Energetic Ion Composition in the Earth's Magnetosphere

Advances in Earth and Planetary Sciences

Energetic Ion Composition in the Earth's Magnetosphere

Edited by

R. G. Johnson
Space Sciences Laboratory, Lockheed Palo Alto Research Laboratory, U.S.A.

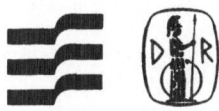

Terra Scientific Publishing Company/Tokyo
D. Reidel Publishing Company / Dordrecht, Boston, London

Library of Congress Cataloging in Publication Data

Main entry under title:

Energetic ion composition in the earth's magnetosphere

    (Advances in earth and planetary sciences)
    " ... presented at the Symposium on the Role of Ion Composition in Under-
standing Magnetospheric Processes, which was held on 11 August 1981 in Edinburgh,
Scotland ... "—Pref.
    1. Magnetosphere—Congress. 2. Ions—Congress. 3. High temperature plasmas
—Congress. I. Johnson, R. G. II. Symposium on the Role of Ion Composition in
Understanding Magnetospheric Processes (1981 : Edinburgh, Lothian) III. Series.
QC809.M35E54   1983   538'.766   83-8629
ISBN-13: 978-94-009-7107-3   e-ISBN-13: 978-94-009-7105-9
DOI: 10.1007/978-94-009-7105-9

Published by Terra Scientific Publishing Company (TERRAPUB),
307 Shibuyadai-haim, 4-17 Sakuragaoka-cho, Shibuya-ku, Tokyo 150, Japan,
in co-publication with D. Reidel Publishing Company, Dordrecht, Holland

Sold and distributed in the U.S.A. and Canada
by Kluwer Boston Inc.,
190 Old Derby Street, Hingham, MA 02043, U.S.A.,
in Japan by Terra Scientific Publishing Company (TERRAPUB),
307 Shibuyadai-haim, 4-17 Sakuragaoka-cho, Shibuya-ku, Tokyo 150, Japan

In all other countries, sold and distributed
by Kluwer Academic Publishers Group,
P. O. Box 322, 3300 AH Dordrecht, Holland

D. Reidel Publishing Company is a member of the Kluwer Academic Publishers Group

Preface

The ten invited papers presented at the Symposium on The Role of Ion Composition in Understanding Magnetospheric Processes, which was held on 11 August 1981 in Edinburgh, Scotland, have formed the principal basis for this book. The symposium was a part of the Fourth Scientific Assembly of the International Association of Geomagnetism and Aeronomy and was convened by the Division III Working Group on Composition of Hot Magnetospheric Plasma under the co-chairmanship of Dr. Gerhard Haerendel and myself.

The symposium papers have been expanded and are supplemented in this book by several additional papers to provide a more comprehensive topical review and a current status report on experimental and theoretical research that relate to our present understanding of the energetic ion composition in Earth's magnetosphere. Thus, the papers include new results as well as a review of published results, with rather extensive reference lists, in the hope that this collection will be useful both for information on current research activities as well as a somewhat lasting topical resource for the space science community.

The definitions of "epithermal," "suprathermal," "warm," "hot," and "energetic" ions and plasmas in this book follow the current convention in this field—that is, each author defines them (or assumes they are defined) in the context of his or her own activities. I have not asked the authors to modify this convention, and the reader is cautioned that "energetic" ions may refer to ion energies in any portion of the range from an eV to several hundred MeV. In particular, the use of "energetic ions" in the title of this book is intended to encompass this wide energy range as well as to indicate that the book is not addressing, as a central issue, the composition of the "cold" ions in the plasmasphere.

As evidenced by most of the papers in this book, a major change has occurred during the past decade in our understanding of the origin and energization of the hot (0.1 to 50 keV) plasmas found in the magnetosphere. Prior to 1972 it was generally believed that the energetic ion population in the magnetosphere was always dominated by H^+ ions (protons) and that the origin of these ions was the sun. This prevalent viewpoint was specifically reflected in the summary of the 1965 conference on Radiation Trapped in the Earth's Magnetic Field (D. Reidel Publishing Company, 1966) which stated: "The solar wind is the source of all electrons and protons of energy less than 100 MeV by some diffusion and acceleration processes which are currently not understood." This earlier viewpoint began to be modified in 1972 with the discovery of large fluxes of energetic (keV) O^+ ions in the magnetosphere, and it may now be contrasted with recent results (discussed herein) which show that during geomagnetic storms, O^+ ions from the Earth's ionosphere often dominate the hot plasma number density, as well as energy density, in broad regions of the outer magnetosphere.

The papers in this book clearly show that we have progressed a long way during the past decade in our understanding of the general morphology of the energetic ion composition in the magnetosphere. However, many (or perhaps most) of the important physical processes responsible for the entry, acceleration, transport, and loss of the ions are still not understood.

I thank the many persons who have contributed to the success of the symposium and the completion of this book. I especially thank the authors who have contributed many weeks of their time to provide well written papers, most often with the necessity of greatly condensing the material to keep the manuscripts to reasonable lengths. The symposium was organized and portions of this book were produced while I was a Visiting Professor at the University of Bern, and I thank Professor J. Geiss and Dr. H. Balsiger for making that visit possible. I am indebted to Professor A. Nishida for his persuasive arguments that this book warranted the collective efforts necessary to bring it to fruition. I thank Mrs. Marian H. Johnson for much valuable assistance with manuscripts, proofs, and data for this book. I also take this opportunity to thank Dr. C.-G. Fälthammar for his early and continuing emphasis on the importance of investigating the energetic ion composition in the magnetosphere and for his efforts in calling international attention to the need for further studies in this area. Last but not least, I especially thank Dr. R. D. Sharp, Dr. E. G. Shelley, Dr. J. B. Reagan, and the many other Lockheed colleagues whose dedication and hard work during the past two decades have made possible many of the exciting discoveries and results discussed in this book and, thus, made possible the book itself.

December, 1982

R. G. Johnson
Lockheed Palo Alto Research Laboratory
Palo Alto, California

Contents

Contents

Energetic Ion Composition in the Earth's Magnetosphere, edited by R. G. Johnson, 1–21.

Principles of Magnetospheric Ion Composition

M. Schulz

*Space Sciences Laboratory, The Aerospace Corporation,
El Segundo, California, U.S.A.*

(Received May 5, 1982)

Ion composition provides an important diagnostic tool for identifying and quantifying source, loss, energization, and transport processes applicable to magnetospheric charged particles. However, the situation is not as simple as had been envisioned a decade ago. Ring-current and radiation-belt ions are believed to come from either the sun or the earth's ionosphere or both. Entry is not direct in either case. Solar-wind velocity and ion composition vary considerably with time. Ion-velocity distributions are partially thermalized in the magnetosheath after crossing the bow shock. Entry of solar-wind ions into the magnetosphere is modulated in part by reconnection efficiency, i.e., by the ratio of cross-tail electric field to "asymptotic" interplanetary electric field. Non-adiabatic motion in the plasma sheet may enable the cross-tail electric field to heat the various ion species by different amounts. Entry of ionospheric ions into the plasma sheet seems to occur primarily through afternoon- and evening-sector auroral arcs, rather than through the polar wind as had previously been postulated. Energization occurs through interaction with the auroral potential structure (which varies with magnetic activity) and through plasma-sheet processes cited above. These are some of the uncertainties that affect the outer boundary condition for the problem of magnetospheric ion composition. Radial transport of magnetospheric ions at ring-current energies ($E \sim 10$–$100 \, \text{keV}$) is achieved primarily by unsteady electrostatic convection. The corresponding diffusion coefficient is independent of charge and mass among particles having drift periods $\gg 2\pi$ times the postulated decay time of an impulse in the convection electric field. Erosion of the ring current occurs primarily through charge exchange and perhaps through wave-particle interactions. Charge exchange favors the survival of O^+ and He^+ over H^+ and He^{++} at energies $\lesssim 50 \, \text{keV}$, but account must be taken of population disparities in the source region and of time scales associated with radial diffusion therefrom. Collisional processes are less important at radiation belt energies ($E \gtrsim 200 \, \text{keV}$), and the mapping of ion distributions from the source region to low L values can be partially understood in terms of a common diffusion profile with M and J (first two adiabatic invariants) conserved. The pitch-angle anisotropy of ring-current ions is likely to generate electrostatic ion-cyclotron waves outside the plasmasphere and electromagnetic ion cyclotron waves inside the plasmasphere. These should result in perferential pitch-angle diffusion of the major ionic constituent and in a possibly weaker (parasitic) pitch-angle diffusion of the minor ionic constituents.

1. Introduction

In more innocent times it was believed that the ion composition of the ring current

1

and radiation belts at a specified energy/nucleon should be the same as the ion composition of the solar wind. The idea was that traversal of the bow shock would separately thermalize each ionic constituent of the solar wind so as to produce a Maxwellian distribution having a temperature ~ 1 keV/nucleon as an outer boundary condition at $L \sim 10$ on the phase-space density, and that the diffusion equation would impose a common profile (defined by the requirement of a divergence-free radial-diffusion current) on all the constituents in the ring current and radiation belts. This concept has largely failed to be supported by the observational data and so is considered obsolete.

One difficulty with the above concept is that the e-folding energy provided by thermalizing the solar wind is too small by an order of magnitude (see below) for assignment to a boundary at $L \sim 10$. Another difficulty is the neglect of ionospheric ions, which can enter the magnetosphere with several keV of energy by virtue primarily of the field-aligned potential drop characteristic of auroral arcs in the afternoon and evening sectors (SHELLEY et al., 1976; GHIELMETTI et al., 1978), and to a lesser extent by processes producing ion acceleration transverse to the magnetic field (GORNEY et al., 1981; SHARP et al., 1977). It seems, however, that entry into the radiation belt or ring current is not direct in either the solar wind or ionospheric case, but rather occurs principally via the plasma sheet. Violation of the first adiabatic invariant of charged-particle motion within the plasma sheet may enable the cross-tail electric field to heat the various ionic species further, but not necessarily by the same amount or by the same amount per nucleon. Moreover, entry of solar wind ions into the plasma sheet is modulated in part by reconnection efficiency, i.e., by the ratio of cross-tail electric field to asymptotic interplanetary electric field. The reconnection efficiency can vary considerably with time, as can the solar-wind velocity and ion-composition. Likewise, the field-aligned auroral potential drop can vary considerably with time. These are some of the uncertainties that can affect the outer boundary condition for the problem of magnetospheric ion composition as a function of energy/nucleon.

Radial transport of magnetospheric ions at ring-current energies ($E \sim 10$–100 keV) is achieved primarily by unsteady electrostatic convection. The corresponding diffusion coefficient is independent of charge and mass among particles having drift periods $\gg 2$ hr if 20 min is the postulated decay time of an impulse in the convection electric field. However, the diffusion current is not divergence-free, since distributed losses constitute an essential ingredient of ring-current dynamics. Erosion of the ring current occurs primarily through charge exchange and perhaps through wave-particle interactions. Charge exchange favors the survival of O^+ and He^+ over H^+ and He^{++} at energies $\lesssim 50$ keV, but account must be taken of population disparities in the plasma sheet and of time scales associated with radial diffusion therefrom. Thus, it is possible for H^+ to predominate over O^+ and He^+ in the early stages of ring-current formation but not in the later stages of a magnetic storm.

Collisional processes are less important at radiation-belt energies ($E \gtrsim 200$ keV), and the mapping of ion distributions from the outer boundary to low L values can be partially understood in terms of a common diffusion profile with M and J (first two adiabatic invariants) conserved. One consequence of this is the appearance of very large ionic pitch-angle anisotropies in the inner radiation zone. Indeed, the pitch-angle anisotropy at low L values is a reflection of the energy spectrum at the outer-boundary location. Inner-zone observations typically reveal that the alpha-particle anisotropy is even larger than the proton anisotropy at the same energy/nucleon. This means that the α/p (alpha/proton) intensity ratio must vary inversely with magnetic latitude along a field line, as indeed it does. The occurrence of a larger alpha-particle than proton anisotropy at fixed energy/nucleon can be understood in terms of an alpha-particle temperature that is less than four times the proton temperature at the outer boundary of trapped radiation.

Much of the observational research emphasis on ion composition in recent years has focused on energies $\lesssim 20$ keV, i.e., on energies so low that they are not truly in the radiation-belt range. Indeed, such energies are so low that they hardly include a significant portion of the storm-time ring-current spectrum. It is therefore quite important to avoid sweeping generalizations in the interpretation of ion-composition results. Meanwhile, progress (e.g., GLOECKLER, 1977) is being made toward the construction of ion-identifying spectrometers that operate in what had been the energy "gap" between ~ 20 keV/charge and ~ 100 keV/nucleon, as required for proper scrutiny of the ring-current spectrum.

The purpose of this work is to outline the basic principles upon which the theoretical study of magnetospheric ion composition is founded. The author helped to compile a much more thorough review of the literature on this subject only a few years ago (CORNWALL and SCHULZ, 1979). Not much would be gained by providing the results of yet another literature search at this time. Thus, the focus of the present review is directed instead on the enduring fundamentals. These relate to the problems of access, energization, transport, and loss of protons and other ions with respect to the earth's magnetosphere and radiation belts.

2. Access

The main sources of magnetospheric ions are the galaxy, the sun, and the ionosphere. The galaxy provides cosmic-ray ions that are themselves too energetic to be trapped by the earth's magnetic field. However, the backscatter (albedo) products of spallation reactions that occur when such cosmic-ray ions strike the upper atmosphere include many neutrons in the 10–100 MeV range. The occasional decay of a neutron from this population contributes a proton in the 10–100 MeV energy range and an electron of energy < 800 keV to the radiation belts. Other sources of electrons are much more important than this one. Other sources of inner-zone ions in the 10–100 MeV energy range are negligible. Thus, inner-belt ion composition is not much of a problem at these very high energies: the ions in this range are almost exlusively protons. The source described in this paragraph is known by the acronym CRAND (cosmic-ray-albedo-neutron decay).

Drift shells at synchronous altitude and beyond are occasionally populated by solar-flare ions (presumably protons) of energy 5–70 MeV (PAULIKAS and BLAKE, 1969). The mode of access in this case is largely direct, in that the drift shells populated are those that pass within a gyro-radius of the magnetopause or neutral sheet.

Of greater interest in the context of magnetospheric ion composition are particles of much lower energy, perhaps $E \sim 1\text{--}10\,\text{keV}$, for which the modes of access are largely indirect. These include magnetosheath ions derived from the shocked solar wind and auroral ions energized primarily by the electric potential drop that occurs along magnetic field lines in the PM (afternoon-evening) sector of the auroral oval. Magnetosheath ions with $E \sim 1\,\text{keV}$ might impinge on the magnetosphere anywhere, but the convection electric field tends to expel them from the dayside region of closed field lines. Conversely, charged particles that enter at the polar cleft (a dayside region of weak magnetic field surrounding the neutral points) are drawn into the tail lobes by the convection electric field, as are particles that impinge on the tail-lobe surfaces. Characteristics of the resulting plasma mantle have been described by SCKOPKE and PASCHMANN (1978).

The convection velocity of plasma in either tail lobe is smaller than the solar-wind velocity by a factor of order $\varepsilon \bar{B}_z / B_t \ll 1$, where B_t is the tail field, $\bar{B}_z$ is the southward component of the interplanetary magnetic field, and ε is the "reconnection efficiency" defined by KENNEL and CORONITI (1975) as the ratio of the east-west component of the cross-trail electric field to that of the interplanetary electric field. Observations of BURTON et al. (1975) suggest a "half-wave rectifier" model such that ε is practically zero for northward $\bar{B}_z$ but at least 0.2 (and perhaps even ~ 1) for southward $\bar{B}_z$. The factor $\varepsilon \bar{B}_z / B_t$ is quite small ($\lesssim 0.1$) in any case, and so only the extreme flanks of the plasma mantle (i.e., regions already within a few earth radii of the current sheet in the tail) are likely to populate the plasma sheet by convection within the first few tens of earth radii downstream from the earth. The reversal of magnetic-field direction across the current sheet and the drawn-to-dusk direction of the cross-tail electric field make the low-latitude AM (pre-dawn) flank an inviting place for magnetosheath ions to enter the magnetosphere in any case.

The access of keV ionospheric ions to the plasma sheet is made possible primarily by the electrostatic potential structure characteristic of the PM (afternoon-evening) sector of the auroral oval. It is in this region (near the boundary between closed and open magnetic field lines) that there develops a virtual discontinuity in the magnetospheric convection electric field if one treats magnetic field lines as electrostatic equipotentials. There is a Pedersen conductance owing to the collisionality of the ionosphere and an Ohmic impedance due to the mirror forces exerted by the inhomogeneous magnetic field on the hot auroral plasma. It seems that these properties conspire with the requirement of current conservation (Ampère's law) to produce a multi-kilovolt potential drop *along* the field lines of the auroral oval. The geometry is such that the corresponding parallel (to $\boldsymbol{B}$) component of the electric field should be upward in the PM sector and downward (if present) in the AM sector. Statistics compiled by GHIELMETTI et al. (1978) from S3-3 data suggest that upgoing ion beams are about six times more probable in the PM sector than in the AM sector, and about 44 times more probable in the dusk quadrant (15–21 hr, magnetic local time) than in the dawn quadrant (03–09 hr, MLT).

The contribution of auroral ion beams to the plasma sheet presumably exceeds that of the polar wind, which had been until recently the favored medium (e.g., AXFORD, 1970) for supplying ionospheric ions to the plasma sheet. The polar wind theoretically consists of protons (along with a few helium ions) having energies

$\lesssim 10$ eV. The auroral ion beams seem to be rich in oxygen as well, with ion energies of several keV at least. It is true that the polar wind delivers ions from the entire polar cap rather than from only the PM half of the auroral oval, but the deficiencies in ion energies and oxygen ions now seem to make the polar wind a less important medium (than the auroral ion beams) for the transfer of ionospheric ions to the plasma sheet. Thus, it seems that the main sources of 1–10 keV ions for the plasma sheet are (1) magnetosheath plasma derived from the shocked solar wind and (2) ionospheric plasma from the PM sector of the auroral oval.

3.   Energization

The importance of the plasma sheet for magnetospheric ion composition resides in the adiabatic and non-adiabatic processes by which ions are further energized there before being transported into the region of closed drift shells. The adiabatic energization process entails the usual gradient-curvature drift of nightside ions across equipotentials of the dawn-to-dusk electric field, or (equivalently) the sunward convection of nightside ions into a spatially increasing magnetic-mirror field. This process occurs in the earthward portion of the nightside plasma sheet, i.e., in a region where the magnetic field lines are topologically dipolar but from which the adiabatic drift shells intersect the magnetopause. The particles in question are "quasi-trapped" in the terminology of ROEDERER (1970), but the failure of their drift shells to close is as much the fault of the convection electric field as it is of the gradient-curvature drift.

The non-adiabatic energization process occurs in the tailward portion of the plasma sheet, where the northward ($\hat{z}$) component of the earth's magnetic field is too weak to enforce conservation of the first two adiabatic invariants. A plasma physicist's prototype for the magnetic field in this distinctly non-dipolar region might be expressed as

$$B = \hat{z} B_z - \hat{x} B_t \tanh (z/z_0), \qquad (1)$$

where $\hat{x}$ is the anti-sunward unit vector, z_0 is the "half-width" of the current sheet, and $B_t \gg B_z$. A model electric field consistent with Eq. (1) is the uniform field $E = -\hat{y}|E_y|$. A theorem of STERN and PALMADESSO (1975) asserts that there is no net gradient-curvature drift in a magnetic field such as Eq. (1) if B_z, B_t, and z_0 are constants. The proof follows from the construction of a bounce-averaged Hamiltonian function that depends on the first two adiabatic invariants M and J but not on the coordinates x_0 and y_0 at which the guiding field line crosses the plane $z = 0$. An immediate corollary is that the electric field $E = -\hat{y}|E_y|$ produces a bounce-averaged drift such that $\dot{x}_0 = -(c/B_z)|E_y|$, but that particle energization does not occur.

The theorem of STERN and PALMADESSO (1975) clearly does not apply if $B_z = 0$. In this case there is a gradient drift in the $+\hat{y}$ direction for ions adiabatic with respect to M, but no curvature drift since the field lines are rectilinear. There is also a net convection in the $\hat{z}$ direction (toward the neutral sheet) for all ions and a net drift in the $-\hat{y}$ direction for ions that fail to conserve M because their trajectories cross the plane $z = 0$. Moreover, the ions that thus fail to conserve M are energized by the uniform electric field. The theorem fails in the case $B_z = 0$ because the second adiabatic invariant J no longer exists, or (equivalently) because the bounce average is no longer defined, even for particles that conserve the first invariant M.

The theorem of STERN and PALMADESSO (1975) also does not prevent particle energization for $B_z \neq 0$ if the particle of interest has too large a rigidity to conserve M and J in the field configuration given by Eq. (1). Thus, the dawn-to-dusk electric field can be expected to add extra energy to ions that have entered the plasma sheet with rigidity $p/q \gtrsim z_0 B_z/c$. Recent results of LYONS and SPEISER (1982) suggest that the energy gained in this way is indeed an increasing function of initial ion energy. This is a geophysical analogy of the rich becoming richer. Moreover, since the energization mechanism involves only single-particle motion in crossed electric and magnetic fields, there is no expectation that any of the energy gained by the favored ions of high initial energy will "trickle down" to the less favored ions of low initial energy.

The above-described idea for charged-particle energization in the tailward portion of the plasma sheet was first worked out in quantitative detail by SPEISER (1965). It is quite remarkable that such a perceptive work has received so little attention over the past seventeen years. In any event, the occurrence of an "ion-heating" process based on single-particle motion makes it premature to invoke plasma turbulence or reconnection as the energization mechanism of the plasma sheet. It appears from the work of LYONS and SPEISER (1982) that non-adiabatic single-particle motion in the crossed electric and magnetic fields is sufficient to account for ion temperatues observed in the tailward portion of the plasma sheet. Still open is the important question of whether the various ionic species are "heated" by different amounts (per particle, per unit charge, or per nucleon) by this mechanism. Sunward convection by the dawn-to-dusk electric field delivers the ions thus energized into the earthward portion of the plasma sheet, where further energization occurs through the conservation of M and J in the crossed *inhomogeneous* magnetic and electric fields that prevail there. Of course, a charged particle can gain even more energy if it is transported (e.g., by guiding-center diffusion) across the boundary between open (plasma-sheet) and closed (ring-current) drift shells, and from there to progressively smaller L values within the region of geomagnetically trapped particles. However, the subject of charged-particle energization by guiding-center diffusion is included below in the section about charged-particle transport across adiabatic drift shells.

4. Transport

For charged particles that conserve their first two adiabatic invariants M and J, the problem of non-adiabatic transport typically entails violation of the third invariant Φ through radial diffusion. At radiation-belt energies it is usual to adopt the dimensionless label L (roughly speaking, the equatorial radius of the drift shell in earth radii) as diffusion coordinate, and this can be done by means of the transformation $L \equiv 2\pi(\mu/a)|\Phi|^{-1}$, where μ is the magnetic moment of the earth and a is the radius of the earth. The diffusion equation for the drift-averaged phase-space density $\bar{f}$ then takes on the form

$$\frac{\partial \bar{f}}{\partial t} = L^2 \frac{\partial}{\partial L}\left[\frac{D_{LL}}{L^2}\frac{\partial \bar{f}}{\partial L}\right], \qquad (2)$$

where D_{LL} is the diffusion coefficient (e.g., SCHULZ and LANZEROTTI, 1974). The functional form of D_{LL} is necessarily model-dependent, but the usual treatment yields results of the form $D_{LL} = D_{LL}^{(m)} + D_{LL}^{(e)}$, where

$$D_{LL}^{(m)} \approx 7 \times 10^{-9} L^{10} [Q(y)/180 D(y)]^2 \, \text{day}^{-1} \qquad (3a)$$

represents the contribution from step-like impulses in the magnetospheric $\mathbf{B}$ field and

$$D_{LL}^{(e)} \approx 1 \times 10^{-10} L^{10} (\gamma Z M_0 y^2 / M)^2 [T(y)/2D(y)]^2 [1 + (\Omega_3 \tau)^{-2}]^{-1} \, \text{day}^{-1} \quad (3b)$$

represents the contribution from exponentially decaying impulses in the electrostatic field that drives magnetospheric convection. The auxiliary functions $Q(y)$, $D(y)$, and $T(y)$ are well approximated (SCHULZ and LANZEROTTI, 1974; DAVIDSON, 1976) by the algebraic forms

$$Q(y) \approx -27.12667 - 45.39913 y^4 + 5.88256 y^8, \qquad (4a)$$

$$D(y) \approx 0.4600577 + 0.1066154 y^{3/4} - 0.1997662 y, \qquad (4b)$$

and

$$T(y) \approx 1.3801730 - 0.6396925 y^{3/4}, \qquad (4c)$$

where y ($= \sin \alpha_0$) is the sine of the equatorial pitch angle. Other symbols appearing in Eq. (3) include the ratio γ of relativistic mass to rest mass, the ionic charge-state number Z, the scale factor $M_0 \equiv 1 \, \text{GeV/gauss}$, the ionic drift frequency $\Omega_3/2\pi$, and the decay time τ ($\sim 1{,}200$ sec) of a model electrostatic impulse. The factor $[1 + (\Omega_3 \tau)^{-2}]$ differs by at most 2.5% from unity for ions having drift periods $2\pi/\Omega_3 \lesssim 20$ min, i.e., for ions of kinetic energy $E \gtrsim 1.2 \, Z/L$ MeV, and so the factor $[1 + (\Omega_3 \tau)^{-2}]$ in Eq. (3b) may be regarded as essentially constant for such particles.

Figures 1a and 1b show the variation of $D_{LL}^{(m)}$ and $D_{LL}^{(e)}$ with L in the limit $\Omega_3 \tau \gg 1$ for nonrelativistic ions having the same energy (and selected values of y) at $L = 7$. The reader may be surprised to see that D_{LL} scales as L^{10} only for $y = 1$. For $\alpha_0 \neq 90°$ account must be taken of the variation of y with L at fixed M and J. According to CHEN and STERN (1975) this variation is well approximated by the algebraic form

$$y^{-2} \approx 1 + 1.35048X - 0.030425 X^{4/3} + 0.10066 X^{5/3} + (X/2.760346)^2, \qquad (5)$$

where $X^2 \equiv (La/\mu)(J^2/8 m_0 M)$ for a particle of rest mass m_0. The limiting case $y = 0$ ($X = \infty$) thus yields $y^2 \propto 1/L$ in Eq. (3b), from which it follows that $D_{LL}^{(e)}$ scales only as L^8 in this limit (see Fig. 1b, dashed line). On the other hand, it seems from Fig. 1a that the deviation of $D_{LL}^{(m)}$ from an L^{10} scaling is almost negligible. Thus, for ion energies sufficiently high that charge exchange and Coulomb drag can truly be neglected as they are in writing Eq. (2), one might anticipate a steady-state diffusion profile of the form

$$\bar{f}(M,J,L) = \left[\frac{1 - (L_1/L)^7}{1 - (L_1/L^*)^7} \right] \bar{f}(M,J,L^*), \qquad (6)$$

where L^* (~ 10) represents the outermost closed drift shell and L_1 represents the innermost drift shell that can trap particles of a given $K^2 \equiv J^2/8 m_0 M$. Thus, the entire energy spectrum of $\bar{f}(M, J, L)$ at $L < L^*$ is determined by the boundary-value

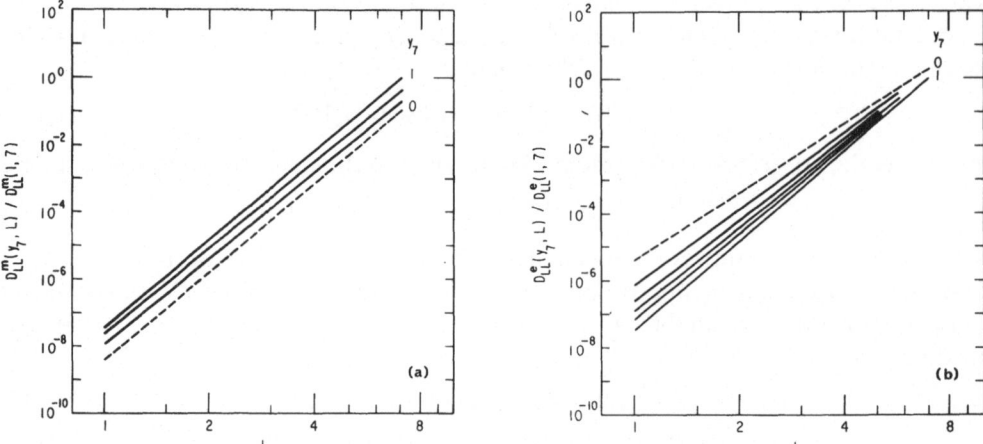

Fig. 1. Radial variation of D_{LL} driven (a) by step-like magnetic impulses and (b) by electrostatic impulses that rise quickly and decay slowly on the time scale of an ion's drift period. Impulses are presumed to conserve M and J so that y varies in accordance with Eq. (5) as a function of L for each value of y_7 (denoting the value of y at $L = 7$), i.e., for $y_7 = 0, 0.4, 0.6$, and 1.0 in (a) and for $y_7 = 0, 0.2, 0.4, 0.6, 0.8$, and 1.0 in (b). Dashed lines ($y_7 = 0$) are not realized in practice, because of the loss cone (SCHULZ and LANZEROTTI, 1974).

spectrum contained in $\bar{f}(M, J, L^*)$. Of particular interest in the case of nonrelativistic ions is the specification

$$\bar{f}(M, J, L^*) = [(y^*)^2(L^*a)^3 E_0^*/\mu M]^{l+1}\bar{f}^* \exp[-\mu M/(y^*)^2(L^*a)^3 E_0^*], \qquad (7)$$

where $l, \bar{f}^*$, and E_0^* are constants. The special case $l = 0$ corresponds to an exponential spectrum, since the particle flux (per unit energy and solid angle) is equal to $2m_0 E \bar{f}$. The substitution $\mu M = L^3 a^3 y^2 E$ in Eq. (7) is valid for all L values in the dipolar field model and yields

$$\bar{f}(M, J, L^*) = [(y^*/y)^2(L^*/L)^3(E_0^*/E)]^{l+1} \bar{f}^* \exp[-(y/y^*)^2(L/L^*)^3(E/E_0^*)], \qquad (8)$$

which can be written more compactly as

$$\bar{f}(M, J, L^*) = (E_0/E)^{l+1}\bar{f}^* \exp(-E/E_0) \qquad (9a)$$

by introducing the parameter

$$E_0 \equiv (y^*/y)^2(L^*/L)^3 E_0^*. \qquad (9b)$$

It follows from Eq. (9) that the index l of a power-law spectrum remains invariant through the diffusion profile given by Eq. (6), whereas the e-folding energy E_0 of an exponential spectrum varies with L and y^* in the same manner as the energy of an individual particle for which M and J are conserved. Moreover, the variation of E_0 with L is stronger for $y^* \sim 1$ than for $y^* \ll 1$, and this property creates anisotropy in the pitch-angle distribution observed at fixed E, over and above the anisotropy contained in the factor $[1 - (L_1/L)^7]/[1 - (L_1/L^*)^7]$, at $L < L^*$.

Application of the foregoing ideas to a body of observational data is illustrated in

Fig. 2 (DUNGEY *et al.*, 1965). The solid curves were constructed by measuring the *e*-folding energies E_0 of exponential proton spectra (DAVIS and WILLIAMSON, 1963) at equatorial pitch angles corresponding (via Eq. (5)) to $\alpha_0 = 10°, 20°, 30°$, and $90°$ at $L = 7$. The dashed curves show the corresponding energy variation of an individual proton but are positioned on the logarithmic ordinate so as to best fit the observed variation of E_0 with L. The dashed curves seem to converge on a common (isotropic) value of $E_0 \sim 12$ keV at $L \sim 10$. It is natural, therefore, to associate the parameters $E_0^* \sim 12$ keV and $L^* \sim 10$ with the outer boundary condition on the proton radiation belt, on which the observations of DAVIS and WILLIAMSON (1963) spanned the energy range $E \approx 0.1–1.6$ MeV. These results for E_0^* and L^* suggest that the adiabatic and non-adiabatic energization processes operative in the plasma sheet must be able to provide a proton temperature ~ 12 keV at $L \sim 10$, at least for the high-energy tail of the proton distribution there. The appearance of a boundary at $L^* \sim 10$ between closed and open drift shells is quite plausible for the (high) proton energies of interest.

The large anisotropy inherent in Eqs. (8) and (9) for $L \ll L^*$ offers a likely explanation for the occurrence of pitch-angle distributions such as those illustrated in Fig. 3 (BLAKE *et al.*, 1973). The proton anisotropy at $L = 1.85$ is an increasing function of energy, as would be expected from the presence of an exponential factor in Eq. (9). The alpha-particle anisotropy is even larger than the proton anisotropy at the same energy/nucleon, as would be expected if the alpha-particle temperature were somewhat less than four times the proton temperature at the outer boundary of trapped radiation, i.e., at $L = L^*$. Such a boundary condition is not really implausible in view

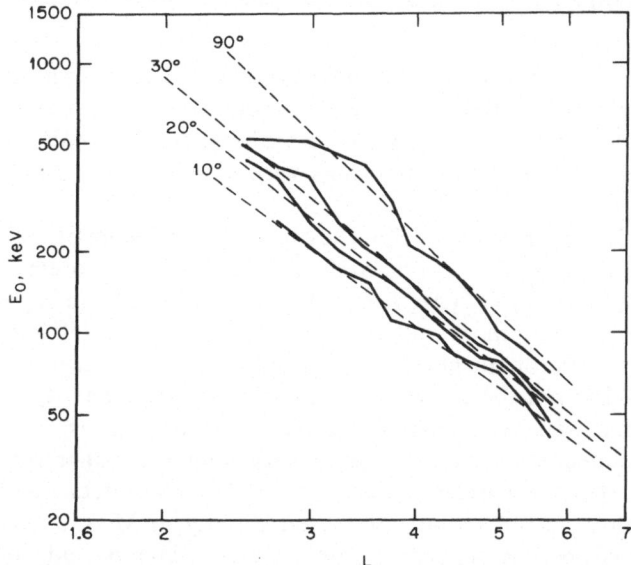

Fig. 2. Empirical *e*-folding energies (solid curves) of observed proton spectra (DAVIS and WILLIAMSON, 1963) at equatorial pitch angles consistent with Eq. (5) and mapping to $\alpha_0 = 10°, 20°, 30°$, and $90°$ (respectively) at $L = 7$; expected variation (dashed curves) of energy with L for individual protons having constant M and J, for selected values of energy and equatorial pitch angle at $L = 7$ (DUNGEY *et al.*, 1965).

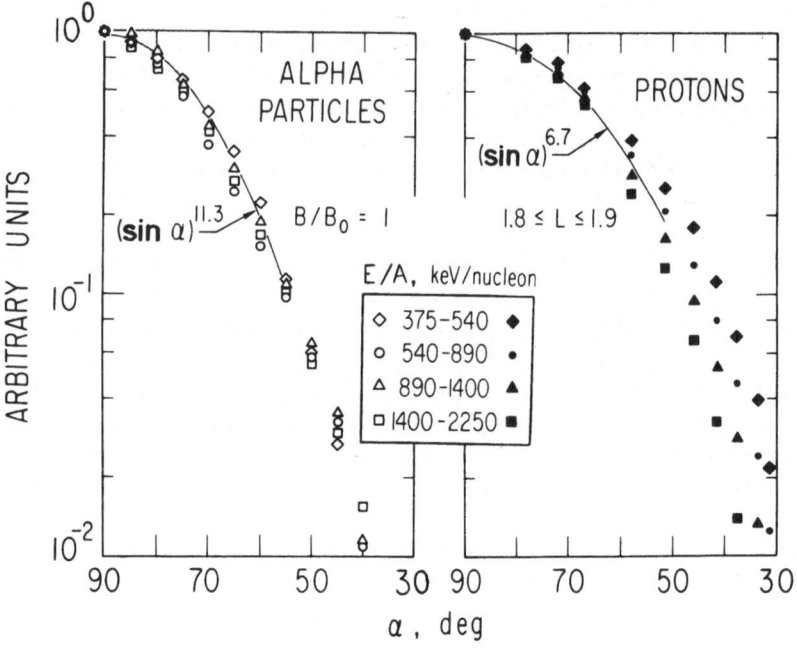

Fig. 3. Equatorial pitch-angle distributions of alpha-particles (left panel) and protons (right panel) in the same four energy/nucleon passbands, as observed on spacecraft OV1-19 (BLAKE *et al.*, 1973). Statistical error bars (not shown) would be about as large as the data-point symbols themselves.

of the adiabatic and non-adiabatic plasma-sheet processes (see above) that must intervene before a solar-wind ion enters the radiation belts. Moreover, there is no expectation that plasma-sheet ions of ionospheric (rather than solar-wind) origin would even have *entered* the plasma sheet with kinetic energy proportional to ionic mass.

It is not really accurate to say that the observed hardening of the proton spectrum with decreasing L in Fig. 2 is a consequence of Liouville's theorem. It is not true, for example, that $\bar{f}(M, J, L) = \bar{f}(M, J, L^*)$. This can be seen from Eq. (6). On the other hand, Liouville's theorem is not violated by the radial-diffusion process. The distinction that must be made here is that Liouville's theorem requires the conservation of $f(M, J, L, \varphi)$ but not the conservation of $\bar{f}(M, J, L)$; the entity that diffuses is the longitude-averaged (i.e., φ-averaged or drift-averaged) phase-space density, whereas the longitude-dependent phase-space density remains conserved. The dilemma confronting the space researcher is that $f(M, J, L, \varphi)$ is difficult to observe with high resolution because particles that differ even slightly in M or J have different drift periods and thus become separated rather quickly in longitude (φ). The primary observable manifestation of drift-phase organization in the phase-space density is the drift-echo phenomenon, and even this is discernible for only a few drift periods following an impulsive change in the magnetospheric *B* field (e.g., LANZEROTTI *et al.*, 1971). Nevertheless, the permanent dispersal of particles in L by such a Hamiltonian process definitely implies their permanent organization in φ for each M and J in the distribution, and such drift-phase organization really *is* a consequence of Liouville's theorem since the drift phase is canonically conjugate to the action variable $(q/c)\Phi$ for a particle of charge q and third invariant Φ (Roberts, C. S., Bell Telephone Labs., personal communication, 1967; SCHULZ and LANZEROTTI, 1974).

The normalization constants 7×10^{-9} day^{-1} and 1×10^{-10} day^{-1} that appear in Eqs. (3a) and (3b), respectively, are those obtained by CROLEY *et al.* (1976) from a variational analysis of inner-zone proton data. It follows that $D_{LL}^{(m)} \gtrsim D_{LL}^{(e)}$ if $M/Z \gtrsim 120$ MeV/gauss (i.e., if $E \gtrsim 40\,Z/L^3$ MeV) for ions that mirror on the equator (i.e., for $y = 1$). For $y \ll 1$ one finds $D_{LL}^{(m)} \gtrsim D_{LL}^{(e)}$ only if $E \gtrsim 180\,Z/L^3$ MeV. These results imply, for example, that ring-current ions (for which $E \approx 10$–100 keV) are transported mainly by electrostatic impulses (unsteady magnetospheric convection). Among ions for which $E \gtrsim 1.2\,Z/L$ MeV (so that $(\Omega_3\tau)^2 \gtrsim 40$ in Eq. (3b)), the diffusion coefficient $D_{LL}^{(e)}$ scales as $(Z/A)^2$ at fixed K^2 $(=J^2/8m_0M)$ and fixed M/A (first invariant per nucleon). This is the scaling to which CORNWALL (1972) appealed in his argument for preferential access of H$^+$ $(Z/A = 1)$ over He^{++} $(Z/A = 1/2)$ and He$^+$ $(Z/A = 1/4)$ to the inner part of the magnetosphere, e.g., to $L \lesssim 3$. Conversely, among ions for which $E \lesssim 30\,Z/L$ keV (so that $(\Omega_3\tau)^2 \lesssim 1/40$ Eq. in (3b)) the diffusion coefficient becomes independent of M, J, Z, and A. In this case one obtains $D_{LL}^{(e)} \approx 8 \times 10^{-5}$ $(\tau/20 \text{ min})^2 L^6$ day^{-1}. Ring-current ions occupy the middle range $(0.1 \lesssim \Omega_3\tau \lesssim 10)$ of the parameter $\Omega_3\tau$ between these two extremes and thus defy efforts to simplify the expression for $D_{LL}^{(e)}$. Perhaps this is not an accident, inasmuch as the decay time τ of an impulse in the convection electric field should naturally reflect the time scale of the underlying particle motion whereby the convection electric field might be discharged.

Although analyses based on radial diffusion alone are quite instructive, additional transport processes must be invoked if one is to account for further details of the prevailing ionic phase-space distributions. For example, one may ask where (in L) the radiation intensity peaks at a given energy E. Differentiation of Eq. (6) with respect to L with Eq. (9a) inserted for $\bar{f}(M, J, L^*)$ yields $(\partial \bar{f}/\partial L)_E = 0$ under the condition

$$\frac{7(L_1/L)^7}{1 - (L_1/L)^7} = -\left[l + 1 + \frac{E}{E_0} \right] \frac{d \ln E_0}{d \ln L}. \tag{10}$$

If the boundary spectrum is assumed to be Maxwellian (so that $l = -1$), then for particles mirroring at the equator one obtains

$$L \approx [(7/3)(L^*/L_1)^3(E_0^*/E)]^{1/10}[1 - (L_1/L)^7]^{1/10} L_1 \tag{11}$$

as the location of peak radiation intensity, since $E_0 = (L^*/L)^3 E_0^*$ for $y^* = 1$ in Eqs. (5) and (9b). The factor $[1 - (L_1/L)^7]^{1/10}$ varies only from 0.99 to 1.00 as L varies from about 1.6 to infinity, and so the expectation based on Eq. (6) is that the location in L of maximum radiation intensity should vary as $E^{-1/10}$. Observations compiled by FRITZ and SPJELDVIK (1981) show clearly that the actual variation closely resembles $E^{-3/16}$, as had been predicted by TVERSKOY (1965), and not $E^{-1/10}$ for $1.5 \lesssim L \lesssim 4.5$; moreover, the peak alpha-particle intensity occurs about 1.3 times as far from the geocenter as the peak intensity for protons of the same energy, and this too had been predicted by TVERSKOY (1965). The essential process considered by TVERSKOY (1965) in addition to radial diffusion was Coulomb drag, which formally causes a particle's M to decrease while K^2 ($=J^2/8m_0M$) remains constant, i.e., causes a particle's energy to decrease while the equatorial pitch angle remains constant. In order to take advantage of the fact the K is conserved by both processes (radial diffusion and Coulomb drag) it is usual to transform from the canonical variables (M, J, Φ) to the "new" variables (M, K, L). The Jacobian of this transformation has an absolute value given by

$$|\partial(M, J, \Phi)/\partial(M, K, L)| = (8m_0M)^{1/2}(2\pi\mu/L^2a), \qquad (12)$$

and so the appropriate Fokker-Planck equation is of the form

$$\frac{\partial \bar{f}}{\partial t} + M^{-\frac{1}{2}}\frac{\partial}{\partial M}\left[M^{\frac{1}{2}}\left(\frac{dM}{dt}\right)_v \bar{f}\right]_{K,L} = L^2\frac{\partial}{\partial L}\left[\frac{D_{LL}}{L^2}\frac{\partial \bar{f}}{\partial L}\right]_{M,K} \qquad (13)$$

Here the transport coefficient $(dM/dt)_v$ is given by

$$(dM/dt)_v = M^{-\frac{1}{2}}(4\pi Z^2q_e^4/m_e)(L^9a^9y^6m_0/2\mu^3)^{\frac{1}{2}}$$

$$\times \left\langle \sum_i N_iZ_i[\gamma^2 - 1 - \gamma^2\ln(4\mu m_eM/m_0y^2L^3a^3I_i)]\right.$$

$$\left. + N_e[\gamma^2 - 1 - \gamma^2\ln(\lambda_D m_e v/\hbar)]\right\rangle, \qquad (14)$$

where $\gamma^2 = 1 + (2\mu M/m_0c^2y^2L^3a^3) = [1 - (v/c)^2]^{-1}$, and where the subscript v denotes "frictional drag" as distinguished from other processes. Other symbols appearing in Eq. (14) include the charge q_e and rest mass m_e of an electron, the densities N_e of free electrons and N_i of neutral atoms or molecules of atomic number Z_i, the mean energy I_i required for atomic excitation or ionization, the Debye length λ_D, and the symbol $\hbar$ (Planck's constant divided by 2π). Thus, for energetic ions of the same M/A and K^2, the transport coefficient $(dM/dt)_v$ scales as $A^{-1/2}Z^2$. This means that the ratio M/A decreases at a rate proportional to $A^{-3/2}Z^2$ for an ion of charge state Z and atomic-mass number A, i.e., as 64:32:8:1 for $H^+ : He^{++} : He^+ : O^+$.

The scaling of $(dM/dt)_v$ with L naturally depends on the models adopted for atmospheric and plasmaspheric density distributions. TVERSKOY (1965) described Coulomb drag in terms of a "lifetime" proportional to $E^{3/2}Z^{-2}A^{-1/2}\langle N_e\rangle^{-1}$ throughout the radiation belts, and thereby replaced the basic transport equation with an equation that would yield some simple results in closed form. The model atmosphere and plasmasphere described by FARLEY and WALT (1971) yielded a "lifetime" $[-d(\ln E)/dt]^{-1}$ roughly proportional to $L^{-1.3}$ for protons of specified M and vanishing K in the interval $1.6 \lesssim L \lesssim 2.5$ (and presumably for higher L values within the plasmasphere). This would correspond to a transport coefficient $(dM/dt)_v$ proportional to $L^{1.3}$ over a region in which hydrogen atoms and free electrons are the dominant agents of Coulomb drag. The ion energy lost to neutral atoms and molecules in Eq. (14) results in elevation of bound electrons to higher energy levels (excitation) or to the spectral continuum (ionization). The energy lost to free electrons (via the term containing N_e and λ_D) results in the Cherenkov emission of Langmuir waves. The logarithms in Eq. (14) are typically quite large ($\gtrsim 10$) for radiation-belt ions.

The angle brackets in Eq. (14) denote the guiding-center average over the bounce and drift motion. Evaluation of the bounce average typically entails a numerical computation, but a few analytical estimates are available in closed form. CORNWALL (1975) suggests that the bounce-averaged atomic-hydrogen density scales as y^{-1} times the equatorial density. SMITH *et al.* (1976) suggest that the bounce-averaged density of hydrogen atoms scales as $\sec^{3.5}\lambda_m$, where λ_m is the mirror latitude. SCHULZ (1977) has described a method of calculating, for any positive integer n, the bounce-averaged value of $(B/B_0)^n$ along a dipolar field line on which $B_0 (\equiv \mu/L^3a^3)$ is the equatorial value of the magnetic-field intensity B. For example, the case $n = 1$ yields the result

$$\langle(B/B_0)\rangle \approx (1.9191y^{-5/4} - 1.1786y^{-1}) \div (1.3802 - 0.6397y^{3/4}), \qquad (15)$$

which might be relevant to a model in which N_e is proportional to B along a plasmaspheric field line. For magnetic drift shells that intersect the model plasmapause, account must be taken of the relative amounts of time spent inside and outside the plasmasphere in evaluating the azimuthal (drift) averages of N_e and $N_e \ln (\lambda_D m_e v/\hbar)$ for use in Eq. (14). The dependence of λ_D on N_e is not a severe complication here, since the argument of the logarithm is typically very large, i.e., since the dependence of the logarithm on N_e is typically quite weak. Thus, it should be sufficient to define $\bar{\lambda}_D$ as the value of λ_D at $B/B_0 = \langle(B/B_0)\rangle$ and to approximate the bounce average of $N_e \ln (\lambda_D m_e v/\hbar)$ by the quantity $\langle N_e\rangle \ln (\bar{\lambda}_D m_e v/\hbar)$.

5. Loss Processes

If Coulomb drag is considered a transport process, as seems appropriate, then the main *loss* mechanism for ring-current and radiation-belt ions appears to be charge exchange. In its simplest manifestation charge exchange causes a singly charged ion to become neutral and thus escape magnetic confinement by the earth's field. This process is described formally by adding a term $-\bar{f}/\tau_q$ to the right-hand side of Eq. (13). The charge-exchange lifetime τ_q is given by $\tau_q^{-1} = \langle N_H\rangle\sigma v$, where N_H is the density of atomic hydrogen in the exosphere, v is the speed of the ion (as in Eq. (14) above), and σ is the ion's cross section for charge exchange with atomic hydrogen. The structure of Eq. (13) makes it desirable to specify τ_q as a function of K, L, and M. Results for H^+ and He^+ ions at $K = 0$ in a model atmosphere of atomic hydrogen are shown in Fig. 4 (CORNWALL, 1971). Lifetimes for H^+ ions are thus typically shorter than for He^+ ions of the same K and M/A. Both $\langle N_H\rangle$ and σv vary with L so as to produce the results shown in Fig. 4; the ion energy (on which σv depends quite strongly) varies as L^{-3} for $K = 0$. The products σv are shown as a function of ion energy for various ions in Fig. 5 (TINSLEY, 1976; CORNWALL and SCHULZ, 1979). The cross sections σ correspond to neutralization for H^+, C^+, N^+, and O^+. Those for helium ions describe this as well as other possibilities, viz., $He^{++} \rightarrow He^+$ and $He^+ \rightarrow He^0$ (dashed curves) and $He^+ \rightarrow He^{++}$ (shown by the $+ + + + +$ curve at $E \gtrsim 150$ keV in Fig. 5).

The possibility of two or more charge states for an ionic species requires that coupled transport equations be introduced. For helium ions the relevant phase-space densities are $\bar{f}_1$ and $\bar{f}_2$, the phase-space densities of He^+ and He^{++}, respectively. The coupled transport equations are of the form

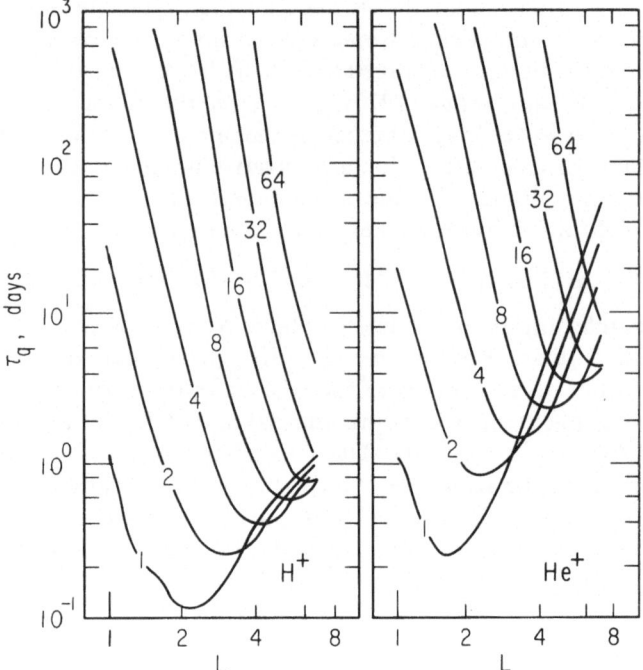

Fig. 4.   Charge-exchange lifetimes against neutralization in model atomic-hydrogen atmosphere for equatorially mirroring ($K = 0$) protons (H^+, left panel) and helium ions (He^+, right panel) at selected values of M/A, measured in MeV/gauss-nucleon (CORNWALL, 1971; SCHULZ and LANZEROTTI, 1974).

$$(\partial \bar{f}_1/\partial t) + C_1\bar{f}_1 = D_1\bar{f}_1 - (\bar{f}_1/\tau_{10}) - (\bar{f}_1/\tau_{12}) + (\bar{f}_2/\tau_{21}) \tag{16a}$$

and

$$(\partial \bar{f}_2/\partial t) + C_2\bar{f}_2 = D_2\bar{f}_2 - (\bar{f}_2/\tau_{21}) + (\bar{f}_1/\tau_{12}), \tag{16b}$$

where C and D are the Coulomb-drag and radial-diffusion operators that appear in Eq. (13), while τ_{ij} denotes the lifetime against transition from charge state i to charge state j. Thus, the lifetime τ_{10} in Eq. (16a) is identical with the τ_q that appears in the helium (He^+) panel of Fig. 4. SPJELDVIK and FRITZ (1978) have generalized Eq. (16) for application to the charge states of oxygen, which are quite numerous but present no difficulty in principle.

Strictly speaking, the source terms $\bar{f}_2/\tau_{21}$ in Eq. (16a) and $\bar{f}_1/\tau_{12}$ in Eq. (16b) should be replaced by nonlocal source terms in which $\bar{f}_i(L')$ is averaged (with appropriate weighting) over a narrow interval $|L' - L| < \Delta L$ about the L value of interest. Such a refinement of Eq. (16) would acknowledge the fact that a change in charge state inherently displaces the guiding center of a geomagnetically trapped ion by instantaneously altering the gyro-radius. For ions of initial charge state Z and initial gyro-radius ρ at the equator, the maximum change in L value from transition to charge state $Z \pm 1$ would be of order $\Delta L = (\rho/a)(Z \pm 1)^{-1}$, where a is the radius of the earth.

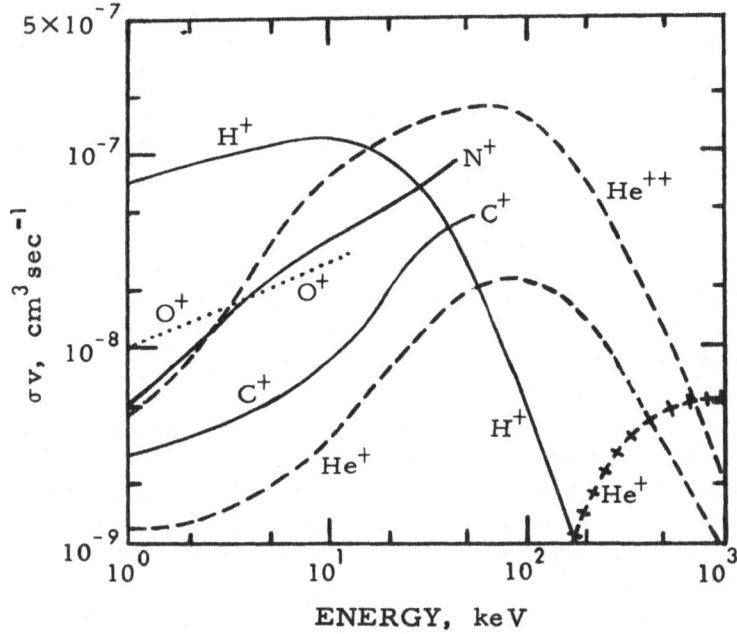

Fig. 5. Products of ion velocity and cross section for charge exchange in atmosphere of atomic hydrogen (CORNWALL and SCHULZ, 1979), based on compilations summarized by CORNWALL (1971) and TINSLEY (1976). Cross sections correspond to neutralization for H^+, C^+, N^+ (solid curves), O^+ (dotted curve), and He^+ (dashed curve); to reduction of charge state (He^{++} to He^+) for He^{++} (dashed curve); and to increase of charge state (He^+ to He^{++}) for He^+ (curve of "plus" signs at lower right).

Such considerations would normally be unimportant, since the adiabatic description of charged-particle motion (upon which. ring-current and radiation-belt theory is founded) requires $\rho/La \ll y \leq 1$ for its validity. However, charge exchange from state i to state j does not remove an ion from the radiation belt unless $j = 0$, nor does charge exchange entail significant pitch-angle diffusion of the trapped ion. Thus, the same ion can experience charge exchange between the same pair of states many times during its residence in the magnetosphere, and the effect of introducing nonlocal source terms to account for this in the equation for $\partial \bar{f}_j/\partial t$ would simulate augmentation of the radial-diffusion coefficient by an amount $\Delta D_{LL}^{(j)} \sim \langle (\rho/a)^2 (Z \pm 1)^{-2} (\tau_{ij} + \tau_{ji}^{-1}) \rangle$. Such an augmentation might (or might not) be quantitatively significant.

Transition from charge state $Z = 1$ to charge state $Z - 1$ ($= 0$) displaces the guiding center to infinity and thus violates the adiabatic description of charged-particle motion. Such an event (i.e., neutralization) removes a fast ion from the ring current but creates a fast neutral whose direction of motion depends on a random variable, viz., the ion's gyrophase at the instant of neutralization. It takes time (~ 20 sec) for the fast neutral to escape from the magnetosphere, and during this time the fast neutral can be "stripped" of one or more electrons so as to re-appear as a born-again ion elsewhere in the magnetosphere. Such reconversion of a fast neutral into an ion is unlikely to occur unless the trajectory of the neutral carries it through the dense atmosphere immediately surrounding the earth. However, MIZERA and BLAKE (1973) have indeed reported evidence of this process in the form of a partial belt of born-again ions of ring-current energy at $L \lesssim 1.1$ near the magnetic equator. Observation of the effect is facilitated by the eccentricity ($\sim$440 km) of the drift shells relative to the earth, since ions reborn at a 100-km altitude just west of the South Atlantic anomaly would gradient-drift to an altitude $\gtrsim$ 900 km in about half a drift period.

Despite the above-described implications for ion transport, charge exchange is primarily a loss mechanism for ring-current ions. It follows from Fig. 5 that H^+ is lost more rapidly than O^+ or He^+ at ion energies $\lesssim 50$ keV. LYONS and EVANS (1976) have emphasized the importance of charge exchange in this context for the ion-composition of the recovery-phase ring current. Their conclusion was that the ring-current ions observed on the S^3 satellite (also known as Explorer 45) decayed much too slowly in intensity (and developed pitch-angle anisotropy much too slowly) to have been H^+ ions. This was a disturbing suggestion at the time, since the predominance of H^+ among the ions of the ring current had by then become an article of faith. The argument of LYONS and EVANS (1976) was not air-tight, since it ignored (a) the likely possibility that radial diffusion would continue during recovery phase and (b) the adiabatic energization that ring-current particles would experience through the azimuthal electric field induced by decay of the ring current itself. However, the work of LYONS and EVANS (1976) alerted space researchers to the likelihood (now generally accepted as fact) that ions heavier than H^+ are major constituents of the ring current at least part of the time. The abundance of O^+ in auroral ion beams (see above under "Access") makes this conclusion seem less surprising in retrospect.

The creation of pitch-angle anisotropy in ring-current ion distributions is an expected consequence of charge exchange, since the bounce-averaged density $\langle N_H \rangle$ of atomic hydrogen in the atmosphere should be an increasing function of mirror latitude λ_m at each L value. It was noted above in connection with Coulomb drag that several functional forms have been proposed for scaling $\langle N_H \rangle$ with equatorial pitch angle ($\alpha_0 \equiv \sin^{-1} y$) or mirror latitude, e.g., y^{-1} by CORNWALL (1975) and $\sec^{3.5} \lambda_m$ by SMITH et al. (1976). The similarity of these two scalings is revealed by the relationship

$$y = (1 + 3\sin^2\lambda_m)^{-1/4}\cos^3\lambda_m \tag{17}$$

that holds between y and λ_m in a dipolar B field. If processes other than charge exchange are neglected for simplicity, then an initially isotropic phase-space density $\bar{f}(E, L; 0)$ evolves into the distribution

$$\bar{f}(E, L, y; t) = \exp(-\langle N_H \rangle \sigma v t)\bar{f}(E, L; 0), \tag{18}$$

which is anisotropic for times $t > 0$ by virtue of the dependence of $\langle N_H \rangle$ on y. CORNWALL (1977) realized that such anisotropy among ring-current ions could lead to instability in electromagnetic ion-cyclotron wave modes. He was able to calculate the growth rate in closed form as a function of wave frequency ($\omega/2\pi$) and elapsed time (t) under the simplifying (but admittedly unrealistic) approximation that $\langle N_H \rangle$ scales as y^{-2} rather than as y^{-1} or as $\sec^{3.5}\lambda_m$. His results are nevertheless revealing and qualitatively consistent with expectation. The growth rate is initially negative at all frequencies $\omega/2\pi > 0$ because the pitch-angle distribution is isotropic, as can be seen from Eq. (18) for $t = 0$. Thereafter the distribution becomes anisotropic, and so the growth rate becomes positive for a band of frequencies $\omega/2\pi$ between zero and some fraction of the ion (in this case, proton) gyrofrequency $\Omega/2\pi$. Experience shows that the fraction should be approximately $A/(A + 1)$, where A is the anisotropy of the distribution. The fraction is exactly $A/(A + 1)$, for example, if the pitch-angle distribution has the form $\sin^{2A}\alpha_0$ or corresponds to a bi-Maxwellian distribution for which the temperature ratio is given by $T_\perp/T_\parallel = A + 1$. Of greater interest than the local growth rate Γ is the path-integrated gain

$$G = \exp\oint (2\Gamma/v_g)ds, \qquad (19)$$

where v_g is the group velocity and ds is the element of arc length along the ray path (usually taken to be a field line in order to simplify the task). It is common in magnetospheric physics to assume that $G \approx \exp(4\Gamma_0 La/v_g)$, where Γ_0 is the equatorial growth rate, in order to simplify the task further. However, LIEMOHN (1967) has demonstrated interesting effects that are overlooked if the path integral is not evaluated more carefully. In any event, the gain G has a maximum with respect to $\omega/2\pi$, and the maximum gain thus determined is not monotonic as a function of t. Indeed, the value of $G_{max}(t)$ is larger than unity for $0 < t < \infty$ but equal to unity for $t = 0$ (when the anisotropy A vanishes) and for $t = \infty$ (when $\tilde{f}$ itself vanishes by virtue of charge exchange). Thus, if $G_{max}(t)$ increases monotonically for $t < t^*$ and decreases monotonically for $t > t^*$, the remaining question is whether $G_{max}(t^*)$ is sufficiently large for instability to occur at all. The usual criterion for instability is $G_{max}(t^*) \gtrsim 400$, i.e., a single-pass gain of at least 20 in wave intensity (or 13 dB in decibel notation) from one foot of the field line to the other, since it is presumed that only about 5% of the wave intensity that reaches either foot of the field line will be reflected back into the magnetosphere. The integral in Eq. (19) refers to propagation once in *each direction* along the field line and so must compensate for two such 5% reflections. If the instability occurs, i.e., if $G_{max}(t^*) \gtrsim 400$, then the resulting ion-cyclotron waves will enhance the loss rate of ring-current ions by causing pitch-angle diffusion into the loss cone. Such diffusion will naturally reduce the anisotropy of the pitch-angle distribution and thus limit the instability, but it will also enhance the effectiveness of charge exchange as a loss mechanism for ring-current ions by increasing the mirror latitude (and hence the magnitude of $\langle N_H \rangle$) for the average ion trapped in the flux tube.

It follows from the above considerations that the probability of having pitch-angle diffusion among ring-current ions increases with the strength of the magnetic storm, i.e., with the intensity of the ring current itself. Recovery from a weak storm might be achieved through charge exchange alone, whereas recovery from a major storm might involve both charge exchange and pitch-angle diffusion (each serving to enhance the effectiveness of the other). The foregoing scenario is incomplete in at least three respects. First, the initial pitch-angle distribution is not likely to be isotropic, since the radial transport that populates the ring current from the plasma sheet creates pitch-angle anisotropy even when charge exchange and Coulomb drag are neglected. This was illustrated above (but for higher-energy ions) in Fig. 2 and in Eqs. (5)–(9). Thus, wave growth occurs even at $t = 0$, although $G_{max}(0)$ is likely to be smaller than $G_{max}(t^*)$ for some $t^* > 0$.

Secondly, the ring current is likely to be populated by two or more ionic species simultaneously. Perhaps H^+ is the major ion initially, i.e., during the main phase of a magnetic storm, with He^{++}, He^+, and O^+ present as (relatively) minor constituents. In this case the electromagnetic mode structure is interrupted by stop-bands above the gyrofrequencies of the heavier ions, and the hydrogen-cyclotron instability is partially suppressed in favor of enhanced excitation at frequencies below the gyrofrequencies of the helium and oxygen ions (e.g., CORNWALL and SCHULZ, 1971; CHIU et al., 1980). Ions having gyrofrequency $\Omega/2\pi$ can experience pitch-angle diffusion through a Doppler-shifted cyclotron resonance with a frequency $\omega/2\pi$ represented in the wave spectrum if the ion energy is sufficient to satisfy the resonance condition

$$kv = (\omega - \Omega)\sec\alpha > 0, \qquad (20)$$

where $2\pi/k$ is the wavelength that corresponds to the wave frequency $\omega/2\pi$. Resonance for $\Omega > \omega$ requires $\cos\alpha < 0$, i.e., particle and wave traveling in opposite directions along the field line. Resonance for $\Omega < \omega$ requires particle and wave to travel in the same direction along the field line so that $\cos\alpha > 0$. As H^+ ceases to be the major ion, the stop bands above the helium gyrofrequencies grow larger and suppress the hydrogen-cyclotron instability altogether (e.g., CHIU et al., 1980), but instability can still occur in the He^{++}, He^+, and O^+ passbands.

Finally, the foregoing scenario is incomplete in that electrostatic waves have been neglected. The usual rule-of-thumb in magnetospheric physics is that electromagnetic waves predominate inside the plasmasphere and electrostatic waves outside. Electromagnetic ion-cyclotron waves are easier to analyze because the cold-plasma dispersion relation can be perturbed in order to extract hot-plasma effects such as wave growth. This does not usually work for electrostatic ion-cyclotron waves, since in this case the cold-plasma dispersion relation does not provide a good approximation of the phase velocity. The usual spectrum of electrostatic ion-cyclotron turbulence is characterized by excitation bands located about midway between harmonics of the ion gyrofrequency, and the usual source of free energy is anisotropy (e.g., $T_\perp > T_\parallel$) in the pitch-angle distribution. Interesting complications should ensue for the ring-current plasma outside the plasmasphere, since (for example) the 3/2 harmonic of the H^+ gyrofrequency coincides with integer harmonics (the third, sixth, and twenth-fourth, respectively) of the He^{++}, He^+, and O^+ gyrofrequencies.

6. Perspective

Implementation of the foregoing principles may impress the reader as an onerous task. It is natural to ask whether a simpler theory of magnetospheric ion composition is not already available. In a certain sense the theory of TVERSKOY (1965) fills this request. TVERSKOY (1965) argued that the maximum in radiation intensity at any energy E should be located at the L value for which the "time scales" associated with radial diffusion and Coulomb drag are equal. This would occur where $M^{-1}(\mathrm{d}\,M/\mathrm{d}t)_v = -L^{-2}D_{LL}$ in the language of the present work. TVERSKOY (1965) neglected the Coulomb drag associated with neutral atoms and molecules in Eq. (14), as well as the radial diffusion associated with electrostatic impulses (unsteady convection) in Eq. (3).

On this basis one would expect to find

$$(4\pi Z^2 q_e^4/m_e)(m_0/2E^3)^{1/2}\langle N_e\rangle\ln(\bar{\lambda}_D m_e v/\hbar) \approx 8.1 \times 10^{-14}L^8[Q(y)/180D(y)]^2 \quad \sec^{-1} \tag{21}$$

at the L value of maximum radiation intensity for specified E, Z, A, and y, since the argument of the logarithm in Eq. (21) is very large ($\gtrsim 10^7$) and the value of γ is of order unity in most situations of interest. Solution of Eq. (21) for L yields

$$L_{\max} \approx 3.00\langle(N_e/10^3 \text{ cm}^{-3})\rangle^{1/8}[(1/16)\ln(\bar{\lambda}_D m_e v/\hbar)]^{1/8}$$
$$\times [180D(y)/Q(y)]^{1/4}Z^{1/4}A^{1/16}(E/1 \text{ MeV})^{-3/16} \tag{22}$$

as the location of peak radiation intensity. The normalization of N_e, E, and $\ln(\bar{\lambda}_D m_e v/\hbar)$ by typical values (10^3 cm^{-3}, 1 MeV, and 16, respectively) of these quantities makes the above expression for $L_{\max}$ easier to evaluate by inspection. It was on the basis of a similar expression that TVERSKOY (1965) anticipated the $E^{-3/16}$ scaling law for $L_{\max}$ and the appearance of the alpha-particle ($Z = 2$, $A = 4$) maximum at 1.3 ($\approx 2^{3/8}$) times as high an L value as the intensity maximum for protons of the same energy.

The scaling laws deduced by TVERSKOY (1965) are amply confirmed by the observations that FRITZ and SPJELDVIK (1981) have recently compiled. Even the leading factor 3.00 in Eq. (22), which is not arbitrary but derived from Eq. (21), seems appropriate in that (for $y = 1$) the 1-MeV proton maximum occurs at $L \sim 3$ and the 1-MeV alpha-particle maximum at $L \sim 4$ (actually at $L = 2.76$ and $L = 3.59$, respectively, upon more careful inspection). Such good agreement seems surprising in view of the fact that Eqs. (21) and (22) have no fundamental basis but only heuristic justification. On closer examination, however, the agreement is less impressive since (for example) the $E^{-3/16}$ scaling law for $L_{\max}$ follows from Eq. (21) only if $\langle N_e\rangle$ is independent of L. The more likely alternative scaling in which $\langle N_e\rangle$ is proportional to L^{-4} would have yielded $L_{\max} \propto Z^{1/6}A^{1/24}E^{-1/8}$, in substantially poorer agreement with the data compiled by FRITZ and SPJELDVIK (1981). Furthermore, the "alpha-particle" observations ordinarily refer to a mixture of $Z = 1$ and $Z = 2$ charge states rather than to $Z = 2$ alone, since the charge-exchange rates that prevail in the earth's magnetosphere make this mixture unavoidable (CORNWALL, 1972). It is not the purpose of this discussion to quibble over details of significant work that is nearly 20 years old, but rather to illustrate how much more sophisticated the treatment of radiation-belt dynamics has become over the intervening period. Just as the heuristic estimates by TVERSKOY (1965) satisfied the needs of an earlier decade of space research, so the work of SPJELDVIK and FRITZ (1978) exemplifies the more fundamental and quantitative approach that is demanded today.

The present work is not intended as a compendium of "who has done what" in the field of magnetospheric ion composition, but rather as an introduction to the principles upon which the modern dynamical theory of geomagnetically trapped ion populations is based. Review articles with extensive bibliographies on the subject have been prepared by CORNWALL and SCHULZ (1979), JOHNSON (1979), SPJELDVIK (1979), and FRITZ and SPJELDVIK (1981), among others. The present assignment seemed to require a different approach, and the foregoing pages represent the outcome of this endeavor.

The author thanks J. M. Cornwall and Y. T. Chiu for providing helpful comments and useful information on the subject reviewed.

This work was supported by the U. S. Air Force Systems Command Space Division (AFSC/SD) under Contract F04701-81-C-0082.

REFERENCES

AXFORD, W. I., On the origin of radiation belt and auroral primary ions, in *Particles and Fields in the Magnetosphere*, edited by B. M. McCormac, pp. 46–59, Reidel, Dordrecht, 1970.

BLAKE, J. B., J. F. FENNELL, M. SCHULZ, and G. A. PAULIKAS, Geomagnetically trapped alpha particles, 2, The inner zone, *J. Geophys. Res.*, **78**, 5498–5506, 1973.

BURTON, R. K., R. L. MCPHERRON, and C. T. RUSSELL, An empirical relationship between interplanetary conditions and D_{st}, *J. Geophys. Res.*, **80**, 4204–4214, 1975.

CHEN, A. J. and D. P. STERN, Adiabatic Hamiltonian of charged particle motion in a dipole field, *J. Geophys. Res.*, **80**, 690–693, 1975.

CHIU, Y. T., J. M. CORNWALL, J. G. LUHMANN, and M. SCHULZ, Argon-ion contamination of the plasmasphere, in *Space Systems and Their Interactions with Earth's Space Environment*, edited by H. B. Garrett and C. P. Pike, pp. 118–147, AIAA, New York, 1980.

CORNWALL, J. M., Transport and loss processes for magnetospheric helium, *J. Geophys. Res.*, **76**, 264–267, 1971.

CORNWALL, J. M., Radial diffusion of ionized helium and protons: A probe for magnetospheric dynamics, *J. Geophys. Res.*, **77**, 1756–1770, 1972.

CORNWALL, J. M., Moment transport equations for wave-particle interactions in the magnetosphere, *J. Geophys. Res.*, **80**, 4635–4642, 1975.

CORNWALL, J. M., On the role of charge exchange in generating unstable waves in the ring current, *J. Geophys. Res.*, **82**, 1188–1196, 1977.

CORNWALL, J. M. and M. SCHULZ, Electromagnetic ion-cyclotron instabilities in multicomponent magnetospheric plasmas, *J. Geophys. Res.*, **76**, 7791–7796, 1971; Correction, *J. Geophys. Res.*, **78**, 6830, 1973.

CORNWALL, J. M. and M. SCHULZ, Physics of heavy ions in the magnetosphere, in *Solar System Plasma Physics*, edited by L. J. Lanzerotti, C. F. Kennel, and E. N. Parker, Vol. 3, pp. 165–210, North-Holland Publ. Co., Amsterdam, 1979.

CROLEY, D. R., JR., M. SCHULZ, and J. B. BLAKE, Radial diffusion of inner-zone protons: Observations and variational analysis, *J. Geophys. Res.*, **81**, 585–594, 1976.

DAVIDSON, G. T., An improved empirical description of the bounce motion of trapped particles, *J. Geophys. Res.*, **81**, 4029–4030, 1976.

DAVIS, L. R. and J. M. WILLIAMSON, Low-energy trapped protons, *Space Res.*, **3**, 365–375, 1963.

DUNGEY, J. W., W. N. HESS, and M. P. NAKADA, Theoretical studies of protons in the outer radiation belt, *Space Res.*, **5**, 399–403, 1965.

FARLEY, T. A. and M. WALT, Source and loss processes of protons of the inner radiation belt, *J. Geophys. Res.*, **76**, 8223–8240, 1971.

FRITZ, T. A. and W. N. SPJELDVIK, Steady-state observations of geomagnetically trapped energetic heavy ions and their implications for theory, *Planet. Space Sci.*, **29**, 1169–1193, 1981.

GHIELMETTI, A. G., R. G. JOHNSON, R. D. SHARP, and E. G. SHELLEY, The latitudinal, diurnal, and altitudinal distributions of upward flowing energetic ions of ionospheric origin, *Geophys. Res. Lett.*, **5**, 59–62, 1978.

GLOECKLER, G., A versatile detector system to measure the charge states, mass compositions and energy spectra of interplanetary and magnetospheric ions (abstract), in *15th International Cosmic Ray Conference, Conference Papers*, Vol. 9, abst. T51, p. 207, Bulg. Acad. Sci., Sofia, Plovdiv meeting, August 1977.

GORNEY, D. J., A. N. CLARK, D. R. CROLEY, J. F. FENNELL, J. G. LUHMANN, and P. F. MIZERA, The distribution of ion beams and conics below 8000 km, *J. Geophys. Res.*, **86**, 83–89, 1981.

JOHNSON, R. G., Energetic ion composition in the earth's magnetosphere, *Rev. Geophys. Space Physics*, 17, 696–705, 1979.

KENNEL, C. F. and F. V. CORONITI, Is Jupiter's magnetosphere like a pulsar's or earth's?, *Space Sci. Rev.*, 17, 857–883, 1975.

LANZEROTTI, L. J., C. G. MACLENNAN, and M. F. ROBBINS, Proton drift echoes in the magnetosphere, *J. Geophys. Res.*, 76, 259–263, 1971.

LIEMOHN, H. B., Cyclotron-resonance amplification of VLF and ULF whistlers, *J. Geophys. Res.*, 72, 39–55, 1967.

LYONS, L. R. and D. S. EVANS, The inconsistency between proton charge exchange and the observed ring current decay, *J. Geophys. Res.*, 81, 6197–6200, 1976.

LYONS, L. R. and T. W. SPEISER, Evidence for current sheet acceleration in the geomagnetic tail, *J. Geophys. Res.*, 87, 2276–2286, 1982.

MIZERA, P. F. and J. B. BLAKE, Observations of ring current protons at low altitudes, *J. Geophys. Res.*, 78, 1058–1062, 1973.

PAULIKAS, G. A. and J. B. BLAKE, Penetration of solar protons to synchronous altitude, *J. Geophys. Res.*, 74, 2161–2168, 1969.

ROEDERER, J. G., *Dynamics of Geomagnetically Trapped Radiation*, pp. 66, 127, Springer, Heidelberg, 1970.

SCHULZ, M., Bounce-averaged synchrotron loss in a dipole field, *J. Geophys. Res.*, 82, 2815–2818, 1977.

SCHULZ, M. and L. J. LANZEROTTI, *Particle Diffusion in the Radiation Belts*, pp. 17–22, 44, 83–95, Springer, Heidelberg, 1974.

SCKOPKE, N. and G. PASCHMANN, The plasma mantle: A survey of magnetotail boundary observations, *J. Atmo. Terr. Phys.*, 40, 261–278, 1978.

SHARP, R. D., R. G, JOHNSON, and E. G. SHELLEY, Observation of an ionosphere acceleration mechanism producing energetic (keV) ions primarily normal to the geomagnetic field direction, *J. Geophys. Res.*, 82, 3324–3228, 1977.

SHELLEY, E. G., R. D. SHARP, and R. G. JOHNSON, Satellite observations of an ionospheric acceleration mechanism, *Geophys. Res. Lett.*, 3, 654–656, 1976.

SMITH, P. H., N. K. BEWTRA, and R. A. HOFFMAN, Decay mechanism for ring current ions (abstract SM11), *EOS, Trans. Am. Geophys. Union*, 57, 982, 1976.

SPEISER, T. W., Particle trajectories in model current sheets, 1, Analytical solutions, *J. Geophys. Res.*, 70, 4219–4226, 1965.

SPJELDVIK, W. N., Expected charge states of energetic ions in the magnetosphere, *Space Sci. Rev.*, 23, 499–538, 1979.

SPJELDVIK, W. N. and T. A. FRITZ, Theory for charge states of energetic oxygen ions in the earth's radiation belts, *J. Geophys. Res.*, 83, 1583–1594, 1978.

STERN, D. P. and P. PALMADESSO, Drift-free magnetic geometries in adiabatic motion, *J. Geophys. Res.*, 80, 4244–4248, 1975.

TINSLEY, B. A., Evidence that the recovery phase ring current consists of helium ions, *J. Geophys. Res.*, 81, 6193–6196, 1976.

TVERSKOY, B. A., Transport and acceleration of charged particles in the earth's magnetosphere, *Geomagnetizm i Aeronomiya*, 7, 226–242, 1965; English translation: *Geomagn. Aeron.*, 7, 177–189, 1965.

Energetic Ion Composition in the Earth's Magnetosphere, edited by R. G. Johnson, 23–41.

Ion Acceleration Mechanisms in the Auroral Regions: General Principles

W. Lennartsson

*Lockheed Palo Alto Research Laboratory,
3251 Hanover Street, Palo Alto, California 94304, U.S.A.*

(Received July 19, 1982)

The theoretical interpretation of particle and field data from the auroral regions has proven to be a complex and rather controversial area of geophysics. Many different models and concepts have emerged in the context of particle acceleration, each emphasizing and idealizing different aspects of the complex interactions between charged particles. This article is a very brief review of representative sections of the literature and is concentrated on general physical principles. The various acceleration mechanisms are grouped into nine categories, each of which can be labelled in terms of idealized spatial and temporal properties of the electric and magnetic fields.

1. Introduction

The last two decades of space exploration have greatly improved our knowledge of the earth's magnetosphere and its various plasma regions. In some respects the observations have been a step ahead of the theory, however, and have raised more questions than they have answered. Many of the new questions have been centered on the acceleration of charged particles and the nature of the electric fields involved, particularly in those regions of the magnetosphere where the solar wind plasma encounters and enters into the earth's magnetic field, that is immediately inside of the magnetopause and along magnetic field lines connected to the auroral regions. Although there is a general consensus among scientists that the solar wind provides both the energy, kinetic and electric, and a good portion of the particles, there is still substantial disagreement as to the relative importance of various acceleration mechanisms. This state of matters has resulted in quite a voluminous literature, and it is impossible to give a comprehensive and fair treatment of all potentially important models in a review of this size. Since the limited discussion here will naturally be some what biased by the author's prejudices, there will be references to other review articles where the reader may find supplementary viewpoints. For general overviews of the solar-wind-magnetosphere interactions the author recommends review articles by Akasofu (1977), Axford (1970, 1976), and Roederer (1979).

The emphasis in this article will be on electric fields that are associated with magnetic field-aligned currents. This, perhaps, may be justified on the grounds that such currents, flowing into and out of the earth's auroral regions, are a fundamental element of the energy transfer from the solar wind to the magnetospheric plasma. Due to the length constraint, the discussion must be limited to general principles of particle acceleration, leaving little room for evaluating the merits of individual models. The arrangement of the different topics is partly adopted from earlier review articles, notably from two articles by FÄLTHAMMAR (1977, 1979). The physical quantities will be expressed in MKS units, and it will be assumed that all particles have only non-relativistic energies.

2. Idealized Single Particle Motion

As a prelude to discussing acceleration mechanisms it is necessary to review a few basic concepts of non-relativistic single particle motion. Thus consider a particle of mass M and charge Q which is acted upon by only a magnetic field $\boldsymbol{B}$ and an electric field $\boldsymbol{E}$, and is moving with velocity components $v_\parallel$ and $\boldsymbol{v}_\perp$ in the directions along and transverse to $\boldsymbol{B}$, respectively. Unless $\boldsymbol{B}$ or $\boldsymbol{E}$ are very inhomogeneous or very rapidly varying in time, it can be shown (e.g., LONGMIRE, 1963) that the transverse motion is a combination of a nearly circular gyro motion at a speed v_g, having an angular frequency $\omega_g = |Q||\boldsymbol{B}|M^{-1}$ and a radius $\rho = M v_g |Q|^{-1}|\boldsymbol{B}|^{-1} = v_g \omega_g^{-1}$, and a drift motion at a velocity which is the sum of a quantity $\boldsymbol{E} \times \boldsymbol{B}|\boldsymbol{B}|^{-2}$ and small perturbations due to inhomogeneities in $\boldsymbol{B}$ and $\boldsymbol{E}$ and to time variations in $\boldsymbol{E}$. The gyro motion is characterized by a nearly constant magnetic moment μ:

$$\mu = M v_g^2 2^{-1} B^{-1} \simeq \text{constant}, \tag{1}$$

where $B = |\boldsymbol{B}|$ and $v_g \simeq |\boldsymbol{v}_\perp - \boldsymbol{E} \times \boldsymbol{B} B^{-2}|$. The perpendicular drift velocity of the gyration center is given by (LONGMIRE, 1963):

$$\boldsymbol{v}_{\perp d} \simeq \boldsymbol{E} \times \boldsymbol{B} B^{-2} + M Q^{-1}[\mathrm{d}E_\perp \mathrm{d}t^{-1} + v_g^2 \boldsymbol{B} \times \nabla B 2^{-1} B^{-1} + v_\parallel^2 \boldsymbol{B} \\ \times (B^{-1}\boldsymbol{B} \cdot \nabla)(B^{-1}\boldsymbol{B})]B^{-2}, \tag{2a}$$

where ∇B is the gradient of B and $B^{-1}\boldsymbol{B} \cdot \nabla$ is the spatial derivative in the direction of $\boldsymbol{B}$. The time derivative of $E_\perp$ is that seen by the moving particle and thus depends on the particle's velocity and on inhomogeneities in $E_\perp$. The last two terms in Eq. (2a) are due to the gradient of $|\boldsymbol{B}|$ and the curvature of the field lines, respectively. The parallel velocity is likewise given by:

$$M (\mathrm{d}\boldsymbol{v} \, \mathrm{d}t^{-1})_\parallel \simeq Q E_\parallel - M v_g^2 (\boldsymbol{B} \cdot \nabla B) 2^{-1} B^{-2}. \tag{2b}$$

The expressions (1) and (2) are valid if and only if $\boldsymbol{B}$ and $\boldsymbol{E}$ vary very little over a distance of one gyro radius and during a time span of one gyro period. Furthermore, the temporal variations in $E_\perp$, even if small, must not be periodic at the gyro frequency.

The considerable difficulties that have been encountered in the theoretical analysis stem from the fact that the particles in reality behave in a collective fashion rather than as single independent particles. This is actually inherent in the above formulas, particularly (2a). Whereas the $\boldsymbol{E} \times \boldsymbol{B}$ drift is identical for ions and electrons, the remaining three drift terms have opposite directions for ions and electrons. Hence, if $v_g \neq 0$ or $v_\parallel \neq 0$ the particle drift can pile up electric charge, thus converting kinetic energy into electric potential energy. This will not happen if the particle density is perfectly constant along the drift trajectories, of course, but that situation is not very realistic. However, in the case of the solar wind interaction with the magnetosphere it is possible for the charges to be conducted through the ionosphere via the auroral magnetic field lines. In fact, such field-aligned currents are a permanent feature of the auroral regions, as inferred from measured perturbations of the magnetic field. A summary of data from weakly disturbed conditions is shown in Fig. 1 (from IIJIMA and POTEMRA, 1976). Such currents can flow not only as a results of local buildup of charge at high altitudes but are probably also driven by the large-scale electric field that already exists in the unobstructed solar wind (ALFVÉN, 1975; JOHNSON, 1978; STERN, 1979; ATKINSON, 1981). In any case, the field aligned currents, in the presence of

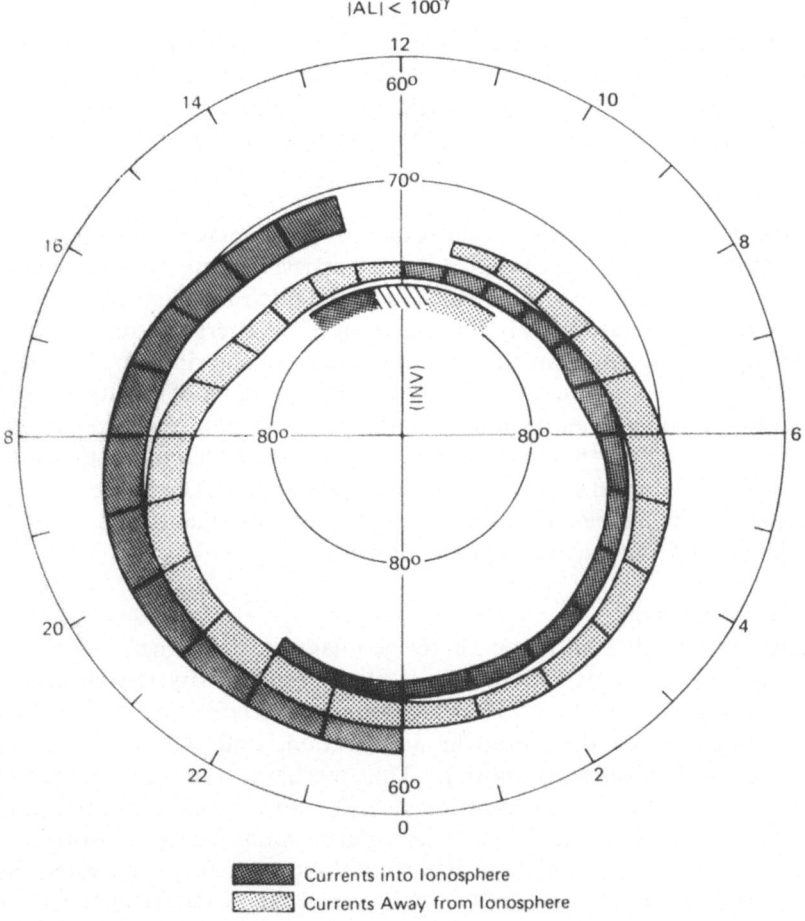

Fig. 1.   A summary of the distribution and flow directions of large-scale field-aligned currents inferred from magnetic field data obtained on 493 passes of Triad at 800 km altitude during weakly disturbed conditions ($|AL| < 100\,\gamma$). The 'hatched' area shown between 1130 and 1230 MLT in the polar cusp region indicates that the current flow directions are often confused. (From IIJIMA and POTEMRA, 1976; copyrighted by the American Geophysical Union).

parallel electric fields (see below), can transfer electric potential energy back into kinetic energy of that subset of particles that carry the current, including upward moving ions from the ionosphere (cf. BLOCK and FÄLTHAMMAR, 1969).

Another likely effect of the differential drift of ions and electrons is a breakdown of the approximations in Eqs. (1) and (2) particularly for the ions which have longer gyro periods and larger gyro radii (for comparable energies). Recent observations, notably by the polar orbiting S3–3 spacecraft, clearly show that Eq. (1) is extensively violated for ions in the auroral regions (see below; see also a companion paper in this book on S3–3 data by R. D. SHARP et al., 1983).

3.   Acceleration by $E_\perp$

The amount of energy that a particle can gain from a perpendicular electric field hinges on the approximation in Eq. (1). As long as Eq. (1) is valid the gyro energy is limited by the increase in the magnetic field strength that a particle encounters during its drift motion. The increase in B may be spatial or temporal.

3.1   Betatron acceleration
The increase in gyro velocity that results from an increase in B is referred to as 'betatron acceleration.' It has the effect that a particle drifting inward in the magnetic field, on closed field lines, will move its equatorial pitch angle closer to 90° and thus move its mirror points successively closer together. The energy increase that is allowed by Eq. (1) has proved sufficient to account for most of the ring current energization, and is discussed in a companion review paper in this book by M. SCHULZ, 1983. It thus appears that betatron acceleration may be the basic mechanism for extracting kinetic energy from $E_\perp$ in the inner magnetosphere, even though the actual particle motion may be more complicated than Eqs. (1) and (2) would suggest. The betatron acceleration due to a perpendicular electric field is insufficient to explain data from the outer magnetosphere and the auroral regions, however.

3.2   Fermi acceleration
A particle that is drifting inward in the geomagnetic field, on closed field lines, will also be forced to move its mirror points closer together by the shrinking spatial dimensions along the field lines. This has the opposite effect on the equatorial pitch angle, as compared to the betatron acceleration, and will divert energy from perpendicular to field-aligned motion. This mechanism is referred to as 'Fermi acceleration,' and is most important for particles with small equatorial pitch angles (not yet within the loss cone). If a particle population has a roughly isotropic velocity distribution prior to acceleration, the total kinetic energy generated by Fermi acceleration will be small compared to that generated by betatron acceleration in a dipole magnetic field (e.g. ALFVÉN and FÄLTHAMMAR, 1963, Chapter 2). On the other hand, the stretched field configuration in the magnetotail may alter that situation significantly, and it has been argued in the literature that Fermi acceleration is indeed a major energization mechanism on closed auroral field lines (e.g. SHARBER and HEIKKILA, 1972).

The approximation (1) is clearly in conflict with some of the energetic ion data from the auroral regions and from regions magnetically connected to the auroral regions (see e.g. SHARP et al., 1977; KLUMPAR, 1979). It is probably also in conflict with low-energy ion data from a large region of the magnetosphere (e.g. BAUGHER et al., 1980; HORWITZ, 1980; HORWITZ et al., 1982). There are basically three different situations in which Eq. (1) breaks down, and these are as follows.

3.3   Singularities in B ($B \simeq 0$)

The differential motion of ions and electrons also has a profound effect on the magnetic field, deforming the geomagnetic field into what is somewhat inappropriately called the 'magnetosphere.' The deformed field is expected to have singularities at the front of the magnetopause and in the distant magnetotail, in the form of neutral points, neutral lines, or neutral surfaces, at which the magnetic field strength approaches zero (ALFVÉN, 1968; COWLEY, 1973; VASYLIUNAS, 1975; SONNERUP, 1979). The perpendicular electric field does not necessarily go to zero there, but may remain essentially the same as in the surrounding plasma, which is assumed to be the case in the so called 'magnetic reconnection' theories. From Eq. (2a) it would seem that the $E \times B$ drift goes to infinity at these locations, but in reality Eq. (1) breaks down because of the finite size of the gyro radii compared with the small dimension of the singularities. Nevertheless, the particles that go through a singularity, or close to it, can move in the direction of the electric field and gain large energies (e.g. DUNGEY, 1953; SPEISER, 1965; HILL, 1975; HAYASHI and SATO, 1978). Whether this is a major energization mechanism appears at present to be a matter of controversy (SONNERUP, 1979). One objection has been that relatively few particles are affected.

This mechanism may be most important during a dynamical situation, when the magnetic field is changing its configuration and may create very large electric fields by induction (HEIKKILA and PELLINEN, 1977; PELLINEN and HEIKKILA, 1978). The effect may be to accelerate a few particles to very high energies, rather than to substantially energize the bulk of the plasma, however.

3.4   Inhomogeneities in $E_\perp$

The acceleration of particles across inhomogeneities in $E_\perp$ has so far received relatively little attention in the literature, and there is no extensive description of it. The following consideration may be of some help.

The first term inside the bracket of Eq. (2a) is called the 'polarization drift.' If this is integrated in time, assuming $B \simeq$ constant, the resulting particle displacement is given by:

$$\Delta r \text{ (polarization)} \simeq M Q^{-1} B^{-2} \Delta E_\perp. \tag{3}$$

Hence, the displacement is limited by the maximum change in the electric field that a particle encounters ($\lesssim 10$ km for a proton if $\Delta E_\perp \lesssim 0.1$ Vm^{-1} and $B \sim 1,000$ nT, for example). The displacements due to the second and third terms in the bracket, the 'gradient-B' and 'curvature drifts,' respectively, are only limited by the spatial dimensions of the magnetosphere, however. Since these two drift terms are in opposite directions for ions and electrons they place rather severe restrictions on a steady-state plasma distribution. Note that the $E \times B$ drift is in the same direction for both ions and electrons and thus cannot neutralize a charge buildup. In order for Eq. (2a) to remain valid, it is necessary that any density gradient in the plasma be perpendicular to the combined gradient-B and curvature drifts, or else that good electric conduction be maintained through the ionosphere via field-aligned currents. But if neither of these two conditions is fully met, Eq. (2a) implies a continued accumulation of charges at various places, until Eq. (2a) breaks down as a description of the particle motion, at least for the ions. The latter is equivalent to a breakdown of Eq. (1), which can be expressed by a formula from COLE (1976) and LENNARTSSON (1980):

$$|V \cdot E_\perp| \gtrsim \omega_g B = |Q| B^2 M^{-1} \tag{4}$$

This is a degenerate form of Eq. (3), when the spatial scale length of $E_\perp$ becomes comparable to or smaller than the displacement of the particles.

Although this consideration is overly simplified, it does suggest the kind of large-amplitude and small-scale fluctuations of $E_\perp$ that are actually observed on a regular basis in the auroral current regions (MOZER et al., 1977, 1980). An example of the observations by S3−3 is shown in Fig. 2 (from MOZER et al., 1977), where $E_\perp$ (as well as $E_\parallel$) has a temporal scale size down to $\lesssim 1$ sec, as seen from the moving spacecraft. This translates to a spatial scale size of a few km or less, if the fluctuations are spatial only,

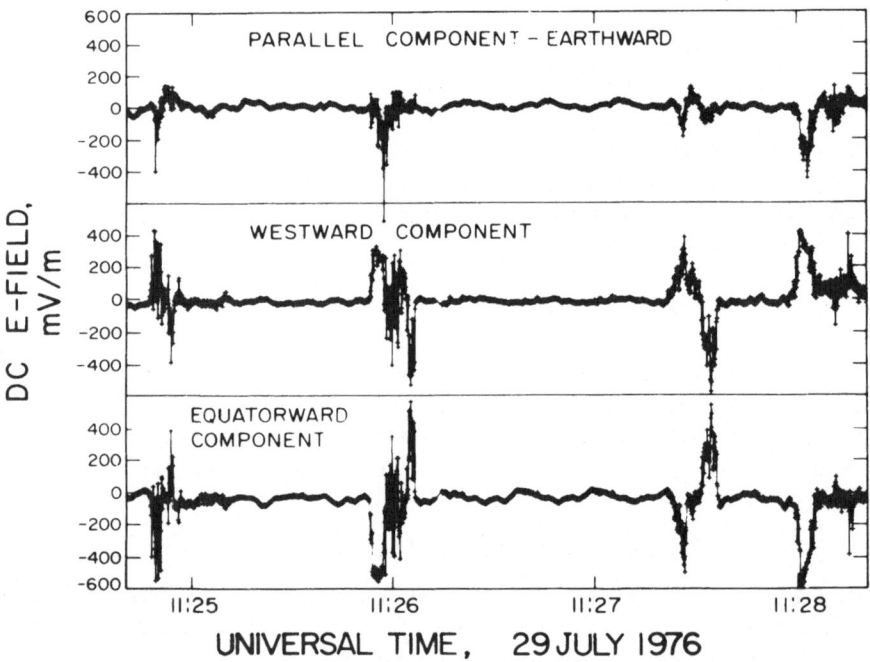

Fig. 2. Electric field measurements made on the S3-3 satellite during its poleward-bound passage through the northern auroral zone at 7,600 km altitude. (From MOZER et al., 1977; copyrighted by the American Physical Society.)

which is sufficient to satisfy Eq. (4) for O^+, at least, at the altitude of these measurements.

A particle traversing a region of enhanced perpendicular electric field under the condition in Eq. (4) can gain part or all of its kinetic energy in the form of gyro energy, provided of course that it enters from the side with the higher potential energy (COLE, 1976; COLE and AGGSON, 1976; SWIFT, 1979; LENNARTSSON, 1980). If the electric field has a 'turbulent' temporal structure, which is perhaps suggested by Fig. 2, the acceleration will probably become a diffusion process. Our understanding of this mechanism will have to await further research, however.

3.5 Cyclotron resonance

According to model calculations by KINDEL and KENNEL (1971) and others, the differential motion of ions and electrons along the magnetic field in the auroral regions is likely to excite electrostatic ion cyclotron waves. These waves may in turn accelerate a subset of the ions to considerable energy via cyclotron resonance (e.g. CLADIS, 1973; UNGSTRUP et al., 1979; OKUDA et al., 1980; OKUDA and ASHOUR-ABDALLA, 1981; DUSENBERY and LYONS, 1981). This mechanism is presently receiving a good deal of attention as the possible key to an understanding of the so called 'conical' ion pitch-angle distribution. It is the special subject of a companion review in this book by ASHOUR-ABDALLA and OKUDA, 1983.

4. Acceleration by $E_\parallel$

Charged particles are more readily accelerated by a parallel component of the electric field than they are by a perpendicular component. This, however, has been the reason for a great deal of controversy about the nature, or even the existence, of parallel electric fields in the magnetosphere. It may seem, in retrospect, that part of this controversy has been due to an emphasis on the parallel mobility of the charge carriers rather than on the effective conductance. If the second term on the right hand side of Eq. (2b) is neglected ($v_g \simeq 0$ or $\boldsymbol{B} \cdot \nabla B \simeq 0$), which may be justified in many cases, it does follow that the particles have a very high mobility. But as the particles are accelerated by $E_\parallel$ their number density decreases, so it does not follow that the parallel current density can increase indefinitely, even if the velocity does. Although the mobility may be hampered by 'anomalous resistivity' the discussion may have underemphasized the fact that a parallel electric field can exist without a reduction of the mobility, as in the classical 'double layer.' Furthermore, it may not have been fully appreciated that the major current carriers in many cases are energetic electrons for which the second term on the right in Eq. (2b), the 'magnetic mirror force,' is of fundamental importance.

4.1 A kinetic perspective

It is easy to show (e.g. LONGMIRE, 1963, Chapter 1) that if the electric and magnetic forces are the only forces on idealized point charges with finite masses they make the phase space (r, $M\boldsymbol{v}$ space) incompressible, although they do not guarantee that a given segment of phase space remains in one piece. If a particle p_1 is observed at a point r_{ob} at time $t = 0$ to have a velocity $\boldsymbol{v}_{p1}$, the incompressibility of phase space of one species of particles can be expressed by

$$f(\boldsymbol{r}_{ob}, \boldsymbol{v}_{p1}, t = 0) = f(\boldsymbol{r}_{p1}(t = -\tau), \boldsymbol{v}_{p1}(t = -\tau), t = -\tau), \tag{5}$$

where f is the density of particles per unit volume of $\boldsymbol{r}$, $\boldsymbol{v}$ space (in the nonrelativistic limit). The left hand member is the density of particles observed near $\boldsymbol{r}_{ob}$ at $t = 0$, having velocities close to and including the velocity of particle p_1. The right hand member is the density of particles close to the position of p_1 at a time τ sec earlier, having the same or nearly the same velocity as p_1 at that time. This equation is alternatively referred to in the literature as a form of 'Liouville's Equation' or 'Vlasov's Equation.'

Now consider Fig. 3. Because of the potential barrier in Fig. 3a, only a part, d^3v_1, of the original velocity region will be populated at $\boldsymbol{r}_{ob}$. Within this region the phase space density will have remained the same, according to Eq. (5), but at the edge of the missing part, d^3v_2, the density has a discontinuity at which it goes to zero. By raising the electric potential of the particle source in Fig. 3b, the entire velocity region will be projected to $\boldsymbol{r}_{ob}$, but it will be displaced,

$$v'^2 = v^2 + 2M^{-1}Q \int_{\boldsymbol{r}_s}^{\boldsymbol{r}_{ob}} \boldsymbol{E} \cdot d\boldsymbol{r} \tag{6}$$

and modified in shape and size. In particular, if $B \neq 0$ and

$$\boldsymbol{E} \parallel \boldsymbol{B} \tag{7}$$

$$B(\boldsymbol{r}_{ob}) = B(\boldsymbol{r}_s), \tag{8}$$

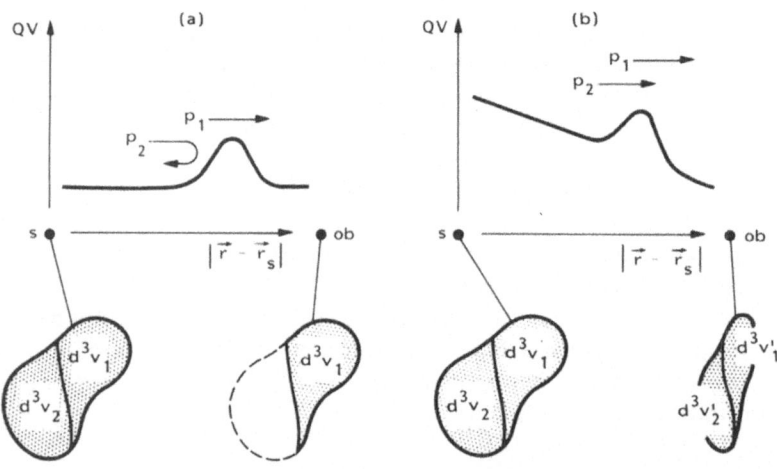

Fig. 3. (a) Due to a potential barrier between the particle source (s) and the observer (ob) only one of the two particles p_1 and p_2 is able to reach the observer. The two particles belong respectively to two adjacent velocity regions, d^3v_1 and d^3v_2, only one of which is populated at $\boldsymbol{r}_{ob}$. (The potential barrier is assumed time-independent here.) (b) By raising the potential of the source, both p_1 and p_2 can reach the observer and both velocity regions are projected to $\boldsymbol{r}_{ob}$ and also distorted (see text).

and $d^3 v_1$ is taken to be an infinitesimal volume element in spherical coordinates, it follows from Eq. (6) that

$$d^3 v_1' = (\cos \alpha_1')^{-1} \cos \alpha_1 (v_1')^{-1} v_1 d^3 v_1, \tag{9}$$

where v_1 and v_1' are the respective velocities and α_1 and α_1' are the respective angles between $\boldsymbol{B}$ and the velocity vectors. Since the phase space density f is still the same in $d^3 v_1'$, according to Eq. (5), the contribution to the field-aligned current density at r_{ob} from particles within $d^3 v_1'$ is independent of the electric potential $V(r_s)$,

$$Q f v_1' \cos \alpha_1' d^3 v_1' = Q f v_1 \cos \alpha_1 d^3 v_1 \tag{10}$$

as long as $Q V(r_s) > Q V(r_{ob})$. This, of course, is just a restatement of the Equation of Continuity. However, the increased potential at r_s has the effect that particles within $d^3 v_2$ will also contribute to the current, so in this case the total current density does increase with increasing potential difference.

Viewed in this perspective it is clear that a plasma volume has a steady-state conductance larger than zero only if the imposed electric field can push certain charge carriers through the entire volume that would otherwise be reflected by internal forces. The final velocity of these charge carriers is irrelevant.

The same reasoning holds true in the case of Coulomb collisions, except that the phase space volume occupied by a given set of particles at time $t = t_0$ will be rapidly fragmented into very small pieces at $t > t_0$, each piece containing but one particle eventually. In that limit Eq. (5) becomes impractical, and the only measurable density will be the random distribution of the fragments.

4.2 The double layer space charge

What Eqs. (6) to (10) demonstrate is the fact that a steady-state electric field cannot produce a higher flux of particles at low potential energy than the flux of particles that enter at high potential energy. Whether the electric field can reach a steady state if it is increased to and beyond the level of current saturation is a more subtle question. Clearly the acceleration and reflection of particles, which are in opposite directions for ions and electrons, produce space charges that create their own electric fields. A transient or fluctuating $E_\parallel$ is actually an intrinsic feature of some models (e.g. HASEGAWA and CHEN, 1975; GOERTZ and BOSWELL, 1979).

It has been proven in laboratory experiments, however, that a plasma, magnetized or not, can under certain conditions support a steady, or nearly steady, electric field at current saturation by redistributing most of the internal space charges into two adjacent layers of opposite polarity, commonly referred to as one 'double layer' (e.g. CRAWFORD and FREESTON, 1963; SCHÖNHUBER, 1963; ANDERSON et al., 1969; QUON and WONG, 1976; COAKLEY et al., 1979; TORVÉN and ANDERSON, 1979; BAKER et al., 1981). The laboratory double layer that has been most extensively studied is a flat disk oriented perpendicular to the current path and connecting radially to the walls of the plasma container. The thickness, L, of the double layer is related to the Debye length $\lambda_D = (\varepsilon_0 k T_e e^{-2} n_e^{-1})^{1/2}$ and the total potential difference V across the layer (FÄLTHAMMAR, 1979),

$$L = \frac{1}{\gamma}\left(\frac{eV}{kT_e}\right)^{1/2} \lambda_D \tag{11}$$

where γ is typically in the range 0.03–0.1. As for the ratio in the bracket, the experiments so far have produced values in the range (TORVÉN, 1979):

$$1 \lesssim \frac{eV}{kT_e} \lesssim 20. \tag{12}$$

Depending on the experimental conditions, the double layer may be steady, intermittent, fixed, or moving (TORVÉN, 1979). For a review of the current theoretical understanding of laboratory double layers, see CARLQVIST (1979).

It is not yet clear how the laboratory double layers can be scaled to magnetospheric conditions, since the influence from the walls of the laboratory plasma containers is not fully understood. It has only been very recently (e.g. BAKER et al., 1981; STENZEL et al., 1981) that the radial confinement of the space charge structure has begun to be replaced by magnetic confinement, as must be the case in the magnetosphere.

A substantial amount of analysis and numerical simulation of auroral conditions has nevertheless been done already (e.g. BLOCK, 1972; KNORR and GOERTZ, 1974; SWIFT, 1976; HUBBARD and JOYCE, 1979; KAN et al., 1979; WAGNER et al., 1980). The more advanced model calculations of auroral double layer structures assume a 2-dimensional electric potential geometry with the shape of a narrow V or U, extended vertically along the magnetic field lines (e.g. SWIFT, 1976, 1979; KAN et al., 1979). These structures have notable similarities to the narrow regions of large perpendicular electric field that are commonly seen at high altitude in the auroral regions (cf. Fig. 2). For a broader review of geophysical applications, see GOERTZ (1979).

4.3 Anomalous resistivity

The classical relation between the electric field and the current density, i, in a fully ionized plasma is

$$E = \eta i = m_e v e^{-2} n_e^{-1} i, \tag{13}$$

where m_e and n_e are the electron mass and density, e is the unit charge and v is a collision frequency which is dominated by long range Coulomb collisions, such that (SPITZER, 1956),

$$v \propto n_i (\bar{v}_e)^{-3},$$

where n_i is the ion density and $\bar{v}_e$ is a typical random velocity of the electrons relative to the ions. The maximum steady-state field that can be supported by Coulomb friction is thus defined by

$$eE = m_e vv_e \simeq m_e v\bar{v}_e,$$

which is a rather negligible field under magnetospheric conditions. If E is increased beyond this limit and if it can be assumed that

$$n_e \simeq n_i \simeq \text{constant} \tag{14}$$

it follows that i approaches infinity, which is contrary to observations.

In order to accommodate a significant parallel electric field within the framework of Eqs. (13) and (14), it has been proposed that a much stronger effective 'friction' may arise between the ions and electrons due to wave-particle interactions, when the electron drift speed exceeds some critical value, and thus produce 'anomalous resistivity' (e.g. SWIFT, 1965; KINDEL and KENNEL, 1971; PAPADOPOULOS and COFFEY, 1974; PALMADESSO et al., 1974; FEDDER, 1976; HUDSON et al., 1978). This approach falls within a research area that has been very active for over 20 years, employing not only theoretical analysis but also numerical simulation and laboratory experiments. For a broad review of the geophysical applications, see HASEGAWA (1974) and PAPADOPOULOS (1977). A recent review of the more fundamental aspects has been given by DUM (1981).

A common analytical approach has been to consider the growth in amplitude of a plasma wave (in particular ion accoustic and ion cyclotron waves and electron plasma oscillations) which is in resonance with a drifting population of electrons (usually a drifting Maxwellian), i.e. electrons drifting relative to the ions. This approach has fostered a popular concept of 'current-driven' instabilities but has not been successful in addressing the non-linear saturated state of the waves. Another approach has been to calculate a resistivity from an assumed saturated wave spectrum. Using the latter method, PAPADOPOULOS (1977) has concluded that only the ion acoustic waves and the parametrically driven waves (driven by an energetic particle beam) are likely to produce a significant steady-state resistivity.

A more direct approach has been to use numerical simulation and plasma experiments. Both methods have revealed highly complex and transient interactions between the waves and the particles, however, and neither method has yet proved that a steady-state 'friction' of the simple form in Eq. (13) can exist between the ions and electrons due to waves (cf. DUM, 1981). It is interesting to note, that the experiments, which can disregard Eq. (14) by nature, often produce 'double layer' structures (e.g. KALININ et al., 1970; DUM, 1981, and references therein).

4.4 Collisionless thermoelectric effect

The enhanced 'frictional force' that may arise between ions and electrons due to wave-particle interactions is often predicted to be the strongest on the bulk portion of the particles, whereas a higher energy particle can move relatively freely (PAPADOPOULOS, 1977, and references therein). This situation is somewhat analogous to classical Coulomb friction, and it has been suggested by HULTQVIST (1971, 1972) that wave-particle interactions can support a substantial thermoelectric field, with a downward direction, along the magnetic field in regions of intense auroral electron precipitation. In this model the electric field is generated by the energetic electrons, depositing negative charge at low altitude, and is supported by the wave-generated 'friction' on the cold electrons moving upwards. The model may explain why precipitating electrons have been observed to be accompanied by energetic ions with a field-aligned velocity distribution (HULTQVIST, 1971, 1972).

4.5   Magnetic confinement

Whatever local process is involved in the transfer of electric potential energy into particle kinetic energy it can be in a steady state only if there is some external 'container' that can support the global electric stresses. The obvious candidate in the geophysical system is the earth's magnetic dipole field. This requires, however, that a substantial portion of the electric space charges be made up of particles for which the second term on the right in Eq. (2b), the 'magnetic mirror force,' is important. It is therefore satisfying that EVANS (1974) was able to demonstrate that the dense low-energy portion of the auroral electron population can be identified, within experimental uncertainties, as being energy-degraded, backscattered and secondary electrons, trapped below the acceleration region. This suggests that the high-energy primary electrons are representative of the particles that make up the negative space charge at high altitude. A three-dimensional representation of a fairly typical velocity distribution of auroral electrons is shown in Fig. 4 (from KAUFMANN and LUDLOW, 1981). The fact that the measured phase space density is a continuous function of $|v|$, with only a plateau separating the low-energy population of atmospheric origin from the high-energy population of magnetospheric origin (on the downward flanks), perhaps reflects a substantial diffusion in phase space due to temporal and spatial fluctuations in the electric field (causing fragmentation and mixing of the original distributions).

The basic nature of the magnetic confinement, under idealized conditions, is illustrated for a hot magnetospheric population of electrons in Fig. 5. This figure is similar to Fig. 3, except that the electrons in this case have to overcome an additional

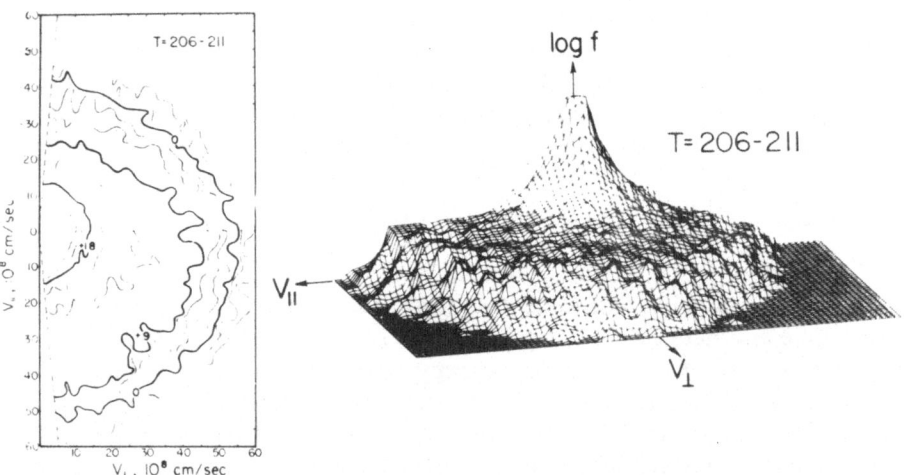

Fig. 4.   Contour and three-dimensional plot of auroral electron distribution function, in the energy range 25 eV–15 keV, measured from a rocket at about 240 km altitude. Downgoing electrons have positive $v_{\parallel}$. Curves of constant $f(v)$ on the contour plot are labeled by the common logarithm of $f(v)$ in $s^3 km^{-6}$. The large central peak on the three-dimensional plot is centered at $v = 0$. This distribution is typical of electrons producing discrete auroral arcs. (From KAUFMANN and LUDLOW, 1981; copyrighted by the American Geophysical Union.)

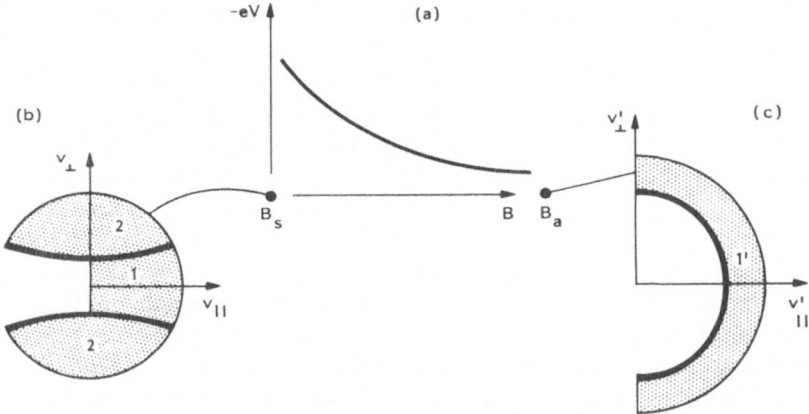

Fig. 5. (a) Schematic distribution of electric potential energy (for electrons) vs. magnetic field strength B between the particle source at B_s and the atmospheric sink at B_a. This distribution is of a type that produces the maximum precipitation for a given $V_a - V_s$ (see text). (b) Velocity regions at B_s (rotationally symmetric around $v_\perp = 0$). Only electrons in Region 1 can reach the atmosphere. The heavy contours represent the hyperboloidal surface: $(B_a B_s^{-1} - 1)v_\perp^2 - v_\parallel^2 = 2m_e^{-1}e(V_a - V_s)$. This surface is projected to $v_\parallel' = 0$ at $B = B_a$. (c) Projection of Region 1 to atmospheric altitude. The heavy contour represents the ellipsoidal surface $(v_\parallel' \geq 0$ only$): (1 - B_s B_a^{-1})v_\perp'^2 + v_\parallel'^2 = 2m_e^{-1}e(V_a - V_s)$. This surface is the projection of the $v_\parallel = 0$ boundary of Region 1. Both surfaces are derived in PERSSON (1966).

'barrier' corresponding to the 'magnetic mirror force' in Eq. (2b). This figure assumes that Eq. (1) is valid and that B is a monotonic function of distance along the field line. Furthermore, it assumes a time-independent electric potential,

$$\frac{\partial}{\partial t} V \equiv 0, \tag{15}$$

such that

$$V(B) - V_s \geq (B - B_s)(B_a - B_s)^{-1}(V_a - V_s). \tag{16}$$

The latter relation ensures that the largest possible velocity region, for a given $V_a - V_s$, is projected from B_s to B_a (LENNARTSSON, 1980) and results in hyperboloidal and ellipsoidal boundaries, as indicated (cf. PERSSON, 1966; BLOCK, 1967; CHIU and SCHULZ, 1978). If Eq. (16) holds true for $B_s B_a^{-1} \to 0$ the ellipsoidal surface at B_a approaches a spherical surface, which means that an isotropic downward flux of electrons at small B_s will be projected as a nearly isotropic flux to B_a, within the downward hemisphere.

If the source population has a Maxwell-Boltzmann distribution with thermal energy $k T_{es}$ (except perhaps for an empty loss region in the upward flux) it thus follows by a simple integration in energy, using Eq. (5), that the current density at B_a is bounded by

$$i_0 \leq i_{\parallel a} \leq (1 + e(V_a - V_s)k^{-1}T_{es}^{-1})i_0 \tag{17}$$

where i_0 is the natural precipitation of charge without acceleration. The complete expression for $i_{\|a}$ from a Maxwell-Boltzmann source as a function of $V_a - V_s$ and B_s, under the maximizing condition Eq. (16), can be found in KNIGHT (1973), and has been evaluated numerically by several authors (e.g. LEMAIRE and SCHERER, 1974; FÄLTHAMMAR, 1977; LYONS et al., 1979; LYONS, 1981; YEH and HILL, 1981; and with a bi-Maxwellian source distribution by FRIDMAN and LEMAIRE, 1980).

What Fig. 5 and the relation in Eq. (17) show under idealized assumptions can be interpreted in a more general sense. The geomagnetic field is indeed a leaky 'container' of charges, but the leakage rate can be substantially increased only by substantially increasing the electric potential energy of the appropriate particles relative to 'ground.' It is tempting to assume that the magnetic confinement, combined with the differential drift of ions and electrons across the magnetic field lines (cf. Eq. (2a)), is the basic cause of parallel electric fields. The virtues of this assumption as a means of ordering auroral particle and field data were discussed briefly by RASSBACH (1973) and LENNARTSSON (1976, 1977a) and at length by LENNARTSSON (1977b). Two recent and more detailed comparisons with specific data sets by LYONS (1981) and YEH and HILL (1981) suggest that the right hand side of Eq. (17) may approximate the major portion of the field-aligned current in intense auroras. If so, the electric charge distribution apparently adjusts itself in such a fashion as to maximize the electron precipitation flux, while leaving the upward flux of ionospheric ions at a moderate level. For a discussion of limits on the upward ion flux, see STERN (1981).

The limits in Eq. (17) are based on Eqs. (1) and (15), leaving the distribution of V along the magnetic field lines as the only parameter determining $i_{\|a}$. In the opposite extreme one might allow the electric field to vary so rapidly in time and space that the particles diffuse across the magnetic field lines. This would not increase the current density, but it would allow the electrons to deposit their negative charges over a larger area, thereby reducing the need for a net parallel acceleration to overcome the 'mirror force.' The problem of determining $E_\| = E_\|(r, t)$ and $E_\perp = E_\perp(r, t)$ theoretically, given the constraints of magnetic confinement, has not yet received widespread attention. One approach, originally employed by ALFVÉN and FÄLTHAMMAR (1963) and PERSSON (1963, 1966) and further developed by LEMAIRE and SCHERER (1974, 1978), WHIPPLE (1977), CHIU and SCHULZ (1978), CHIU and CORNWALL (1980), CHIU et al. (1981), and STERN (1981), has been to seek a time-independent potential distribution $V = V(r)$ that can be matched with suitably selected particle distributions to satisfy quasi-neutrality. The electron and ion number densities at arbitrary points in space are calculated from specified velocity distributions at specified points, using Eqs. (1) and (6) and a time-independent form of Eq. (5), as $n_e = n_e(B, V)$ and $n_i = n_i(B, V)$. The 'solution' is reached when $n_e \simeq n_i$ everywhere. This approach seems to require substantially different pitch-angle distributions for the positive and negative components of the energetic particles, and the resulting $V(r)$ is popularly referred to as the 'effect of pitch-angle anisotropies.' Although it is a constructive approach, it does have intrinsic difficulties. One is the question of uniqueness of the solutions (LENNARTSSON, 1977b; STERN, 1981), another is the question of the validity of the underlying assumptions (LENNARTSSON, 1980).

The ultimate test of any theory is, of course, a comparison with observations. Given that we know of large amplitude plasma wave 'noise' or 'turbulence' in the auroral plasma (e.g. FREDRICKS et al., 1973; GURNETT and FRANK, 1976), along with narrow 'double layer' or 'electrostatic shock' structures (e.g. MOZER et al., 1977; TEMERIN et al., 1982), as well as particle distributions that show signs of diffusion in velocity space (e.g. BURCH et al., 1974; HOFFMAN and LIN, 1981; cf. also Fig. 4), we have no reason to anticipate a simple theory. A study of the magnetic confinement may be the right starting point, however.

5. Concluding Remarks

Even though certain mechanisms have received more attention than the others in the literature, it is probably fair to say that the 'basic' mechanisms have yet to be agreed upon. In the global context it is conceivable that several mechanisms have essential functions in redistributing kinetic energy from one group of particles to another. From the point of view of ion composition data it may be particularly interesting to study those mechanisms that are intrinsically mass dependent. There are many examples of data in the accompanying review articles in this book that show evidence of selection by ion species in the acceleration processes, such as the differences in energy and pitch angle distributions between the terrestrial O^+ and H^+ ions in the S3–3 data.

Several of the referenced models, if taken at face value, are indeed mass dependent. For example, any acceleration mechanism that involves a violation of Eq. (1), as discussed in Subsections 3.3 to 3.5, is likely to affect the ions of different mass in different fashions. If the electric field is nearly independent of time, the kinetic energy gain is only determined by the product of a particle's charge and the traversed potential difference, of course, but the resulting pitch angle may still be mass dependent. If, on the other hand, the electric field is time-dependent, i.e. 'turbulent,' for example, the energy gained by different ion species may differ considerably. The latter is an explicit feature in models of cyclotron resonance (see references in Subsection 3.5), but may have to be considered in any realistic model of the electric interaction between a large number of charged particles.

The author is pleased to thank Dr. R. G. Johnson for suggesting the topic of this review and Dr. G. T. Davidson for helpful discussions. This work was supported by the Office of Naval Research under contract N00014–78–C–0479 and the Lockheed Independent Research Program.

REFERENCES

AKASOFU, S.-I., Physics of the Magnetospheric Substorm, D. Reidel, Boston, 1977.
ALFVÉN, H., Some properties of magnetospheric neutral surfaces, J. Geophys. Res., 73, 4379–4381, 1968.
ALFVÉN, H., Electric durrent structure of the magnetosphere, in Physics of the Hot Plasma in the Magnetosphere, edited by B. Hultqvist and L. Stenflo, p. 1, Plenum Press, New York, 1975.
ALFVÉN, H. and C.-G. FÄLTHAMMAR, Cosmical Electrodynamics, Fundamental Principles, Clarendon Press, Oxford, 1963.
ANDERSSON, D., M. BABIC, S. SANDAHL, and S. TORVÉN, On the maximum current carrying capacity of a low pressure discharge, Proc. 9th Int. Conf. on Phenomena in Ionized Gases, Bucharest, Rumania, p. 142, 1969.
ASHOUR-ABDALLA M. and H. OKUDA, Transverse acceleration of ions on auroral field lines, this volume, 43–72, 1983.

ATKINSON, G., Duality of the magnetic flux tube and electric current description of magnetospheric plasma and energy flow, *Rev. Geophys. Space Phys.*, **19**, 617, 1981.

AXFORD, W. I., The origin of radiation belt and auroral particles. in *Particles and Fields in the Magnetosphere*, ed. B. C. McCormac, pp. 46–59, d. Reidel, Dordrecht, Holland, 1970.

AXFORD, W. I., Flow of mass and energy in the solar system, in *Physics of Solar Planetary Environments*, edited by D. J. Williams, p. 270, Am. Geophys. Union, 1976.

BAKER, K. D., L. P. BLOCK, R. KIST, W. KAMPA, N. SINGH, and H. THIEMAN, Studies of strong laboratory double layers and comparison with computer simulation, *J. Plasma Phys.*, **26**, 1, 1981.

BAUGHER, C. R., C. R. CHAPPELL, J. L. HORWITZ, E. G. SHELLEY, and D. T. YOUNG, Initial thermal plasma observations fromo ISEE 1, *Geophys. Res. Lett.*, 7, 657, 1980.

BLOCK, L. P., Coupling between the outer magnetosphere and the high-latitude ionosphere, *Space Sci. Rev.*, 7, 198, 1967.

BLOCK, L. P., Potential double layers in the ionosphere, *Cosm. Electrodyn.*, **3**, 349, 1972.

BLOCK, L. P. and C.-G. FÄLTHAMMAR, Field aligned currents and auroral precipitation, in *Atmospheric Emissions*, eds. B. M. McCormac and A. Omholt, p. 285, Van Nostrand Reinhold, New York, 1969.

BURCH, J. L., S. A. FIELDS, W. B. HANSON, R. A. HEELIS, R. A. HOFFMAN, and R. W. JANETZKE, Characteristics of auroral electron acceleration regions observed by Atmophsere Explorer-C, *J. Geophys. Res.*, **81**, 2223, 1976.

CARLQVIST, P., Some theoretical aspects of electrostatic double layers, in *Wave Instabilities in Space Plasmas*, edited by P. J. Palmadesso and K. Papadopoulos, p. 83, D. Reidel Publ. Co., 1979.

CHIU, Y. T. and M. SCHULZ, Self-consistent particle and parallel electrostatic field distributions in the magnetospheric-ionospheric auroral region, *J. Geophys. Res.*, **83**, 629, 1978.

CHIU, Y. T. and J. M. CORNWALL, Electrostatic model of a quiet auroral arc, *J. Geophys. Res.*, **85**, 543, 1980.

CHIU, Y. T., J. M. CORNWALL, and M. SCHULZ, Effects of auroral-particle anisotropies and mirror forces on high-latitude electric fields, in *Physics of Auroral Arc Formation*, edited by S.-I. Akasofu and J. R. Kan, Am. Geophys. Union, p. 234, 1981.

CLADIS, J. B., Effect of magnetic field gradient on motion of ions resonating with ion cyclotron waves, *J. Geophys. Res.*, **78**, 8129, 1973.

COAKLEY, P., L. JOHNSSON, and N. HERSHKOWITZ, Strong laboratory double layers in the presence of a magnetic field, *Phys. Lett.*, **70A**, 425, 1979.

COLE, K. D., Effects of crossed magnetic and (spatially dependent) electric fields on charged particle motion, *Planet. Space Sci.*, **24**, 515, 1976.

COLE, K. D. and T. L. AGGSON, $V E \times B$ effects on heavy ions in the magnetosphere, *Space Research XVI, 1975 COSPAR Proceedings*, edited by M. J. Rycroft, p. 639, Akademie-Verlag, Berlin, 1976.

COWLEY, S. W. H., A self-consistent model of a simple magnetic neutral sheet system surrounded by a cold, collisionless plasma, *Cosmic Electrodynamics*, **3**, 448, 1973.

CRAWFORD, F. W. and I. L. FREESTON, *Proc. VIth Int. Conf. on Phenomena in Ionized Gases*, Paris, I, p. 461, 1963.

DUM, C. T., Anomalous resistivity and plasma dynamics, in *Physics of Auroral Arc Formation*, edited by S.-I. Akasofu and J. R. Kan, Am. Geophys. Union, p. 408, 1981.

DUNGEY, J. W., Conditions for the occurrence of electrical discharges in astrophysical systems, *Philos. Mag. Ser.*, 7, 44, 725, 1953.

DUSENBERY, P. B. and L. R. LYONS, Generation of ion-conic distribution by upgoing ionospheric electrons, *J. Geophys. Res.*, **86**, 7627, 1981.

EVANS, D. S., Precipitating electron fluxes formed by a magnetic field-aligned potential difference, *J. Geophys. Res.*, **79**, 2853, 1974.

FÄLTHAMMAR, C.-G., Problems related to macroscopic electric fields in the magnetosphere, *Rev. Geophys. Space Phys.*, **15**, 457, 1977.

FÄLTHAMMAR, C.-G., Non-resistive electric potential drops in cosmical plasmas, in *Particle Acceleration Mechanisms in Astrophysics* (La Jolla Institute-1979), edited by Arons, McKee, and Max p. 27, American Inst. of Physics Conference Proceedings, Vol. 56, 1979.

FEDDER, J. A., Effects of anomalous resistivity on auroral Birkelund current systems, *Ann. Geophys.*, **32**, 1975, 1976.

FREDRICKS, R. W., F. L. SCARF, and C. T. RUSSELL, Field-aligned currents, plasma waves, and anomalous resistivity in the disturbed polar cusp, *J. Geophys. Res.*, **78**, 2133, 1973.

FRIDMAN, M. and J. LEMAIRE, Relationship between auroral electron fluxes and field-aligned electric potential difference, *J. Geophys. Res.*, **85**, 664, 1980.

GOERTZ, C. K., Double layers and electrostatic shocks in space, *Rev. Geophys. Space Phys.*, **17**, 418, 1979.

GOERTZ, C. K. and R. W. BOSWELL, Magnetopsphere-ionosphere coupling, *J. Geophys. Res.*, **84**, 7239, 1979.

GURNETT, D. A. and L. A. FRANK, A region of intense plasma wave turbulence on auroral field lines, *J. Geophys. Res.*, **82**, 1031, 1977.

HASEGAWA, A., Instabilities and nonlinear processes in geophysics and astrophysics, *Rev. Geophys. Space Phys.*, **12**, 273, 1974.

HASEGAWA, A. and L. CHEN, Kinetic process of plasma heating due to Alfvén wave excitation, *Phys. Rev. Lett.*, **35**, 370, 1975.

HAYASHI, T. and T. SATO, Magnetic reconnection: acceleration, heating, and shock formation, *J. Geophys. Res.*, **83**, 217, 1978.

HEIKKILA, W. J. and R. J. PELLINEN, Localized induced electric field within the magnetotail, *J. Geophys. Res.*, **82**, 1610, 1977.

HILL, T. W., Magnetic merging in a collisionless plasma, *J. Geophys. Res.*, **80**, 4689, 1975.

HOFFMAN, R. A. and C. S. LIN, Study of inverted-*V* auroral precipitation events, in *Physics of Auroral Arc Formation*, edited by S.-I. Akasofu and J. R. Kan, Am. Geophys. Union, p. 80, 1981.

HORWITZ, J. L., Conical distributions of low-energy ion fluxes at synchronous orbit, *J. Geophys. Res.*, **85**, 2057, 1980.

HORWITZ, J. L., C. R. BAUGHER, C. R. CHAPPELL, E. G. SHELLEY, and D. T. YOUNG, Conical pitch angle distributions of very low-energy ion fluxes observed by ISEE 1, *J. Geophys. Res.*, **87**, 2311, 1982.

HUBBARD, R. F. and G. JOYCE, Simulation of auroral double layers, *J. Geophys. Res.*, **84**, 4297–4304, 1979.

HUDSON, M. K., R. L. LYSAK, and F. S. MOZER, Magnetic field-aligned potential drops due to electrostatic ion cyclotron turbulence, *Geophys. Res. Lett.*, **5**, 143, 1978.

HULTQVIST, B., on the production of a magnetic-field-aligned electric field by the interaction between the hot magnetospheric plasma and cold ionosphere, *Planet Space Sci.*, **19**, 749, 1971.

HULTQVIST, B., On the interaction between the magnetosphere and the ionosphere, in *Solar-Terrestrial Physics*, Proceedings of the International Symposium, Leningrad, USSR, May 11–19, 1970, E. R. Dyer (Gen. Ed.), pp. 176–198, D. Reidel, Dordrecht, 1972.

IIJIMA, T. and T. A. POTEMRA, Field-aligned currents in the dayside cusp observed by Triad, *J. Geophys. Res.*, **81**, 5971, 1976.

JOHNSON, F. S., The driving force for magnetospheric convection, *Rev. Geophys. Space Phys.*, **16**, 161–167, 1978.

KALININ, Yu. G., D. N. LIN, L. I. RUDAKOV, V. D. RYUTOV and V. A. SKORYUPIN, Observation of current convective instability during turbulent heating of plasma by a current, *Zh. Eksp, Teor, Fiz.*, **59**, 1056, 1970 (*Sov. Phys. JETP*, **32**, 573, 1971).

KAN, J. R., L. C. LEE, and S.-I. AKASOFU, Two-dimensional potential double layers and discrete auroras, *J. Geophys. Res.*, **84**, 4305, 1979.

KAUFMANN, R. L. and G. R. LUDLOW, Auroral electron beams: stability and acceleration, *J. Geophys. Res.*, **86**, 7577, 1981.

KINDEL, J. M. and C. F. KENNEL, Topside current instabilities, *J. Geophys. Res.*, **76**, 3055, 1971.

KLUMPAR, D. M., Tranversely accelerated ions: an ionspheric source of hot magnetospheric ions, *J. Geophys. Res.*, **84**, 4229, 1979.

KNIGHT, S., Parallel electric fields, *Planet. Space Sci.*, **21**, 741, 1973.

KNORR, G. and C. K. GOERTZ, Existence and stability of strong potential double layers, *Astrophys. Space Sci.*, **31**, 209, 1974.

LEMAIRE, J. and M. SCHERER, Ionosphere-plasmasheet field-aligned currents and parallel electric fields, *Planet. Space Sci.*, **22**, 1485, 1974.

LEMAIRE, J. and M. SCHERER, Field aligned distribution of plasma mantle and ionospheric plasma, *J. Atmos. Terr. Phys.*, **40**, 337, 1978.

LENNARTSSON, W., On the magnetic mirroring as the basic cause of parallel electric fields, *J. Geophys. Res.*, **81**, 5583, 1976.

LENNARTSSON, W., On high-latitude convection field inhomogeneities, parallel electric fields and inverted-*V*

40   W. LENNARTSSON

precipitation events, *Planet Space Sci.*, **25**, 89, 1977a.

LENNARTSSON, W., On the role of magnetic mirroring in the auroral phenomena, *Astrophys. Space Sci.*, **51**, 461, 1977b.

LENNARTSSON, W., On the consequences of the interaction between the auroral plasma and the geomagnetic field, *Planet. Space Sci.*, **28**, 135, 1980.

LONGMIRE, C. L., *Elementary Plasma Physics*, John Wiley and Sons, Inc., New York, London, 1963.

LYONS, L. R., Discrete aurora as the direct result of an inferred, high-altitude generating potential distribution, *J. Geophys. Res.*, **86**, 1, 1981.

LYONS, L. R., D. S. EVANS, and R. LUNDIN, An observed relation between magnetic field aligned electric fields and downward electron energy fluxes in the vicinity of auroral forms, *J. Geophys. Res.*, **84**, 457, 1979.

MOZER, F. S., N. K. HUDSON, R. B. TORBERT, B. PARADY, and J. YATTEAU, Observations of paired electrostatic shocks in the polar magnetosphere, *Phys. Rev. Lett.*, **38**, 292, 1977.

MOZER, F. S., C. A. CATTELL, M. K. HUDSON, R. L. LYSAK, M. TEMERIN, and R. B. TROBERT, Satellite measurements and theories of low altitude auroral particle acceleration, *Space Sci. Rev.*, **27**, 155, 1980.

OKUDA, H. and M. ASHOUR-ABDALLA, Formation of a conical distribution and intense ion heating in the presence of hydrogen cyclotron waves, *Geophys. Res. Lett.*, **8**, 811, 1981.

OKUDA, H., C. Z. CHENG, and W. W. LEE, Anomalous diffusion and ion heating in the presence of electrostatic ion cyclotron instabilities, *Phys. Rev. Lett.*, **46**, 427, 1980.

PAPADOPOULOS, K. and T. COFFEY, Anomalous resistivity of the auroral plasma, *J. Geophys. Res.*, **79**, 1558, 1974.

PAPADOPOULOS, K., A review of anomalous resistivity for the ionosphere, *Rev. Geophys. Space Phys.*, **15**, 113, 1977.

PALMADESSO, P. J., T. P. COFFEY, S. L. OSSAKOV, and K. PAPADOPOULOS, Topside ionosphere ion heating due to electrostatic ion cyclotron turbulence, *Geophys. Res. Lett.*, **1**, 105, 1974.

PELLINEN, R. and W. HEIKKILA, Energization of charged particles to high energies by an induced substorm electric field within the magnetotail, *J. Geophys. Res.*, **83**, 1544, 1978.

PERSSON, H., Electric field along a magnetic line of force in a low-density plasma, *Phys. Fluids*, **6**, 1756, 1963.

PERSSON, H., Electric field parallel to the magnetic field in a low-density plasma, *Phys. Fluids*, **9**, 1090, 1966.

QUON, B. M. and A. Y. WONG, Formation of potential double layers in plasmas, *Phys. Rev. Lett.*, **37**, 1393, 1976.

RASSBACH, M. E., Upward Birkeland currents, *J. Geophys. Res.*, **78**, 7553, 1973.

ROEDERER, J. G., Earth's magnetosphere: global problems in magnetospheric plasma physics, in *Solar System Plasma Phys.*, Vol. II, edited by C. F. Kennel, L. J. Lanzerotti, and E. N. Parker, p. 1, North-Holland Publ. Co., 1979.

SHARBER, J. R. and W. J. HEIKKILA, Fermi acceleration of auroral particles, *J. Geophys. Res.*, **77**, 3397, 1972.

SHARP, R. D., R. G. JOHNSON, and E. G. SHELLEY, Observations of an ionospheric acceleration mechanism producing energetic (keV) ions primarily normal to the geomagnetic field direction, *J. Geophys. Res.*, **82**, 3324, 1977.

SHARP, R. D., A. G. GHIELMETTI, R. G. JOHNSON, and E. G. SHELLEY, Hot plasma composition results from the S3–3 spacecraft, this volume, pp. 167–193, 1983.

SCHÖNHUBER, M.-J.,Über das Auftreten und Verhalten von Raumladungs-Doppelschichten in Hg-Niederdruckgasentladungen, *Zeit. F. angew. Phys.*, **15**, 454. 1963.

SCHULZ, M., Principles of magnetosphere ion composition, this volume, pp. 1–21, 1983.

SONNERUP, B. U. Ö., Magnetic field reconnection, *Solar System Plasma Phys.*, Vol. III, edited by L. T. Lanzerotti, C. F. Kennel and E. N. Parker, p. 45, North-Holland Publ. Co., 1979.

SPEISER, T. W., Particle trajectories in a model current sheet based on the open model of the magnetosphere, with applications to auroral particles, *J. Geophys. Res.*, **70**, 1717–1728, 1965.

SPITZER, L., *Physics of Fully Ionized Gases*, Chapter 5, Interscience, New York, 1956.

STENZEL, R. L., M. OOYAMA, and Y. NAKAMURA, Potential double layers in strongly magnetized plasmas, in *Physics of Auroral Arc Formation*, edited by S.-I. Akasofu and J. R. Kan, p. 226, Am. Geophys. Union, 1981.

STERN, D. P., The electric field and global electrodynamics of the magnetosphere, *Rev. Geophys. Space Phys.*, **17**, 626, 1979.

STERN, D. P., One-dimensional models of quasi-neutral parallel electric fields, *J. Geophys. Res.*, **86**, 5839, 1981.

SWIFT, D. W., A mechanicm for energizing electrons in the magnetosphere, *J. Geophys. Res.*, **70**, 3061, 1965.

SWIFT, D. W., An equipotential model for auroral arcs-II. Numerical solutions. *J. Geophys. Res.*, **81**, 3935, 1976.

SWIFT, D. W., An equipotential model for auroral arcs: the theory of two-dimensional laminar electrostatic shocks, *J. Geophys. Res.*, **84**, 6427, 1979.

TEMERIN, M., K. CERNY, K. W. LOTKO, and F. S. MOZER, Observations of double layers and solitary waves in the auroral plasma, in publication in *Phys. Rev. Lett.*, 1982.

TORVÉN, S., Formation of double layers in laboratory plasmas, in *Wave instabilities in Space Plasmas*, edited by P. J. Palmadesso and K. Papadopoulos, pp. 109–128, D. Reidel Publ. Co. 1979.

TORVÉN, S. and D. ANDERSSON, Observations of electric double layers in a magnetized plasma column, *J. Phys. D.: Appl. Phys.*, **12**, 717, 1979.

UNGSTRUP, E., D. M. KLUMPAR, and W. J. HEIKKILA, Heating of ions to superthermal energies in the topside ionosphere by electrostatic ion cyclotron waves. *J. Geophys. Res.*, **84**, 4289, 1979.

WAGNER, J. S., T. TAJIMA, J. R. KAN, J. N. LEBOEUF, S.-I. AKASOFU, and J. M. DAWSON, *V*-potential double layers and the formation of auroral arcs, *Phys. Rev. Lett.*, **45**, 803, 1980.

VASYLIUNAS, V. M., Theoretical model of magnetic field line merging, *Rev. Geophys. Space Phys.*, **13**, 303, 1975.

WHIPPLE, E. C. JR., The signature of parallel electric fields in a collisionless plasma, *J. Geophys. Res.*, **82**, 1525, 1977.

YEH, H.-C. and T. W. HILL, Mechanism of parallel electric fields inferred from observations, *J. Geophys. Res.*, **86**, 6706, 1981.

Energetic Ion Composition in the Earth's Magnetosphere, edited by R. G. Johnson, 43–72.
Copyright © 1983 by Terra Scientific Publishing Company (TERRAPUB), Tokyo.

Transverse Acceleration of Ions on Auroral Field Lines

M. Ashour-Abdalla* and H. Okuda**

*Institute of Geophysics and Planetary Physics, University of California,
Los Angeles, California, U.S.A.
**Plasma Physics Laboratory, Princeton University,
Princeton, New Jersey, U.S.A.*

This paper examines transverse ion heating on auroral field lines associated with current-driven electrostatic ion cyclotron waves theoretically and by numerical simulations. The auroral plasma is assumed to consist of drifting electrons and stationary hydrogen and oxygen ions. Depending on the ratio of the electron drift speed to the thermal speed and the ratio of hydrogen to oxygen concentrations, preferential heating of either hydrogen or oxygen ions can take place. It is found that unless the oxygen ions are a minority species, oxygen transverse heating generally exceeds that of hydrogen ions. Theory and numerical simulations are in good agreement.

1. Introduction

It has only recently been recognized that heavy ions are important in the dynamics and stability of different regions in the earth's magnetosphere. Heavy ions have been observed in different regions of the earth's magnetosphere. Cold ($\approx$ eV) He^+ was first observed in the plasmasphere on OGO-1 (Taylor *et al.*, 1965) and OGO-5 (Harris *et al.*, 1970), and a typical He^+/H^+ ratio of 1% was given. Furthermore both He^+ and O^+ have been found to be important constituents of the thermal and suprathermal plasma populations outside the plasmasphere (Young *et al.*, 1977; Young, 1979).

Since the surprising discovery by Shelley *et al.* (1972) of large fluxes of precipitating O^+ having energies up to 12 keV (the upper limit of their instrument) much has been learned about heavy ions in auroral plasmas (see Johnson, 1982). Shelley *et al.* (1976) and Ghielemetti *et al.* (1978) reported observations of intense fluxes of O^+ streaming up high latitude auroral field lines near $1 R_E$. Measurements from the S3–3 Satellite and from the ISIS-1 and ISIS-2 satellites at altitudes $\approx 1,500$ km show the existence of ion distributions with enhanced particle fluxes at pitch angles between 90–130 degrees (Sharp *et al.*, 1977; Klumpar, 1979; Ungstrup *et al.*, 1979). Such distributions are commonly termed ion conic distributions. Gorney *et al.* (1981) have completed a statistical study from the S3–3 particle data of upflowing ions as a function of local time, latitude, altitude and magnetic activity. They found that ion conics were observed uniformly at altitudes above 2,000 km, suggesting a low altitude generation region for conics. At such low altitudes, the recent S3–3 data show that the ion composition can vary from more than 10% oxygen to 90% oxygen (Mizera *et al.*, 1980).

The presence of ion conics requires a mechanism which causes preferential acceleration in the direction perpendicular to the earth's magnetic field. Waves interacting with the ions is a plausible mechanism for preferentially heating the ions at large pitch angles. Hydrogen cyclotron waves at harmonics of the gyrofrequency have been observed at altitudes of $\simeq 6,000$ km on board the S3–3 satellite. The presence of hydrogen cyclotron waves correlates well with the presence of ion beams (KINTNER et al., 1979). At lower altitudes, accurate measurements of ion cyclotron waves are difficult to make due to large Doppler shifting effects which obscure the wave frequency. Thus, if conics are indeed generated at low altitudes, ion cyclotron waves could be present, but they would be difficult to resolve.

In this paper, we will address the formation of the conical distribution function and the acceleration of different ion species on auroral field lines. We assume that the return flow of the electrons from the ionosphere is responsible for exciting ion cyclotron waves, and we ask what are the nonlinear consequences of the instability and in particular if the waves produce the observed conical distributions.

There has been a great deal of interest in the acceleration of heavy ions along auroral field lines by ion cyclotron waves. Assuming the presence of large amplitude electrostatic hydrogen cyclotron waves, test particle orbits of heavy ions have been calculated by several authors to study the heating. LYSAK et al. (1980) considered the ion orbits in a set of fixed amplitude waves with definite phase relation. They found that the heating of hydrogen and oxygen ions was comparable. PAPADOPOULOS et al. (1980) argued that the heavy ions form a minority constituent of many magnetospheric multi-ion plasmas. As such, the heavy ions do not affect the collective mode structure of the plasma due to their small abundance ratios. They thus examined the acceleration of large M/Q (where M is the mass and Q is the charge) ions in the presence of a coherent large amplitude electrostatic hydrogen cyclotron wave. They found that when the wave amplitude exceeds a certain critical value the particle orbits become stochastic and therefore can be accelerated by the wave. The most important result was that the maximum energy achieved by an ion scaled as $(M/M_H)^{5/3}$, where M_H is the mass of the hydrogen and M is the mass of the heavy ion under consideration. From the above study, it was concluded that heavy ions are thus preferentially accelerated. SINGH et al. (1981) investigated the acceleration of ionospheric ions, by ion cyclotron waves, as a function of their initial temperature. They found that O^+ ions are heated only when their initial temperature satisfies the condition $T_{O+} > T_H$. On the other hand, He^+ and He^{++} were accelerated even when they were initially very cold. Recently, DUSENBERY and LYONS (1981), considered heating of ions due to ion cyclotron waves excited by the upgoing thermal electrons in the low altitude auroral zone. Their calculations were performed assuming the initial ion distributions consisted solely of H^+ or solely of O^+. The resulting ion flux versus energy spectra was in good agreement with the low altitude conic observations from ISIS-2 (DUSENBERY and LYONS, 1981). ASHOUR-ABDALLA et al. (1981) considered the oxygen heating associated with the current-driven electrostatic ion cyclotron waves in the presence of hydrogen and oxygen ions. In their simulations, they found that oxygen ions can be heated preferentially by the oxygen cyclotron waves. They attributed this to the fact that oxygen cyclotron waves required a lower electron drift speed threshold than hydrogen cyclotron waves.

In this paper we examine the transverse acceleration of ions on auroral field lines associated with the current-driven electrostatic ion cyclotron waves in a plasma consisting of hydrogen and oxygen ions. In Section 2, we present results from a marginal stability analysis for ion cyclotron waves. The simulation results are presented in Section 3. Finally, in Section 4, we conclude by discussing the heating of the ions and the evolution of the ion cyclotron waves on auroral field lines.

2.   Theoretical Analysis

2.1   Marginal Stability Analysis of EHC Waves

We shall consider the dispersion relation for the electrostatic ion cyclotron waves in the presence of drifting Maxwellian electrons with respect to hydrogen and oxygen background ions. Assuming bi-Maxwellian distributions for both hydrogen and oxygen ions, the dispersion relation can be written as follows (DRUMMOND and ROSENBLUTH, 1962; KINDEL and KENNEL, 1971; K. F. LEE, 1972; ASHOUR-ABDALLA *et al.*, 1981; OKUDA and ASHOUR-ABDALLA, 1981)

$$
1 + \frac{1}{k^2 \lambda_e^2} \left[1 + \frac{\omega - k_\parallel v_{de}}{k_\parallel v_{te}} Z\left[\frac{\omega - k_\parallel v_{de}}{k_\parallel v_{de}} \right] \right]
$$
$$
+ \frac{1}{k^2 \lambda_{H\parallel}^2} \left[1 + \sum_n \frac{\omega - n\Omega_H(1 - T_{H\parallel}/T_{H\perp})}{k_\parallel v_{H\parallel}} \Gamma_n(\mu_H) Z\left(\frac{\omega - n\Omega_H}{k_\parallel v_{H\parallel}} \right) \right]
$$
$$
+ \frac{1}{k^2 \lambda_{O\parallel}^2} \left[1 + \sum_n \frac{\omega - n\Omega_O(1 - T_{O\parallel}/T_{O\perp})}{k_\parallel v_{O\parallel}} \Gamma_n(\mu_O) Z\left(\frac{\omega - n\Omega_O}{k_\parallel v_{O\parallel}} \right) \right] = 0 \qquad (1)
$$

The subscripts H and O denote quantities related to the hydrogen and oxygen ions, respectively. The subscripts $\parallel$ and $\perp$ refer to quantities parallel and perpendicular to the external magnetic field. v_{de} is the electron drift speed along the magnetic field and v_{te} is the electron thermal speed. λ_e is the electron Debye length defined by $\lambda_e = v_{te}/\sqrt{2}$ ω_{pe}. $\Gamma_n(\mu) = I_n(\mu)\exp(-\mu)$ where I_n is the modified Bessel function of order n and $\mu = k_\perp^2 v_\perp^2/2\Omega^2 = k_\perp^2 \rho_\perp^2$ and Z is the plasma dispersion function. Note we have assumed $\omega_{pe}^2/\Omega_e^2 \ll 1$ which is valid on low altitude auroral field lines.

Let us first consider the electrostatic hydrogen cyclotron (EHC) waves when the conditions $\omega \approx n\Omega_H$ and $k_\perp \rho_H \approx n$ are satisfied. Under such conditions, since $\omega \gg \Omega_H$ and $k_\perp \rho_O \gg 1$ the motion of the oxygen ions can be described by the straight line approximation. The dielectric constant is then given by $(1/k^2 \lambda_O^2)Z'(\omega/k v_O)$ (KINDEL and KENNEL, 1971). Near the fundamental hydrogen cyclotron frequency, $\omega \approx \Omega_H$ and $k_\perp \rho_H \approx 1$, equation (1) is reduced to

$$
1 + \frac{\omega - k_\parallel v_{de}}{k_\parallel v_{te}} Z\left[\frac{\omega - k_\parallel v_{de}}{k_\parallel v_{te}} \right]
$$
$$
+ \frac{T_e}{T_{H\parallel}} \frac{n_H}{n_e} \left[1 + \frac{\omega - \Omega_H(1 - T_{H\parallel}/T_{H\perp})}{k_\parallel v_{H\parallel}} \Gamma_1 Z\left(\frac{\omega - \Omega_H}{k_\parallel v_{H\parallel}} \right) \right]
$$
$$
- \frac{T_e}{T_{H\parallel}} \frac{n_H}{n_e} \left[(1 - \Gamma_1)\left(1 - \frac{T_{H\parallel}}{T_{H\perp}} \right) + \frac{T_{H\parallel}}{T_{H\perp}} \frac{1 - \Gamma_0}{\mu_H} \right] = 0 \qquad (2)
$$

where the oxygen contribution has been neglected. The oxygen contribution to the hydrogen cyclotron waves is significant only when $(m_H/m_O)(n_O/n_H) > 1$ or $n_O/n_H > 16$ in which case the oxygen contribution should be retained in Eq. (2). It should be stressed, however, that in Eq. (2) the effects of oxygen ions are implicitly included in the ratio of hydrogen density to electron density, n_H/n_e. As we shall see later, this density ratio plays an important role in determining the stability of EHC waves.

We have shown earlier, for a hydrogen plasma, that one of the nonlinear consequences of the hydrogen cyclotron wave is the perpendicular heating $(T_{H_\perp} \gg T_{H_\parallel})$ which enhances the cyclotron damping by causing a downward frequency shift of the hydrogen cyclotron waves (OKUDA and ASHOUR-ABDALLA, 1981, 1982). If the cyclotron damping overcomes the inverse electron Landau damping, the instability shuts off by the perpendicular heating of the hydrogen ions. The same process must occur for the hydrogen wave and hydrogen heating in a hydrogen-oxygen plasma. The maximum perpendicular heating of the hydrogen ions can be calculated from the marginal stability analysis for a given electron drift speed, v_{de}/v_{te}.

For a small electron drift, $v_{de}/v_{te} < 1$, the electron contribution in Eq. (2) may be approximated by $i\sqrt{\pi}(\omega - k_\parallel v_{de})/k_\parallel v_{te}$ so that for marginal stability (real ω), we find, for the real part,

$$
\frac{\omega - \Omega_H}{\Omega_H} =
$$
$$
-\frac{\Gamma_1 \dfrac{T_{H_\parallel}}{T_{H_\perp}} \dfrac{n_H}{n_e} \rho Z_R(\rho) \dfrac{T_e}{T_{H_\parallel}}}{\rho \Gamma_1(\mu_H) Z_R(\rho) \dfrac{n_H}{n_e} \dfrac{T_e}{T_{H_\parallel}} + 1 - \dfrac{T_e}{T_{H_\parallel}} \dfrac{n_H}{n_e} \left[(1 - \Gamma_1)\left(1 - \dfrac{T_{H_\parallel}}{T_{H_\perp}}\right) + \dfrac{T_{H_\parallel}}{T_{H_\perp}} \dfrac{1 - \Gamma_0}{\mu_H} - 1 \right]}
$$

$$(3)$$

and for the imaginary part,

$$
\frac{v_{de}}{v_{te}} = \frac{(1 + \Delta)\rho}{\Delta}\left(\frac{m_e}{m_H}\right)^{1/2}\left(\frac{T_{H_\parallel}}{T_{H_\perp}}\right)^{1/2} + \frac{T_e}{T_{H_\parallel}} \frac{n_H}{n_e} \rho \frac{(\Delta + T_{H_\parallel}/T_{H_\perp})}{\Delta} \Gamma_1 \exp(-\rho^2) \quad (4)
$$

where

$$
\rho \equiv \frac{\omega - \Omega_H}{k_\parallel v_{H_\parallel}} \tag{5}
$$

$$
\Delta \equiv \frac{\omega - \Omega_H}{\Omega_H}. \tag{6}
$$

Note that in Eqs. (3) and (4) we neglected the oxygen contribution to the EHC waves. It is clear from Eq. (3) that the frequency of EHC waves approaches Ω_H as n_H/n_e is reduced.

v_{de}/v_{te} can be minimized numerically with respect to Δ and ρ or μ_H and ρ using Eqs. (3) and (4) for given values of $T_{H_\perp}/T_{H_\parallel}$ and n_H/n_e. For a given set (μ_H, ρ), equation (3) is first used to find Δ which is then used to calculate v_{de}/v_{te} from Eq. (4).

For $v_{de}/v_{te} \gtrsim 1$, the small argument of the electron Z function breaks down and one must retain the full Z function for the electrons. One then finds, for the real part,

$$\frac{\omega - \Omega_H}{\Omega_H} = -\frac{\Gamma_1 \dfrac{T_{H\parallel}}{T_{H\perp}} \dfrac{n_H}{n_e} \rho\, Z_R(\rho) \dfrac{T_e}{T_{H\parallel}}}{\rho \Gamma_1 Z_R(\rho) \dfrac{n_H}{n_e} \dfrac{T_e}{T_{H\parallel}} + \left[1 + \dfrac{\omega - k_\parallel v_{de}}{k_\parallel v_{te}} Z_R\left(\dfrac{\omega - k_\parallel v_{de}}{k_\parallel v_e}\right)\right] - \dfrac{T_e}{T_{H\parallel}} \dfrac{n_H}{n_e}}$$

$$\left[(1 - \Gamma_1)\left(1 - \frac{T_{H\parallel}}{T_{H\perp}}\right) + \frac{T_{H\parallel}}{T_{H\perp}} \frac{1 - \Gamma_0}{\mu_H} - 1\right]. \tag{7}$$

and, for the imaginary part,

$$\frac{k_\parallel v_{de} - \omega}{k_\parallel v_{te}} \exp\left[-\left(\frac{k_\parallel v_{de} - \omega}{k_\parallel v_{te}}\right)^2\right] = \frac{T_e}{T_{H\parallel}} \frac{n_H}{n_e} \frac{\omega - \Omega_H(1 - T_{H\parallel}/T_{H\perp})}{k_\parallel v_{H\parallel}} \Gamma_1 Z_I\left(\frac{\omega - \Omega_H}{k_\parallel v_{H\parallel}}\right).$$

Noting that for $v_{de}/v_{te} \gtrsim 1$, the exponential electron term is then the dominant term, the critical drift speed is given by

$$\frac{v_{de}}{v_{te}} = \frac{1 + \Delta}{\Delta} \rho \frac{v_{H\parallel}}{v_{te}} + \left[\ln \frac{\dfrac{v_{de}/v_{te} - \dfrac{1 + \Delta}{\Delta} \rho(v_{H\parallel}/v_{te})}}{(T_e/T_{H\parallel})(n_H/n_e) \rho \dfrac{\Delta + T_{H\parallel}/T_{H\perp}}{\Delta} \Gamma_1 \exp(-\rho^2)}\right]^{1/2} \tag{8}$$

where Δ and ρ are defined as before by Eqs. (5) and (6). Equations (7) and (8) can be solved numerically to find the minimum drift v_{de}/v_{te} for a given $T_{H\parallel}/T_{H\perp}$ and $n_H/T_{H\perp}$ and n_H/n_e which in turn should give the marginal stability for a given drift speed v_{de}/v_{te}.

Figure 1 shows the values of $T_{H\perp}/T_{H\parallel}$ for a given v_{de}/v_{te} as a function of n_H/n_e found from Eqs. (3) and (4) or Eqs. (7) and (8) for $T_e/T_{H\parallel} = 1$. Since the hydrogen parallel temperature changes only slightly as a result of hydrogen cyclotron instability, $T_{H\perp}/T_{H\parallel}$ represents the maximum hydrogen perpendicular heating for a given electron drift speed (OKUDA and ASHOUR-ABDALLA, 1982). For the pure hydrogen plasma, $n_H/n_e = 1$, the maximum heating agrees with the authors' earlier work (OKUDA and ASHOUR-ABDALLA, 1982; ASHOUR-ABDALLA and OKUDA, 1982).

As the hydrogen concentration decreases, or the oxygen concentration increases, the maximum attainable hydrogen temperature decreases as shown in Fig. 1. This is because the frequency of the hydrogen wave approaches Ω_H as n_H/n_e decreases as shown in Eq. (7) so that a small increase of $T_{H\perp}/T_{H\parallel}$ is enough to stabilize the EHC waves. Note also for a sufficiently large electron drift, there is no solution for marginal stability indicating that EHC waves always remain unstable regardless of the hydrogen temperature anisotropy. For such a large drift, generation of temperature anisotropy is not sufficient to stabilize the hydrogen cyclotron instability and other effects such as plateau formation and heating of the electrons must come into play for the stabilization. Note, however, that we have assumed the ion velocity distributions to be

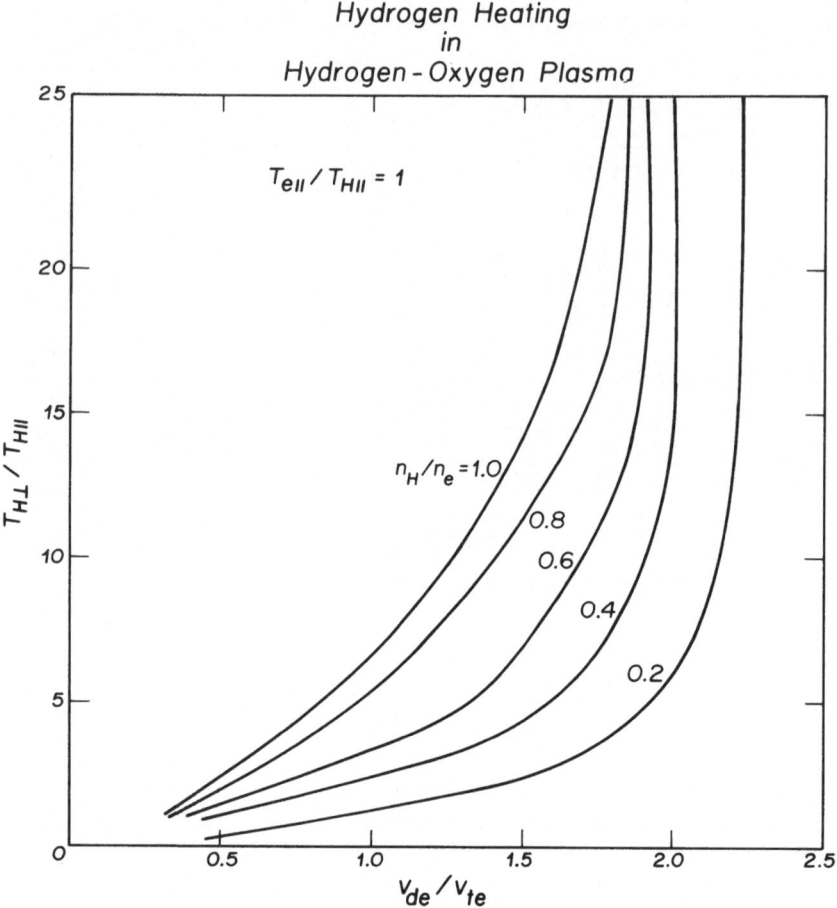

Fig. 1.   Maximum temperature anisotropy, $T_{H\perp}/T_{H\parallel}$, for hydrogen ions determined from the marginal stability analysis for different hydrogen concentrations, $T_{e\parallel} = T_{H\parallel} = T_{O\parallel}$.

bi-Maxwellian throughout the calculations so that if the distributions deviate significantly from bi-Maxwellian, the marginal stability analysis given by Fig. 1 may have to be modified.

   Figure 2 shows marginal stability plots for $T_e/T_{H\parallel} = 2$. Note the maximum attainable hydrogen temperature anisotropy is larger in this case, than for the case $T_e = T_{H\parallel}$ (Fig. 1). Equations (3) and (7) show that the frequency of the EHC waves increases as $T_e/T_{H\parallel}$ increases. Therefore, for $T_e/T_{H\parallel} > 1$, a large hydrogen temperature anisotropy is required to lower the frequency and stabilize the waves. Since the maximum temperature anisotropy is a sensitive function of the value of $T_e/T_{H\parallel}$, even modest electron heating can cause a large increase in ion perpendicular heating.

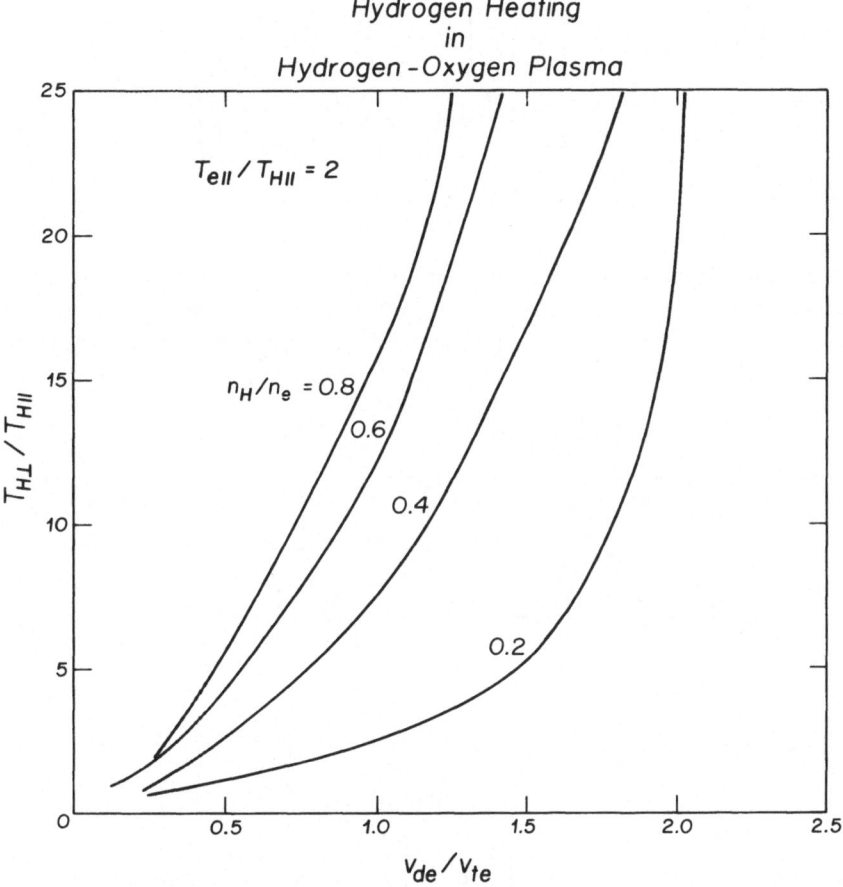

Fig. 2.   Same as Fig. 1 except for $T_{e\,\|}/T_{H\,\|} = T_{e\,\|}/T_{O\,\|} = 2$.

2.2   Marginal stability analysis of EOC waves

In considering the stability of EHC waves, we have neglected the effect of oxygen ions on hydrogen waves. As long as the hydrogen waves grow much faster than the oxygen waves, and the heating of hydrogen ions occurs much quicker than the oxygen heating, the effects of oxygen ions on the hydrogen heating can be neglected since most of the electron energy is exclusively fed into the hydrogen ions. When considering oxygen waves, we must include the effect of hydrogen ions. For EOC waves at $\omega \simeq \Omega_O$ and $k_\perp \rho_O \simeq 1$, the behavior of the hydrogen ions can be described by the guiding center approximation since $\omega \ll \Omega_H$ and $k_\perp \rho_O \ll 1$. We also can neglect the hydrogen dielectric constant ω_{pH}^2/Ω_H^2 as compared to $(k^2 \lambda_e^2)^{-1}$. Making the above approximation and setting $\omega \simeq \Omega_O$ and $k_\perp \rho_O \simeq 1$, equation (1) can be written as

$$1 + \frac{\omega - k_\parallel v_{de}}{k_\parallel v_{te}} Z\left(\frac{\omega - k_\parallel v_{de}}{k_\parallel v_{te}}\right) + \frac{T_e}{T_{H_\parallel}}\left[1 + \frac{\omega}{k_\parallel v_{H_\parallel}} Z\left(\frac{\omega}{k_\parallel v_{H_\parallel}}\right)\right]$$

$$+ \frac{T_e}{T_{O_\parallel}} \frac{n_O}{n_e}\left[1 + \frac{\omega - \Omega_O(1 - T_{O_\parallel}/T_{O_\perp})}{k_\parallel v_{O_\parallel}}\Gamma_1(\mu_O)Z\left(\frac{\omega - \Omega_O}{k_\parallel v_{O_\parallel}}\right)\right]$$

$$+ \frac{T_e}{T_{O_\parallel}} \frac{n_O}{n_e}\left[(1 - \Gamma_1(\mu_O))\left(1 - \frac{T_{O_\parallel}}{T_{O_\perp}}\right) + \frac{T_{O_\parallel}}{T_{O_\perp}}\frac{1 - \Gamma_0(\mu_O)}{\mu_O}\right] = 0. \tag{9}$$

Let us first calculate the marginal stability for the oxygen waves in order to find the maximum temperature anisotropy $T_{O_\perp}/T_{O_\parallel}$ for a given electron drift speed. Assuming ω is real for marginal stability and separating Eq. (17) into real and imaginary parts, we find

$$\frac{\omega - \Omega_O}{\Omega_O} = \cfrac{- \Gamma_1 \dfrac{n_O}{n_e} \dfrac{T_e}{T_{O_\parallel}} \dfrac{T_{O_\parallel}}{T_{O_\perp}} \rho Z_R(\rho)}{1 + \dfrac{\omega - k_\parallel v_{de}}{k_\parallel v_{te}} Z_R\left[\dfrac{\omega - k_\parallel v_{de}}{k_\parallel v_{te}}\right] - \dfrac{T_e}{T_{O_\parallel}} \dfrac{n_O}{n_e}\left[(1 - \Gamma_1)\left(1 - \dfrac{T_{O_\parallel}}{T_{O_\perp}}\right) + \dfrac{T_{O_\parallel}}{T_{O_\perp}}\right.}$$

$$\left.\dfrac{1 - \Gamma_O}{\mu_O}\right] + \dfrac{T_e}{T_{H_\parallel}} \dfrac{n_H}{n_e}\left[1 + \dfrac{\omega}{k_\parallel v_{H_\parallel}} Z_R\left(\dfrac{\omega}{k_\parallel v_{H_\parallel}}\right)\right] + \dfrac{n_O}{n_e} \dfrac{T_e}{T_{O_\parallel}}(1 + \rho\Gamma_1 Z_R(\rho)) \tag{10}$$

and

$$\frac{v_{de}}{v_{te}} = \frac{1 + \Delta}{\Delta}\rho\frac{v_{O_\parallel}}{v_{te}} +$$

$$\left[\cfrac{\dfrac{v_{de}}{v_{te}} - \dfrac{1 + \Delta}{\Delta}\rho\dfrac{v_{O_\parallel}}{v_{te}}}{\ln\dfrac{T_e}{T_{H_\parallel}} \dfrac{n_H}{n_e}\left[\dfrac{m_H}{m_O}\dfrac{T_{e_\parallel}}{T_{H_\parallel}}\right]^{1/2}(1 + \Delta)\dfrac{\rho}{\Delta}\exp\left[-\dfrac{\rho^2}{\Delta^2}\dfrac{v_{O_\parallel}^2}{v_{H_\parallel}^2}\right] + \dfrac{T_{e_\parallel}}{T_{O_\parallel}}\dfrac{n_O}{n_e}\rho\dfrac{\Delta + T_{O_\parallel}/T_{O_\perp}}{\Delta}\Gamma_1\exp(-\rho^2)}\right]^{1/2} \tag{11}$$

where $\rho = (\omega - \Omega_O)/k_\parallel v_{O_\parallel}$ and $\Delta = (\omega - \Omega_O)/\Omega_O$ as before.

Equations (10) and (11) can be solved numerically to find the minimum v_{de}/v_{te} for a given temperature anisotropy $T_{O_\perp}/T_{O_\parallel}$ and density concentration n_O/n_e. The maximum temperature anisotropy for the oxygen ions as a function of electron drift is shown in Figs. (3) and (4). Figure 3 is for case $T_e/T_{O_\parallel} = 1$, whereas figure 4 is for the case $T_e/T_{O_\parallel} = 2$. Comparing Figs. 3 and 4 with Figs. 1 and 2, we note that the oxygen perpendicular heating attainable is in general larger than that of the hydrogen ions. This is because the critical drift speed for the oxygen cyclotron waves is in general lower than that of the hydrogen cyclotron waves so that the oxygen ions must be heated more than the hydrogen ions in order to reach saturation.

2.3   Ion heating on auroral field lines

In the following, we shall assume that the electron drift speed is sufficiently large so that both EHC and EOC waves are excited. Clearly, there are situations, depending on

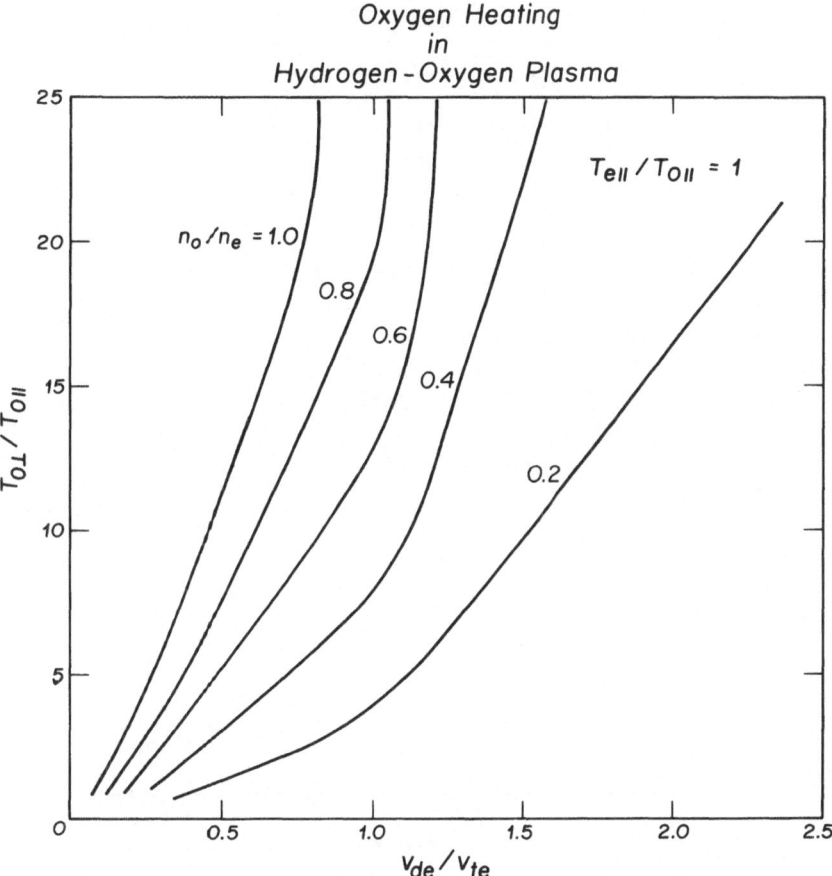

Fig. 3.   Maximum temperature anisotropy, for oxygen ions determined from the marginal stability analysis for different oxygen concentrations, $T_{e\parallel} = T_{H\parallel} = T_{O\parallel}$.

the density ratio of hydrogen and oxygen and the electron drift speed, where either EHC waves or EOC waves can be destabilized seperately. Obviously, under such conditions, only the corresponding ion species will be substantially heated. Here we consider the more interesting case where both EHC and EOC waves are unstable.

Our theoretical model is shown in Fig. 5 where the ionosphere, located at $x < 0$ is the reservoir for the drifting Maxwellian electrons streaming upwards along auroral field lines. As the electrons stream through hydrogen and oxygen ions for $x > 0$, both hydrogen and oxygen cyclotron waves are excited. As a result of wave excitation, the electron distribution develops a plateau at $x = l_H$ so that no instability can occur for $x > l$. Since the hydrogen waves grow much faster than the oxygen waves, unless the hydrogen ions are a minority species, initially, only the hydrogen heating takes place with little oxygen heating.

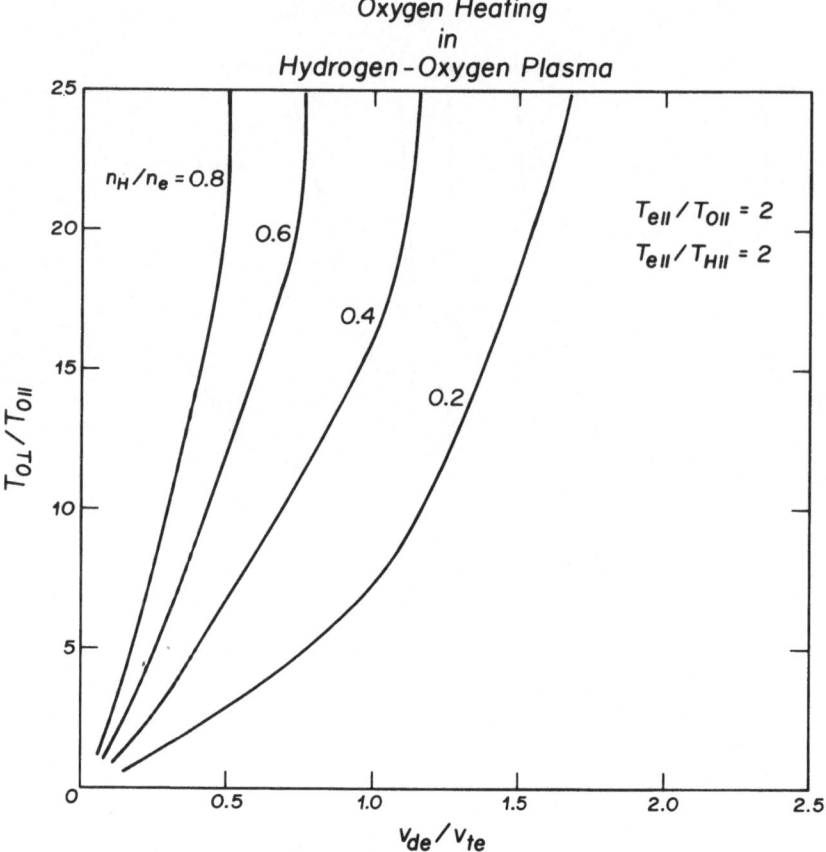

Fig. 4. Same as Fig. 3 except for $T_{e\perp}/T_{O\perp} = T_{e\perp}/T_{H\perp} = 2$.

The length l_H may be estimated from $l_H = \langle v_{de} \, \tau_D \rangle$ where τ_D is the diffusion time of the electron velocity space to establish a plateau (Ashour-Abdalla and Okuda, 1982; Okuda and Ashour-Abdalla, 1982). Since $\tau_H = \gamma_H^{-1}$ where γ_H is the linear growth rate for the hydrogen cyclotron waves, we find from Eq. (2)

$$\frac{\gamma_H}{\Omega_H} = \Gamma_1 \frac{v_{de}}{v_{te}} \frac{n_H}{n_e} \frac{T_e}{T_H} \tag{12}$$

which is smaller than the growth rate for a pure hydrogen plasma by n_H/n_e. Therefore l_H is given by

$$l_H = \rho_H \left[\frac{m_H}{m_e}\right]^{1/2} \left[\frac{T_H}{T_e}\right]^{1/2} \frac{n_e}{n_H} \tag{13}$$

which is larger than the l_H for a pure hydrogen plasma by n_e/n_H. This clearly is because the growth rate is now reduced by n_H/n_e so that it takes a longer time for the electron

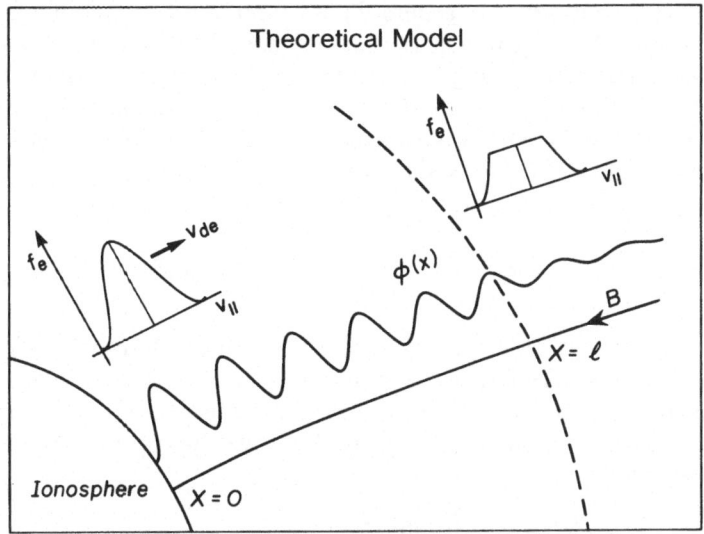

Fig. 5. A schematic plot of the auroral plasma model used in the theoretical analysis. The ionosphere located at $x < 0$ represents a reservoir of electrons which injects drifting Maxwellian electrons along auroral field lines at $x > 0$.

velocity distribution to establish a plateau.

In order to estimate the length l_O over which the electrons develop a plateau due to the growth of EOC waves, let us temporarily assume that only the EOC waves are unstable. Then l_O can be written

$$l_O = \langle v_{de}\, \tau_O \rangle$$

where τ_O is the electron diffusion time in velocity space associated with EOC waves. From Eq. (9), we find the growth rate of the EOC waves near Ω_O to be

$$\frac{\gamma_O}{\Omega_O} = \Gamma_1 \frac{n_O}{n_e} \frac{T_e}{T_O} \left[\frac{v_{de}}{v_{te}} - \frac{T_e}{T_H} \frac{n_H}{n_e} \frac{\Omega_O}{k_\parallel v_{H_\parallel}} \exp\left(-\frac{\Omega_O^2}{k_\parallel^2 v_{H_\parallel}^2} \right) \right]$$

$$\approx \Gamma_1 \frac{n_O}{n_e} \frac{T_e}{T_O} \frac{v_{de}}{v_{te}} \tag{14}$$

so that

$$l_O = \langle v_{de}\tau_O \rangle = v_{de}\gamma_O^{-1}$$

$$= \rho_O \left[\frac{m_O}{m_e}\right]^{1/2} \left[\frac{T_O}{T_H}\right]^{1/2} \frac{n_e}{n_O}. \tag{15}$$

Comparing l_H and l_O, we find

$$\frac{l_O}{l_H} = \frac{\rho_O}{\rho_H} \left[\frac{m_O}{m_e}\right]^{1/2} \left[\frac{T_O}{T_H}\right]^{1/2} \frac{n_H}{n_O}$$

$$= \frac{m_O}{m_H} \frac{T_O}{T_H} \frac{n_H}{n_O} \tag{16}$$

Equation (16) shows that in general $l_O > l_H$ and that $l_H > l_O$ holds only for a very low concentration of hydrogen ions. In that later case, EOC waves grow at a faster rate than EHC waves. In the following we assume that the growth rate of the EHC waves is larger than that of the EOC waves or equivalently $l_H < l_O$. In this case, free energy from the electrons is fed into the EHC waves almost exclusively and very little is available for EOC waves until the EHC waves saturate.

The energy lost from the electrons ΔE^e during τ_H may be estimated from

$$\Delta E^e = \frac{m_e n_e}{2} \int (f_O^e - f_\infty^e) v_\parallel^2 \, dv_\parallel l_H S \tag{17}$$

where S is the cross-sectional area of auroral field lines, f_O^e and f_∞^e are the electron distributions at the source region $(x < 0)$ and at $x > 0$ where a plateau is formed. Equating the energy loss from the electrons to the energy gain of the hydrogen ions (Okuda and Ashour-Abdalla, 1982), the hydrogen perpendicular heating during time τ_H is given by

$$\Delta T_{H_\perp} = \Delta E^e / n_H l_H S$$

$$= \frac{m_e}{30} \frac{v_{de}^5}{v_{te}^3} \frac{n_e}{n_H} \quad \text{for} \quad v_{de} < v_{te}$$

$$= \frac{m_e}{3} (v_{de}^2) \frac{n_e}{n_H} \quad \text{for} \quad v_{de} > v_{te} \tag{18}$$

Note that $\Delta T_{H_\perp}$ is larger than that for a pure hydrogen plasma by n_e/n_H since the number of hydrogen ions is now smaller enabling each hydrogen ion to gain more energy during τ_H.

Assuming the hydrogen energy loss is smaller than the energy input from the electrons given by Eq. (18), the hydrogen temperature in $0 < x < l_H$ increases in time until the hydrogen cyclotron instabilities saturate by generating a sufficiently large temperature anisotropy.

The time for the saturation of the hydrogen perpendicular heating, τ_{sH}, for $0 < x < l_H$ is estimated to be

$$\tau_{sH} = \tau_H \frac{T_{H_\perp}^f}{\Delta T_{H_\perp}} \tag{19}$$

where $T_{H_\perp}^f$ is the final hydrogen perpendicular temperature at saturation determined from the marginal stability curves given by Figs. 1 and 2.

Since the length, l_O, over which the electron velocity distribution develops a plateau due to the growth of EOC waves is generally much longer than l_H, we expect EHC waves and hydrogen heating to extend upward along auroral field lines until the hydrogen energy loss across the magnetic field becomes comparable to the energy input to the hydrogen ions from the electrons. Therefore the maximum length l_{tH} on which hydrogen cyclotron turbulence exists is estimated by balancing the energy input from the electrons to the energy loss due to cross-field diffusion (Ashour-Abdalla and Okuda, 1982).

$$l_{tH} = \rho_H \left[\frac{m_H}{m_e}\right]^{1/2} \left[\frac{a}{\rho_H}\right]^2 \left[\frac{D_{BH}}{D_{H_\perp}}\right] \qquad (20)$$

which takes the same form as the pure hydrogen plasma. Here a is the transverse dimension of the auroral arc ($a^2 = S$), $D_{BH} \equiv \rho_H^2 \Omega_H$ is the Bohm diffusion coefficient for hydrogen ions and $D_{H_\perp}$ is the crossfield hydrogen diffusion coefficient on auroral field lines.

So far we have neglected the effects of oxygen ions on hydrogen waves and hydrogen heating. So long as the hydrogen waves grow much faster than the oxygen waves and consequently, the heating of hydrogen ions occurs much faster than the oxygen heating, the effects of oxygen ions on the hydrogen heating can be neglected since most of the electron energy is exclusively fed into the hydrogen ions. As the turbulence regions of the EHC waves extends along auroral field lines, more oxygen heating takes place since the saturated hydrogen ions can no longer absorb energy from the electrons. In particular, if the distance l_O over which the electron velocity distribution develops a plateau due to EOC, is shorter than l_{tH}, the maximum length on which the hydrogen cyclotron turbulence exists,

$$l_O < l_{tH}, \qquad (21)$$

then the oxygen ions can be heated to their marginal stability limit, which is more than the hydrogen ions due to the larger temperature anisotropy required for saturation of EOC waves for a given electron drift as seen in Figs. 1 and 3. If condition Eq. (21) is satisfied, hydrogen ions are heated to their maximum temperature determined from the marginal stability of the hydrogen waves for $0 < x < l_O$. Since the saturated EHC waves are no longer able to absorb energy from the electrons, a significant amount of electron energy streaming from $x < 0$ is now fed into the EOC waves resulting in the heating of oxygen ions. If equation (21) is satisfied, then the oxygen ions are heated to the maximum temperature determined from the marginal stability analysis for the EOC waves for $0 < x < l_O$.

If, on the other hand,

$$l_O > l_{tH} \qquad (22)$$

is satisfied on auroral field lines, oxygen heating will never reach the maximum value determined from the saturation of the EOC waves. In this case, the electrons lose most of their free energy to the EHC waves resulting in hydrogen heating in the region $0 < x < l_{tH}$.

When the hydrogen concentration is small enough so that the growth rate for the EOC waves is larger than that for EHC waves, most of the energy from the electrons is fed into the oxygen ions. Noting the oxygen heat loss along field lines is smaller than the oxygen heating from the electrons for $v_{de} > v_{te}$, the maximum length l_{tO} over which the EOC waves exist is estimated by balancing the energy input with the energy loss associated with the perpendicular heat conduction,

$$l_{tO} = \rho_O \left[\frac{m_O}{m_e}\right]^{1/2} \left[\frac{a_\perp}{\rho_O}\right]^2 \left[\frac{D_{BO}}{D_{O_\perp}}\right] \qquad (23)$$

where $D_{BO} = \rho_O^2 \Omega_O$ is the Bohm diffusion coefficient for oxygen ions and $D_{O_\perp}$ is the oxygen cross-field diffusion coefficient. Comparing Eq. (23) with Eq. (20), we find

$$\frac{l_O}{l_H} = \left[\frac{m_O}{m_H}\right]^{1/2} \frac{\rho_O}{\rho_H} \frac{D_{H_\perp}}{D_{O_\perp}} \sim 16 \qquad (24)$$

for $T_{O_\parallel} = T_{H_\parallel}$ and $D_{BH} = D_{BO}$. Therefore, the localization length of the EOC waves along auroral field lines is much longer than that of the EHC waves. Note, however, this can happen only when the oxygen ions are the majority species ($n_O/n_H > 16$) which is not always the case on auroral field lines.

3. Results of Simulations from the Uniform Recycling Model

In this section, we describe the results obtained from the uniform recycling model in which a fraction of the electron distribution is replaced by the initial drifting Maxwellian at a constant rate in time and uniformly in space. This model assumes the presence of a uniformly distributed plasma source, which is different from the ionospheric situation where a flow of fresh electrons enter at one end of the system. Nevertheless, as pointed out previously (Okuda and Ashour-Abdalla, 1981) the nonlinear saturation of the instability and ion heating for both cases should be similar, provided the ion energy loss is much smaller than the energy input from the drifting electrons. However, this model treats the instability as absolute, and the ion heat loss along auroral field lines is not included in the model. The ion heating found in the uniform recycling model should correspond to the theoretical model given in Section 2 in which a maximum $T_\perp/T_\parallel$ was calculated from the marginal stability analysis when the electron heating remains negligible.

In Section 2, we have seen that, for a given electron drift speed, oxygen heating generally exceeds that of hydrogen unless the hydrogen ions are the dominant species ($n_H/n_e > 0.8$). This is because the oxygen cyclotron waves have in general a lower threshold drift speed for the electrons so that oxygen ions must be heated more to achieve the marginal stability. Here, we present results from two cases with different hydrogen and oxygen ion concentrations. In the first case, hereafter referred to as 'case 1,' we set $\bar{n}_O = \bar{n}_H$, where $\bar{n}_O$ is the average oxygen density and $\bar{n}_H$ is the average hydrogen density. In case 2, hereafter referred to as 'case 2,' we set $\bar{n}_O/\bar{n}_e = 0.9$ and $\bar{n}_H/\bar{n}_e = 0.1$, where $\bar{n}_e$ is the average electron density. The other parameters relevant to both cases are: the system length $L = 256\,\Delta$ where Δ is equal to the electron Debye length $\lambda_{e_\parallel}$, $T_{e_\parallel} = T_{H_\parallel} = T_{O_\parallel} = T_{H_\perp} = T_{O_\perp}$, $v_{de} = v_{te}$, $\Omega_e/\omega_{pe} = 10$, $m_H/m_e = 1837$, $m_O/m_H = 16$, $\Omega_H \Delta t = 0.1$ and $k_\parallel/k_\perp = 0.05$. The recycling time is chosen to be $\Omega_H t = 10$ for both cases.

Figure 6 shows simulation results for case 1 where equal hydrogen and oxygen densities are used. The left panel shows the results at an early time $\Omega_H t = 220$, whereas the right panel shows the results at a later time $\Omega_H t = 2,000$. In the upper panel we plot the electrostatic potential $e\phi/T_e$ (a, b). In the middle panel we show the hydrogen density profile $n_H(x)/\bar{n}_H$ (c) and the oxygen density profile $n_O(x)/\bar{n}_O$ (d). The electrostatic energy for each Fourier mode, or equivalently $k_\perp \rho_i$ is shown in panels (e), (f) at two different times. We note that at $\Omega_H t = 220$, the amplitude of the electrostatic potential

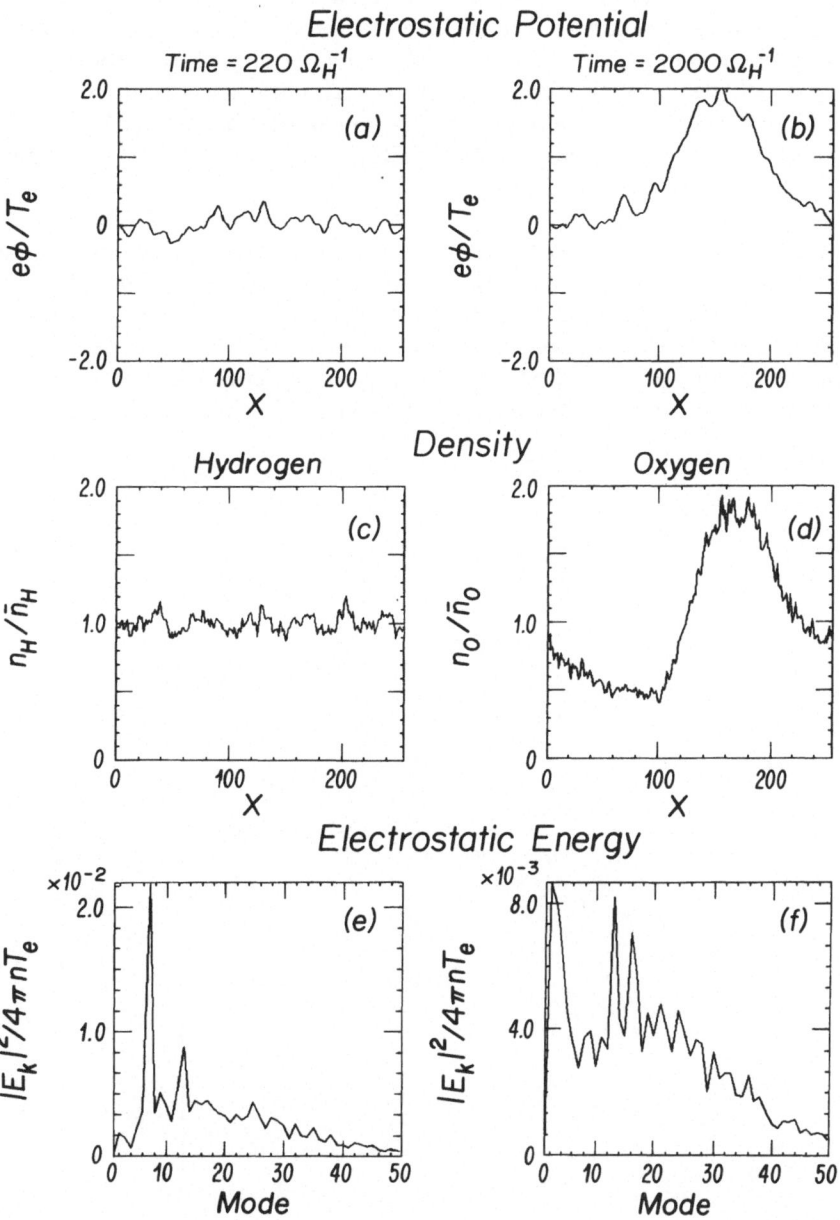

Fig. 6.   Shown on the left column are (a) electrostatic potential $e\phi(x)/T_e$, (c) hydrogen density $n_H(x)/\bar{n}_H$, (e) electrostatic field energy $|E_k|^2/4\pi nT_e$ versus mode number at time $\Omega_H t = 220$. Shown on the right column are (b) electrostatic potential $e\phi(x)/T_e$, (d) oxygen density $n_O(x)/\bar{n}_O$ and (f) electrostatic field energy $|E_k|^2/4\pi nT_e$ versus mode number at time $\Omega_H t = 2,000$.

is $e\phi/T_e \approx 0.2$ and is dominated by short wavelength perturbations. Similar short wavelength perturbations can be seen in the hydrogen density profile (c) at $\Omega_H t = 220$ whose amplitude is $\delta n_H/\bar{n}_H \approx 0.2$. This can be seen more clearly from the electrostatic energy (e) which peaks at a mode number ≈ 7 and the wavelength of this mode corresponds to $k_\perp \rho_H \approx 0.74$. Therefore, at this time of the simulation, $\Omega_H t = 220$, the observed perturbations are dominated by the EHC waves and the EOC waves have not grown to a large amplitude yet.

Looking at a later time $\Omega_H t = 2,000$ or $\Omega_O t = 125$, we note the presence of large amplitude waves $e\phi/T_e \approx 1.0$ (b) and the oxygen density profile (d) is modulated by a long wavelength mode of large amplitude. The electrostatic energy peaks at mode 2 whose wave number corresponds to $k_\perp \rho_O \approx 0.85$. Thus, it is clear that at an early stage hydrogen cyclotron waves are growing whereas at a later stage oxygen cyclotron waves are dominant, owing to the difference in the gyrofrequencies for both species. To investigate the nature of the wave activity more precisely, we plot in Fig. 7 the time

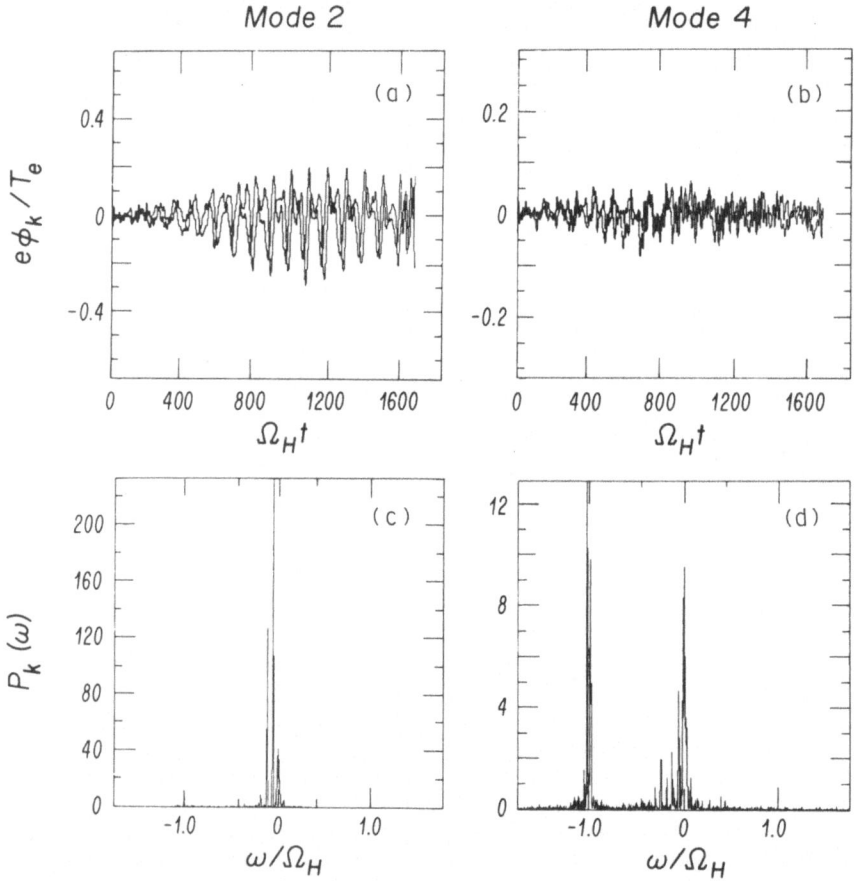

Fig. 7.   Time history (top panel) and the power spectrum (lower panel) for two Fourier modes, mode 2 in (a), (c) and mode 4 in (b), (d).

history and the frequency spectrum of mode 2 panel (a, c) and mode 4 panel (b, d). The upper panel shows the time history whereas the lower panel shows the power spectrum versus frequency normalized to the hydrogen cyclotron frequency. Figure 7 (a) shows that the amplitude of mode 2 starts increasing substantially at around $\Omega_H t = 450$, reaches a peak value of $e\phi/T_e \approx 0.3$ at $\Omega_H t = 1,100$ and then starts slowly decreasing. The corresponding power spectrum (c) indicates clearly the presence of coherent narrow band emission at a frequency slightly above the oxygen cyclotron frequency and its harmonics confirming the long wavelength mode is the EOC wave. Figures 7b and 7d show similar plots for mode 4, corresponding to $k_\perp \rho_H = 0.42$. It is clear from Fig. 7d that this mode is a mixed mode in which both EHC and EOC waves are excited to a comparable degree. The maximum value of $e\phi/T_e \approx 0.05$ as shown in Fig. (7b). The fact that the EOC wave saturates at a small amplitude is because the wavelength of mode 4 is shorter $(k_\perp \rho_O = 1.7)$ than the wavelength of mode 2 $(k_\perp \rho_O = 0.85)$.

The effect of these waves on the perpendicular particle distribution is shown in Fig. 8 in which the perpendicular velocity distributions of hydrogen (a, b) and the oxygen ions (c, d) are shown at two different times. The hydrogen distribution is heated substantially up to a factor of 4 at time $\Omega_H t = 220$ (a), while the oxygen distribution is hardly affected. At a much later time $\Omega_H t = 2,000$, the perpendicular oxygen distribution has undergone significant heating and is much hotter than the hydrogen distribution. Both the oxygen and hydrogen perpendicular distributions resemble Maxwellians. Comparing Figs. 8 (a, b) with (c, d), we observe that the hydrogen heating has saturated by the time $\Omega_H t = 220$ whereas oxygen heating has just begun at that time.

The time development of the perpendicular temperature of the hydrogen and oxygen ions is shown in Fig. 9. The hydrogen perpendicular temperature increases rapidly and then saturates at a value of six times the initial temperature at time $\Omega_H t = 600$. In contrast, the oxygen ions are hardly heated initially. Only when the hydrogen cyclotron waves have saturated at time $\Omega_H t = 600$ does the oxygen temperature increase drastically. The oxygen temperature at time $\Omega_H t = 2,000$ is about 14 $T_\perp(0)$ and is still increasing without saturating during the course of this simulation. The fact that the oxygen heating is much larger than that of hydrogen ions is consistent with our theoretical prediction given in Section 2.

The parallel electron distribution (a, b) and parallel hydrogen distribution (c, d) are shown at two different times in Fig. 10. It is interesting to note that even in the presence of a continuous source of fresh electrons, the electrons are heated and have diffused out (a, b). How much diffusion can take place depends on the diffusion coefficient associated with the EHC waves and the recycling rate. Note the recycling time used in these simulations is longer than the growth time of the EHC waves. Therefore, at later times after the saturation of the EHC instability, the velocity space diffusion of the electrons diminishes and the electron distribution should resemble the original drifting Maxwellian. The hydrogen parallel distribution (c, d) shows the heating as well as the net drift along the magnetic field. The parallel heating is much smaller than the perpendicular heating since $k_\parallel/k_\perp \ll 1$ for EHC waves. The net drift of hydrogen ions along the magnetic field is caused by the anomalous resistivity caused by the EHC waves (OKUDA and ASHOUR-ABDALLA, 1981). The time history of the parallel

Perpendicular Distribution Functions

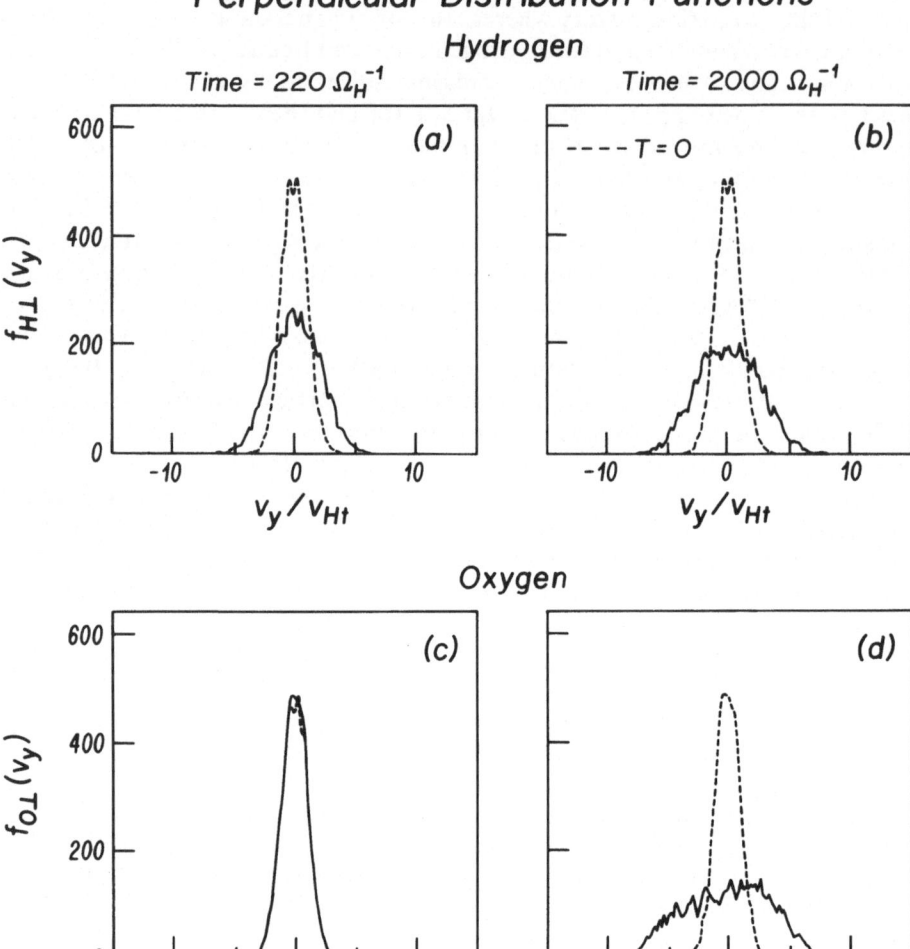

Fig. 8.   Hydrogen (top panel) and oxygen (bottom panel) perpendicular velocity distributions at two different time steps, $\Omega_H t = 220$ (left column) and $\Omega_H t = 2,000$ (right column).

electron, hydrogen and oxygen temperatures is shown in Fig. 11. The parallel electron temperature increases rapidly at first until the saturation of the hydrogen cyclotron waves, then fluctuates about a constant value. This is because the electron heating due to EOC waves balances with the cooling due to recycling at this stage. The parallel hydrogen temperature also increases about the same amount. It is interesting to note that while the hydrogen perpendicular heating stopped at $\Omega_H t \approx 600$, parallel heating continues to the end of the simulation. This is probably caused by the hydrogen Landau damping of the EOC waves in which both hydrogen and oxygen parallel

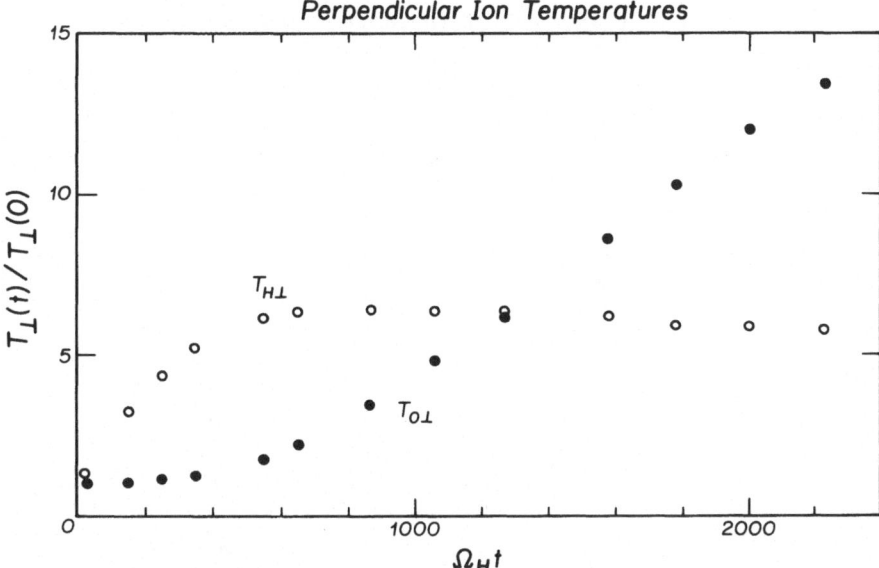

Fig. 9.   Time history of perpendicular temperatures of hydrogen and oxygen ions for case 1.

temperatures increase in time at about the same rate. As discussed in Section II, the maximum heating of both hydrogen and oxygen ions depends sensitively on the ratios of $T_{e\parallel}/T_{H\parallel}$ and $T_{e\parallel}/T_{O\parallel}$, respectively, which are larger than unity as shown in Fig. 11. We therefore expect that both ion species would be heated more than the values given in Figs. 1 and 3 since they are for $T_{e\parallel} = T_{H\parallel} = T_{O\parallel}$.

Now we turn to the second case, where the relative concentration of oxygen is $\bar{n}_O/\bar{n}_e = 0.9$ and hydrogen $\bar{n}_H/\bar{n}_e = 0.1$. As expected from linear theory, the EOC waves should play a dominant role in this case. Figure 12 shows that the electrostatic potential reaches a huge amplitude at $\Omega_H t = 650$, $e\phi/T_e \approx 2$, (a) and is modulated by the growth of the second mode of the system. The corresponding oxygen density profile (b) also shows large spatial modulation $\delta n_O/\bar{n}_O \approx 0.5$ and the electrostatic energy (c) peaks at $k_\perp \rho_O = 0.84$. Hydrogen density, not shown in the figure, shows very little modulation in this case.

The growth of the waves can be seen in Fig. 13 where we plot the time history of modes 2 (a, c) and 4 (b, d). In the left panel (a, c) we see the growth of a large amplitude $e\phi_k/T_e \approx 1.0$ oxygen cyclotron wave. The wave reaches a peak amplitude at $\Omega_H t = 650$. The right panel (b, d) corresponding to mode 4, shows the growth of both oxygen and hydrogen waves, but at a smaller amplitude. Note the initial rapid growth of the wave amplitude, up to $\Omega_H t = 200$ in (b) is associated with the EHC waves.

The effect of the waves on the particle distributions is shown in Fig. 14. Here we plot the hydrogen perpendicular distribution (a, b), oxygen perpendicular distribution (c, b), and electron parallel distribution at two different times. At an early time $\Omega_H t = 650$, we note that the hydrogen heating (a) exceeds that of oxygen (b). At a later time $\Omega_H t = 2,200$, the oxygen distribution (d) is hotter than the hydrogen perpendicular

Parallel Distribution Functions

Electrons

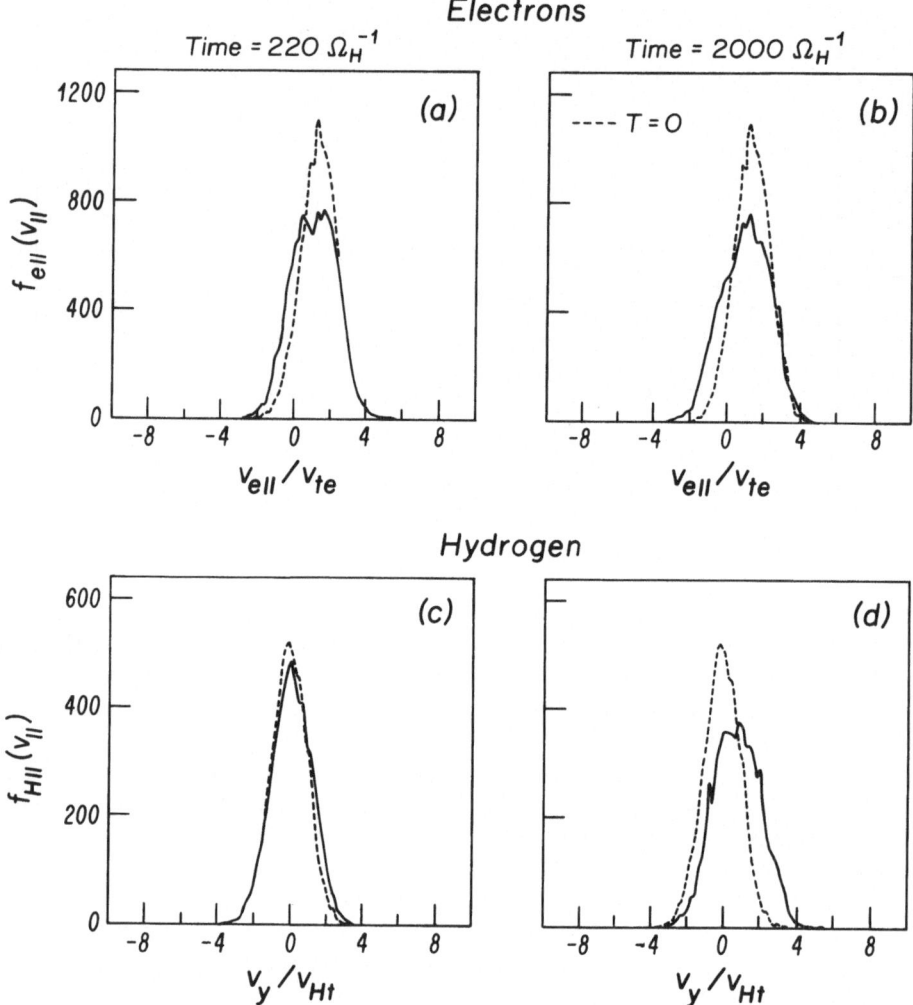

Fig. 10.   Electron (top panel) and hydrogen (lower panel) parallel velocity distributions at two different times, $\Omega_H t = 220$ (left column) and $\Omega_H t = 2{,}000$ (right column).

distribution. This behavior is similar to what we observed in case 1 where $\bar{n}_H = \bar{n}_O$ was used. It is interesting to observe the behavior of the parallel electron temperature shown in Fig. 14 (e) and (f). At time $\Omega_H t = 650$ (e), the electrons have been heated by a factor of 2, then cool off later at $\Omega_H t = 2{,}200$ (f). This heating of the electron parallel distribution is significantly larger than that in case 1 and can be explained by quasilinear diffusion caused by the growth of both EHC and EOC waves. In fact, this large increase in electron temperature seen in Fig. 15 is due to the combined growth of

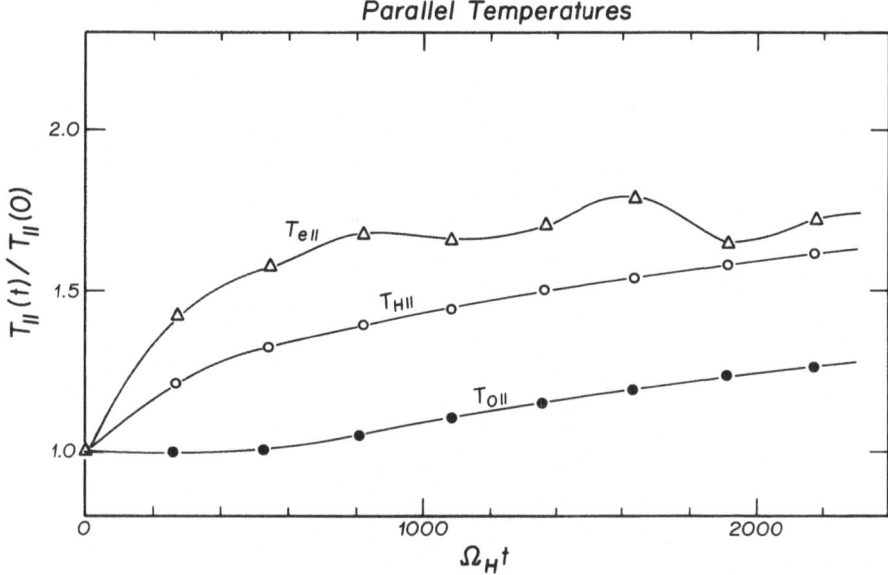

Fig. 11.   Time history of parallel temperatures for electrons, hydrogen ions and oxygen ions for the case 1.

the hydrogen and oxygen cyclotron waves. The fact that the electron heating in this case is larger than in case 1, is because both EHC and EOC waves grow on the same time scale here while in case 1, EHC waves grow much faster than the EOC waves. The temperature of the electrons is expected to be restored to its initial value once the waves have saturated. This can be seen in Fig. 15, where the parallel electron temperature has decreased substantially after $\Omega_H t = 800$. The parallel hydrogen and oxygen temperatures also increase, but only slightly at early times.

Figure 16 shows the increase in the perpendicular ion temperatures. The hydrogen temperature increases and reaches a maximum of 6 times the initial value, whereas the oxygen reaches a maximum of about 9 times the initial value. In the present case, where the oxygen ions are a majority species, the time scale in which the oxygen temperature rises is comparable to that for the hydrogen temperature. This is because the growth rates of EHC and EOC waves are comparable in this case.

It is illuminating to compare the maximum temperature achieved in the simulations, for different density concentrations, with those obtained from marginal stability analysis in Section 2. From Fig. 1, it appears that the hydrogen temperature anisotropy required to stabilize hydrogen cyclotron waves is much greater for case 1, $\bar{n}_H = \bar{n}_O$, than for case 2, $\bar{n}_H/\bar{n}_e = 0.1$. Yet, results from the simulations show that $T_{H_\perp} \approx 6 T_{H_\perp}(t = 0)$ for both cases (Figs. 9, 16). The theoretical temperature anisotropy assuming the distribution remains a bi-Maxwellian is $T_{H_\perp}/T_{H_\parallel} \approx 3$ for case 1 and $T_{H_\perp}/T_{H_\parallel} \approx 1$ for case 2. This apparent contradiction between theory and simulation is due to the different behavior of the parallel electron temperature. As we have seen in Fig. 15, for case 2, the parallel electron temperature increases rapidly due to the

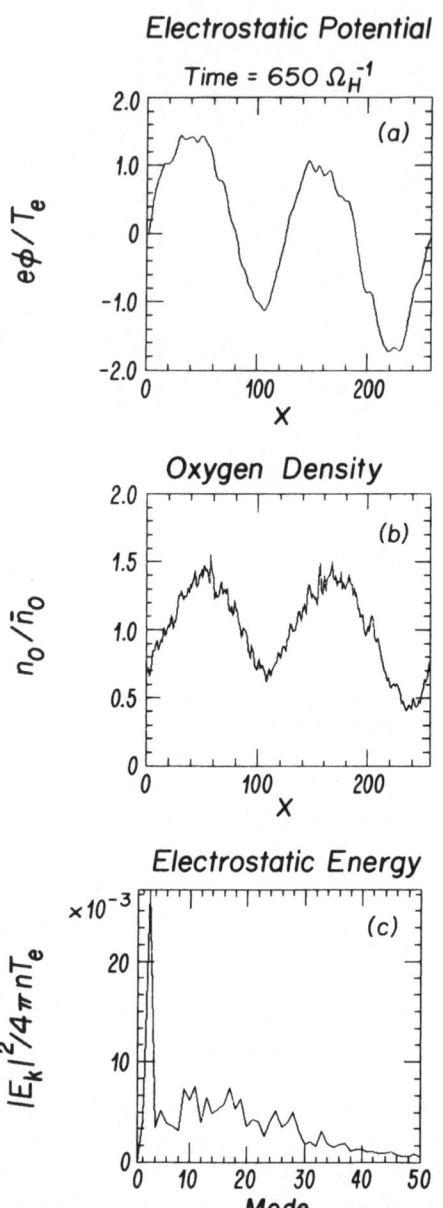

Fig. 12.   Shown are (a) electrostatic potential $e\phi(x)/T_e$, (b) oxygen density $n_O(x)/\bar{n}_O$ and (c) electrostatic field energy $|E_k|^2/4\pi n T_e$ versus mode number, at time $\Omega_H t = 650$ for case 2.

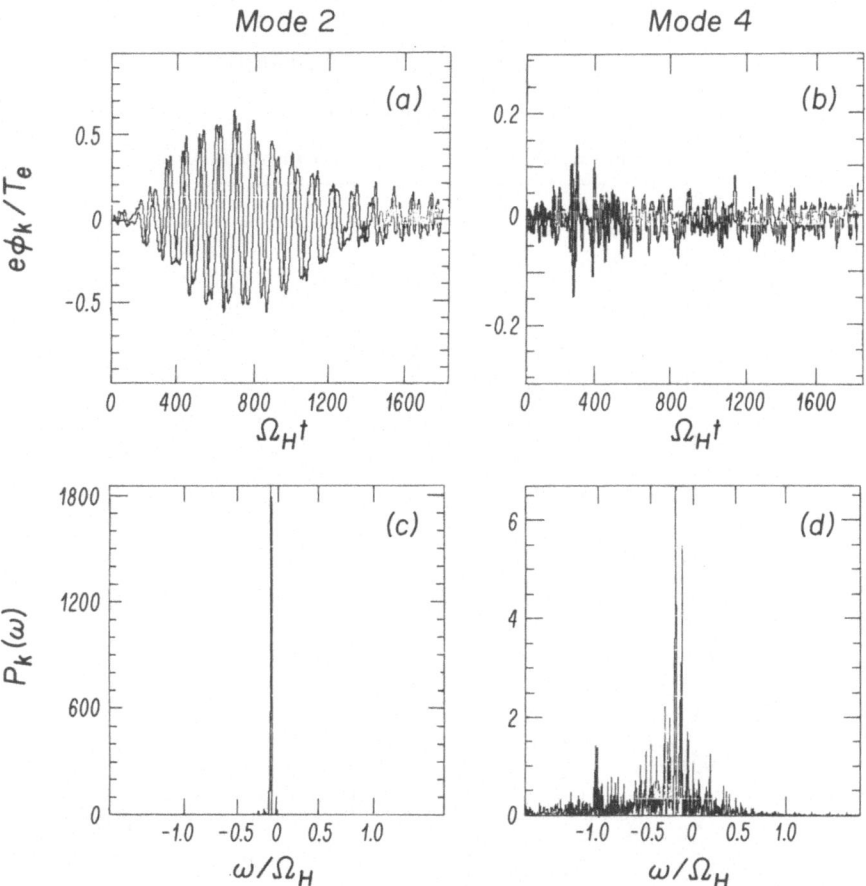

Fig. 13.   Time history (top panel) and power spectrum (lower panel) for two different Fourier modes, mode 2 in (a), (c) and mode 4 in (b), (d).

combined growth of both hydrogen and oxygen cyclotron waves, which cause the electrons to diffuse in velocity space resulting in $T_{e\,\|}/T_{H\,\|} = 2$ at an early stage. Recall that as the electron temperature is increased, ion cyclotron damping diminishes and large temperature anisotropies are required to stabilize the waves as shown in Fig. 3. In the marginal stability analysis used to obtain Figs. 1–4, the parallel electron temperature was kept constant, while it was allowed to vary in the simulation. If we compare the simulation results with Figs. 3 and 4 for $T_{e\,\|}/T_{H\,\|} = 2$, then we find reasonable agreement between simulation and theory.

4.   Summary and Discussion

In this paper we have addressed the problem of transverse acceleration of ions associated with the current-driven electrostatic ion cyclotron waves in a plasma

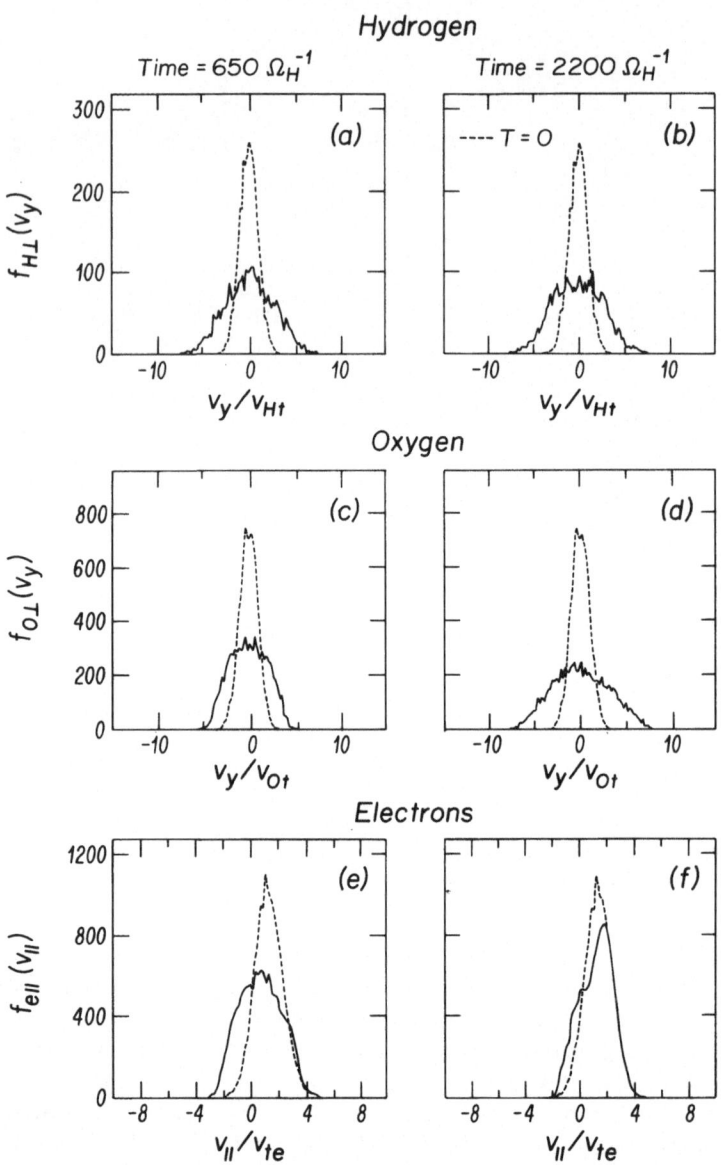

Fig. 14.   Hydrogen perpendicular velocity distributions (top panel) (a), (b), oxygen perpendicular velocity distributions (middle panel) (c), (d), and electron parallel velocity distributions (lower panel) (e), (f) at two different times. Left column is at $\Omega_H t = 650$ and right column is at $\Omega_H t = 2,200$.

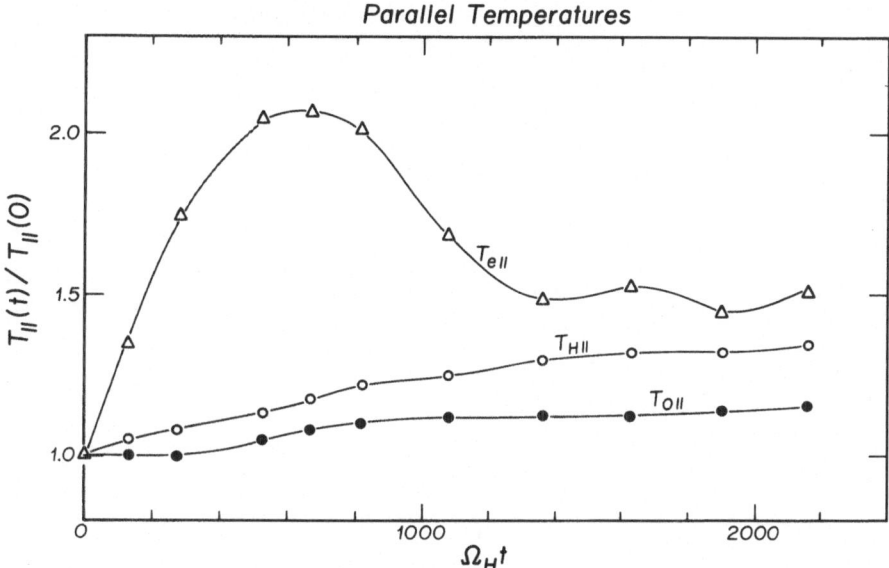

Fig. 15.   Time history of parallel temperature for electrons, hydrogen and oxygen ions for the case 2.

consisting of hydrogen and oxygen ions. We have studied this problem using a theoretical approach (Section 2) as well as simulation techniques (Section 3). First, in Section 2, we performed a marginal stability analysis for both EHC and EOC waves. From this marginal stability analysis we were able to determine the maximum temperature anisotropy $T_\perp/T_\parallel$ generated by ion cyclotron waves. This maximum temperature anisotropy was then used to determine the heating of hydrogen and oxygen ions, assuming the ion energy loss remains small. Similar to the case of a pure hydrogen plasma, (ASHOUR-ABDALLA and OKUDA, 1982), we found that ion heat losses along and across the field lines were small provided $v_{de} \approx v_{te}$. The fact that ion heat losses were small permitted us to perform our simulations using a uniform recycling model, in which no ion heat loss takes place. In general, theoretical predictions and simulation results are in good agreement. Let us summarize our findings.

4.1   Conclusions from theory

(1)   The maximum temperature anisotropy, $T_\perp/T_\parallel$, for a given electron drift speed is generally larger for oxygen ions than for hydrogen ions. This can be attributed to the fact the EOC waves require a lower electron drift speed threshold than EHC waves. As such, for a given electron drift speed, oxygen ions must be heated more than hydrogen ions to attain marginal stability.

(2)   For equal densities of oxygen and hydrogen ions $n_O = n_H$ and $v_{de} = v_{te}$, $T_{e\parallel} = T_{H\parallel} = T_{O\parallel}$, the perpendicular temperature of hydrogen ions increases by as much as a factor of 3 compared to as much as a factor of 10 for oxygen ions. (Figs. 1 and 3). When the parallel electron temperature is larger than the ion parallel temperatures ($T_{e\parallel} = 2T_{H\parallel} = 2T_{O\parallel}$) and all the other parameters are kept constant, the hydrogen

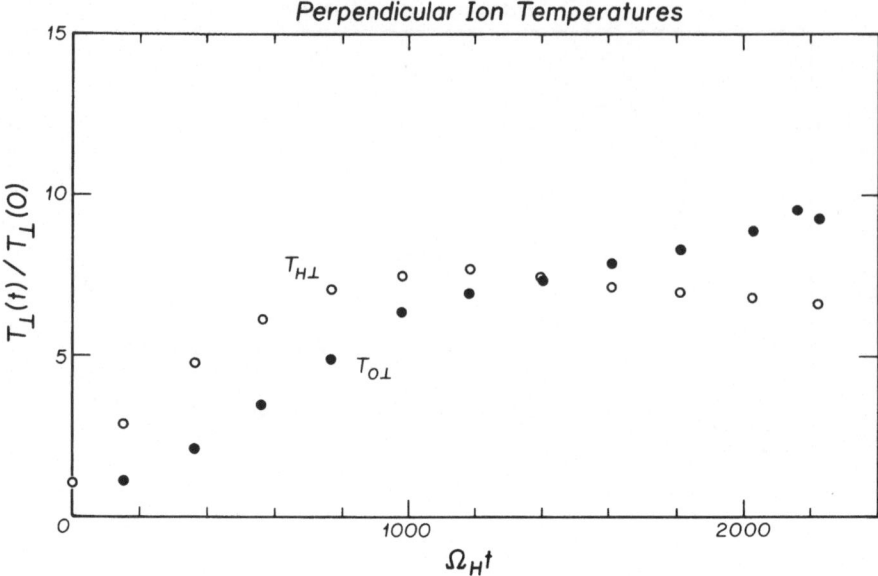

Fig. 16.   Time history of perpendicular temperatures of hydrogen and oxygen ions for case 2.

ions can be heated by more than a factor of 20.

(3)   If the electron drift speed is sufficiently large compared to the thermal speed, then there is no marginal stability for either EHC or EOC waves. In this case, the instabilities are expected to saturate by processes not included in the marginal stability analysis. Examples of such processes are the formation of an electron plateau or electron heating due to quasilinear diffusion which reduces the ratio v_{de}/v_{te}.

(4)   Hydrogen ions are heated much more rapidly than oxygen ions, unless the oxygen ion concentration is very large ($n_O/n_H > 16$). This is because EHC waves generally grow on a faster time scale than EOC waves. Thus, most of the free energy from the electrons is fed into hydrogen ions and very little energy is available for oxygen heating initially.

(5)   Oxygen ion heating takes place after the hydrogen heating has saturated, since saturated hydrogen ions can no longer absorb energy from the electrons. The oxygen ions can be heated more than the hydrogen ions if the distance l_O over which the electron velocity distribution develops a plateau due to EOC being shorter than l_{tH}, the maximum length over which the hydrogen cyclotron turbulence exists.

(6)   The localization length for the EHC waves on auroral field lines, l_{tH}, is determined by balancing the energy input from the electrons with the hydrogen energy loss across the magnetic field. For typical auroral plasma parameters, the localization length l_{tH} is estimated to be $\approx 500-1,000$ kms.

(7)   The localization length of the EOC waves is also of the order of l_{tH}. At distance $x > l_{tH}$, there will be no free energy available from the electrons for the growth of the EOC waves.

4.2   Conclusions from simulations

(1)   For case 1, equal oxygen and hydrogen densities, the EHC waves initially grow much faster than the EOC waves. As a result, the hydrogen temperature initially rises more rapidly than that of the oxygen ions. After the saturation of the EHC waves, we find that the oxygen ions are heated to larger values than the hydrogen ions. The heating of both ion species agrees well with theoretical predictions when the parallel heating of electrons and ions is taken into account.

(2)   The hydrogen parallel temperature increases even after the saturation of the hydrogen perpendicular temperature as shown in Fig. 10. The heating rate of the hydrogen parallel temperature is about the same as that of the oxygen parallel temperature. Therefore, it seems that the parallel heating of hydrogen ions must be due to hydrogen Landau damping of the EOC waves.

(3)   When a large concentration of oxygen ions is used as in case 2, both the EHC and EOC waves grew on the same time scale, because the waves had approximately the same growth rates.

(4)   The main difference between case 1 and 2, where oxygen was a majority species, was the behavior of the electrons. In the latter case, the parallel electron temperature was considerably larger than in the former case. This can be explained by the fact that both the EHC and EOC waves grew on the same time scale, thereby significantly heating the electrons. This increase in parallel electron temperature resulted in substantial heating of the hydrogen ions.

(5)   The oxygen ions in case 2 were heated by a factor of 9. The oxygen ion heating was larger than hydrogen ion heating. Nevertheless, the oxygen heating is smaller than theoretical prediction for $n_O/n_e = 0.9$. It is possible that the choice of simulation parameters for case 2 was not close to that for marginal stability. In particular, the angle of propagation $k_\parallel/k_\perp$ should have been taken smaller for case 2.

As stated previously, the simulation model employed here assumes the presence of a uniformly distributed plasma source along field lines, thereby neglecting the wave convection and heat losses. Clearly the model is valid in regions of auroral field lines near the ionosphere where fresh electrons stream into the plasma. When considering the global nature of auroral field lines, the localized source at the ionosphere must be taken into account. Nevertheless, we believe that the nonlinear saturation of the instability and the associated ion heating for both cases should be similar, provided the energy loss is much smaller than the energy input from the electrons. This is because once the cyclotron waves have saturated, they do not interact with the flow of fresh electrons. Thus, the electrons can reach higher altitudes without significant distortion of their velocity distribution, which is essentially equivalent to having a plasma source at a higher altitude. In order to test the above assumption and include ion heat losses, we developed a model, the constant flux model, in which drifting electrons enter at only one end of the field line (ASHOUR-ABDALLA and OKUDA, 1982). In the case of a pure hydrogen plasma, we found that hydrogen heating was similar for both simulation models provided $v_{de} > v_{te}$. In this case, as we have seen in Section 2, ion heat losses are negligible as compared to hydrogen heating by EHC waves. Our results using the constant flux model, for equal densities of oxygen and hydrogen and $v_{de} = v_{te}$ are shown in Fig. 17. Figure 17 shows the hydrogen perpendicular velocity distribution (a)

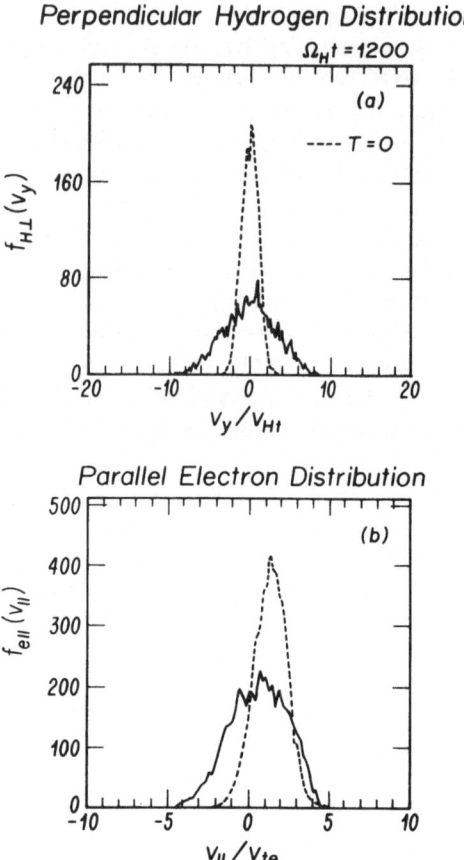

Fig. 17.   Hydrogen perpendicular velocity distribution (a) and electron parallel velocity distribution (b) for the constant flux model in the region $0 < x < 100$.

and the parallel electron distribution (b) at $\Omega_H t = 1,200$, in the region $0 < x < 100$. The heating of hydrogen ions agrees with the theoretical predictions in Section 2. In fact, the maximum perpendicular temperature in the constant flux model ($T_\perp = 9T_\perp(t = 0)$) is larger than that for the uniform recycling model ($T_\perp = 6T_\perp(t = 0)$). This is because the parallel electron temperature is larger in the constant flux model than in the uniform recycling model. In the uniform recycling model, the electron heating is constantly compensated by the cooling associated with the fresh electron distribution whereas no such cooling exists in the constant flux model.

Thus we feel confident that ion cyclotron waves, EHC and EOC, are important for heating ions in the auroral ionosphere and that the estimated heating rates are appropriate.

We would like to thank R. G. Johnson for his support and encouragement of our theoretical work. We are very grateful to R. J. Walker for his careful reading of this manuscript. It is a pleasure to acknowledge helpful discussions with H. Balsiger, F. V. Coroniti, D. Gorney, R. G. Johnson, C. F. Kennel, P. L. Pritchett and R. J. Walker. This work was supported by Solar Terrestrial Theory Grant NAGW-78 and Air Force Contract F19628-82-K-0019 and NSF Grant ATM81-15257. In addition, H. Okuda would like to acknowledge DOE grant DE-AC02-76-CH03073.

REFERENCES

Ashour-Abdalla, M., H. Okuda, and C. Z. Cheng, Acceleration of heavy ions on auroral field lines, *Geophys. Res. Lett.*, **8**, 795, 1981.

Ashour-Abdalla, M. and H. Okuda, Plasma physics on auroral field lines: The formation of ion conic distributions, *High Latitude Magnetospheric/Ionospheric Plasma Physics*, edited by B. Hultquist, Plenum Publ. Corp., New York, to be published in 1982.

Drummond, W. E. and M. N. Rosenbluth, Anomalous diffusion arising from microinstabilities in a plasma, *Phys. Fluids*, **5**, 1507, 1962.

Dusenbery, P. B. and L. R. Lyons, Generation of ion-conic distribution by upgoing ionospheric electrons, *J. Geophys. Res.*, **86**, 7627, 1981.

Ghielmetti, A. G., R. G. Johnson, R. D. Sharp, and E. G. Shelley, The latitudinal, diurnal and altitudinal distributions of upward flowing energetic ions of ionospheric origin, *Geophys. Res. Lett.*, **5**, 59, 1978.

Gorney, D. J., A. Clarke, D. Croley, J. Fennell, J. Luhmann, and P. Mizera, The distribution of ion beams and conics below 8,000 km, *J. Geophys. Res.*, **86**, 83, 1981.

Harris, K. K., G. W. Sharp, and C. R. Chapell, Observations of the plasmapause from OGO-5, *J. Geophys. Res.*, **75**, 219, 1970.

Johnson, R. G., The hot ion composition, energy and pitch angle characteristics above the auroral zone ionosphere, *High Latitude Magnetospheric/Ionospheric Plasma Physics*, edited by B. Hultquist, Plenum Publ. Corp., New York, to be published in 1982.

Kindel, J. M. and C. F. Kennel, Topside current instabilities, *J. Geophys. Res.*, **76**, 3055, 1971.

Kintner, P. M., M. C. Kelley, R. D. Sharp, A. Ghielmetti, M. Temerin, C. A. Cattell, P. F. Mizera, and J. F. Fennell, Simultaneous observations of energetic (keV) upstreaming ions and electrostatic hydrogen cyclotron waves, *J. Geophys. Res.*, **84**, 7201, 1979.

Klumpar, D. M., Transversely accelerated ions: an ionospheric source of hot magnetospheric ions, *J. Geophys. Res.*, **84**, 4229, 1979.

Lee, K. F., Ion cyclotron instability in current-carrying plasmas with anisotropic temperatures, *J. Plasma Phys.*, **8**, 379, 1972.

Mizera, P. F., J. F. Fennell, D. R. Croley, A. L. Vampola, F. S. Mozer, R. B. Torbert, M. Temerin, R. Lysak, M. Hudson, C. A. Cattell, R. G. Johnson, R. D. Sharp, A. G. Ghielmetti, and P. M. Kintner, The aurora inferred from S3-3 particles and fields, *J. Geophys. Res.*, **86**, 2329, 1981.

Okuda, H. and M. Ashour-Abdalla, Formation of a conical distribution and intense ion heating in the presence of hydrogen cyclotron waves, *Geophys. Res. Lett.*, **8**, 811, 1981.

Okuda, H. and M. Ashour-Abdalla, Acceleration of hydrogen ions and conic formation along auroral field lines, submitted to *J. Geophys. Res.*, 1982.

Papadopoulos, K., J. D. Gaffey, Jr., and P. J. Palmadesso, Stochastic acceleration of large M/Q ions by hydrogen cyclotron waves in the magnetosphere, *Geophys. Res. Lett.*, **7**, 1014, 1980.

Sharp, R. D., R. G. Johnson, and E. G. Shelley, Observation of an ionospheric acceleration mechanism producing energetic (keV) ions primarily normal to the geomagnetic field direction, *J. Geophys. Res.*, **82**, 3324, 1977.

Shelley, E. G., R. G. Johnson, and R. D. Sharp, Satellite observations of energetic heavy ions during a geomagnetic storm, *J. Geophys. Res.*, **77**, 6104, 1972.

Shelley, E. G., R. D. Sharp, and R. G. Johnson, Satellite observations of an ionospheric acceleration mechanism, *Geophys. Res. Lett.*, **3**, 654, 1976.

Singh, N., R. W. Schunk, and J. J. Sojka, Energization of ionospheric ions by electrostatic hydrogen cyclotron waves, *Geophys. Res. Lett.*, **8**, 1249, 1981.

Taylor, H. A., H. C. Brinton, and C. R. Smith, Positive ion composition in the magnetosphere obtained from the OGO-A satellite, *J. Geophys. Res.*, **70**, 5769, 1965.

Ungstrup, E., D. M. Klumpar, and W. J. Heikkila, Heating of ions to superthermal energies in the topside ionosphere by electrostatic ion cyclotron waves, *J. Geophys. Res.*, **84**, 4289, 1979.

Young, D. T., Ion composition measurements in magnetospheric modeling, AGU Geophysical Monograph Series, **21**, 340, 1979.

Young, D. T., J. Geiss, H. Balsinger, P. Eberhardt, A. G. Ghielmetti, and H. Rosenbauer, Discovery of He^{2+} and O^{2+} ions of terrestrial origin in the outer magnetosphere, *Geophys. Res. Lett.*, **4**, 561, 1977.

Energetic Ion Composition in the Earth's Magnetosphere, edited by R. G. Johnson, 73–98.

What Magnetospheric Workers Should Know about Solar Wind Composition

S. J. Bame, W. C. Feldman, J. T. Gosling,
D. T. Young, and R. D. Zwickl

University of California,
Los Alamos National Laboratory,
Los Alamos, New Mexico, U.S.A.

(Recived March 1, 1982)

The solar wind is a dynamic plasma flow which interacts with the magnetosphere through a variety of processes. Ions of the more common solar elements and isotopes are normally in the flow in various charge states, e.g. $^1H^+$, $^4He^{++}$, O^{7+} and O^{6+}, $Si^{9+}-Si^{7+}$, and $Fe^{13+}-Fe^{7+}$. These ions have abundances high enough that, in principle, they can be used for determining how the particles gain access to the magnetosphere, what fractions of the various ions enter, how they are accelerated and transported within the magnetosphere, and how they are ultimately lost. The present state of knowledge of the solar wind composition is reviewed with emphasis on those features which might contribute to magnetospheric studies. The average elemental and isotopic abundances are given along with a discussion of how they vary with time and with solar wind conditions. Two types of short-lived solar wind impulsive events which contain unusual ions are described. In one type, ions such as Fe^{16+} and Si^{12+} are found, and in the other, ions of $^4He^+$ and somtimes O^{2+} and O^{3+} are detected. The potential, as well as the limitations of using solar wind composition as a test probe for magnetospheric studies, are briefly considered.

1. Introduction

Historically the solar wind has been looked upon as the principal source of magnetospheric plasma populations above a few eV. This view has changed in recent years with the discovery that the ionosphere is a prolific source of kilovolt particles, primarily O^+, but also H^+ and He^+ (Shelley *et al.*, 1972, 1976; Johnson *et al.*, 1974). Roughly speaking, it is now thought that on average the solar wind and terrestrial ionosphere each contribute about 50% to magnetospheric plasmas below ~ 20 keV (Balsiger *et al.*, 1980), which is the current limit of mass spectrometer measurements. At radiation belt energies ($\gtrsim 0.3$ MeV/nucleon), on the other hand, ion distributions

By acceptance of this article, the publisher recognizes that the U.S. Government retains a nonexclusive, royalty-free license to publish or reproduce the published form of this contribution, or to allow others to do so, for U.S. Government purposes.

The Los Alamos National Laboratory requests that the publisher identify this article as work performed under the auspices of the U.S. Department of Energy.

near the equatorial plane exhibit elemental abundances clearly indicative of a solar origin (WILLIAMS, 1981). In the energy interval 20 keV to 0.3 MeV/nucleon, where the bulk of the energy in the terrestrial ring current lies, there are at present no composition measurements (cf. WILLIAMS, 1981). However it is reasonable to assume that when such measurements are made, the solar wind will be found to be a major contributor to this population as well.

Since the solar wind contains ions covering a broad range of ion mass, charge and energy, it provides a wide choice of test particles for probing processes related to particle access and entry, acceleration, transport and loss within the magnetosphere. Several tests of these processes have been suggested (AXFORD, 1969; KRIMIGIS, 1972; BLAKE, 1973). Most fundamental, of course, is a determination of the mechanisms and subsequent rates at which solar wind particles enter the magnetosphere. Numerous attempts have been made to establish a solar wind origin for magnetospheric ions to answer this question. Chief and technically easiest among magnetospheric measurements is the $^4He^{++}/H^+$ ratio, first observed in precipitating auroral fluxes (REASONER, 1973; SHARP et al., 1974; LYNCH et al., 1976). Prior to the discovery of kilovolt H^+ of terrestrial origin (in association with O^+ in upward flowing ion beams, (SHELLEY et al., 1976) the detection of He^{++}/H^+ ratios of one to several percent was taken as indicating a solar wind origin for all precipitating ions. However, the presence of a large and variable fraction of terrestrial H^+ at kilovolt energies vitiates the use of He^{++}/H^+ alone in identifying the solar wind contribution to magnetospheric plasmas.

A second test, proposed by AXFORD (1969), involves measuring the $^4He/^3He$ ratio. Because ^3He fluxes are expected to be low, this measurement has thus far been carried out only with the foil trapping technique. BÜHLER et al. (1976) found a ratio of 2950 ± 250, close to the average for the solar wind (Table 1), in two fairly intense post-midnight auroras. LIND et al. (1979) used a foil exposure technique on Skylab with

Table 1. Average elemental and isotopic abundances in various solar system reservoirs.

Z	Element	A	Solar system (a)	Photosphere (b)	Corona (b)	Solar wind	Terrestrial atomosphere (d)
1	H	1	1×10^{12}	1×10^{12}	1×10^{12}	1×10^{12}	—
2	He	3	1.2×10^7	—	—	2.2×10^7	5.8×10^4
2	He	4	6.9×10^{10}	—	—	4.7×10^{10}	4.7×10^{10}
8	O	16	6.7×10^8	6.8×10^8	5.6×10^8	5.2×10^8	1.9×10^{15}
10	Ne	20	9.6×10^7	—	—	8.3×10^7	1.5×10^{11}
10	Ne	21	2.9×10^5	—	—	2.0×10^5	4.1×10^8
10	Ne	22	1.2×10^7	—	—	6.0×10^6	1.5×10^{10}
10	Ne		1.1×10^7	2.8×10^7	3.5×10^7	9.0×10^7	1.6×10^{11}
14	Si	28	2.9×10^7	3.6×10^7	4.5×10^7	7.5×10^7	—
18	Ar	36	3.1×10^6	$3.8 \times 10^{6(c)}$	$4.7 \times 10^{6(c)}$	2.9×10^6	2.8×10^{12}
26	Fe	56	2.4×10^7	2.5×10^7	4.7×10^7	5.3×10^7	—

$^{a)}$ CAMERON (1973). $^{b)}$ WITHBROE (1971). $^{c)}$ Reduced from value tabulated in WITHBROE (1971) according to the relative abundance of Ar^{36} (.842) tabulated in CAMERON (1973). $^{d)}$ BANKS and KOCKARTS (1973), scale is set such that ^4He is 4.2×10^{10}.

exposure times of 47 to 181 days. They found a weighted average $^4He/^3He$ ratio of 3,100 ± 200. The particles measured by Lind et al. were trapped at low L values (typically $L \lesssim 4$) with characteristic energies of ≈ 30 keV. Thus the helium isotope data strongly suggest a solar wind source for helium in the ~ 30 keV range, with perhaps, a small admixture of terrestrial ^4He.

A similar conclusion can be drawn from very high energy measurements at 1–4 MeV which show C/O flux ratios of ~ 0.5 (FRITZ and WILKEN, 1976; HOVESTADT et al., 1978; KONRADI et al., 1980). Since these represent the extreme high energy tail of the magnetospheric particle distribution it is virtually impossible to infer from these measurements the origins of the bulk of ring current and radiation belt particles. Both the $^4He/^3He$ and C/O data would seem to indicate, however, that the solar wind source becomes increasingly important with increasing ion energy. This is consistent with the fact that solar wind ions enter the magnetosphere at relatively high magnetic moment per nucleon in comparison to terrestrial ion beams which are injected with kilovolt energies rather deep within the magnetosphere.

High charge state ions (e.g. O^{6+}) would at first seem to be an unambiguous indicator of the presence of solar wind ions. SPJELDVIK and FRITZ (1978) have shown, however, that the charge state of energetic trapped oxygen will be altered through charge changing collisions with neutral hydrogen in the earth's geocorona. When taken in combination with radial diffusion, Spjeldvik and Fritz have argued that the presence of high charge state oxygen in the trapped MeV radiation is no guarantee of solar wind origin. In fact the presence of high charge state oxygen from a strong terrestrial source may dominate any solar wind source, at least for the more energetic inner zone ions. At lower energies (e.g. tens of keV at geostationary orbit) high charge state oxygen might provide a reasonable tracer since oxygen at these energies is relatively fresh and therefore unaltered by charge transfer and diffusion processes. At the present time, however, there are no instruments capable of detecting O^{6+} at the expected flux levels although this may change with the coming of NASA's OPEN (Origin of Plasmas in the Earth's Neighborhood) program. Other candidate ions which could serve as solar wind tracers (e.g. Si, S, Fe) are much more abundant in the wind than their terrestrial counterparts (Table 1), but unfortunately their abundances are low enough that they cannot be detected with existing magnetospheric plasma instruments which are restricted by rather high background and limited aperture.

At higher energies (0.5 MeV/nucleon) HOVESTADT et al. (1978) have reported an Fe/O ratio of 10^{-2} in the trapped radiation. They find in addition some evidence to suggest that violation of the first adiabatic invariant may limit the stable trapping of heavier solar wind ions, which have large gyroradii at these energies.

A somewhat different approach to investigating the contribution of solar wind ions to magnetospheric populations is to combine composition with solar wind dynamics, for example the use of variations in solar wind composition. A second approach is to employ longer term statistical studies in order to reduce problems of single point measurements made within the admittedly inhomogeneous magnetosphere. Preliminary studies by YOUNG et al. (1981) covering $2\frac{1}{2}$ years of data obtained by GEOS-1 and 2 near geostationary orbit show that long term (1 month) averages of the He^{++}/H^{+} ratio tend to remain constant with an average value of (7.7 ±

3.4) $\times 10^{-3}$. Although energy per charge coverage of the GEOS data is restricted to < 16 keV/Q, there is a clear indication that He^{++} is deficient at energies of a few tens of keV in the region of geostationary orbit.

From this brief introductory review it can be seen that use of solar wind composition signatures as tracers within the magnetosphere is an active field and will become even more so as larger numbers of improved instruments are flown on future missions. The next step in these studies will be to investigate not only "how much" solar wind is present in the magnetosphere, but also to discover the processes mediating its entry through the semi-permeable membrane of the magnetopause. In order to do this it is necessary to understand not only the average composition of the solar wind, but its variation with other solar wind parameters which may affect particle entry and transport. The intent of this paper is therefore to give the reader a detailed overview of the composition and composition-related dynamics of the solar wind. We begin with a consideration of what ions are expected and measurement techniques in Section 2 and then discuss in Section 3 the average elemental and isotopic abundances. How these abundances vary is the topic of Section 4. Ionization states of solar wind heavy ions are taken up in Section 5.

2.   Heavy Ions Expected in the Solar Wind and Techniques for Measuring Them

A number of techniques have been used to measure the composition of the solar wind. In this section these will be briefly considered along with a discusion of what ion species are expected in the wind and what is required to improve our ability to resolve them. Most of the measurements have been made with electrostatic analyzers, ESAs, which measure energy per unit charge, E/Q, spectra of the ions, but have no intrinsic ability to discriminate between ions. As will be seen, under favorable conditions it is possible to assign mass per unit charge, M/Q, values to resolved E/Q peaks and to identify self-consistently the more abundant ion constituents forming the peaks. Various kinds of ESAs have been flown: cylindrical curved plate analyzers, spherical section curved plate analyzers, ion traps, and gridded plasma or Faraday cups. The ions transmitted through ESAs are detected in two ways: the current of analyzed ions is measured with electrometers or the ions are counted individually with open electron multipliers. In past experiments the ESA capability has been augmented in two ways. First, a coarse pulse height analysis of ions detected with electron multipliers has been used to discriminate between H$^+$, He^{++}, and ions heavier than He^{++} (BAME et al., 1968a). Second, because ion current depends on the ion charge level, Q, it has been possible to identify specific peaks in E/Q spectra by comparing the ion current measured with an electrometer ESA with simultaneously measured ion counting rates determined with another ESA (SCHWENN et al., 1980). More detailed descriptions of ESA techniques as well as some of the other measurement techniques mentioned in this section can be found in several reviews (e.g. HUNDHAUSEN, 1968; VASYLIUNAS, 1971; NEUGEBAUER, 1981).

The widths of individual spectral peaks in solar wind E/Q spectra are proportional to the local kinetic or thermal temperatures of the ions comprising the peaks. At times these temperatures are so high that individual ion peaks cannot be resolved. To overcome this problem, E/Q analysis can be combined with crossed field velocity selection (Wien filter) so that M/Q is determined. Some recent experiments using this technique are described in BALSIGER *et al.* (1976), SHELLEY *et al.* (1978), and COPLAN *et al.* (1978).

Even with M/Q analysis some ambiguities remain, as will be discussed later in this section. A new generation of instruments is required to eliminate these ambiguities. One instrument, in preparation for flight on the International Solar Polar Mission spacecraft, uses a technique in which ions, after electrostatic analysis, are post accelerated by about 30 kV into and through a thin foil which forms the entrance to a drift space (GLOECKLER and HSIEH, 1979). Secondary electrons ejected by the ion's passage through the foil are detected, generating a "start" pulse. After traversing the drift space the ion is stopped in a solid state counter which measures its total energy, E_T, i.e., roughly its original energy plus the charge dependent energy gained by post acceleration. Detection of the secondary electrons ejected by the ion at the solid state counter surface produces a "stop" pulse, which combined with the "start" pulse gives the ion time-of-flight and hence its velocity, v. In principle, the combination of the ion E/Q with the post acceleration values of v and E_T uniquely identify the ion. Other advanced instruments with similar features in part, but significant differences in detail, have been under development.

Other methods more limited in scope have been used to measure solar wind abundance ratios. Aluminum foils were exposed on the lunar surface during some of the Apollo missions and then returned. Solar wind gases trapped in the foils were released and analyzed with a mass spectrometer revealing abundance ratios of helium relative to other noble gases, neon isotope ratios, and the ^{4}He/^{3}He ratio (e.g. GEISS *et al.*, 1970b). Other isotopic ratio determinations have been made by analyzing gases from lunar materials returned in the Apollo program (e.g. EBERHARDT, 1974; BOGARD *et al.*, 1974). Because of the poorly known exposure and turnover times, diffusion, and other effects, solar wind abundance results derived from the analysis of lunar materials should be used with caution. A determination of the solar wind iron abundance for a limited exposure time has been made by analysis of pits etched in mica which was exposed on the lunar surface (ZINNER *et al.* 1974). Iron ions have also been detected in high speed solar wind streams with solid state detectors (MITCHELL and ROELOF, 1980; MITCHELL *et al.*, 1981).

Until recently most measurements of the solar wind have been made with ESAs, which are well suited for measuring the dynamic properties of the wind with good energy and time resolution. The early observations established that a solar wind ion E/Q spectrum generally displays a prominent peak followed by a lower intensity peak at about twice the E/Q position of the first (e.g. NEUGEBAUER and SNYDER, 1966). An example of such a spectrum is shown in Fig. 1(a). On the basis of self-consistent arguments, the prominent peak was identified as hydrogen ions, H^+, and the lower intensity peak as helium ions, He^{++}, because hydrogen and helium are by far the most abundant solar elements. The solar wind emanates from the solar corona, which ordinarily has a temperature of $1-2 \times 10^6$K. At these temperatures virtually all of the hydrogen is ionized and the helium is doubly ionized except for a miniscule fraction of He^+. Furthermore, it seemed reasonable to suppose that the H^+ and He^{++} ions

S. J. BAME *et al.*

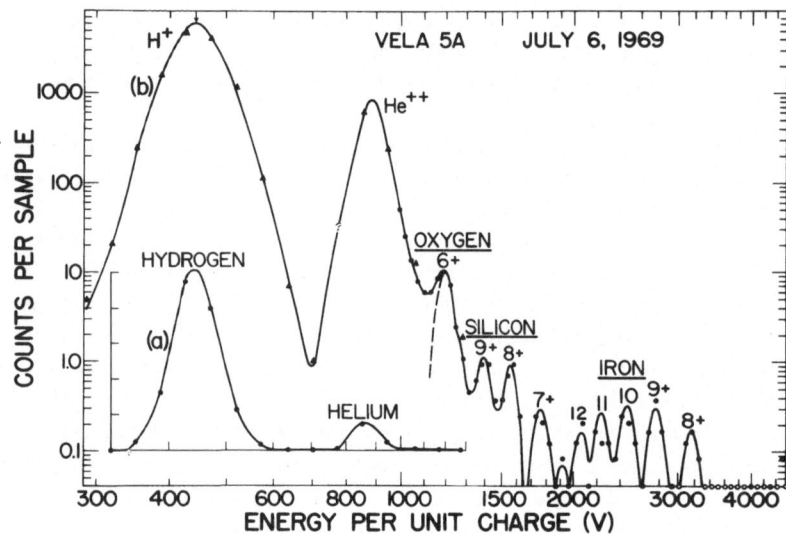

Fig. 1. (a) A Vela 5A solar wind spectrum shown with relative intensity plotted on a linear scale. Such spectra, which were commonly obtained with electostatic analyzers during 1962–1965, were interpreted as showing that the solar wind is predominantly composed of H^+ and He^{++} ions. (b) The same spectrum with intensity plotted with triangles on a logarithmic scale. Joined to it is a higher resolution heavy ion spectrum plotted with circular dots. Self-consistent identifications of the multiple peaks are shown above the individual peaks.

should be locked together in the flow with a nearly common bulk velocity. Thus, He^{++} ions with four times the energy and twice the charge would be expected to appear in a spectrum at twice the E/Q value of H^+ ions. This widely accepted interpretation of the early results was consistently borne out by the observations and was later confirmed with measurements from an electrostatic analyzer combined with a Wien filter velocity selector (OGILVIE *et al.*, 1968a, b).

Although the ability of an ESA to separate He^{++} from H^+ is limited when the flow speed is low and the local kinetic temperature high, usually the distributions are separable. Most of the present knowledge of solar wind He^{++}, discussed later in this paper, has been developed using data from ESAs.

The earliest observations of minor ions other than He^{++} were made in solar wind plasma with low kinetic temperature using high resolution, hemispherical plate electrostatic analyzers on Vela 3A and 3B. Small ion peaks at other E/Q positions in the Vela spectra were self-consistently identified as O^{6+}, O^{7+}, $^4He^+$, and $^3He^{++}$ ions travelling at the same speed as the protons (BAME *et al.*, 1968a). The experimental determination of the near commonality of all ion bulk velocities furnished an indispensable tool for subsequent identifications of other heavy ions, such as those of Si and Fe (BAME *et al.*, 1970; BAME, 1972).

To illustrate the complexity associated with measuring the solar wind composition, a composite spectrum from Vela 5A is presented in Fig. 1(b). This spectrum displays the usual prominent peaks, H^+ and He^{++}, measured with a standard solar wind analyzer. Joined to the He^{++} peak is a simultaneously measured spectrum from a second analyzer, optimized for heavy ions, showing a number of resolved peaks beyond He^{++}. Peaks beyond He^{++} are principally due to O^{6+}, Si^{9+} to Si^{7+} (with some admixture of sulfur ions), and Fe^{13+} to Fe^{8+}. In this particular case the thermal width of the He^{++} peak is too great for O^{7+} ions to be resolved.

Figure 2 has been prepared to show what a 'typical' low speed, low temperature solar wind E/Q spectrum should look like, so as to illustrate the difficulties associated with resolving all of the ions expected to be in the solar wind. The continuous, heavy line spectrum was synthesized by assuming representative amounts of H^+ and He^{++} and by adding together the contributions of all ions with abundances above a selected threshold level. Ion abundances were based on estimates of abundances of the elements in the solar corona (WITHBROE, 1971). The distributions of ion charge levels, or ionization states of the various elements were calculated using the density independent ionization equilibria of JORDAN (1969) supplemented with the ionization equilibria of MEWE (1972). For the illustrative purposes of this 'typical' spectrum, an isothermal coronal freezing-in temperature of $2.0 \times 10^6 K$ was chosen for ions from carbon through neon and $1.58 \times 10^6 K$ was used for the heavier ions. These temperatures are based on a study of the temperature and temperature gradient of the corona (BAME et al., 1974) and reflect the fact that ionization states for the heavier elements 'freeze' in at greater heights in the corona. It was further assumed that every ion species beyond He^{++} produces the same peak width and shape (near instrumental) and that the peaks

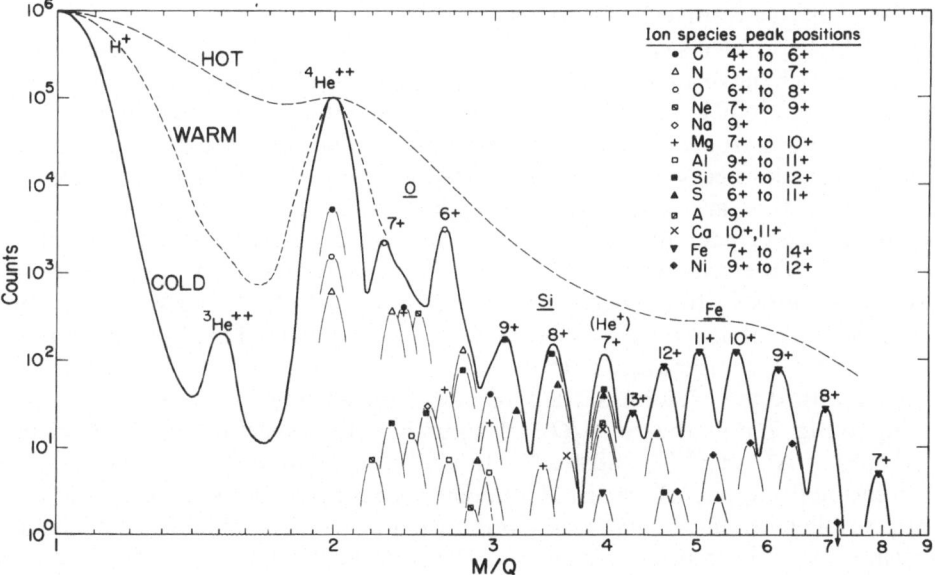

Fig. 2. A solar wind ion spectrum synthesized using assumptions described in the text (solid, heavy curve). Beyond the He^{++} peak, the spectrum was generated by summing the contributions of all ions. Peak positions of all of the various ions having intensities above a selected threshold level are shown with the symbols given in the table. The dashed curves show representative spectra for 'warm' and 'hot' solar wind kinetic, or thermal temperatures.

appear at E/Q positions appropriate for a solar wind in which all ion species have a common bulk velocity. This latter assumption makes it possible to replace the E/Q scale with an M/Q scale. This synthesized spectrum bears a remarkable resemblance to heavy ion spectra from Vela such as that shown in Fig. 1(b), contributing to the confidence that can be placed in the identification of the various ion species. Similar E/Q spectra have been reported by other experimenters (GRÜNWALDT, 1976; ZASTENKER and YERMOLAEV, 1981).

A number of problems associated with composition determinations become immediately apparent on inspection of Fig. 2. Note that in some ranges, e.g. between O^{7+} and O^{6+}, a number of ion species are so closely spaced that they cannot be resolved, even at low solar wind thermal temperatures. In other cases such as for the O^{6+} and Fe ions, the abundances of the primary ions are so much higher than those of the competing ions that their resolution is assured as long as the thermal or kinetic temperature of the ions is low and the E/Q instrumental resolution and sensitivity are adequate. (Of course, this same feature insures that the lesser species cannot be identified with an E/Q analysis alone.) However, noting the dashed spectra labelled 'warm' and 'hot' in the figure, the importance of the local ion thermal temperatures is immediately evident. Successful E/Q abundance determinations require relatively low temperatures, which are ordinarily found only in low speed solar wind. At intermediate temperatures $^3He^{++}$ is submerged in the tail of the H^+ peak and O^{7+} falls in the tail of the He^{++} peak. As the other minor ion peaks broaden, the ability to resolve them is lost, even when it is still possible to resolve He^{++} from H^+. At the highest temperatures even the ability to resolve He^{++} is lost.

The above problems can be alleviated substantially by using instruments referred to earlier, which have the capability to determine M/Q directly. Such an analysis removes the requirement for low ion temperatures, but there are some serious problems associated with instruments which have been used so far. Because of the complicated and lengthy cycles necessary for their operation, the instruments have limited time resolution and thus are unsuited for measuring the dynamic properties of the solar wind. Also, due principally to package size and weight restrictions, it has not been possible to push the instrumental M/Q capabilities into the interesting range occupied by Si and Fe ions in high speed solar wind streams. Indeed, the velocity filtering provided by current instruments is not as good as that provided by the solar wind itself when the thermal temperature is low.

Even with direct determinations of M/Q, not all ions can be resolved. Again referring to Fig. 2, note that at $M/Q = 4$, a number of ion species lie in the same peak, i.e. Si^{7+}, S^{8+}, A^{9+}, Ca^{10+}, and Fe^{14+}. In addition, as discussed later in this paper, He^+ occasionally appears here also. Untangling the contributions of these various ions is not possible without an advanced composition instrument capable of independent determinations of both mass and charge. The same problem holds for the interesting case of the fully ionized ions of C, N, and O. As shown in the figure, those ions all fall at the same M/Q position occupied by the much more abundant He^{++}. Again, only an advanced instrument will be capable of separating those fully ionized species. It is to be hoped that successful measurements will be made with such instruments on future missions.

3.   Average Abundance

The solar wind consists primarily of protons and electrons with a small admixture of heavier "minor" ions. Nearly all parameters characterizing the solar wind near 1 AU vary over a large range. One of the most notable is the number density of protons and electrons, which has been observed at 1 AU to vary over a range spanning nearly four decades—from a low of $\cong 0.02$ cm^{-3} to a high of $\cong 150$ cm^{-3}. Similarly, the abundances of heavy ions in the solar wind also vary over a large range (see, for example, Fig. 3). Nevertheless, long-term average abundances can be derived from existing measurements and provide a useful measure of the abundances of the minor ions which may enter the earth's magnetosphere from the solar wind.

The average abundances of solar wind heavy ions measured using several different experimental techniques are collected in Table 1. Also shown are the abundance values for the same elements in the solar system, photosphere, corona, and terrestrial atmosphere. Where possible, the separate measured elemental and isotopic abun-

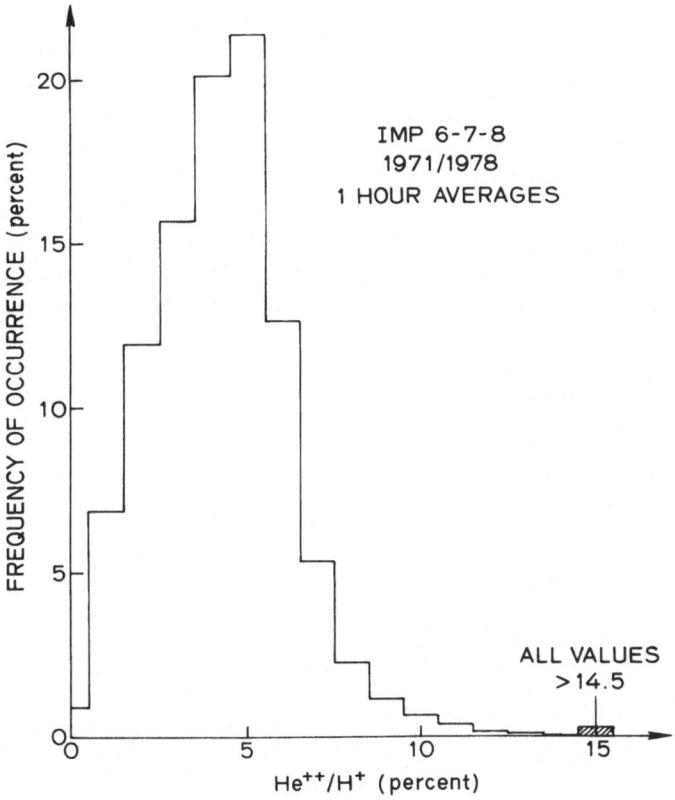

Fig. 3.   Histogram of hourly averages of helium abundance relative to hydrogen using Los Alamos IMP data obtained between March 1971 and December 1978. A total of 39,441 hourly averages are included in the analysis. From BORRINI *et al.* (1981a).

dances have been normalized to hydrogen whose value is set to 10^{12} for ease of comparison with other abundance compilations. Factors affecting the use of Table 1 for magnetospheric and perhaps other studies are discussed below for various groups of elements. Organization into groups is made according to experimental technique because various uncontrolled factors of the measurements lead to uncertainties which are peculiar to each technique.

3.1 ^1H, ^4He, and ^3He

As discussed in the last section, measurements of solar wind ^{1}H and ^{4}He have been made routinely from a variety of space platforms for nearly 20 years using electrostatic analyzers. The accuracy of this technique is about 10% for individual samples. However, as will be discussed in Section 4, the ^{4}He abundance derived from individual samples varies over a range exceeding a factor of 100. The ^{4}He abundance listed in Table 1 is the average of all measurements made using the Los Alamos IMP plasma analyzers between 1971 and 1976 (FELDMAN *et al.*, 1977). This value was chosen because the IMP data represent the largest homogeneous set for which the abundance has been routinely estimated. It is used as the connecting link between measurements of the abundances of the heavy noble gases which are made relative to ^{4}He (GEISS *et al.*, 1970b, 1972; GEISS, 1973a, b) and the abundance scale relative to ^{1}H in Table 1.

Although ^{3}He is occasionally resolved in E/Q spectra from electrostatic analyzers (BAME *et al.*, 1968a, 1979; GRÜNWALDT, 1976), its abundance has been measured most completely using three other experimental techniques. The most-extensive set of measurements has been made using a combined electrostatic analyzer and magnetic velocity filter (COPLAN *et al.*, 1978); the results of this investigation (OGILVIE *et al.*, 1980) were used in Table 1. Although the individual measurements of ^{4}He/^{3}He varied over a range spanning two orders of magnitude (see Fig. 4), it is useful to note that the average ratio agrees very well with that obtained using the Al foil collection techniques

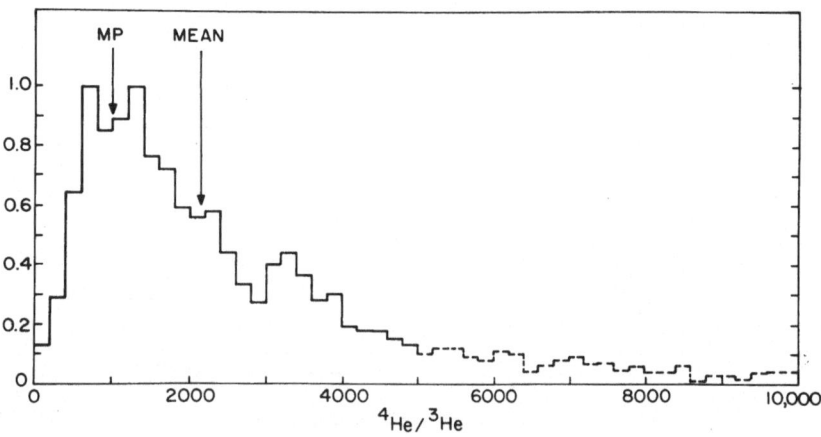

Fig. 4.   Histogram of ^{4}He^{++}/^{3}He^{++} density ratios obtained by the ion composition experiment on ISEE 3 between August and November 1978 and between March and August 1979. A total of 4334 individual spectra are included in the histogram. From OGILVIE *et al.* (1980).

(GEISS et al., 1972; GEISS, 1973b) and from analyses of surface implanted ions in Surveyor material and lunar fines (GEISS, 1973a, b). This fact lends confidence to attempts to relate the abundances of the heavier noble gases to hydrogen through the ^{4}He abundance adopted in Table 1.

3.2 O, Si, and Fe

Measurements of the abundances of O, Si, and Fe in the solar wind having the least experimental uncertainties have been made using specially designed electrostatic analyzers (BAME et al., 1970). However, because relatively few heavy ion spectra were available in the data sample (BAME et al., 1975), the statistical accuracy is limited. Because the spectra analyzed were sampled during low speed, low temperature solar wind flows, the averages given here refer to those flows only. Whether or not the abundances in high speed flows are different is not known with certainty, but there is some evidence that the average abundances are similar in high and low speed flows, as mentioned below.

A special note concerning the use of the Si abundance listed in Table 1 is in order. Because the ions of S and Si cannot be resolved from each other in heavy ion spectra measured using an electrostatic analyzer, the abundance of S (the less abundant component) relative to Si needs to be assumed before the Si abundance can be determined. The solar system abundances tabulated in CAMERON (1973) were used for this purpose (BAME et al., 1974). Any deviation from these values in the solar wind will affect directly the Si abundance in Table 1 (see e.g., Table 1 of BAME et al., 1974).

As mentioned earlier, estimates of the solar wind Fe abundance have also been made using two other techniques (ZINNER et al., 1974; MITCHELL and ROELOF, 1980). Although both these measurements suffer from possibly large systematic uncertainties related both to experimental technique and limitations in the sampling, it is worth noting that their values agree with the average obtained using electrostatic analysis alone which is listed in Table 1. This point is especially significant since, as noted above, the electrostatic technique determines only the abundances in low speed solar wind, whereas the mica nuclear track detectors deployed during the Apollo 17 mission were exposed during passage of a stream-stream interaction driven by a modest high speed stream (ZINNER et al., 1974), and the IMP 7/8 low energy solid state detectors are only sensitive when sampling the very highest speed solar wind flows (MITCHELL and ROELOF, 1980).

3.3 Ne and Ar

Measurements of the solar wind abundances of the Ne and Ar isotopes having the least uncertainties have been made using the foils exposed on the moon during various Apollo missions. Because of limited exposure time, the resultant abundances should be used with caution when taken as representative of the average solar wind. However, the fact that the average ^{4}He/^{3}He abundance ratio from the Apollo measurements (GEISS et al., 1970b, 1972; GEISS, 1973b) agrees well with the average from a much more complete set of measurements (OGILVIE et al., 1980) lends confidence that the Ne and Ar isotopic and elemental abundances are also representative of average conditions. The absolute scale of these abundances in Table 1 is based on the ^{4}He abundance as determined with the Los Alamos IMP 6, 7, and 8 plasma analyzers between 1971 and 1976.

3.4   C, N, Mg, Kr, and Xe

Estimates of the solar wind elemental and isotopic composition of C, N, Mg, Kr, and Xe have been possible only by measuring the amounts of those particles trapped in the surface layers of lunar dust and rocks. This technique is potentially very important because it can provide information, especially about the history of the sun's outer atmosphere, which is unavailable using any other technique. However, becuse of the rather varied and largely unknown environment experienced by lunar surface material over the past history of the moon, abundance determinations based on amounts of surface correlated material are subject to very large uncertainties (see e.g. EBERHARDT, 1974). The magnitude and effect of these uncertainties are not generally agreed upon by researchers in the field (see e.g. EPSTEIN and TAYLOR, 1972; GEISS, 1973a; GOEL and KOTHARI, 1972; BOGARD et al., 1974). It is the opinion of the present authors that although the present experimental difficulties may be resolved in the future, the results obtained to date are not sufficiently certain to be useful by themseles as a guide for magnetosphere tracer experiments. In consequence, no abundances for these elements are included in Table 1. However, it may be of some use to report that all current determinations are consistent with the solar system abundances tabulated in CAMERON (1973) to within about 20%.

4.   Abundance Variations

The solar wind heavy ion abundance values quoted in the previous section are average values of quantities which, in fact, are highly variable. Figure 3, which is a histogram of 1-hour averaged values of the abundance of $^4He^{++}$, obtained over an ~8-year interval, demonstrates this variability for $^4He^{++}$, the most abundant and most easily measured solar wind heavy ion. Because solar wind helium is almost 100% $^4He^{++}$ with few exceptions (mentioned in Section 5), it is customary to refer to the measured $^4He^{++}/^1H^+$ number ratio as the 'helium' abundance. Remarkably, measured values of the helium abundance vary from less than 0.1% to greater than 35%, a factor of 350. This great variability was not expected prior to in situ observations, but has been reported by virtually all solar wind observing groups (e.g. NEUGEBAUER and SNYDER, 1966; OGILVIE and WILKERSON, 1969; ROBBINS et al., 1970; FORMISANO et al., 1970; BAME, 1972; OGILVIE, 1972; BOLLEA et al., 1972; NEUGEBAUER, 1981). Although generally attributed to some fractionation process occurring in the solar atmosphere (e.g. JOKIPII, 1966; NAKADA, 1969), neither the origin of this variability nor its dependence upon abundance values in the lower and intermediate corona are clearly understood (e.g. GEISS et al., 1970a; JOSELYN and HOLZER, 1978; BORRINI and NOCI, 1979). Despite this, a certain ordering of the observed variability has been achieved and will be discussed in later paragraphs.

It is probable that a similar degree of variability exists for the other heavy elements and isotopes present in the solar wind. For example, the histogram of observed $^4He^{++}/^3He^{++}$ density ratios shown in Fig. 4 (OGILVIE et al., 1980) demonstrates that the ^{3}He isotope is present in solar wind spectra at 1 AU in variable amounts relative to ^{4}He (see also BAME et al., 1968a; GRÜNWALDT, 1976). However, because the number of measurements of these other heavy elements and isotopes is small (certainly, less than 100 individual measurements have been reported for all elements other than H and He), our knowledge of this variability is less secure. As mentioned before, measurements of solar wind O, Si, and Fe have for the most part been restricted to intervals when the solar wind flow speed and local thermal temperature are low and the density is high. Thus most of our information of the variability of these species pertains to flows between high speed streams, i.e. the interstream plasma (e.g. BAME et al., 1979), to noncompressive density enhancements (e.g. GOSLING et al., 1977; FENIMORE, 1980), and to plasma driving interplanetary shocks (e.g. BAME et al., 1979). Even with these limitations, present evidence (BAME et al., 1975, 1979; GRÜNWALDT, 1976; ZASTENKER and YERMOLAEV, 1981) indicates that the O/H ratio varies from 1.9×10^{-4} to 1.8×10^{-2}, the Si/H ratio varies from 2.8×10^{-5} to 1.18×10^{-3}, and the Fe/H ratio varies from 1.4×10^{-5} to 1.2×10^{-3}. These correspond to variations over ranges of 95, 42, and 86 respectively.

There is some expectation that the abundances of elements heavier than H should vary in concert with one another. Indeed, a limited statistical study (BAME et al., 1975) provided evidence that Fe and Si abundance variations are often coupled, and specific examples of simultaneous enhancements of He, O, Si, and Fe relative to H have been presented in the literature (ZASTENKER and YERMOLAEV, 1981; ZWICKL et al., 1982). On the other hand, the same study which provided statistical evidence for a coupling of Fe and Si abundance variations (BAME et al., 1975) did not find unambiguous evidence for any other element abundance correlations. In fact, it was shown that the abundance of O, Si, and Fe relative to He could vary by a factor of 10 from event to event. That is, the abundances of heavy elements relative to H need not vary in concert with one another. A better understanding of the degree of coupling of heavy element abundances in the solar wind will not be achieved until larger data bases are obtained. The present, sparse data base for elements heavier than He is too limited to permit definitive statements on this important topic.

No such limitation exists for helium abundance variability; on the contrary, measurements of it are nearly as complete as measurements of H itself. Although the observed variability is often perplexing and difficult to interpret, it is now clear that much of it is closely associated with the large scale structure of the interplanetary plasma (i.e. high speed streams, stream interfaces, sector structure, noncompressive density enhancements, and transient disturbances), and consequently is related to the structure of the corona where the solar wind originates (i.e. coronal holes, coronal streamers, and coronal transients). Within the cores of most quasi-stationary high speed streams the helium abundance is relatively constant with an average value of 0.048 ± 0.005 (BAME et al., 1977). By way of contrast, great variability of the helium abundance is a characteristic feature of low speed flows (BAME et al., 1977; FELDMAN et al., 1977). Periods of very low (say $< \sim 0.02$) abundance occur preferentially at low flow speeds (e.g. HIRSHBERG et al., 1972b); in particular, a minimum often occurs $\sim 1/2$ day prior to the onset of a high speed stream (HIRSHBERG et al., 1974; GOSLING et al., 1978), and an abrupt increase is commonly observed at the interface that separates what was originally dense slow gas from what was originally less dense fast gas at the leading edge of a high speed stream. These features can be seen in the plots of Fig. 5 which show the average helium abundance, proton density, and flow speed patterns observed near 23 discontinuous stream interfaces on two different temporal scales.

The helium abundance minimum which often precedes a high speed stream on the average is centered upon a polarity reversal in the interplanetary magnetic field (BORRINI et al., 1981a; GOSLING et al., 1981). Figure 6, which shows the results of a superposed epoch analysis of solar wind plasma data obtained in the vicinity of 23 magnetic field sector boundaries, demonstrates the coupled variations in helium abundance, proton density flow speed, and proton temperature which are characteristic of interplanetary field polarity reversals. Importantly, such variations are often

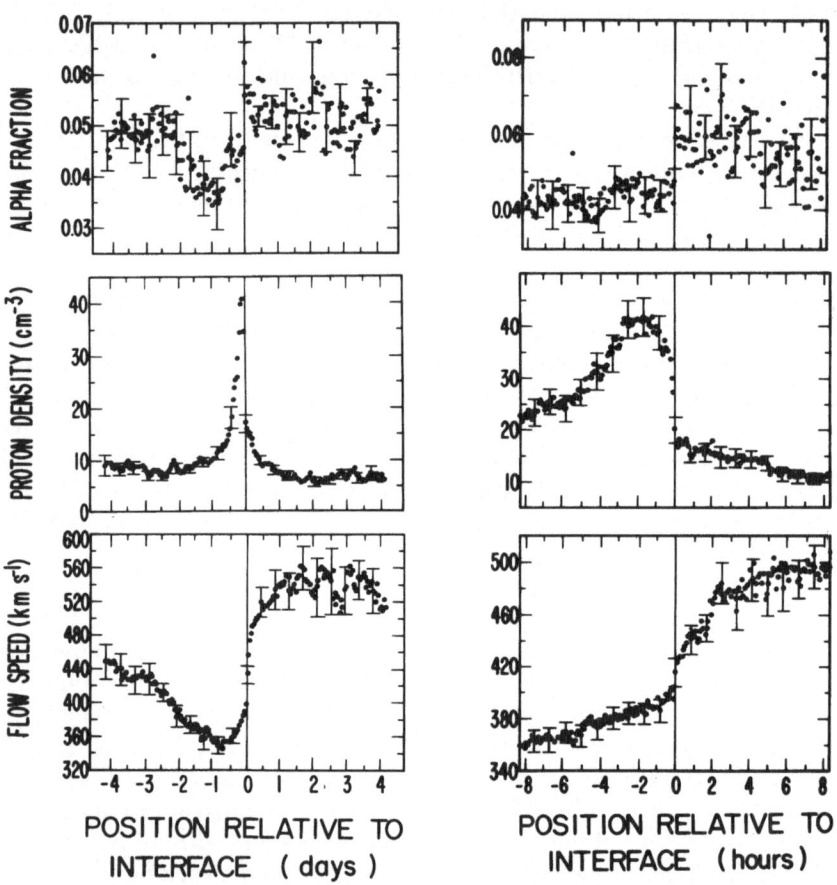

Fig. 5. Superposed epoch plots of the solar wind proton density, flow speed, and helium abundance observed near 23 discontinuous stream interfaces. The data are shown on two different time scales to show the long and short term variations more clearly. Note the minimum in helium abundance ~1/2 day prior to the interface and the discontinuous jump in abundance which occurs at the interface. Adapted from GOSLING et al. (1978).

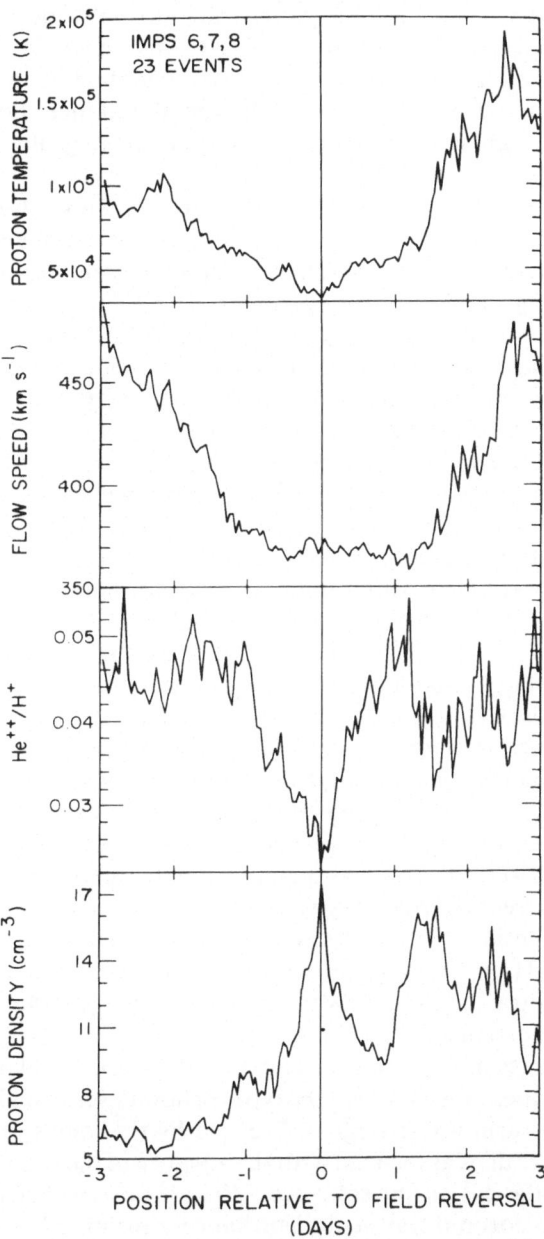

Fig. 6. Superposed epoch plots of the solar wind proton temperature, flow speed, helium abundance, and proton density for 23 well defined sector boundaries. The minimum in A(He) is a characteristic signature of such boundaries. Collectively, the coupled variations illustrated here signal passage of a coronal streamer at 1 AU. From GOSLING *et al.* (1981) as adapted from BORRINI *et al.* (1981a).

observed near polarity reversals that would not normally be identified as sector boundaries. The coupled variations shown in Fig. 6, which include a helium abundance minimum as an important component, have been identified as the 1 AU signal of a coronal streamer. It is not presently clear why the abundance should be low within such structures, although suggestions have been offered (e.g. BORRINI *et al.*, 1981a; HOLLWEG, 1981).

Helium enhancements greater than ~0.10 are relatively infrequent (see Fig. 3) and account for less than ~1% of all helium abundance measurements. It has long been recognized that enhancements of this magnitude or larger are frequently observed 10–20 hours following passage of interplanetary shocks (e.g. GOSLING *et al.*, 1967; BAME *et al.*, 1968b; OGILVIE *et al.*, 1968; LAZARUS and BINSACK, 1969; HIRSHBERG *et al.*, 1970, 1971). Statistical studies of such enhancements reveal that ~50% are preceded by shocks, but that the rest occur independent of shock phenomena (HIRSHBERG *et al.*, 1972a; FENIMORE, 1980; BORRINI *et al.*, 1982a). Further, only about 50% of all shocks are followed by enhancements (BORRINI *et al.*, 1981b). Independent of their association with interplanetary shocks, enhancements are strongly correlated with abnormal local thermal depressions (GOSLING *et al.*, 1973; MONTGOMERY *et al.*, 1974), unusual ionization temperatures (BAME *et al.*, 1979; FENIMORE, 1980; SCHWENN *et al.*, 1980; GOSLING *et al.*, 1980) and enhanced magnetic field strength (BORRINI *et al.*, 1981b, 1982a). Collectively, these associations indicate that a large enhancement signals the arrival at 1 AU of plasma ejected from low in the corona during a disturbance such as a large solar flare or eruptive prominence. When the ejection speed is sufficiently high, a shock forms in front of the ejecta.

Figure 7, which shows average values of proton density, bulk flow speed, proton temperature and magnetic field strength as a function of helium abundance, helps summarize our present understanding of its variability. Low abundance, which is preferentially observed near polarity reversals in the interplanetary field where the proton density is high and the flow speed and proton temperature are low, signals the intersection of the spacecraft trajectory with a coronal streamer. Average abundance (~5%) occurs preferentially within the cores of high speed streams, which originate from coronal holes. High abundance is observed at average proton densities and speeds, and at low proton temperatures and high field strengths, and is associated with transient coronal disturbances.

Long-term averages of the helium abundance in the ecliptic plane at 1 AU (Fig. 8) reflect the relative frequency with which the types of flow described above are observed. These frequencies, in turn, are strongly influenced by the long-term evolution of the corona at low solar latitudes associated with the advance of the solar activity cycle. The highest average abundances during solar cycle 20 were observed near and shortly after solar maximum when coronal transient disturbances presumably were most frequent. Certainly the abrupt rise of the average during 1977 and 1978 was associated with an abrupt rise in the frequency of such transients (BORRINI *et al.*, 1982b). Abundance minimums near solar minimum in 1965 and 1975–1976 can be ascribed to a preponderance of coronal streamers at low solar latitudes during that epoch of the solar cycle (e.g. HUNDHAUSEN *et al.*, 1981). During the years between solar maximum and solar minimum, the flows in the ecliptic at 1 AU originated from a mixture of

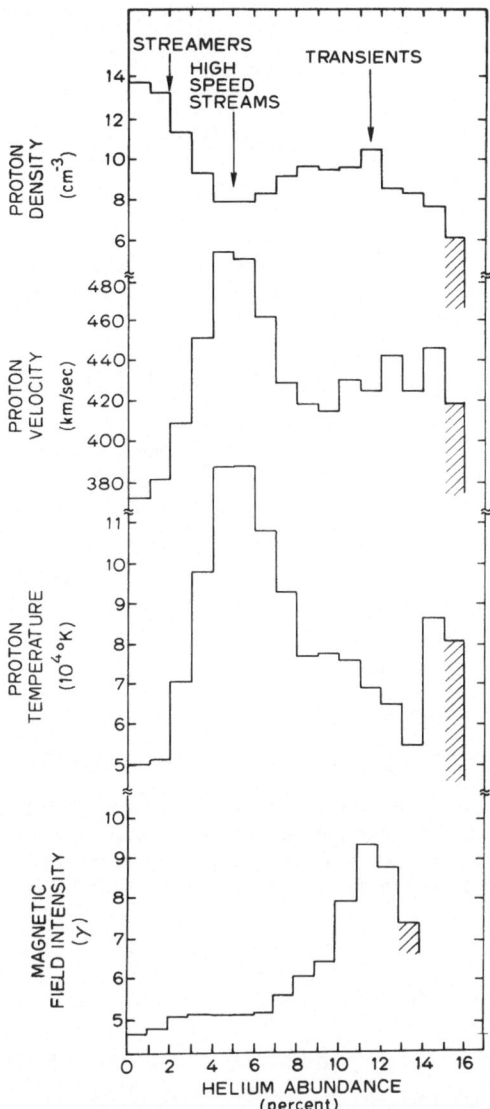

Fig. 7.   Average values of solar wind proton density, flow speed, proton temperature and magnetic field strength as a function of helium abundance. Adapted from BORRINI *et al.* (1982b).

coronal holes, coronal streamers, and transient disturbances, and the intermediate abundance values observed during these epochs reflect this mix.

Altogether then, solar wind helium abundance variability has been extensively studied and proven to be rich in detail on a variety of temporal and spatial scales. When combined with suitable magnetospheric measurements of the abundance variability, these measurements can be exploited in studies investigating the origin of magnetospheric plasma.

S. J. BAME *et al.*

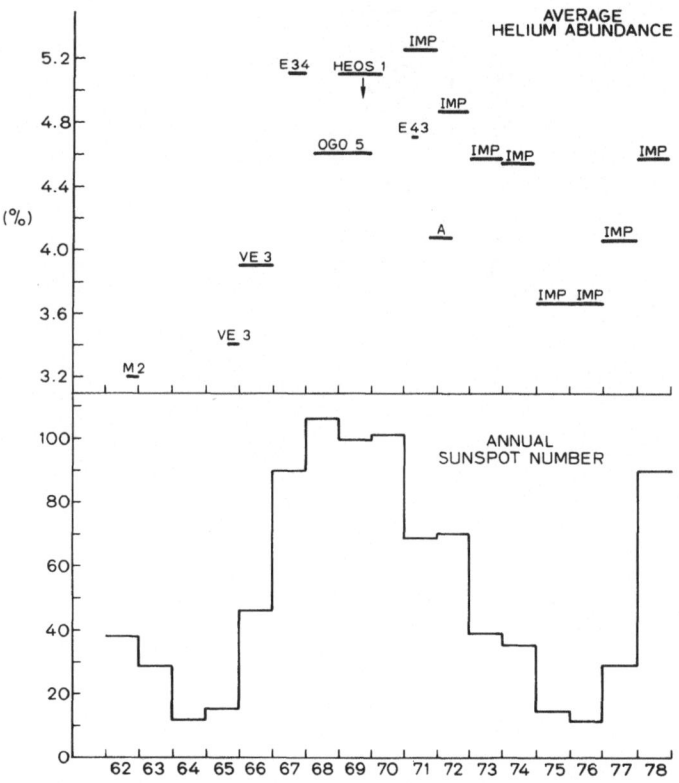

Fig. 8. Long term averages of helium abundance and yearly sunspot numbers as a function of time. The different symbols employed refer to different spacecraft. All of the averages except the IMP data points were originally compiled by OGILVIE and HIRSHBERG (1974). The addition of the IMP points through 1976 was published by FELDMAN *et al.* (1978), and the final two IMP points are included in a study by BORRINI *et al.* (1982b).

5. Ionization States of Solar Wind Ions

The ionization state of the expanding corona is thought to remain constant beyond a heliocentric distance of ∼5 solar radii (BRANDT and HODGE, 1964; HUNDHAUSEN *et al.*, 1968). This 'freezing-in' of the ionization state occurs because the mean free path for collisions producing ionization and recombination changes of the charge levels of ions becomes very long in comparison to the expansion scale height as the plasma expands outward from the sun (HUNDHAUSEN *et al.*, 1968). Hence, this early 'freezing-in' of ionization states allows quantitative determinations of the thermal state of the corona to be made with measurements at 1 AU. In Section 2 we discussed the ionization states that occur at normal coronal temperatures.

Until recently, most studies of ionization states have been based on solar wind heavy ion measurements from simple high resolution electrostatic analyzers (e.g. BAME *et al.*, 1968a, 1970, 1974, 1979; FENIMORE, 1980). However, as mentioned earlier, these studies have been largely limited to periods of low solar wind velocity and low kinetic temperatures. Recently, measurements over a greater range of temperatures have become possible with the introduction of multi-element instruments such as mass spectrometers. For example, the O^{7+}/O^{6+} number ratio of oxygen has now been measured over a large range of solar wind speeds (OGILVIE and VOGT, 1980).

Of particular interest, both for the study of transient coronal phenomena and for potential application in tracer studies of the magnetosphere, are certain short-lived solar wind events in which the ionization state impulsively changes to an altogether new state which persists for times of a few hours to the better part of a day. The new state may undergo evolutionary changes, and usually disappears rather quickly. Two broad categories of these unusual events, which have been related to energetic events in the corona, have been observed. In the first, a normal heavy ion spectrum such as that in Fig. 1 is replaced with a spectrum having a form indicative of an unusually high freezing-in coronal temperature (BAME et al., 1979; FENIMORE, 1980). Ions which are not found in normal spectra appear, such as Fe^{16+} and Si^{12+}. It can be inferred with confidence that at such a time the fully ionized ions C^{6+}, N^{7+}, and O^{8+} have unusually high abundances. They all have the same M/Q as He^{++}, and so are not resolved in E/Q spectra. The brief appearances of these unusual ions opens the possibility for future experiments which can trace the entry of solar wind ions into the magnetosphere. However, as discussed in the introduction and last section of this paper, effective implementation of this technique must await development of new instruments with higher sensitivity and greater resolving power for both solar wind and magnetospheric measurements. Such experiments will require nearly full time monitoring, both in the solar wind and in a multiplicity of spatial regions in the magnetosphere.

In the second category of unusual impulsive events, spectra occasionally appear which show evidence that the ionization state of at least a fraction of the plasma was established at unusually low freezing-in coronal temperatures. Ions of He^+, generally unobservable in normal solar wind (FELDMAN et al., 1974), appear at relatively high intensity levels (SCHWENN et al., 1980; GOSLING et al., 1980; ZWICKL et al., 1982). The potential utility of these He^+ appearances for magnetospheric tracer experiments seems more limited than for the heavier ion impulsive events, since He^+ ions of ionospheric origin are commonly observed in the magnetosphere, whereas ions of Si and Fe are below present limits of detectability.

Considering these impulsive events in greater detail, a spectrum taken during an event indicative of a high temperature coronal origin is shown in Fig. 9, along with a 'normal' spectrum, given for comparative purposes (BAME et al., 1979). The upper normal spectrum was observed in low speed, low temperature 'interstream' solar wind. Assuming a common bulk speed, the vertical lines and arrows indicate the expected positions of various ion species of elements with significant solar abundances. As shown earlier, it is usually possible to identify a group of peaks as due to a particular element, and thus determine its ionization state. Comparing the normal spectrum of Fig. 9 to the spectra shown in Figs. 1 and 2, the groups of peaks due to O, Si, and Fe can be identified easily. Ionization state temperatures for the Fe and O ions, estimated with isothermal fits to the peaks, are given in Fig. 9.

The lower spectrum in Fig. 9 was obtained from plasma in a post-shock,

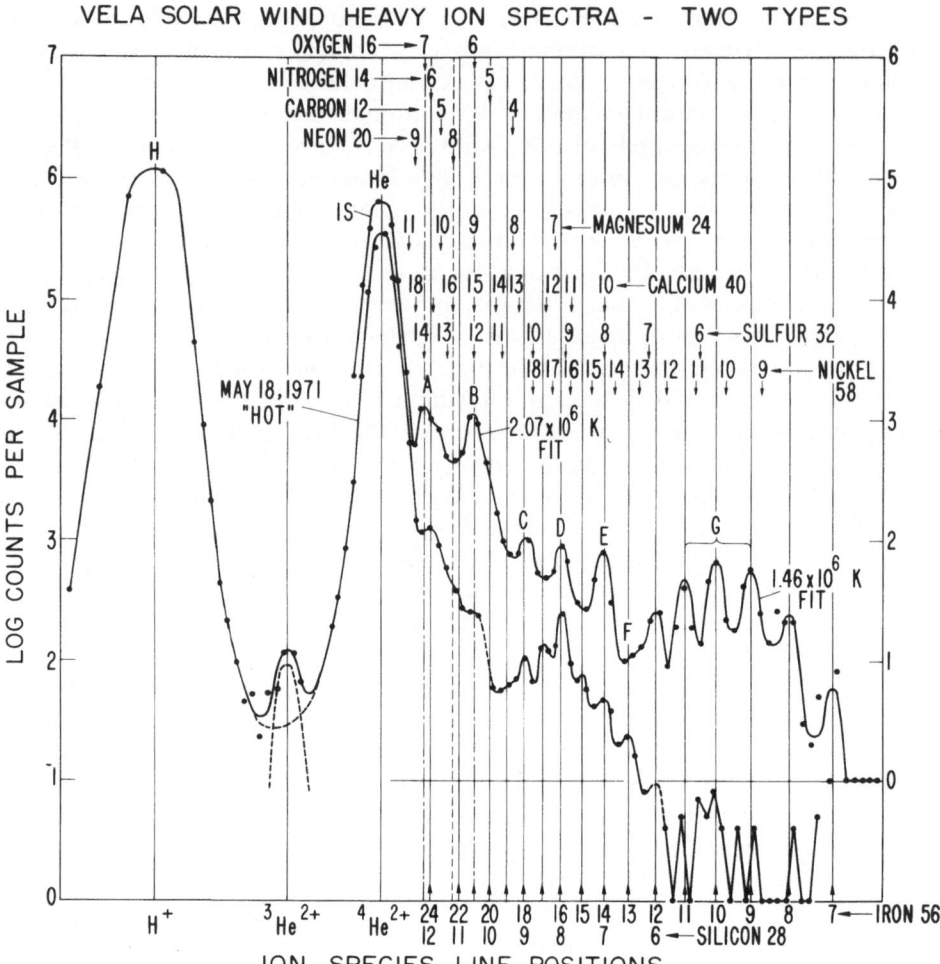

Fig. 9. A heavy ion spectrum from low speed, low temperature solar wind measured with Vela 5A from 0326 to 0422 UT on June 23, 1969 (upper spectrum) and a "hot" spectrum measured with Vela 6B from 0238 to 0429 UT on May 18, 1971 (lower spectrum). The upper spectrum is representative of 'normal' spectra which usually contain well defined O, Si, and Fe peaks. The Fe peaks in the hot spectrum indicate a much hotter corona than is observed for 'normal' spectra. From BAME *et al.* (1979).

temperature-depressed solar wind flow following a solar flare. The spectrum displays a large number of peaks in the range of the C–D–E label. These peaks can be identified as Fe^{13+} through Fe^{18+} with Fe^{16+} the most prominent. An isothermal fit gives a coronal freezing-in temperature of 3.0×10^6K, a factor of two higher than the temperature determined for Fe in the upper spectrum. A fit to the oxygen peaks also indicates a higher than normal freezing-in temperature. In other examples of impulsive events, 'hot' spectra have been found showing freezing-in coronal temperatures up to and above 10×10^6K (FENIMORE, 1980).

An impulsive event spectrum indicative of a low temperature coronal origin of some of the plasma is given in Fig. 10. This spectrum shows that a mixture of normal (O^{7+}, O^{6+}) and very cold (He^+, O^{2+}, O^{3+}) plasma can be found together. The occurrence of such a mixed spectrum is relatively rare and appears to be related to transient coronal phenomena such as disappearing filaments (SCHWENN et al., 1980; GOSLING et al., 1980; ZWICKL et al., 1982). Coronal freezing-in temperatures for the He^+, O^{2+}, and O^{3+} ions in this type of spectrum are unusually low (i.e. near $10^5 K$), while the O^{6+} and O^{7+} ion abundances are indicative of freezing-in temperatures in the $1-2 \times 10^6 K$ range. Currently, there is no altogether certain explanation for these observed mixed ionization states.

6.   Limits of Current Knowledge with Respect to Magnetospheric Applications

In the introduction we summarized those particle measurements within the magnetosphere which were aimed at delineating the contributions of solar wind plasma to magnetospheric populations. That there is not total complementarity between what can realistically be measured in the magnetosphere vs what can be measured in the solar wind, is not too surprising. For one thing, in the solar wind ions are confined to a fairly small region of particle velocity space, which eases the problems of their detection

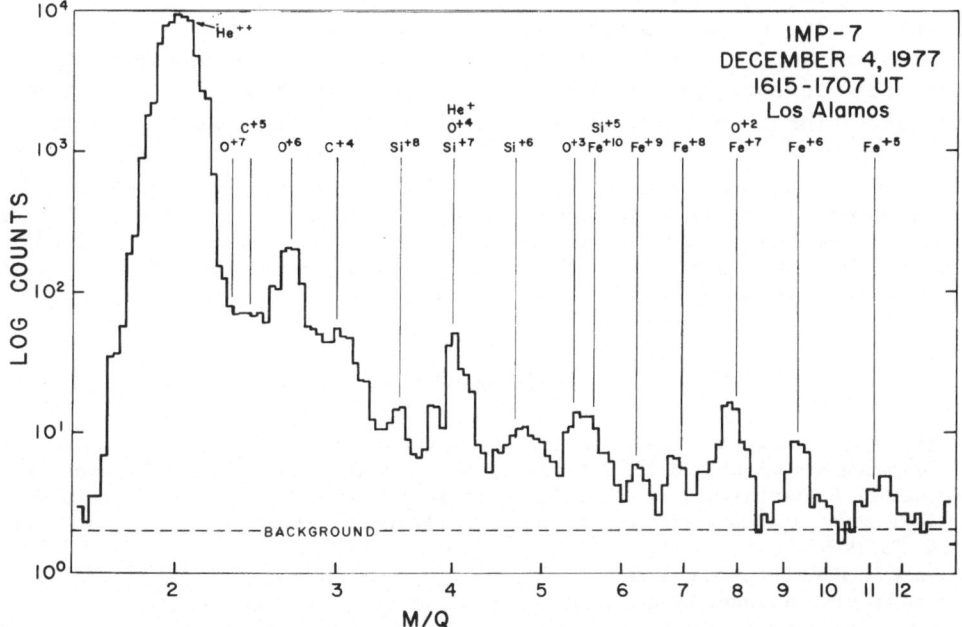

Fig. 10.   An unusual summed spectrum from low speed, low temperature solar wind within a non-compressive density enhancement measured with IMP 7 from 1615–1707 UT on December 4, 1977. The enhanced peak at $M/Q = 4$ has been seen only 3 times in 8 years of normal Los Alamos IMP and ISEE solar wind data. The presence of a peak at O^{6+} along with He^+, O^{2+}, and O^{3+} indicates the presence of a mixed plasma representing simultaneously warm and cold coronal plasma. From ZWICKL et al. (1982).

somewhat. Once inside the bow shock and magnetosphere, however, solar wind ions are thermalized and their velocity space density decreases. This effect increases the problems of detection, particularly in the $10 \sim 100$ keV range where the bulk of the ring current lies. Setting aside the problems of detection, we explore briefly the limits of current solar wind knowledge and ask which additional measurements could be useful in the future for studies of magnetospheric composition.

Elemental abundances, particularly the $C/N/O$ content, but also heavier ions, are of interest for further studies of radiation belt ions such as those reported by FRITZ and WILKEN (1976), HOVESTADT et al. (1978), and KONRADI et al. (1980). Elemental composition, except for H and He, can only rarely be obtained from present day solar wind measurements (cf. Section 3) and certainly not on a routine basis. The ^{3}He abundance, on the other hand, is now measured routinely in the solar wind (OGILVIE et al., 1980); but its measurement in the magnetosphere is most difficult and verges on the impossible with present magnetospheric instruments. As we have seen, charge state measurements, while not routine, are certainly feasible in the solar wind under appropriate conditions. They are currently of rather limited use in the magnetosphere, however, due to detection problems at low energy and confusion caused by charge state redistribution at higher energies.

We may summarize by saying that solar wind heavy ion abundances (charge state and elemental) with the possible exception of ^{4}He^{++} are not presently measured on the kind of routine basis needed for detailed tracer studies. There is a complementary problem on the magnetospheric side. Presently planned future instruments and missions may improve this situation considerably for both solar wind and magnetospheric measurements. This would seem to require a similar improvement in solar wind techniques if the use of composition as a probe is to become a viable research tool.

This review was prepared under the auspices of the U.S. Department of Energy. Portions of the Los Alamos data reported here were derived from the Vela nuclear test detection program jointly conducted by the Department of Defense and Department of Energy. Other Los Alamos data come from the National Aeronautics and Space Agency IMP and ISEE scientific satellite programs which have provided partial support for our space plasma work.

REFERECES

AXFORD, W. I., On the origin of radiation belt and auroral primary ions, in *Particles and Fields in the Magnetosphere*, edited by B. M. McCormac, p. 46, D. Reidel, Dordrecht, The Netherlands, 1969.
BALSIGER, H., P. EBERHARDT, J. GEISS, A. GHIELMETTI, H. P. WALKER, D. T. YOUNG, H. LOIDL, and H. ROSENBAUER, A satellite-borne ion mass spectrometer for the energy range 0 to 16 keV, *Space Sci. Inst.*, **2**, 499, 1976.
BALSIGER, H., P. EBERHARDT, J. GEISS, and D. T. YOUNG, Magnetic storm injection of 0.9 to 16-keV/e solar and terrestrial ions into the high-altitude magnetosphere, *J. Geophys. Res.*, **85**, 1645, 1980.
BAME, S. J., Spacecraft observations of the solar wind composition, in *Solar Wind*, edited by C. P. Sonett, P. J. Coleman, and J. M. Wilcox, p. 535, NASA SP-308, 1972.
BAME, S. J., A. J. HUNDHAUSEN, J. R. ASBRIDGE, and I. B. STRONG, Solar wind ion composition, *Phys. Rev. Lett.*, **20**, 393, 1968a.
BAME, S. J., J. R. ASBRIDGE, A. J. HUNDHAUSEN, and I. B. STRONG, Solar wind and magnetosheath observations during the Jan. 13–14, 1967 geomagnetic storm, *J. Geophys. Res.*, **73**, 5761, 1968b.

BAME, S. J., J. R. ASBRIDGE, A. J. HUNDHAUSEN, and M. D. MONTGOMETRY, Solar wind ions: $^{56}Fe^{+8}$ to $^{56}Fe^{+12}$, $^{28}Si^{+7}$, $^{28}Si^{+8}$, $^{28}Si^{+9}$, and $^{16}O^{+6}$, *J. Geophys. Res.*, **75**, 6360, 1970.

BAME, S. J., J. R. ASBRIDGE, W. C. FELDMAN, and P. D. KEARNEY, The quiet corona: Temperature and temperature gradient, *Solar Phys.*, **35**, 137, 1974.

BAME, S. J., J. R. ASBRIDGE, W. C. FELDMAN, M. D. MONTGOMERY, and P. D. KEARNEY, Solar wind heavy ion abundances, *Solar Phys.*, **43**, 463, 1975.

BAME, S. J., J. R. ASBRIDGE, W. C. FELDMAN, and J. T. GOSLING, Evidence for a structure-free state at high solar wind speeds, *J. Geophys. Res.*, **82**, 1487, 1977.

BAME, S. J., J. R. ASBRIDGE, W. C. FELDMAN, E. E. FENIMORE, and J. T. GOSLING, Solar wind heavy ions from flare-heated coronal plasma, *Solar Phys.*, **62**, 179, 1979.

BANKS, P. M. and G. KOCKARTS, *Aeronomy*, Part A, p. 30, Academic press, New York, 1973.

BLAKE, J. B., Experimental test to determine the origin of geomagnetically trapped radiation, *J. Geophys. Res.*, **78**, 5822, 1973.

BOGARD, D. D., W. C. HIRSCH, and L. E. NYQUIST, Noble gases in Apollo 17 fines: mass fractionation effects in trapped Xe and Kr, *Proc. Fifth Lunar Sci. Conf.*, **2**, 1975, 1974.

BOLLEA, D., V. FORMISANO, P. C. HEDGECOCK, G. MORENO, and F. PALMIOTTO, Heos 1 helium observations in the solar wind, in *Solar Wind*, edited by P. J. Coleman, C. P. Sonett, and J. M. Wilcox, p. 588, NASA SP-308, Washington D.C., 1972.

BORRINI, G. and G. NOCI, Dynamics and abundance of ions in coronal holes, *Solar Phys.*, **64**, 367, 1979.

BORRINI, G., J. T. GOSLING, S. J. BAME, W. C. FELDMAN, and J. M. WILCOX, Solar wind helium and hydrogen structure near the heliospheric current sheet: a signal of coronal streamers at 1 AU, *J. Geophys. Res.*, **86**, 4565, 1981a.

BORRINI, G., J. T. GOSLING, S. J. BAME, and W. C. FELDMAN, An analysis of shock wave disturbances observed at 1 AU from 1971 through 1978, *J. Geophys. Res.*, **87**, 4365, 1981b.

BORRINI, G., J. T. GOSLING, S. J. BAME, and W. C. FELDMAN, Helium abundance enhancements in the solar wind, *J. Geophys. Res.*, **87**, 7370, 1982a.

BORRINI, G., J. T. GOSLING, S. J. BAME, and W. C. FELDMAN, Helium abundance variations in the solar wind, *Solar Phys.*, 1982b (in press).

BRANDT, J. C. and P. W. HODGE, Solar System Astrophys., pp. 210, McGraw-Hill, New York, 1964.

BÜHLER, F., W. I. AXFORD, H. J. A. CHIVERS, and K. MARTI, Helium isotopes in an aurora, *J. Geophys. Res.*, **81**, 111, 1976.

CAMERON, A. G. W., Abundances of the elements in the solar system, *Space Sci. Rev.*, **15**, 121, 1973.

COPLAN, M. A., K. W. OGILVIE, P. A. BOCHSLER, and J. GEISS, Ion composition experiment, *IEEE Trans. Geosci, Electron.*, **GE-16**, 185, 1978.

EBERHARDT, P., The solar wind as deduced from lunar samples, in *Solar Wind Three*, edited by C.T. Russell, p. 58, Inst. Geophys. Plan. Phys., UCLA, 1974.

EPSTEIN, S. and H. P. TAYLOR, Jr., O^{18}/O^{16}, Si^{30}/Si^{28}, C^{13}/C^{12}, and D/H studies of Apollo 14 and 15 samples, *Proc. 3rd Lunar Sci. Conf.*, **3**, 1429, 1972.

FELDMAN, W. C., J. R. ASBRIDGE, S. J. BAME, and P. D. KEARNEY, Upper limits for the solar wind He^+ content at 1 AU, *J. Geophys. Res.*, **79**, 1808, 1974.

FELDMAN, W. C., J. R. ASBRIDGE, S. J. BAME, and J. T. GOSLING, Plasma and magnetic fields from the sun, in *The Solar Output and its Variation*, edited by O. R. White, p. 351, Colorado Associated University Press, Boulder, CO, 1977.

FELDMAN, W. C., J. R. ASBRIDGE, S. J. BAME, and J. T. GOSLING, Long-term variations of selected solar wind properties: Imp 6, 7, and, 8 results, *J. Geophys. Res.*, **83**, 2177, 1978.

FENIMORE, E. E., Solar wind flows associated with hot heavy ions, *Astrophys. J.*, **235**, 245, 1980.

FORMISANO, V., F. PALMIOTTO, and G. MORENO, α-particle observations in the solar wind, *Solar Phys.*, **15**, 479, 1970.

FRITZ, T. A. and B. WILKEN, Substorm generated fluxes of heavy ions at the geostationary orbit, in *Magnetospheric Particles and Fields*, edited by B. M. McCormac, p. 171, D. Reidel, Dordrecht, The Netherlands, 1976.

GEISS, J., Abundances in the solar wind and some applications to astrophysics and geophysics, in *Solar Cosmic Rays and Their Penetration into the Earth's Magnetosphere*, Leningrad, 1973a.

GEISS, J., Solar wind composition and implications about the history of the solar system, *Proc. of the 13th Inter. Cosmic Ray Conf.*, **5**, 3375, 1973b.

GEISS, J., P. HIRT, and H. LEUTWYLER, On acceleration of ions in corona and solar wind, *Solar Phys.*, **12**, 458, 1970a.

GEISS, J. P. EBERHARDT, F. BÜHLER, J. MEISTER, and P. SIGNER, Apollo 11 and 12 solar wind composition experiments: fluxes of He and Ne isotopes, *J. Geophys. Res.*, **75**, 5972, 1970b.

GEISS, J., F. BÜHLER, H. CERUTTI, P. EBERHARDT, and C. FILLEUX, Apollo 16 preliminary science report, *NASA SP-315*, Nat. Aeronaut. and Space Admin., p. 14–1, 1972.

GLOECKLER, G. and K. C. HSIEH, Time-of-flight technique for particle identification at energies from 2–400 keV/nucleon, *Nuclear Instrum. Methods*, **165**, 537, 1979.

GOEL, P. S. and B. K. KOTHARI, Total nitrogen contents of some Apollo 14 lunar samples by neutron activation analysis, *Proc. 3rd Lunar Sci. Conf.*, **3**, 2041, 1972.

GOSLING, J. T., J. R. ASBRIDGE, S. J. BAME, A. J. HUNDHAUSEN, and I. B. STRONG, Measurements of the interplanetary solar wind during the large geomagnetic storm of April 17–18, 1965, *J. Geophys. Res.*, **72**, 1813, 1967.

GOSLING, J. T., V. PIZZO, and S. J. BAME, Anomalously low proton temperatures in the solar wind following interplanetary shock waves—Evidence for magnetic bottles?, *J. Geophys. Res.*, **78**, 2001, 1973.

GOSLING, J. T., E. HILDNER, J. R. ASBRIDGE, S. J. BAME, and W. C. FELDMAN, Noncompressive density enhancements in the solar wind, *J. Geophys. Res.*, **82**, 5005, 1977.

GOSLING, J. T., J. R. ASBRIDGE, S. J. BAME, and W. C. FELDMAN, Solar wind stream interfaces, *J. Geophys. Res.*, **83**, 1401, 1978.

GOSLING, J. T., J. R. ASBRIDGE, S. J. BAME, W. C. FELDMAN, and R. D. ZWICKL, Observations of large fluxes of He$^+$ in the solar wind following an interplanetary shock, *J. Geophys. Res.*, **85**, 3431, 1980.

GOSLING, J. T., G. BORRINI, J. R. ASBRIDGE, S. J. BAME, W. C. FELDMAN, and R. T. HANSEN, Coronal streamers in the solar wind at 1 AU, *J. Geophys. Res.*, **86**, 5438, 1981.

GRÜNWALDT, H., Solar wind composition from the Helios 2 plasma experiment, *Space Res.*, **XVI**, 681, 1976.

HIRSHBERG, J., A. ALKSNE, D. S. COLBURN, S. J. BAME, and A. J. HUNDHAUSEN, Observation of a solar flare induced interplanetary shock and helium enriched driver gas, *J. Geophys. Res.*, **75**, 1, 1970.

HIRSHBERG, J., J. R. ASBRIDGE, and D. E. ROBBINS, The helium-enriched plasma from the proton flares of August/September 1966, *Solar Phys.*, **18**, 313, 1971.

HIRSHBERG, J., S. J. BAME, and D. E. ROBBINS, Solar flares and solar wind helium enrichments: July 1965–July 1967, *Solar Phys.*, **23**, 467, 1972a.

HIRSHBERG, J., J. R. ASBRIDGE, and D. E. ROBBINS, Velocity and flux dependence of the solar-wind helium abundance, *J. Geophys. Res.*, **77**, 3583, 1972b.

HIRSHBERG, J., J. R. ASBRIDGE, and D. E. ROBBINS, The helium component of solar wind velocity streams, *J. Geophys. Res.*, **79**, 934, 1974.

HOLLWEG, J. V., Minor ions in the low corona, *J. Geophys. Res.*, **86**, 8899, 1981.

HOVESTADT, D., G. GLOECKLER, C. Y. FAN, L. A. FISK, F. M. IPAVICH, B. KLECKER, J. J. O'GALLAGHER, and M. SCHOLER, Evidence for solar wind origin of energetic heavy ions in the earth's radiation belt, *Geophys. Res. Lett.*, **5**, 1055, 1978.

HUNDHAUSEN, A. J., Direct observations of solar wind particles, *Space Sci. Rev.*, **8**, 690, 1968.

HUNDHAUSEN, A. J., H. E. GILBERT, and S. J. BAME, Ionization state of the interplanetary plasma, *J. Geophys. Res.*, **73**, 5485, 1968.

HUNDHAUSEN, A. J., R. T. HANSEN, and S. F. HANSEN, Coronal evolution during the sunspot cycle: Coronal holes observed with the Mauna Loa K-coronameters, *J. Geophys. Res.*, **86**, 2079, 1981.

JOHNSON, R. G., R. D. SHARP, and E. G. SHELLEY, The discovery of energetic He$^+$ ions in the magnetosphere, *J. Geophys. Res.*, **79**, 3135, 1974.

JOKIPII, J. R., Effects of diffusion on the composition of the solar corona and the solar wind, in *The Solar Wind*, edited by R. J. Mackin and M. Neugebauer, p. 215, Pergammon, New York, 1966.

JORDAN, C., The ionization equilibrium of elements between carbon and nickel, *Mon. Not. Roy. Astron. Soc.*, **142**, 501, 1969.

JOSELYN, J. A. and T. E. HOLZER, A steady three-fluid coronal expansion for nonspherical geometries, *J. Geophys. Res.*, **83**, 1019, 1978.

KONRADI, A., T. A. FRITZ, and S.-Y. SU, Time-averaged fluxes of heavy ions at the geostationary orbit, *J. Geophys. Res.*, **85**, 5149, 1980.

KRIMIGIS, S. M., The charge composition aspect of energetic trapped particles, presented at Solar Terrestrial Relations Conf., Calgary, Canada, August 1972.

LAZARUS, A. J. and J. H. BINSACK, Observations of the interplanetary plasma subsequent to the July 7, 1966 proton flare, *Ann. IQSY*, **3**, 378, 1969.

LIND, D. L., J. GEISS, and W. STETTLER, Solar and terrestrial noble gases in magnetospheric precipitation, *J. Geophys. Res.*, **84**, 6435, 1979.

LYNCH, J., D. PULLIAM, R. LEACH, and F. SCHERB, The charge spectrum of positive ions in a hydrogen aurora, *J. Geophys. Res.*, **81**, 1264, 1976.

MEWE, R., Calculated solar X-radiation from 1 to 60 Å, *Solar Phys.*, **22**, 459, 1972.

MITCHELL, D. G. and E. C. ROELOF, Thermal iron ions in high speed solar wind streams: Detection by the IMP 7/8 energetic particles experiment, *Geophys. Res. Lett.*, **7**, 661, 1980.

MITCHELL, D. G., E. C. ROELOF, W. C. FELDMAN, S. J. BAME, and D. J. WILLIAMS, Thermal iron ions in high speed solar wind streams, 2. Temperatures and bulk velocities, *Geophys. Res. Lett.*, **8**, 827, 1981.

MONTGOMERY, M. D., J. R. ASBRIDGE, S. J. BAME, and W. C. FELDMAN, Solar wind electron temperature depressions following some interplanetary shock waves: Evidence for magnetic merging?, *J. Geophys. Res.*, **79**, 3103, 1974.

NAKADA, M. P., A study of the composition of the lower solar corona, *Solar Phys.*, **7**, 302, 1969.

NEUGEBAUER, M., Observations of solar wind helium, *Fundamentals Cosmic Phys.*, **7**, 131, 1981.

NEUGEBAUER, M. and C. W. SNYDER, Mariner 2 observations of the solar wind, 1, Average properties, *J. Geophys. Res.*, **71**, 4469, 1966.

OGILVIE, K. W., Helium abundance variations, *J. Geophys. Res.*, **77**, 4227, 1972.

OGILVIE, K. W. and T. D. WILKERSON, Helium abundance in the solar wind, *Solar Phys.*, **8**, 435, 1969.

OGILVIE, K. W. and J. HIRSHBERG, The solar cycle variation of the solar wind helium abundance, *J. Geophys. Res.*, **75**, 4595, 1974.

OGILVIE, K. W. and C. VOGT, Variation of the average 'freezing-in' temperature of oxygen ions with solar wind speed, *Geophys. Res. Lett.*, **7**, 577, 1980.

OGILVIE, K. W., N. MCILWRAITH, and T. D. WILKERSON, A mass-energy analyzer for space plasmas, *Rev. Sci. Inst.*, **39**, 441, 1968a.

OGILVIE, K. W., L. F. BURLAGA, and T. D. WILKERSON, Plasma observations on Explorer 34, *J. Geophys. Res.*, **73**, 6809, 1968b.

OGILVIE, K. W., M. A. COPLAN, P. BOCHSLER, and J. GEISS, Abundance ratios of $^4He^{++}/^3He^{++}$ in the solar wind, *J. Geophys. Res.*, **85**, 6021, 1980.

REASONER, D. L., Auroral helium precipitation, *Rev. Geophys. Space Phys.*, **11**, 169, 1973.

ROBBINS, D. E., A. J. HUNDHAUSEN, and S. J. BAME, Helium in the solar wind, *J. Geophys. Res.*, **75**, 1178, 1970.

SCHWENN, R., H. ROSENBAUER, and K.-H. MÜHLHAUSER, Singly-ionized helium in the driver gas of an interplanetary shock wave, *Geophys. Res. Lett.*, **7**, 201, 1980.

SHARP, R. D., R. G. JOHNSON, and E. G. SHELLEY, Satellite measurements of auroral alpha particles, *J. Geophys. Res.*, **79**, 5167, 1974.

SHELLEY, E. G., R. G. JOHNSON, and R. D. SHARP, Satellite observations of energetic heavy ions during a geomagnetic storm, *J. Geophys. Res.*, **77**, 6104, 1972.

SHELLEY, E. G., R. D. SHARP, and R. G. JOHNSON, Satellite observations of an ionospheric acceleration mechanism, *Geophys. Res. Lett.*, **3**, 654, 1976.

SHELLEY, E. G., R. D. SHARP, R. G. JOHNSON, J. GEISS, P. EBERHARDT, H. BALSIGER, G. HAERENDEL, and H. ROSENBAUER, Plasma composition experiment on ISEE-A, *IEEE Trans Geosci. Electron.*, **GE-16**, 266, 1978.

SPJELDVIK, W. N. and T. A. FRITZ, Theory for charge states of energetic oxygen ions in the earth's radiation belts, *J. Geophys. Res.*, **83**, 1583, 1978.

VASYLIUNAS, V. M., Deep space plasma measurements, in *Methods of Experimental Physics, Plasma Physics*, Vol. 9-B, edited by R. H. Lovberg and H. R. Griem, p. 49, Academic Press, New York, 1971.

WILLIAMS, D. J., Ring current composition and sources: an update, *Planet. Space Sci.*, **29**, 1195, 1981.

WITHBROE, G. L.; The chemical composition of the photosphere and the corona, in *Menzel Symposium on Solar Physics, Atomic Spectra, and Gaseous Nebulae*, p. 127, N.B.S. Special Pub. No. 353, 1971.

YOUNG, D. T., H. BALSIGER, and J. GEISS, Observed increase in the abundance of kilovolt O^+ in the magnetosphere due to solar cycle effects, *Adv. Space Res.*, 1, 309, 1981.

ZASTENKER, G. N. and Yu. I. YERMOLAEV, Observations of solar wind stream with high abundance of heavy ions and relation with coronal conditions, *Planet. Space Sci.*, 29, 1235, 1981.

ZINNER, E., R. M. WALKER, J. BORG, and M. MAURETTE, Apollo 17 lunar surface cosmic ray experiment — measurement of heavy solar wind particles, *Proc. Fifth Lunar Sci. Conf.*, 3, 2975, 1974.

ZWICKL, R. D., J. R. ASBRIDGE, S. J. BAME, W. C. FELDMAN, and J. T. GOSLING, He^+ and other unusual ions in the solar wind: a systematic search covering 1972–1980, *J. Geophys. Res.*, 87, 7379, 1982.

Energetic Ion Composition in the Earth's Magnetosphere, edited by R. G. Johnson, 99–141.
Copyright © 1983 by Terra Scientific Publishing Company (TERRAPUB), Tokyo.

Composition and Characteristics of the Polar Wind

W. J. RAITT and R. W. SCHUNK

Department of Physics and Center for Atmospheric and Space Sciences,
Utah State University, Logan, Utah 84322, U.S.A.

(Received March 1, 1982)

We have discussed in general terms the mechanism of the polar wind, describing a pressure gradient driven flow of light ions (H^+, He^+) from a source region around 300 km altitude to the outer regions of the magnetosphere along geomagnetic field lines. Although it is pointed out that the dynamical nature of the high-latitude ionospheric source will result in an outward flux of light ions which is both spatially and temporally variable, the average value is generally accepted to be of the order of 3×10^8 cm^{-2}sec^{-1} for H^+ ions and 2×10^7 cm^{-2}sec^{-1} for He^+ ions.

The results of detailed theoretical models of H^+ and He^+ outflow are presented. The resulting density, temperature and velocity profiles are examined under a variety of conditions to bracket the range of values which might be expected to be encountered in the terrestrial environment. Recent developments of the theoretical modeling of light ion outflow are discussed, in which changes are predicted in the distribution function of the polar wind species in terms of temperature differences parallel and perpendicular to the magnetic field direction.

The final section collects together the published experimental evidence for the polar wind. The measurements exist only in the upper ionosphere and consist of either direct observations of the flow velocity, or indirect observations in terms of differences in the light ion density at a given altitude between low latitude regions of little or no outflow, and polar regions where a strong outflow is expected.

The principal conclusion presented in the paper is that polar wind studies represent a field of geophysical research in which theoretical modeling has far outstripped experimental observations. It is hoped that programs such as the existing ISEE and Dynamics Explorer projects and the proposed OPEN project will go some way to alleviate this deficiency.

1. Introduction

Early work of AXFORD and HINES (1961) and DUNGEY (1961) on the configuration of the earth's distant geomagnetic field due to the interaction of the solar wind and the earth's geomagnetic field led to the now generally accepted view of the gross features of the magnetosphere. The effect of the interaction is to generate current systems which result in the earth's geomagnetic field on the sunward side being compressed and on the anti-sunward side being drawn out into a long tail-like structure. The magnetic field lines which form the tail of the magnetosphere originate from the high latitude (polar) regions of the earth's surface. A recent schematic diagram of a meridional cross-section

W. J. RAITT and R. W. SCHUNK

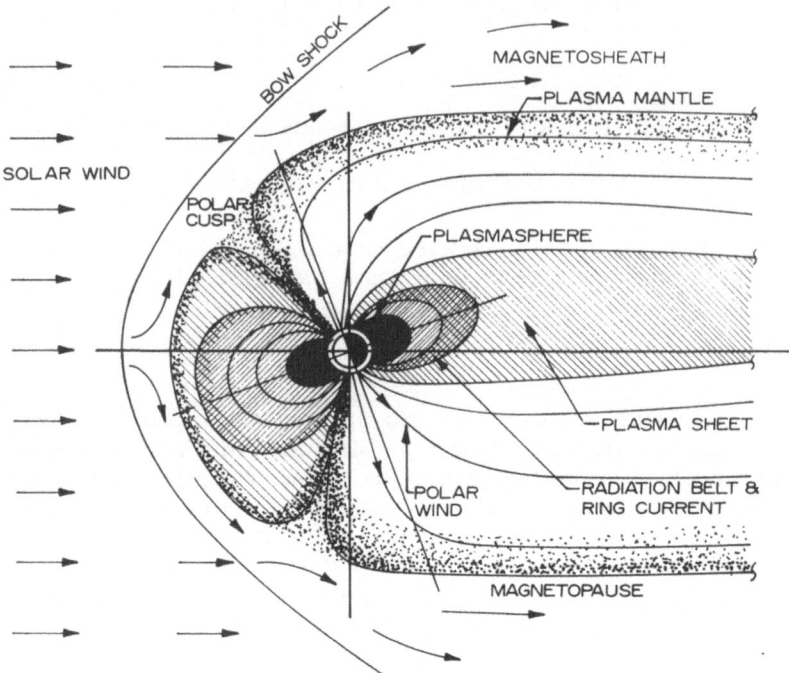

Fig. 1. Schematic illustration of the earth's bow shock and magnetosphere (from BAHNSEN, 1978). Reprinted by permission of Pergamon Press.

of the magnetosphere is shown in Fig. 1. One result of the extended tail of the magnetosphere is that there will be a continual plasma pressure gradient away from the earth parallel to the magnetic field lines between the relatively dense ionospheric plasma at about 300 km altitude and the distant geomagnetic tail.

Initial studies of the effects of plasma density gradients on the diffusion dominated regions of the ionosphere had concentrated on applying plasma transport equations to solve for the case of the diffusive equilibrium distribution of multi-ion species (cf. MANGE, 1960). In this case, the pressure gradient existing between the F-region peak and higher altitudes results in the ions distributing themselves such that the polarization electric field set up by the pressure gradient is just balanced by the gravitational force on the ions. Under these circumstances the outward flux of ions is zero. When the concept of open or extended field lines became accepted, it was realized that the possibility existed for ions to continuously flow out from the earth into distant regions of the magnetosphere linked by geomagnetic field lines to the polar ionosphere. This condition is one of the solutions of the ion density distribution under diffusion dominated conditions and is known as 'dynamic equilibrium' as opposed to 'diffusive equilibrium' discussed earlier. Both types of equilibrium represent steady state solutions to the plasma transport equations resulting in the requirement that the outward flux is a constant, zero in the case of diffusive equilibrium, and some finite value in the case of dynamic equilibrium (BANKS and KOCKARTS, 1973; BAUER, 1973). Boundary conditions on the rate of production of the outflowing ions and Coulomb collisions with stationary ions result in an upper limit to the outward flux. The most dramatic ionospheric effect of dynamic equilibrium is to produce a marked reduction in ion density of the outflowing species compared with the diffusive equilibrium situation. For a single ion species in its parent neutral gas, with electron, ion and neutral temperatures equal, the scale height is decreased by a factor of two for the change from diffusive to dynamic equilibrium.

For a mixture of light and heavy ions, such as H^+ and O^+, detailed solutions of the plasma transport equations, which will be discussed later in this paper, show that the gravitational attraction on the heavier O^+ ion is such that the polarization electric field is not sufficient to permit the dynamic equilibrium condition and the ions form a diffusive equilibrium altitude-density profile. However, H^+ is light enough that once the inhibiting effect of H^+-O^+ collisions is reduced, as the O^+ density falls off, the polarization electric field can drive the ions into a continuous outflow parallel to the geomagnetic field (BANKS and HOLZER, 1968). This outflow of light ions from the polar regions of the earth's ionosphere was termed the polar wind by Axford. Further studies (BANKS and HOLZER, 1969a, b) have shown that He^+ is also light enough to flow out along open geomagnetic field lines, although the characteristics of the outflow differ considerably from that of H^+.

As the polar wind ions increase their outflow velocity when they escape the effects of Coulomb collisions with O^+, the conventional plasma transport equations eventually become invalid due to the transition from collision dominated motion to collisionless motion. To accommodate the collisionless region, kinetic models of the polar wind have been developed by DESSLER and CLOUTIER (1969) and by LEMAIRE and SCHERER (1970, 1971, 1972). Comparisons between the hydrodynamic and kinetic models of the collisionless polar wind have shown that a good match occurs between the two techniques (HOLZER et al., 1971).

The studies of both the diffusive and dynamic equilibrium distribution of ion density with altitude which have been published are for steady state conditions. At present, no detailed studies of the time dependent nature of the outflow have been made, although some speculations on the characteristics of a shock front propagating along a suddenly depleted flux tube have been made (BANKS et al., 1971).

It is apparent that this outflow of plasma will provide a source of magnetospheric plasma in the distant magnetospheric tail with its origin in the plasma generated in the polar regions by ionization of the earth's neutral atmosphere by solar ultraviolet light in the sunlit pole and by particle ionization over the whole of the polar cap but concentrated in the auroral regions. The outflowing plasma into the magnetospheric tail is generated as a relatively cool thermal plasma, but as will be discussed later in this paper, recent studies have shown that quite drastic modifications occur to the distribution functions of both the ions and electrons as the plasma flows away from its source region in the ionosphere to great distances along the geomagnetic field lines.

As was mentioned above, the ultimate source of the polar wind is the polar ionospheric plasma which has a peak density at around 300 km altitude. The source of H^+ to form the main element of polar wind is the accidentally resonant charge exchange reaction

W. J. RAITT and R. W. SCHUNK

$$O^+ + H \rightleftharpoons H^+ + O.$$

Thus, the amount of H^+ available to flow away from the earth is directly controlled by the ambient density of O^+ ions in the topside ionosphere. Recent studies of the distribution of ion densities over the polar cap as a function of position and time have shown that the O^+ density is by no means uniform. Hence, different flux tubes will have different initial plasma pressures to drive the polar wind, resulting in varying outward fluxes. The large variability of O^+ density is a result of mapping the convection electric field about the magnetic pole which rotates diurnally about the geographic pole in the geographic reference frame (SOJKA *et al.*, 1979, 1981a, b). Some idea of the complexity of the variation of the O^+ density over the polar ionosphere can be seen in Fig. 2, which

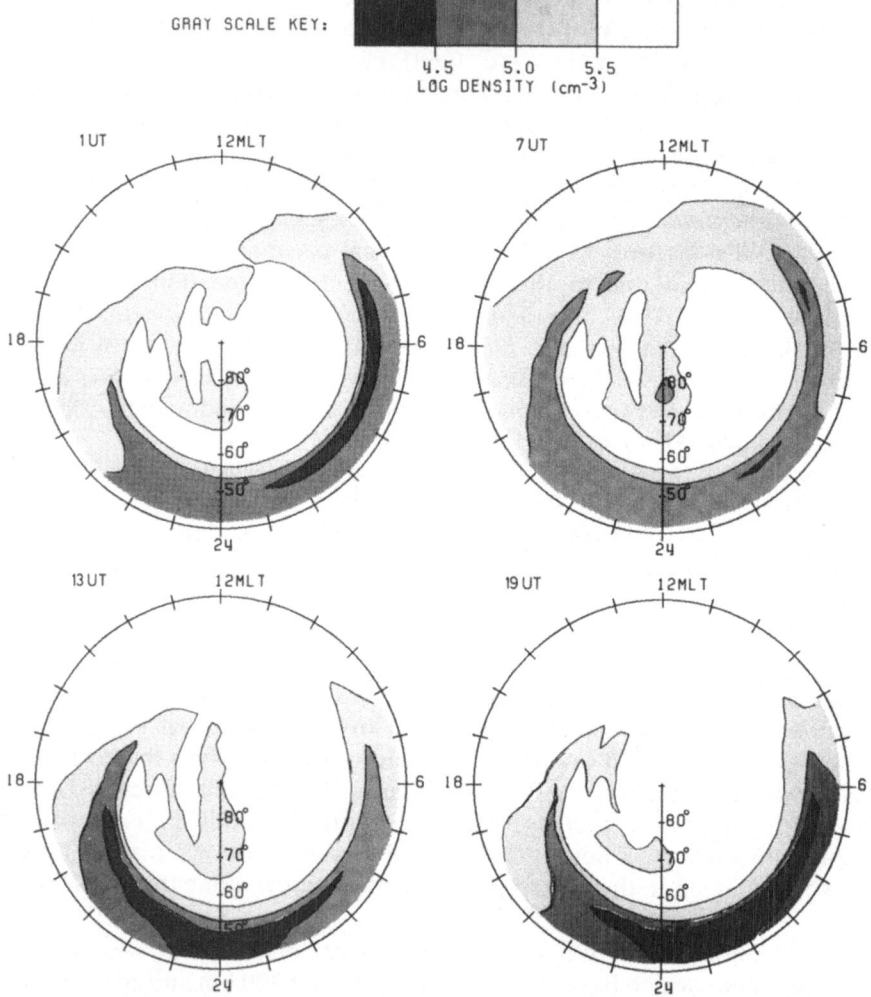

Fig. 2.   Contours of the electron density at 300 km for four universal times displayed in the magnetic quasi-inertial frame. The gray scaling corresponds to different density levels, as indicated in the key (from SOJKA, *et al.* (1981b)).

shows the distribution of O^+ ion density over the polar regions at four universal times. These plots were made for conditions of solar maximum and low geomagnetic activity.

The source of He^+ ions flowing away from the earth's ionosphere is less variable than the H^+ source, since it is produced primarily by chemical balance between photoionization of He and charge exchange between He^+ and the molecules O_2 and N_2. However, local heating of the neutral atmosphere by plasma drift or by particle precipitation will increase the loss processes for He^+ in localized regions. The sources and Coulomb collision processes in the topside ionosphere for the H^+ and He^+ ions result in a flux limited outflow having values as large as 3×10^8 cm^{-2}sec^{-1} for H^+ ions and 2×10^7 cm^{-2}sec^{-1} for He^+ ions.

Closed magnetic field lines in the vicinity of the plasmapause also have plasma removed from the region at the equatorial plane resulting in outflow at lower latitudes and also interhemispheric plasma flow. However, this process is related to the configuration and dynamics of the plasmasphere rather than the population of the distant magnetosphere with low energy plasma, and so the process will not be further considered here.

In this paper we will discuss the results of detailed modelling of steady state outflow of H^+ and He^+ ions from the polar ionosphere into the tail regions of the magnetosphere. We will then survey the experimental evidence for the existence of this phenomenon in the altitude region where most of the modelling has been done; that is to say, below about 3,000 km. Finally, we will summarize the present state of polar wind studies and suggest areas for further modelling and experimental observations.

2. Theoretical Models of the Polar Wind

Numerous theoretical models have been developed over the last decade to describe the polar wind, including hydrodynamic models (BANKS and HOLZER, 1968, 1969a, b; MARUBASHI, 1970; BANKS, 1973; BAILEY and MOFFETT, 1974; BANKS et al., 1974, 1976; STROBEL and WEBER, 1972; RAITT et al., 1975, 1977, 1978a, b; SCHUNK et al., 1978; OTTLEY and SCHUNK, 1980), hydromagnetic models (HOLZER et al., 1971), kinetic models (LEMAIRE and SCHERER, 1970, 1971, 1972, 1973; LEMAIRE, 1972) and models based on generalized transport equations (SCHUNK and WATKINS, 1979, 1981, 1982). The hydrodynamic, hydromagnetic and generalized transport equations are obtained by taking velocity moments of the Boltzmann equation in an effort to derive conservation equations for the physically significant moments of the distribution function, such as density, drift velocity, temperature, stress tensor and heat flow vector. The hydrodynamic equations result if the plasma is assumed to be collision-dominated, while the hydromagnetic equations correspond to the collisionless moment equations. The generalized transport equations are similar to the hydromagnetic equations, except that collisional terms are retained, and therefore these equations provide a continuous transition from the collision-dominated to the collisionless regimes. The kinetic models, on the other hand, are obtained by directly integrating the collisionless Boltzmann equation.

All of the different polar wind models have limitations. For example, there are mathematical difficulties connected with the moment equation approach because of the need to truncate the infinite set of moment equations (cf. SCHUNK, 1977). The kinetic solutions, on the other hand, satisfy the full hierarchy of moment equations deduced from the collisionless Boltzmann equation. However, in this case the reliability of a given velocity moment depends on the expression adopted for the velocity distribution function at the boundary.

Our aim is not to compare the different mathematical models of the polar wind, but rather to present the results that have been obtained from the various models in order to elucidate the basic characteristics of the polar wind. In the subsections that follow, we discuss the outflow of both H^+ and He^+ and the effect that various high-latitude processes have on this outflow. In our discussion, we will emphasize the more recent results.

2.1 H^+ outflow

In the initial studies of the polar wind, the hydrodynamic continuity and momentum equations for H^+ and O^+ were solved by assuming an isothermal ionosphere (BANKS and HOLZER, 1968, 1969a, b; MARUBASHI, 1970) and the basic characteristics of the density and flow velocity structure of the polar wind were elucidated. Subsequently, the polar wind energy balance and the effects of convection electric fields have been taken into account (BANKS et al., 1974; RAITT et al., 1975, 1977, 1978a, b).

Typical polar wind results are shown in Figs. 3a and 3b for the case when convection electric field effects are neglected. Figure 3a shows the effect on the H^+ and O^+ densities and H^+ field-aligned drift velocity of different H^+ escape velocities at 3,000 km. Curve (a) represents near diffusive equilibrium, with H^+ becoming the dominant ion at 900 km (the O^+ density in this case follows the lower curve of the shaded region). As the upper boundary velocity is increased, the H^+ density is progressively reduced with a peak in the H^+ density profile appearing near 600–700 km altitude. Curve (h), which represents a flow velocity of 20 km s^{-1} at 3,000 km, corresponds to a supersonic flow of H^+ with an escape flux of 8.5 $\times 10^7$ cm^{-2}s^{-1}.

The H^+ and O^+ temperatures are both affected by the H^+ flow, as shown in Fig. 3b. For O^+, the behavior is fairly simple in that as H^+ flows out of the topside ionosphere with an increasing velocity, the O^+ temperature at high altitudes decreases. This behavior results because as H^+ becomes a minor ion, the O^+-neutral thermal coupling becomes stronger. For H^+, on the other hand, the variation of temperature with escape velocity is more complicated. As the H^+ escape velocity is increased, the H^+ temperature at high altitudes first decreases, then increases, and then decreases again. This behavior is related to the relative contributions made to the H^+ thermal balance by convection, advection, thermal conduction, frictional heating, and collisional cooling, and we refer the reader to the paper by RAITT et al. (1975) for a more detailed discussion of the H^+ energy balance.

A separate peculiarity of H^+ flow is its flux limiting character. As the outward flow velocity increases, the H^+ flux rapidly rises to a saturation limit, first described by HANSON and PATTERSON (1964) and given analytic form by GEISLER (1967). Figure 4 illustrates this behavior in terms of the H^+ density at 3,000 km. At sufficiently high densities, the flux is inward (negative) but as the H^+ density is lowered, the outward

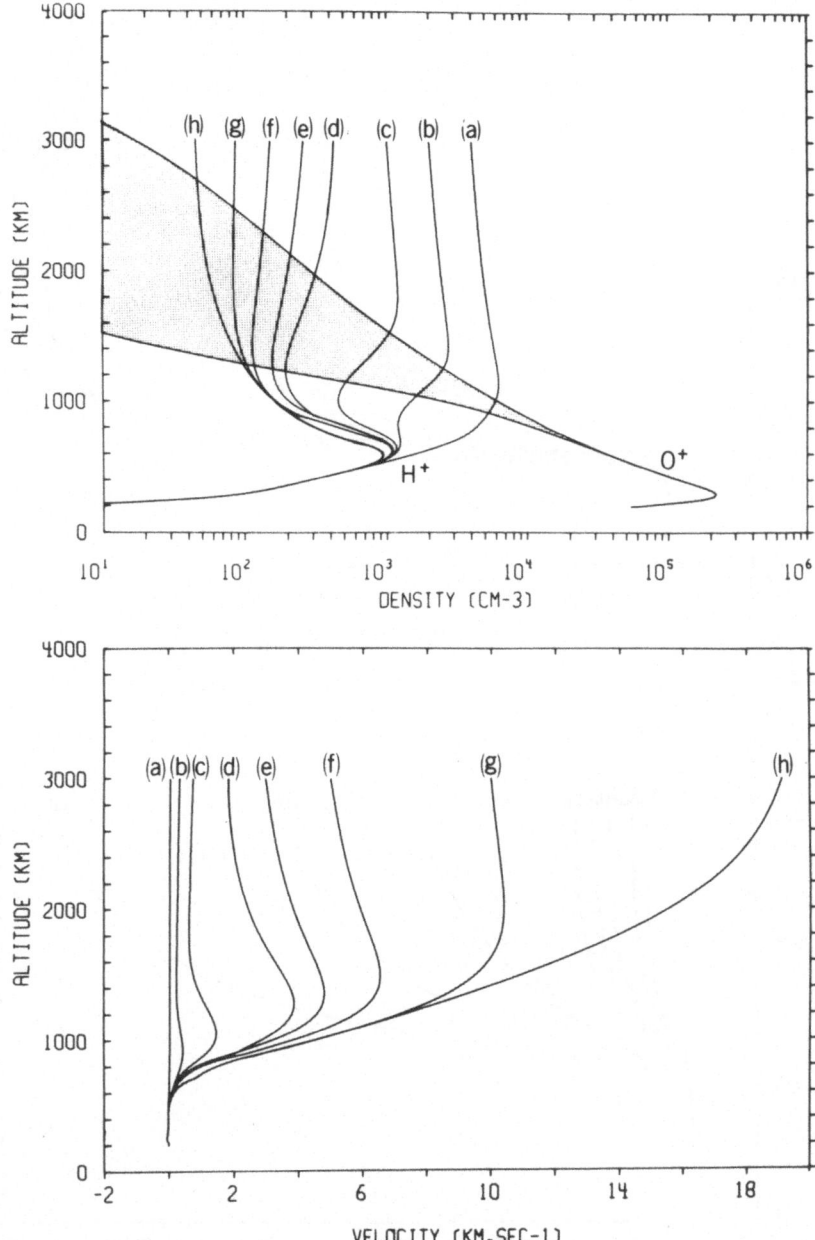

Fig. 3a.   Theoretical H^+ density and field-aligned drift velocity profiles for the Earth's daytime high-latitude ionosphere. The different curves correspond to different H^+ escape velocities at 3000 km: (a) 0.06, (b) 0.34, (c) 0.75, (d) 2.0, (e) 3.0, (f) 5.0, (g) 10.0, (h) 20.0 km s^{-1}. The shaded region shows the range of O^+ densities (from RAITT *et al.* (1975)).

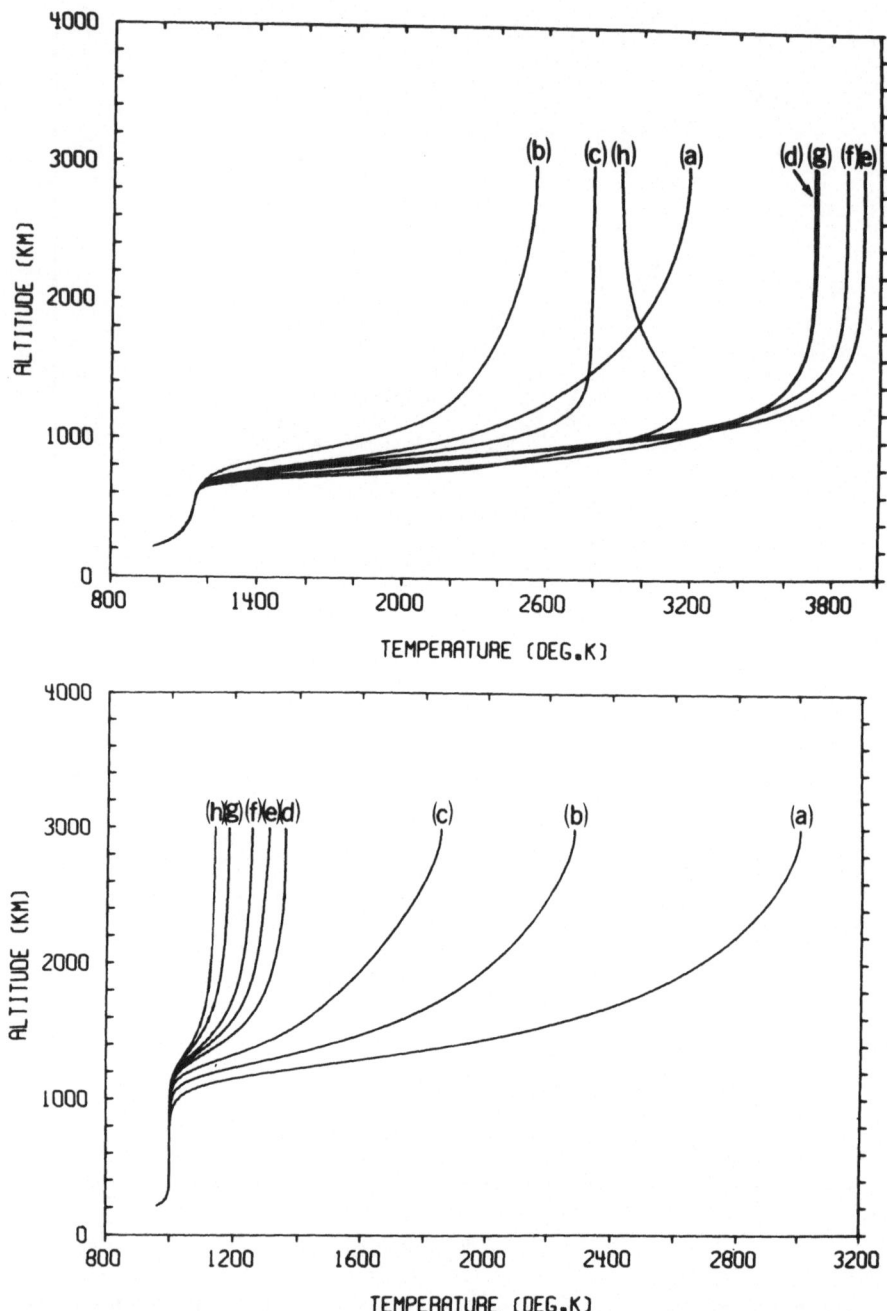

Fig. 3b. Theoretical H$^+$ (top) and O$^+$ (bottom) temperature profiles for the Earth's daytime high-latitude ionosphere. These profiles correspond to the density and drift velocity profiles shown in Fig. 3a (from RAITT et al. (1975)).

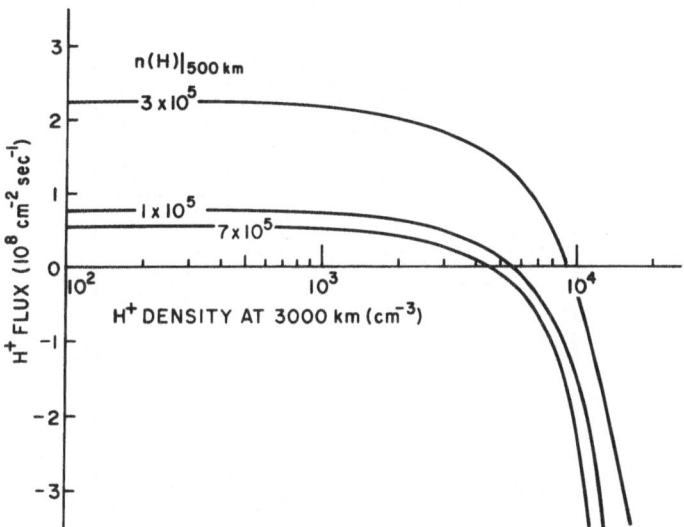

Fig. 4. H^+ flux for different H^+ boundary densities and neutral hydrogen densities (taken from BANKS (1972)).

flux is quickly established and soon saturates in magnitude. Thus, while arbitrarily large plasma inflow can occur in response to high densities within the protonosphere, the outflow is limited by various atmospheric constraints. These include the plasma temperatures, the neutral atomic hydrogen concentration, the O^+ density and the atmosphere composition. The limiting fluxes for several atomic hydrogen densities are included in Fig. 4, showing that the H^+ flux changes proportionately to the density of H.

2.1.1 Effect of viscous stress

In the initial studies of the polar wind by BANKS and HOLZER (1968, 1969b) and MARUBASHI (1970), the effect of viscous stress on the H^+ gas was neglected. Consequently, when H^+ was a minor ion, the H^+ momentum equation possessed singularities at $M = \pm 1$; that is, at the point of transition from subsonic to supersonic flow. For such a case, the different solutions to the Mach number equation are shown schematically in Fig. 5. For $M > 0$, corresponding to polar wind outflow, there are both subsonic ($M < 1$) and supersonic ($M > 1$) regions. All of the solutions that remain subsonic at all altitudes are valid, physical solutions. However, for supersonic outflow only the critical solution A is a physical solution. For this solution, the H^+ flow is subsonic at low altitudes, passes through the singular point $M = 1$, and then is supersonic at high altitudes.

The effect of H^+ stress on the density and temperature structure of polar wind flow has been studied by RAITT et al. (1975). From the mathematical point of view, H^+ stress introduces a term in the H^+ momentum equation that is proportional to d^2M/dS^2 and, consequently, eliminates the singularity that appeared in the original studies of the polar wind. The quantitative effect of viscous stress on the H^+ density,

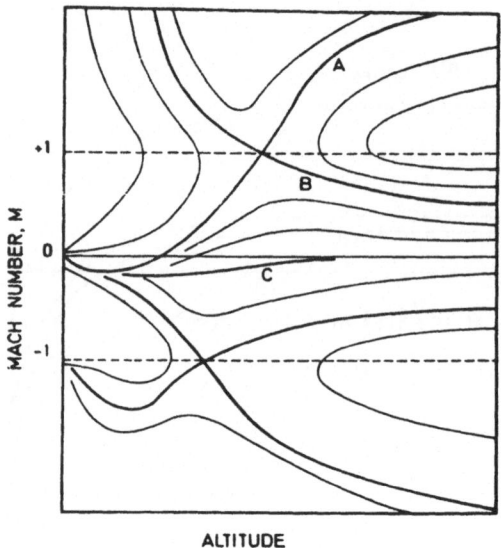

ALTITUDE

Fig. 5.   Ion Mach number diagram showing the transition from subsonic to supersonic flow for both outward and inward solutions (from BANKS and KOCKARTS (1973)).

drift velocity and temperature profiles is shown in Fig. 6 for the case of a boundary H^+ outflow velocity of 3 km sec^{-1} at 3,000 km (curve e in Figs. 3a and 3b). The profiles labelled 1 were obtained with the standard coefficient of viscosity, while those labelled 2 were calculated for a factor of 3 reduction in this coefficient. Decreasing viscous stress acts to decrease the H^+ density while increasing the flow speed and temperature.

   2.1.2   Effect of thermal diffusion
   Thermal diffusion occurs when there is a temperature gradient in a plasma and is a manifestation of the effect that heat flow has on the momentum balance. Figures 7a and 7b show H^+ density and temperature profiles calculated with and without allowance for thermal diffusion. With allowance for thermal diffusion, there is approximately a 50% increase in the H^+ density below 1,500 km and approximately a 1,400°K decrease in the H^+ temperature above 1,500 km. This behavior results because in the altitude region where H^+ is a minor ion, thermal diffusion acts to drive the lighter H^+ ions toward cooler regions, i.e., towards lower altitudes (SCHUNK and WALKER, 1969, 1970). The downward thermal diffusion force therefore acts to reduce the upward flow of H^+ ions and this, in turn, results in enhanced H^+ densities and reduced H^+ temperatures.

   2.1.3   Effect of diffusion-thermal heat flow
   Diffusion-thermal heat flow occurs whenever there is a relative drift between two or more ion gases (ST.-MAURICE and SCHUNK, 1977; CONRAD and SCHUNK, 1979). Depending on the direction of the relative drift, this process can either assist or oppose ordinary thermal conduction. Since there is a large H^+-O^+ relative drift along polar wind field lines, this process is potentially important for the H^+ energy balance. Figure 8 shows H^+ temperature profiles calculated with and without allowance for diffusion-

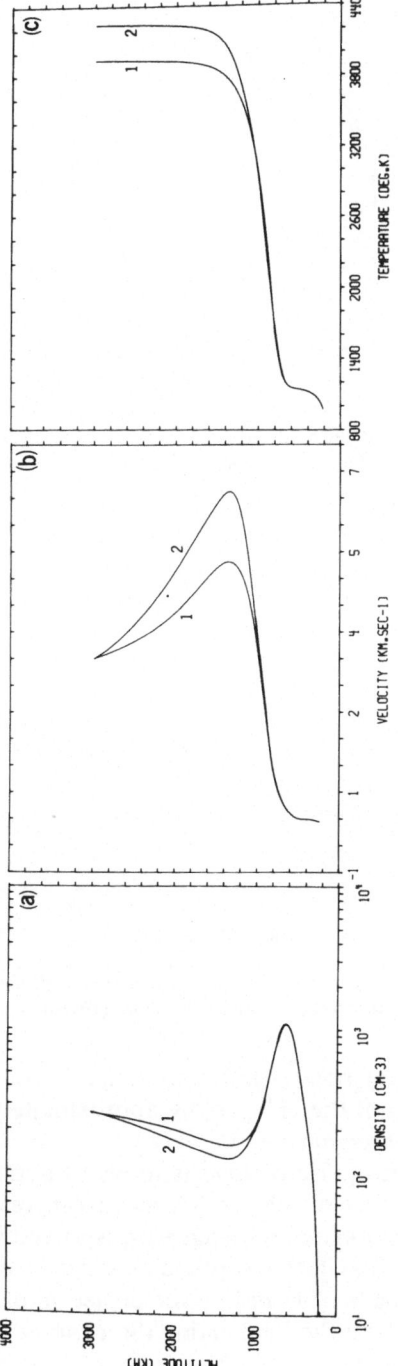

Fig. 6. Profiles of H$^+$ density, drift velocity and temperature for an outflow velocity of 3 km sec^{-1} at 3,000 km. Profile 1 corresponds to the standard coefficient of viscosity used in the results of Figs. 3a and 3b, while for profile 2 the coefficient of viscosity was reduced by a factor of 3 (from RAITT *et al.* (1975)).

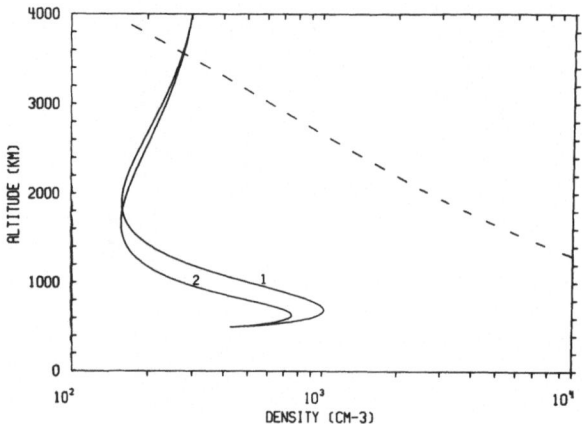

Fig. 7a. H$^+$ density profiles calculated with (1) and without (2) allowance for thermal diffusion. The dashed curve is the O$^+$ density profile (from BANKS *et al.* (1974)).

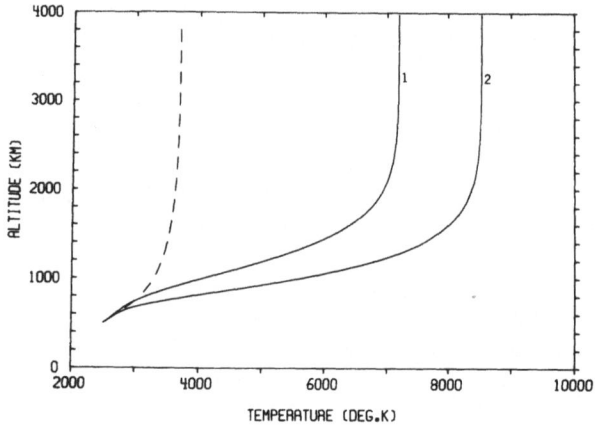

Fig. 7b. H$^+$ temperature profiles calculated with (1) and without (2) allowance for thermal diffusion. The dashed curve is the O$^+$ temperature profile (from BANKS *et al.* (1974)).

thermal heat flow for typical polar wind conditions. In the polar wind, diffusion-thermal heat flow acts to cool the H$^+$ gas, by approximately 600°K above 900 km.

2.1.4 *Effect of the convection electric field*

The effect of a convection electric field is to heat the plasma through frictional interaction between ions and neutrals, which acts to increase the plasma pressure differential between the ionosphere and magnetosphere and to enhance the outward flux of H$^+$ ions. In practice, this effect is opposed by a higher topside O$^+$ density, which reduces the H$^+$ diffusion coefficient, and by the depletion of O$^+$ in the F_2-region by chemical effects associated with the ion heating (SCHUNK *et al.*, 1975), which reduces the production rate of H$^+$ ions.

RAITT *et al.* (1977) studied the effect that convection electric fields have on the

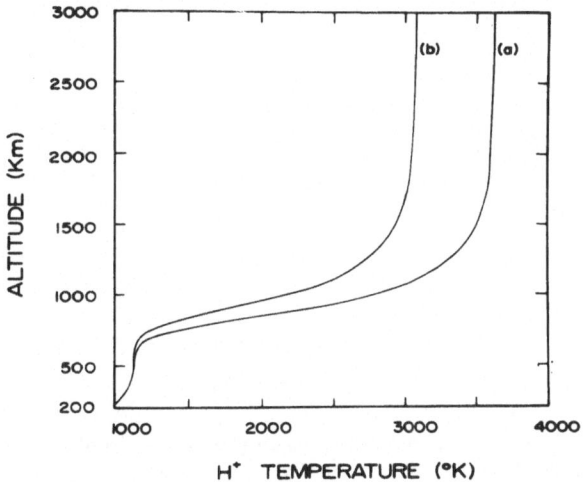

Fig. 8. H$^+$ temperature profiles calculated without (curve a) and with (curve b) allowance for diffusion-thermal heat flow. For these calculations, the H$^+$ outflow velocity at 3,000 km was set equal to 10 km/s (from SCHUNK *et al.* (1978)).

polar wind, but they separately studied the electric field heating and O$^+$ depletion effects. Figures 9a–c show H$^+$ density, drift velocity, and temperature profiles for an H$^+$ outflow velocity of 10 km/sec at 3,000 km, a F_2-peak O$^+$ density of 2.1 $\times 10^5$ cm^{-3}, and for convection electric fields of 0, 25, 50, and 100 mVm^{-1}. As the convection electric field increases, the general trend is for an increased H$^+$ density, a decreased H$^+$ drift velocity and an enhanced H$^+$ temperature. The electric field also acts to increase the O$^+$ temperature and hence topside O$^+$ density (not shown in Figs. 9a–c). The increased O$^+$ density, in turn, acts to slow the H$^+$ outflow and thereby

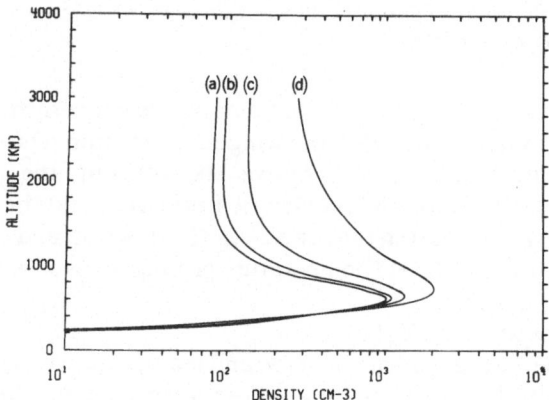

Fig. 9a. H$^+$ density profiles as a function of altitude for a 10 km/sec H$^+$ outflow velocity at 3,000 km and for different perpendicular electric fields: (a) 0, (b) 25, (c) 50, and (d) 100 mVm^{-1} (from RAITT *et al.* (1977)).

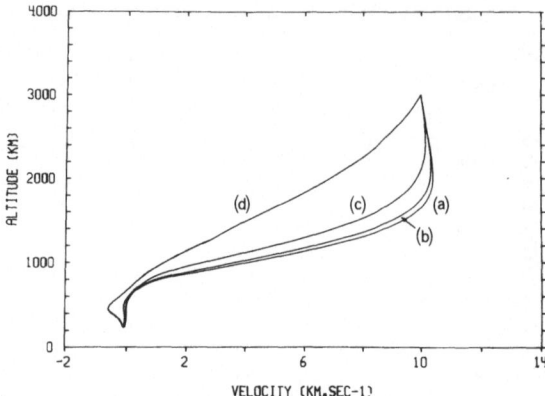

Fig. 9b.   H$^+$ drift velocity profiles as a function of altitude. Curves (a)–(d) correspond to the H$^+$ density profiles (a)–(d) in Fig. 9a (from RAITT *et al.* (1977)).

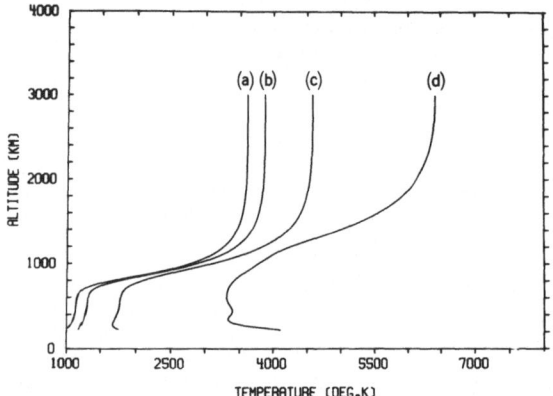

Fig. 9c.   H$^+$ temperature profiles as a function of altitude. Curves (a)–(d) correspond to curves (a)–(d) in Figs. 9a and 9b (from RAITT *et al.* (1977)).

increase the H$^+$ density. As far as the H$^+$ temperature is concerned, the electric field provides a direct heat source below 600 km owing to the frictional interaction between H$^+$ ions and the neutral atmosphere. Above 600–800 km, the H$^+$ temperature increases rapidly with altitude for all four electric field cases, and this results from the additional H$^+$-O$^+$ frictional heating that occurs as H$^+$ flows up and out of the topside ionosphere. At high altitudes, the H$^+$ temperature profiles are isothermal owing to the dominance of thermal conduction.

   2.1.5   Effect of changes in N_mF_2

   The effect on the H$^+$ density, drift velocity, and temperature profiles of a change in the F_2-peak O$^+$ density is shown in Figs. 10a–c, respectively. These profiles were calculated for an upper boundary H$^+$ outflow velocity of 10 km/sec, a 50 mVm^{-1} convection electric field, and N_mF_2 values of 2.1×10^5, 7.0×10^4, and 2.1×10^4 cm^{-3}.

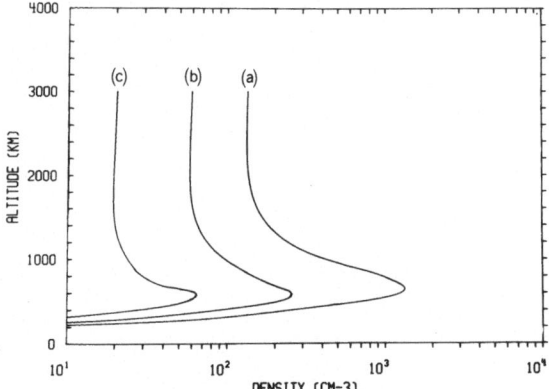

Fig. 10a.   H$^+$ density profiles as a function of altitude for a perpendicular electric field of 50 mVm^{-1}, a 10 km/sec H$^+$ outflow velocity at 3,000 km, and a range of F-region peak O$^+$ densities: (a) 2.1 × 10^5, (b) 7.0 × 10^4, and (c) 2.1 × 10^4 cm^{-3} (from RAITT *et al.* (1977)).

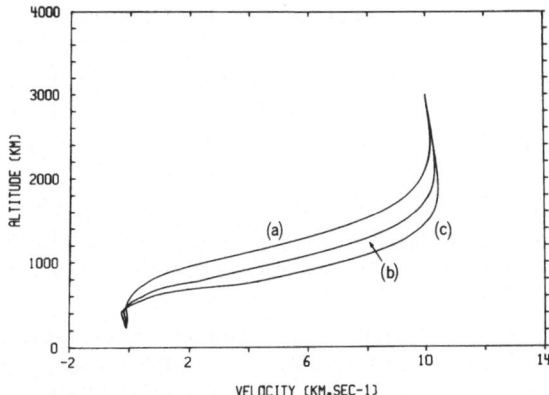

Fig. 10b.   H$^+$ drift velocity profiles as a function of altitude. Curves (a)–(c) correspond to curves (a)–(c) in Fig. 10a (from RAITT *et al.* (1977)).

A decrease in the O$^+$ density results in a decrease in the H$^+$ density and an increase in the H$^+$ outflow velocity at all altitudes. The increased H$^+$ velocities simply result from the reduced resistance that is associated with the lower O$^+$ densities. The reduced H$^+$ densities, on the other hand, result from both an increased H$^+$ escape velocity and from a reduction in H$^+$ production from the charge exchange reaction O$^+$ + H → H$^+$ + O. The effect of a reduction in the F_2-peak O$^+$ density is more complicated for the H$^+$ temperature than for the density and drift velocity. Above 1,000 km, a reduced N_mF_2 produces lower H$^+$ temperatures, which results from a decreased H$^+$-O$^+$ frictional heating rate. Below 1,000 km, the more complex behavior is a result of more heating and cooling terms becoming important in the H$^+$ energy balance (cf. RAITT *et al.*, 1977). In particular, there is a large change between case (a) and case (c) at the altitude where thermal convection switches from being a heat source to a heat sink.

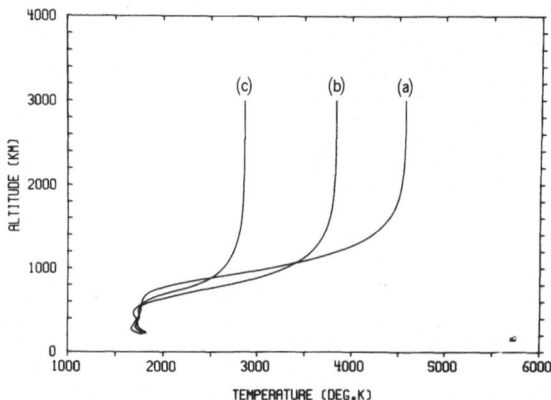

Fig. 10c. H$^+$ temperature profiles as a function of altitude. Curves (a)–(c) correspond to curves (a)–(c) in Figs. 10a and 10b (from RAITT *et al.* (1977)).

2.1.6   Summary of H$^+$ outflow parameters

The variation of the H$^+$ temperature at the upper boundary with both the convection electric field and the H$^+$ outflow velocity is shown in Table 1. The general trend is for an increase in $T(H^+)$ with increasing electric field and a decrease in $T(H^+)$ with increasing outflow velocity. This latter behavior is a result of the H$^+$ flow being supersonic for 10 and 20 km/sec, causing a reduction in frictional heating by the velocity dependence of the H$^+$-O$^+$ collision frequency, and a reduction in thermal convective heating owing to the nonlinear acceleration term, which acts to reduce the velocity gradient along the magnetic field. Note that the maximum H$^+$ temperature predicted for the range of electric fields and outflow velocities considered is about 6,500°K.

Table 2 shows the H$^+$ escape flux at the upper boundary as a function of convection electric field and upper boundary H$^+$ outflow velocity. For a given electric field, the H$^+$ escape flux remains essentially constant as the H$^+$ outflow velocity is increased, which indicates that the H$^+$ flow was in a flux limited condition for each of the outflow velocities considered. The magnitude of the limited H$^+$ flux does, however, increase with increasing convection electric field. The increased outward flux results

Table 1.   The variation in the upper boundary H$^+$ temperature in °K as a function of perpendicular electric field and upper boundary H$^+$ outflow velocity. The peak O$^+$ density is 2.1×10^5 cm^{-3}.

$E_\perp$ mV/m	$W_\parallel$ (H$^+$) km/sec		
	5	10	20
0	3,775	3,617	3,050
25	4,018	3,880	3,345
50	4,632	4,583	4,022
100	5,991	6,428	5,936

from the increased H^+ production from the reaction $O^+ + H \rightleftharpoons H^+ + O$, which in turn is a consequence of the elevated O^+ temperature and enhanced O^+ column density that is associated with an enhanced electric field.

Table 3 shows the variation of the upper boundary H^+ escape flux as a function of convection electric field and peak O^+ density. The H^+ escape flux increases with increasing convection electric field for each peak O^+ density shown, but the relative increase between the 0 and $100\ mVm^{-1}$ electric fields is less for the lower O^+ peak density cases. For a given electric field, the H^+ escape flux also increases with increasing O^+ peak density, with the relative increase being greater for the higher electric field.

2.2 He^+ outflow

RAITT et al. (1978a) have made extensive calculations of the characteristics of He^+ outflow from the high-latitude topside ionosphere. Steady state solutions to the hydrodynamic continuity, momentum and energy equations were obtained self-consistently, yielding He^+ density, drift velocity, and temperature profiles over the altitude range from 200 to 2,000 km. To cover the wide variation expected at high latitudes, several parameters were varied, including the He^+ and H^+ outflow velocities, the F-region peak electron density, the convection electric field, and the He

Table 2. The variation of the upper boundary H^+ outward flux in $cm^{-2}\ sec^{-1}$ as a function of perpendicular electric field and upper boundary H^+ outflow velocity. The peak O^+ density is $2.1 \times 10^5\ cm^{-3}$.

$E_\perp$ mV/m	$W_\parallel (H^+)$ km/sec		
	5	10	20
0	8.03,7	8.25,7	8.30,7
25	9.28,7	9.52,7	9.68,7
50	1.31,8	1.34,8	1.36,8
100	2.66,8	2.72,8	2.76,8

Note: $8.03,7 \equiv 8.03 \times 10^7$ etc.

Table 3. The variation of the upper boundary H^+ outward flux in $cm^{-2}\ sec^{-1}$ as a function of perpendicular electric field and peak O^+ density. The upper boundary H^+ outflow velocity is 10 km/sec.

$E_\perp$ mV/m	Peak $n(O^+)\ (cm^{-3})$		
	2.1 ,5	7.0 ,4	2.1 ,4
0	8.25,7	4.34,7	1.68,7
25	9.52,7	4.72,7	1.67,7
50	1.34,8	5.95,7	2.02,7
100	2.72,8	1.01,8	3.16,7

Note: $8.25,7 \equiv 8.25 \times 10^7$ etc.

and N_2 neutral density profiles. However, for all cases He^+ was assumed to be a minor ion at all altitudes.

2.2.1 Effect of the convection electric field and He^+ outflow velocity

Figures 11a, 11b, and 11c show corresponding He^+ density, drift velocity, and temperature profiles for convection electric fields of 0, 50, and 100 mVm^{-1} and for upper boundary He^+ outflow velocities of 0.1, 0.5, and 2.5 km s^{-1}. For these calculations the O^+ density at the F_2 peak was 2.1×10^5 cm^{-3} and the H^+ outflow velocity at 3,000 km was 10 km s^{-1}.

The general characteristics of the He^+ density profiles are similar to those of H^+. There is a region below about 600 km where equilibrium between production and loss dominates, whereas at higher altitudes diffusion is more important, resulting in a peak He^+ density in the vicinity of 600 km. The higher He^+ outflow cases qualitatively resemble the flux limited profiles computed by BANKS and HOLZER (1969). For the 0 and 50 mVm^{-1} perpendicular electric fields, this peak is unaffected by the upper boundary He^+ outflow velocity. When the perpendicular electric field is 100 mVm^{-1} there is a slight reduction in the He^+ peak density as the He^+ upper boundary outflow velocity increases from 0.1 to 2.5 km s^{-1}.

The reason for this behavior is a consequence of the increasing ion temperature as the perpendicular electric field increases. The higher ion temperature causes a decreased Mach number for a given outflow at the upper boundary. The result is that the profiles become nearer to diffusive equilibrium profiles above the peak, this effect being particularly noticeable in the most extreme case of $E_\perp = 100$ mVm^{-1} and 0.1 km s^{-1} He^+ outflow velocity (Fig. 11a, bottom panel). For this case the increased peak density and much higher upper boundary density results from the He^+ outward flux being below the saturated flux of 3.8×10^6 cm^{-2}s^{-1} at 3.2×10^6 cm^{-2}s^{-1}. This again is due to the high ion temperature and the inhibiting effect on the He^+ diffusion coefficient of the higher H^+ and O^+ densities for $E_\perp = 100$ mVm^{-1}.

With regard to the He^+ drift velocity profiles, the zero perpendicular electric field case is similar in characteristics to the H^+ velocity profiles in that the ion velocity parallel to B is negative at low altitudes, becomes positive, then increases to reach a peak value before falling back to the upper boundary value. A notable difference between He^+ and H^+ in the detail of the profiles is that the rapid increase in the He^+ outflow velocity occurs at a higher altitude than for H^+; the He^+ increase occurring at about 1,300 km while the H^+ increase occurs around 800 km. This difference is primarily due to the smaller diffusion coefficient for He^+, resulting in production and loss being more important to higher altitudes for He^+ than for H^+.

In comparing the velocity profiles for increasing perpendicular electric fields it can be seen that at higher altitudes, above about 1,000 km, there is a decrease in the He^+ outflow velocity. This is a result of the flux limited condition for all but the extreme case of $E_\perp = 100$ mVm^{-1} and 0.1 km s^{-1} He^+ outflow boundary velocity. The increased scale height results in increased number densities at these altitudes and since the outward flux $(N(He^+)W_\parallel(He^+))$ is constant $W_\parallel(He^+)$ must be lower.

As far as the He^+ temperature profiles are concerned, the zero electric field case shows that below about 1,300 km the boundary He^+ outflow velocity has very little effect on the He^+ temperature and the profile is similar to the O^+ temperature profile.

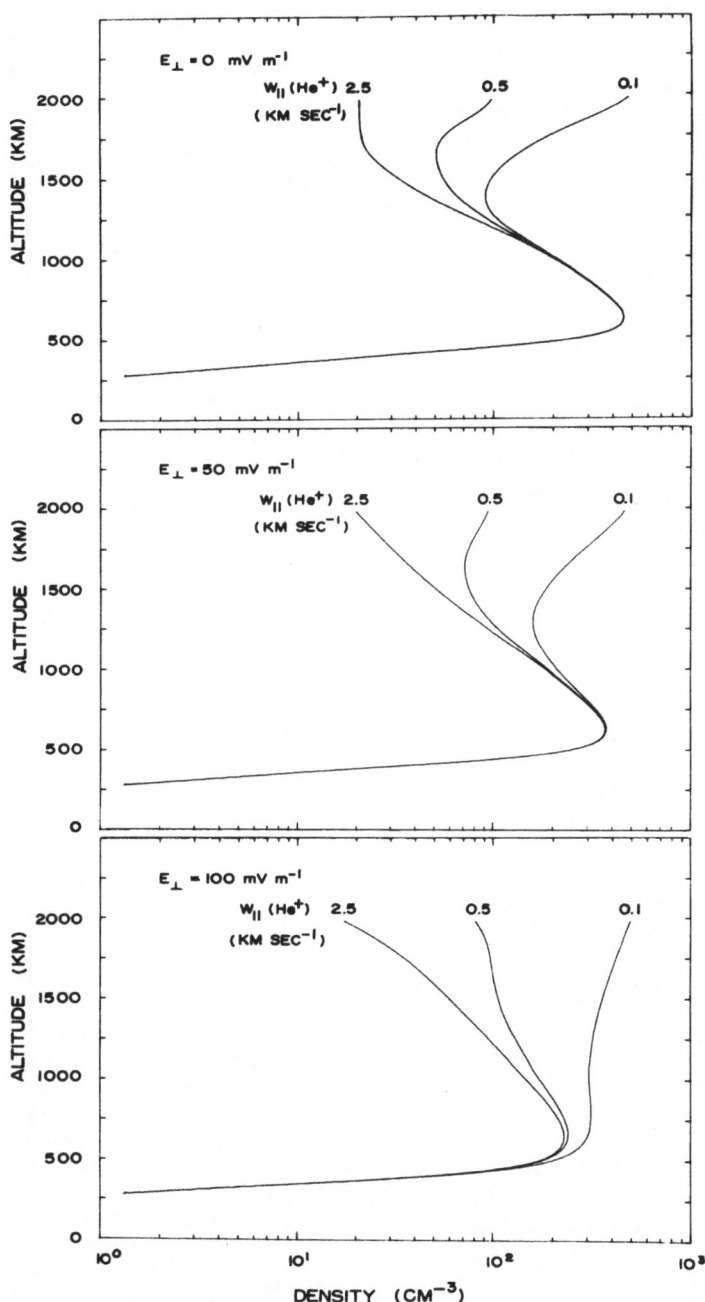

Fig. 11a. He$^+$ density profiles as a function of altitude for He$^+$ upper boundary outflow velocities of 0.1, 0.5, and 2.5 km s^{-1}. The upper, middle and lower panels show the profiles for convection electric fields of 0, 50, and 100 mVm^{-1}, respectively (from RAITT *et al.* (1978a)).

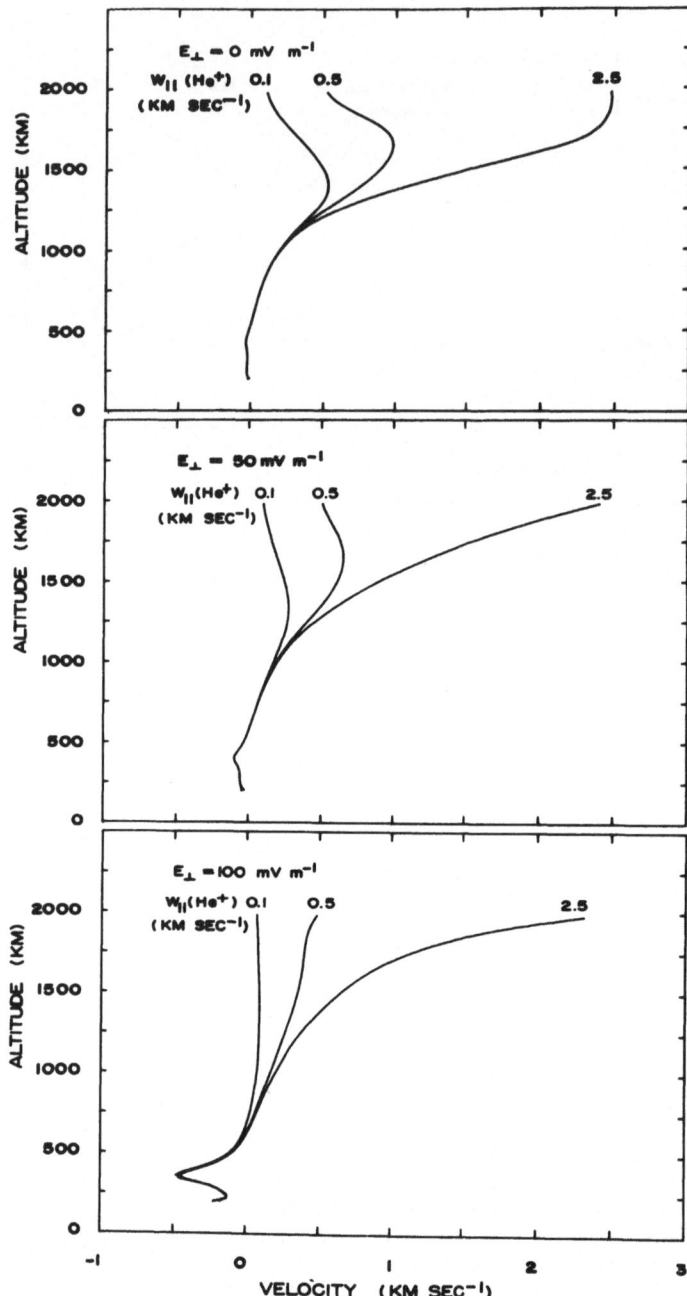

Fig. 11b.   He$^+$ drift velocity profiles as a function of altitude for He$^+$ upper boundary velocities of 0.1, 0.5, and 2.5 km s^{-1}. The upper, middle and lower panels correspond to convection electric fields of 0, 50, and 100 mVm^{-1}, respectively (from RAITT et al. (1978a)).

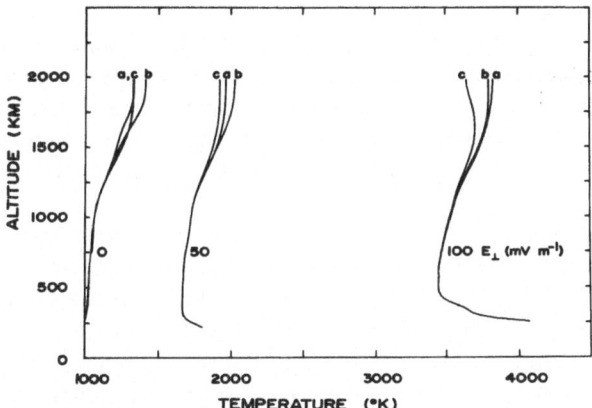

Fig. 11c. He^+ temperature profiles as a function of altitude for He^+ upper boundary velocities of : (a) 0.1, (b) 0.5, and (c) 2.5 km s^{-1}. Each family of curves corresponds to a different convection electric field, as indicated by the labels (from RAITT *et al.* (1978a)).

Above 1,300 km a more rapid increase in velocity occurs, which results in flow terms such as advection, convection, thermal conduction, and Joule heating by collisions with O^+ being more important in the energy balance. These terms do not, however, dominate the energy balance and coupling to O^+ remains an important term up to the top boundary. This prevents a large variation of He^+ temperature with He^+ outflow velocity in contrast to the marked affect for H^+ ions. The main reason for the small effect of frictional heating is that the He^+ velocity shows a rapid increase only above 1,300 km, while for H^+ the rapid increase in outflow velocity begins at about 800 km. At 1,300 km the O^+ density is typically a factor of 5 lower than at 800 km and the Joule heating is consequently reduced. It should be noted, however, that the behavior of the Joule heating agrees with that observed for H^+ in that the subsonic-transonic cases of 0.1 and 0.5 km s^{-1} outflow velocity show an increase in upper boundary temperature, while the supersonic outflow case of 2.5 km s^{-1} shows a reduction. This can be attributed to the same reason as that for H^+ temperatures, a reduction in the velocity dependent collision frequency as the Mach number exceeds unity.

The groupings of the families of He^+ temperature profiles for different He^+ outflow velocities for each of the three perpendicular electric fields shows that this latter parameter has the strongest influence on the He^+ temperature. The increase in frictional heating of He^+ ions near the lower boundary by collisions with the neutral gas as $E_\perp$ increases can be seen. This frictional heating rapidly declines with altitude as the neutral gas density decreases until the strong coupling to the O^+ ions takes over. Like the zero perpendicular electric field case, the effect of the He^+ outflow is minimal until about 1,300 km when the flow, conduction, and He^+-O^+ frictional heating terms become significant.

2.2.2 *Solar cycle, seasonal, and geomagnetic activity variations*

Motivated by the measurements of HOFFMAN and DODSON (1980), which indicated that the He^+ escape flux exhibits a large seasonal variation, RAITT *et al.* (1978b) did a detailed quantitative study of the solar cycle, seasonal, and geomagnetic activity variations of the limiting He^+ escape flux. These authors found that the limiting He^+ escape flux is not sensitive to the convection electric field, the O^+ peak density, or the H^+ outflow velocity, and that the large variation in escape flux is related to the variation of the neutral atmosphere, which affects the He^+ production and loss rates. Helium ions are produced by photoionization of neutral He and lost primarily in reactions with N_2.

Figure 12 shows He^+ density profiles as a function of altitude for solar maximum and solar minimum, for summer and winter, and for high and low geomagnetic activity. For these calculations, the He^+ outflow velocity at the upper boundary was set equal to 0.5 km s^{-1}, which produced a saturated He^+ escape flux for all cases. It is apparent that the seasonal variation of the He^+ density is the largest, but there is also a significant variation with geomagnetic activity. There is very little difference between solar maximum and solar minimum as far as the peak He^+ densities are concerned. The main solar cycle effect is the enhanced scale height above the peak at solar maximum, which is a consequence of the generally higher neutral temperatures at solar maximum. The variation of the He^+ density shown in Fig. 12 closely follows the variation of the neutral He density below 600 km, as pedicted by the MSIS atmospheric model. Therefore, the enhanced He^+ densities in winter are a consequence of the 'winter helium bulge.'

The limiting He^+ escape fluxes corresponding to the He^+ density profiles shown in Fig. 12 are given in Table 4. The He^+ escape flux shows a variation of about 2 orders of magnitude over the range of geophysical conditions studied, which encompassed extreme solar cycle, seasonal, and geomagnetic activity variations. The He^+ escape flux varies from a low value of 0.99×10^5 cm^{-2}s^{-1} for solar minimum, summer, high magnetic activity conditions to a maximum of $1-2 \times 10^7$ cm^{-2}s^{-1} for solar maximum, winter, low magnetic activity conditions. The ratios between winter and summer He^+ escape fluxes vary from 32 to 36 for high magnetic activity and from 23 to 25 for low magnetic activity. The He^+ escape fluxes during solar maximum are $1.5-2.0$ times greater than those for solar minimum.

2.2.3   Effect of topside He^+ heat flux

Although He^+ probably exists in a state of outflow at high latitudes most of the time, there are occasions when He^+ can become the dominant ion in the polar cap at altitudes above about 600 km (SHEPHERD et al., 1976). To study this situation, OTTLEY

Table 4.   Limiting He^+ escape fluxes (cm^{-2} sec^{-1}). The labels high and low correspond to high and low geomagnetic activity, respectively.

	Solar maximum		Solar minimum	
	Summer	Winter	Summer	Winter
High	1.61(5)	5.07(6)	0.99(5)	3.54(6)
Low	5.97(5)	1.35(7)	3.02(5)	7.40(6)

Note: $1.61(5) \equiv 1.61 \times 10^5$ etc.

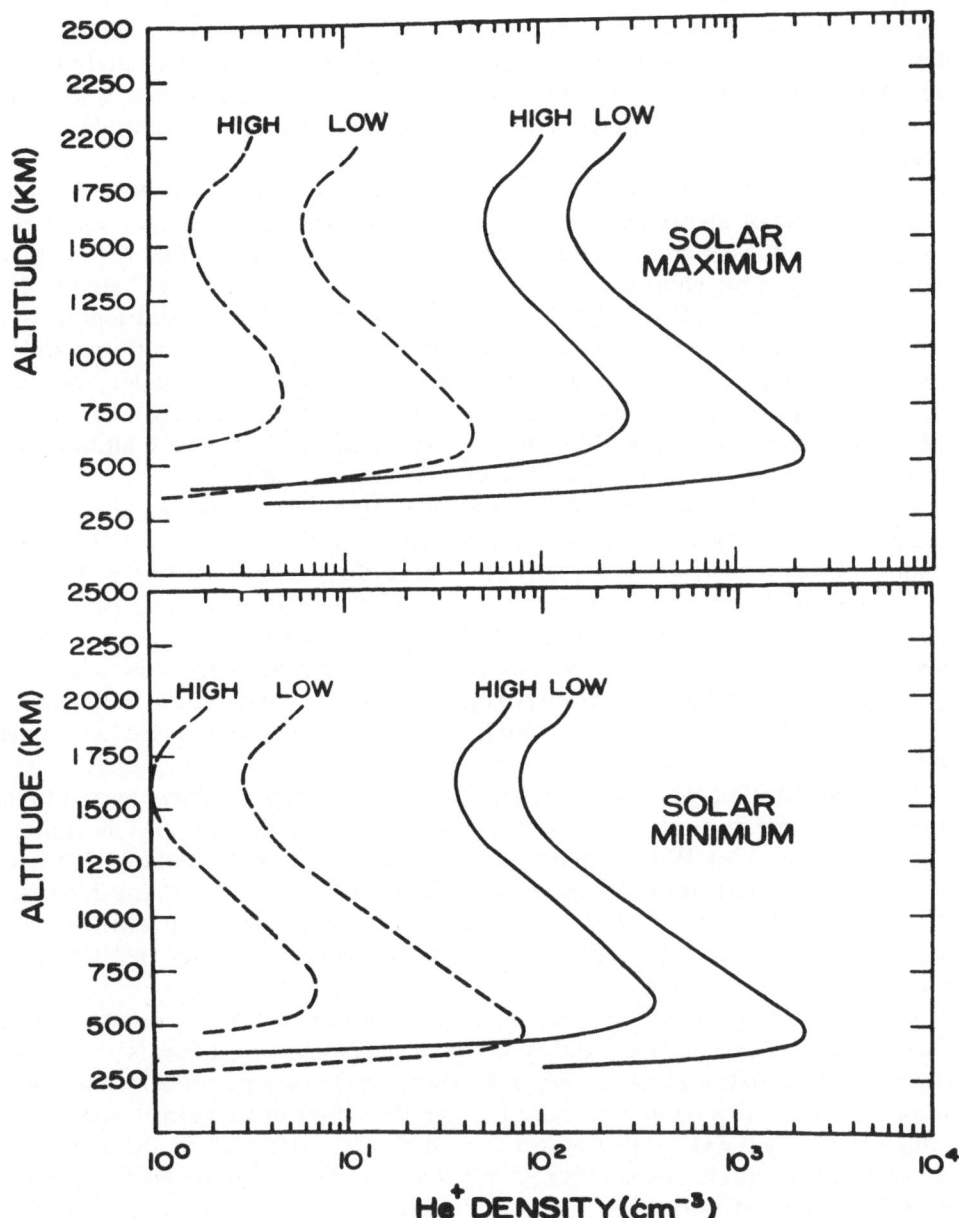

Fig. 12.   He$^+$ density profiles as a function of altitude for solar maximum and solar minimum, for summer (dashed curves) and winter (solid curves), and for high and low geomagnetic activity. For all cases the He$^+$ outflow velocity at the upper boundary was set equal to 0.5 km s^{-1} (from RAITT *et al.* (1978b)).

and SCHUNK (1980) calculated the density and temperature structure of He$^+$ for subsonic outflows, which result in He$^+$ being a major ion or an important minor ion. In addition to He$^+$ outflow velocity, they studied the effect of convection electric fields, different electron temperature distributions, and different assumed He$^+$ and O$^+$ heat fluxes at high altitudes. The latter effect had not been considered in previous studies of the high-latitude topside ionosphere, but it could be important if high-altitude ion heat sources exist.

One of the interesting results to emerge from the OTTLEY and SCHUNK (1980) study was that because of thermal diffusion the change in the 'diffusive equilibrium' He$^+$ density profile with an increase in the topside He$^+$ heat flux is very similar to the change obtained for 'dynamic equilibrium' with an increase in the topside He$^+$ escape flux. This is shown in Figs. 13a and 13b, where He$^+$ and O$^+$ density and temperature profiles are plotted as a function of altitude for diffusive equilibrium conditions and for three upper boundary He$^+$ heat fluxes. The top panels correspond to no He$^+$ heat flux at 2,500 km, while for the middle and bottom panels a downward He$^+$ heat flux at 2,500 km was selected such that the He$^+$/O$^+$ temperature ratio at high altitudes was approximately 2 for the middle panels and 3 for the bottom panels.

The variation of the ion and electron temperatures with an increasing topside He$^+$ heat flux is straightforward. Without a topside He$^+$ heat flux, $T(\text{He}^+) \approx T(\text{O}^+)$ at all altitudes. As the downward topside He$^+$ heat flux is increased, the He$^+$/O$^+$ temperature ratio at most altitudes increases, with the exact increase depending on altitude and the magnitude of the topside He$^+$ heat flux. Associated with this increased temperature ratio is a slight increase in $T(\text{O}^+)$ at high altitudes from about 2,000°K (top panel) to about 2,500°K (bottom panel). The electron temperature profile, on the other hand, exhibits a negligibly small change even though $T(\text{He}^+)$ changes significantly.

The small effect of He$^+$ on the O$^+$ and electron temperatures is due in part to the fact that He$^+$ remains a minor ion to higher altitudes as the topside He$^+$ heat flux is increased (see Fig. 13a). Without a topside He$^+$ heat flux (top panel) the He$^+$ density increases with altitude until it becomes the major ion at about 1,200 km. However, with a topside He$^+$ heat flux the He$^+$ density first increases with altitude, then decreases, and then becomes nearly constant; the net effect being that He$^+$ remains a minor ion to higher altitudes.

The behavior of the He$^+$ and O$^+$ densities shown in Fig. 13a is due to thermal diffusion, which is the additional diffusion that results when a heat flow is present in a plasma. For the situation shown in Fig. 13b, where the He$^+$ temperature increases with altitude, thermal diffusion acts to drive the light He$^+$ ions downward toward cooler regions and the heavy O$^+$ ions upward toward hotter regions (cf. ST.-MAURICE and SCHUNK, 1977; CONRAD and SCHUNK, 1979). As the upper boundary He$^+$ heat flux is increased, thermal diffusion becomes more effective in driving He$^+$ downward and O$^+$ upward, which accounts for the large increase in the He$^+$/O$^+$ transition altitude shown in Fig. 13a.

2.3   Collisionless polar wind

The polar wind results that have been discussed in the previous subsections are

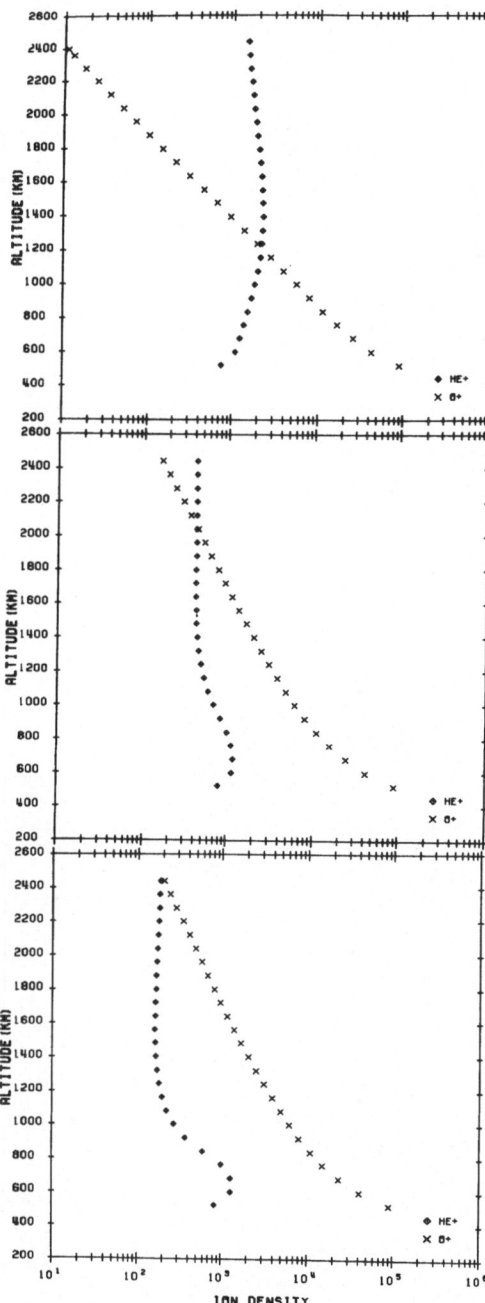

Fig. 13a. Ion density profiles as a function of altitude for no He$^+$ outflow and three He$^+$ heat fluxes at the upper boundary. The upper boundary He$^+$ heat fluxes in units of eV cm^{-2} s^{-1} are 0 (top), -4.69×10^7 (middle), and -5.63×10^7 (bottom) (from OTTLEY and SCHUNK (1980)).

W. J. RAITT and R. W. SCHUNK

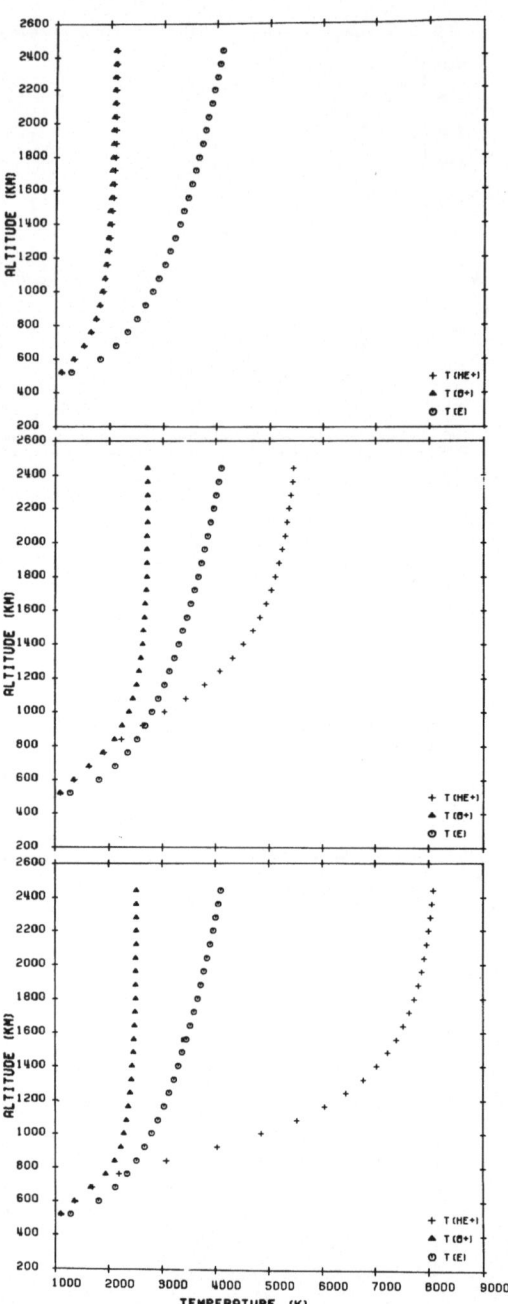

Fig. 13b. Ion and electron temperature profiles as a function of altitude for no He⁺ outflow and three He⁺ heat fluxes at the upper boundary. The upper boundary He⁺ heat fluxes in units of eV cm⁻² s⁻¹ are 0 (top), -4.69×10^7 (middle), and -5.63×10^7 (bottom) (from OTTLEY and SCHUNK (1980)).

valid at the altitudes where the H^+ and He^+ gases are collision-dominated. As a rough guide, the ion gases are effectively collision-dominated when

$$U_i/H_i v_i \ll 1$$

where U_i is the ion field-aligned drift velocity, H_i is the ion density scale height, and v_i is the appropriate ion collision frequency. For H^+, this condition generally begins to break down at 1,000 km and is clearly violated at 2,000 km (RAITT et al., 1975). When the plasma is not collision-dominated, the H^+ pressure distribution becomes anisotropic and the H^+ heat flow vector is not simply related to the gradient in the H^+ temperature.

As noted earlier, the collisionless characteristics of the polar wind can be described by kinetic, hydromagnetic and generalized transport models. For supersonic flow, these models produce density and drift velocity profiles that are similar to those obtained from the hydrodynamic equations. However, the ion temperature distributions are different, with the collisionless models yielding large temperature anisotropies at high altitudes. Typical results are shown in Fig. 14, where the H^+ and O^+ temperatures parallel and perpendicular to the geomagnetic field are plotted as a function of altitude for collisionless, supersonic H^+ outflow. The ion temperature distributions were calculated with both kinetic and hydromagnetic models and the

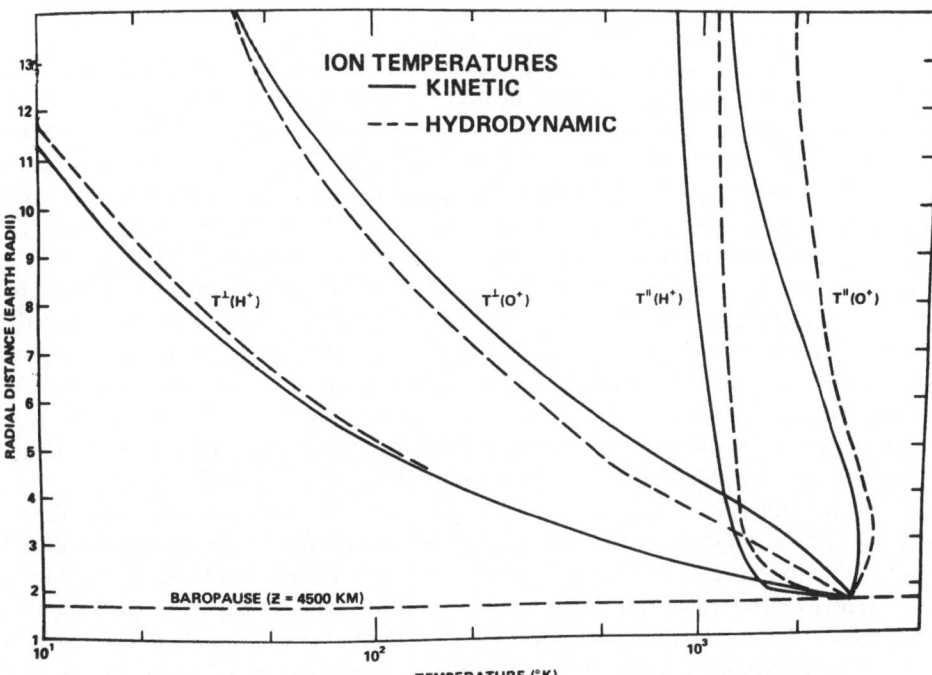

Fig. 14. O^+ and H^+ temperatures parallel and perpendicular to the geomagnetic field obtained from kinetic (solid curves) and hydrodynamic (dashed curves) models of the collisionless, supersonic polar wind (from HOLZER et al. (1971)).

results are similar. The parallel ion temperatures are essentially constant with altitude at high altitudes, while the perpendicular ion temperatures decrease monotonically with altitude. The net result is a parallel-to-perpendicular temperature anisotropy that grows with altitude, reaching nearly a factor of 50 for H^+ ions at a distance of 10 earth radii.

SCHUNK and WATKINS (1981, 1982) used generalized transport equations and obtained polar wind solutions that were continuous through the transition from the collision-dominated to the collisionless regimes. Both subsonic and supersonic H^+ outflows were considered. The dramatic result to emerge from these studies was that the character of the solution to the generalized transport equations was different for subsonic and supersonic H^+ outflows. Figure 15 shows the H^+ temperatures parallel and perpendicular to the geomagnetic field for supersonic H^+ outflow and for both cold (top panel) and hot (bottom panel) electron temperature distributions. For both cases, there is an appreciable H^+ temperature anisotropy with $T_{p\parallel} > T_{p\perp}$ at all altitudes above 1,500 km. The perpendicular H^+ temperature displays a rapid decrease with altitude at low altitudes and then tends to go constant at high altitudes. The variation of $T_{p\parallel}$ with altitude, on the other hand, is more complicated. At low altitudes, $T_{p\parallel}$ displays a decrease with altitude that is similar to the $T_{p\perp}$ decrease. Above about 3,000 km, $T_{p\parallel}$ exhibits little variation with altitude all the way to the top boundary for the low electron temperature case. However, for the high electron temperature case, $T_{p\parallel}$ is roughly constant with altitude between 3,000 and 5,000 km and then decreases with altitude above this altitude range.

The behavior of the parallel and perpendicular H^+ temperatures shown in the top panel of Fig. 15 for the cold T_e distribution is in good qualitative agreement with that obtained from the kinetic and hydromagnetic models (see Fig. 14). The main qualitative difference is that the generalized transport equations produce an H^+ temperature anisotropy that tends to go constant at high altitudes, whereas the anisotropy obtained from the kinetic and hydromagnetic models does not display this tendency. Also, solutions to the generalized transport equations for supersonic H^+ outflow can be obtained only with an upward H^+ heat flow from the lower ionosphere.

For subsonic H^+ outflow, it is possible to obtain solutions to the generalized transport equations with a downward H^+ heat flow, which implies that a high altitude heat source exists above the altitude range of interest. Figure 16 shows that for subsonic H^+ outflow with a downward heat flow the variation of the parallel and perpendicular H^+ temperatures is different from that found for supersonic outflow. For subsonic outflow the anisotropy is opposite to that found for supersonic outflow, with $T_{p\parallel} > T_{p\perp}$ below about 3,500 km and $T_{p\perp} > T_{p\parallel}$ above this altitude. Also, in contrast to what was found for supersonic outflow, for this case the parallel and perpendicular H^+ temperatures increase with altitude above 3,500 km, owing to the high-altitude heat source. However, as was found for supersonic outflow, the H^+ temperature anisotropy tends to go constant at high altitudes, i.e., the temperature anisotropy is regulated.

The basic difference between the subsonic and supersonic results can be traced to the increased importance of heat flow in relation to convection for the subsonic case. For subsonic H^+ outflow, the downward heat flow acts as a source of energy for the H^+ gas and is the largest term in the energy balance. Below 3,500 km this source competes

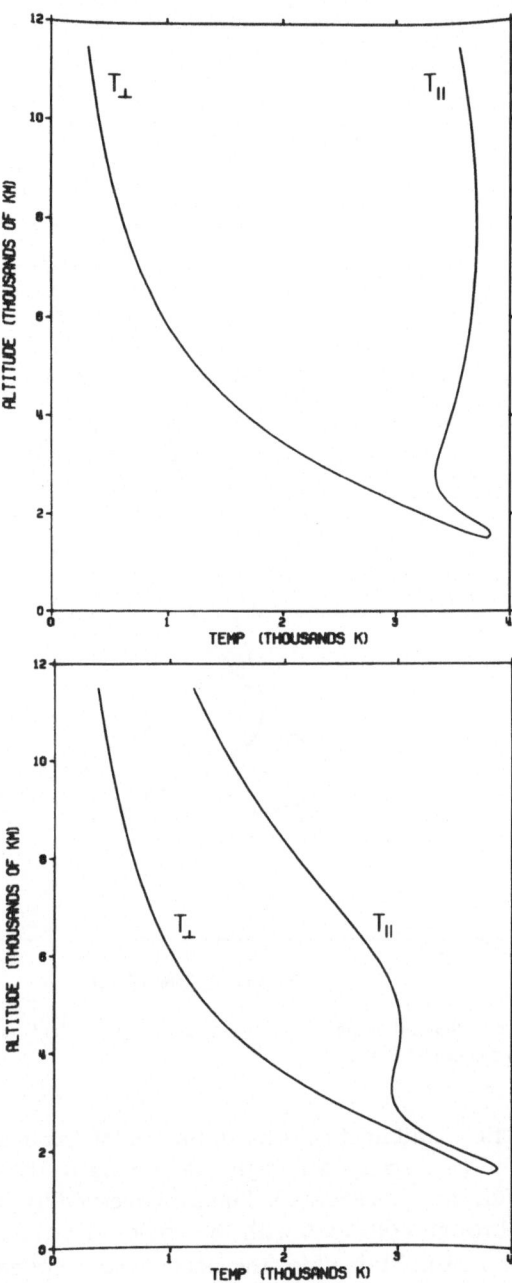

Fig. 15. H$^+$ temperatures parallel and perpendicular to the geomagnetic field versus altitude for supersonic flow and for both cold (top panel) and hot (bottom panel) electron temperature distributions (from SCHUNK and WATKINS (1982)).

W. J. RAITT and R. W. SCHUNK

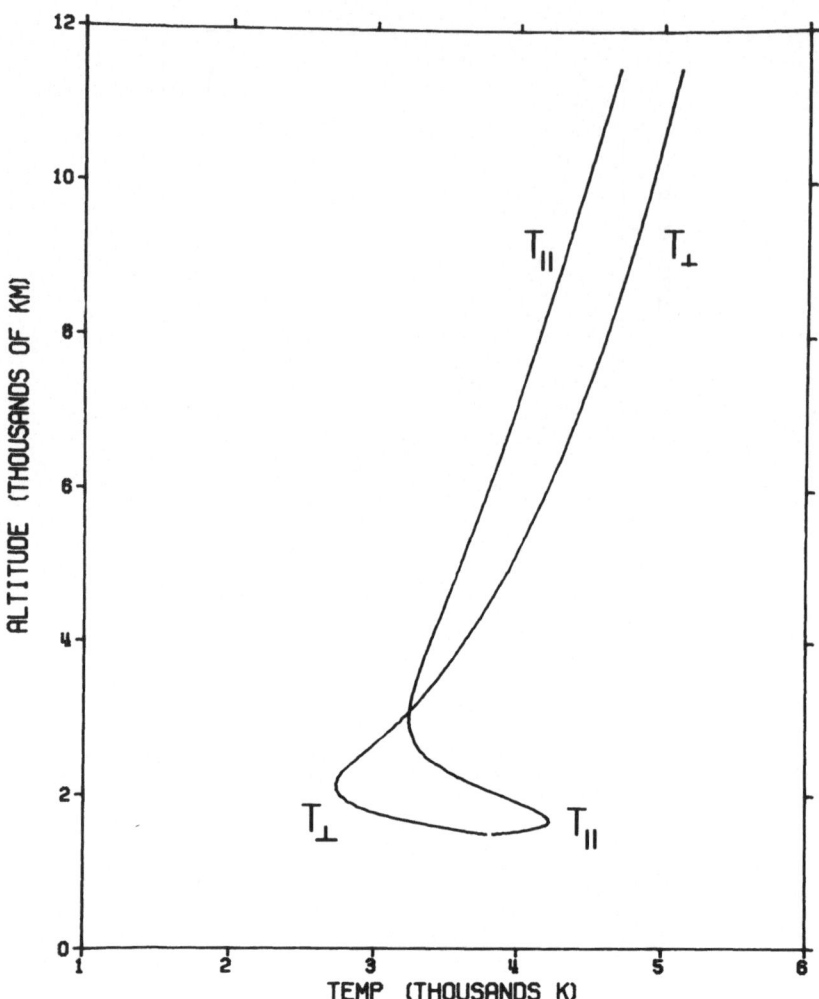

Fig. 16. H$^+$ temperatures parallel and perpendicular to the geomagnetic field versus altitude for subsonic flow (from SCHUNK and WATKINS (1982)).

with advection, convection, and heat transfer to the colder electrons. However, above 3,500 km, advection and convection are negligible, owing to the decrease of the H$^+$ drift velocity with altitude, and the energy balance is governed by energy input via heat flow and energy loss through collisions with the colder electrons.

Different results are also obtained for the electron gas depending on whether the H$^+$ outflow is subsonic or supersonic. For subsonic H$^+$ outflow, the electron gas remains collision dominated to high altitudes; that is, the electron temperature anisotropy is small, the electron temperatures are low, the electron temperature profile is isothermal at high altitudes, and the electron heat flow is given by the classical collision-dominated expression at all altitudes. For supersonic H$^+$ outflow, on the other hand, the electron gas becomes collisionless above about 3,000 km. In the collisionless regime an electron temperature anisotropy develops such that the electron temperature perpendicular to the geomagnetic field, $T_{e\perp}$, is greater than the temperature parallel to the geomagnetic field, $T_{e\parallel}$. This anisotropy increases with altitude, and at 12,000 km $T_{e\perp}/T_{e\parallel} \sim 2$. Also, for supersonic H^+ outflow the magnitude of the downward electron heat flux at 1,500 km has a dramatic effect on the individual parallel and perpendicular electron temperatures. For small lower-boundary electron heat fluxes, $T_{e\parallel}$ and $T_{e\perp}$ are low at all altitudes ($T_{e\perp} < 7,000°K$), but for large electron heat fluxes at the lower boundary, $T_{e\perp}$ can exceed 100,000°K at 12,000 km. This behavior is shown in Fig. 17a, where the electron temperatures parallel and perpendicular to the geomagnetic field are plotted as a function of altitude for supersonic H^+ outflow and for lower boundary electron temperature gradients of 0.1° (top panel), 1° (middle panel) and 2°K km^{-1} (bottom panel).

The electron heat flow profiles that are associated with the three cases shown in Fig. 17a are given in Fig. 17b. An increase in the boundary T_e gradient results in an increase in the magnitude of the downward heat flow at all altitudes. A study of the balance of the electron heat flow equation indicates that below about 2,000 km the balance is determined by the usual collision-dominated terms. However, above 4,700 km for the $\nabla T_e = 0.1°K$ km^{-1} case and above 3,000 km for the $\nabla T_e = 1°$ and 2°K km^{-1} cases the standard collision-dominated heat flow expression is not valid. Above these altitudes the electron temperature gradient is not balanced by the collision term, as in the case of collision dominance, but instead is balanced by an electron temperature anisotropy term. In essence, the electron heat flow is collisionless above these altitudes, and hence it is possible to obtain altitude profiles of q_e that are completely different from those predicted by the standard collision-dominated heat flow expression. Note that for the $\nabla T_e = 1°K$ km^{-1} boundary condition the electron heat flow profile is essentially constant with altitude above about 3,000 km.

3. Experimental Observations

Published data on the outflow of plasma from the earth's ionosphere fall into two general categories; direct measurements of plasma flow parallel to the geomagnetic field, and measurements of ionospheric plasma parameters indirectly related to the effects of outflow. Direct measurements of the outflow velocity are particularly sparse, but even the published, indirect measurements related to light ion outflow are limited in effects of outflow. Direct measurements of the outflow velocity are particularly sparse, but even the published, in direct measurements related to light ion outflow are limited in directly related to the polar wind, other than the existence of thermal and suprathermal plasma in the magnetospheric tail.

Direct measurements of the upward flow of light ions from the earth's polar ionosphere have been made using instruments to measure the offset of the flow direction of light ions from the direction opposite to a satellite velocity vector, due to a bulk motion of the ions parallel to the geomagnetic field. Since the satellite velocity is generally in the range 7 to 8 km/s and models indicate that the bulk upward flow can be

W. J. Raitt and R. W. Schunk

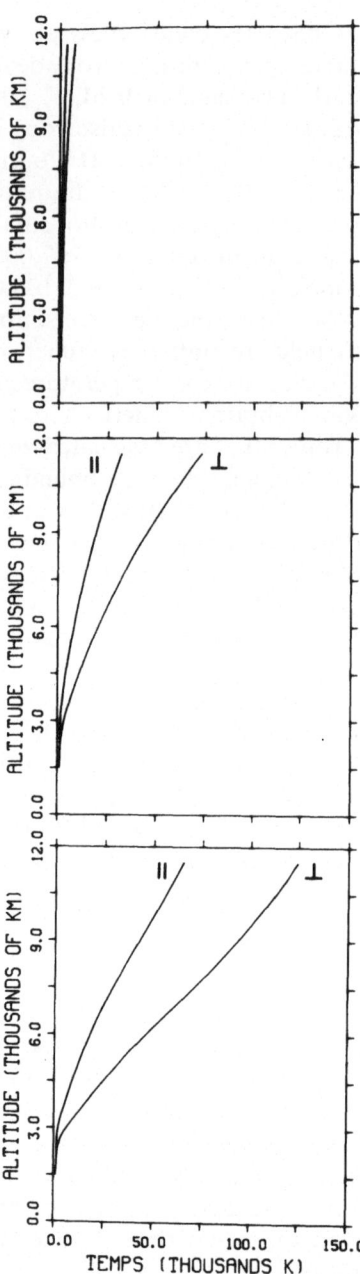

Fig. 17a. Electron temperatures parallel and perpendicular to the geomagnetic field versus altitude for supersonic flow and for three boundary electron temperature gradients. The boundary electron temperature gradients are 0.1° (top panel), 1° (middle panel), and 2°K km^{-1} (bottom panel) (from Schunk and Watkins (1981)).

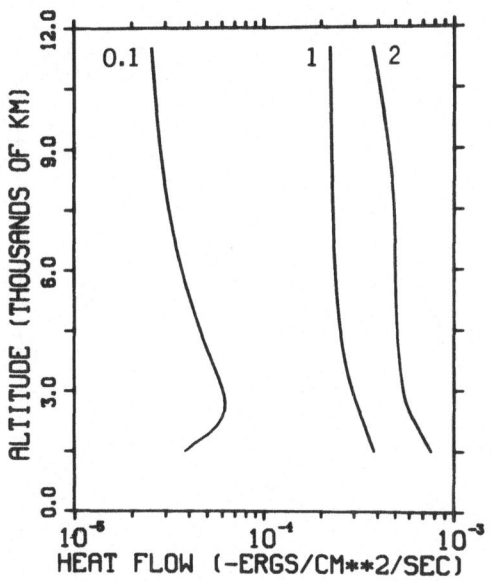

Fig. 17b. Electron heat flow profiles for supersonic flow and for three boundary electron temperature gradients. The numbers next to the curves correspond to the different boundary electron temperature gradients in degrees Kelvin per kilometer (from SCHUNK and WATKINS (1981)).

in the range 2 to 20 km/s, the resultant velocity vector in the frame of reference of the satellite should show a significant shift from the anti-ram direction.

The existence of the polar wind was first determined directly by HOFFMAN (1970) using data from the Explorer 31 satellite. He studied the phase shift between O^+ and H^+ ions detected by an ion mass spectrometer mounted with a radial view direction on a satellite orbiting in a cartwheel mode (spin axis normal to the orbit plane). The orbital characteristics of Explorer 31 enabled measurements to be made in the altitude range 500 to 3,000 km. The measurements showed the existence of H^+ ion flow velocities of up to 10 km/s parallel to the geomagnetic field and flux levels of the order of 10^8 cm^{-2}sec^{-1} above 2,500 km.

A similar ion mass spectrometer instrument to analyze ionospheric ions was flown on the Isis-2 satellite in a 1,400 km altitude circular, polar orbit. During the lifetime of the satellite, a large data base of observations over both the southern and northern polar regions was obtained. The roll modulation of the ion mass spectrometer signals for O^+, He^+ and H^+ was again analyzed for phase shift, details of the technique are described by HOFFMAN et al. (1974). A more refined technique for obtaining the relative positions of the O^+, He^+ and H^+ peaks during a roll period is described by HOFFMAN and DODSON (1980). Figure 18 shows an example of the roll modulation of the three ions O^+, He^+, and H^+ and the fitted curves defining the peak measured ion concentrations. Data of this type have been used to show the latitudinal dependence of ion concentration, light ion flux, and light ion outflow velocity for H^+ and He^+ for conditions of summer and winter solstice and equinox for geomagnetically quiet times

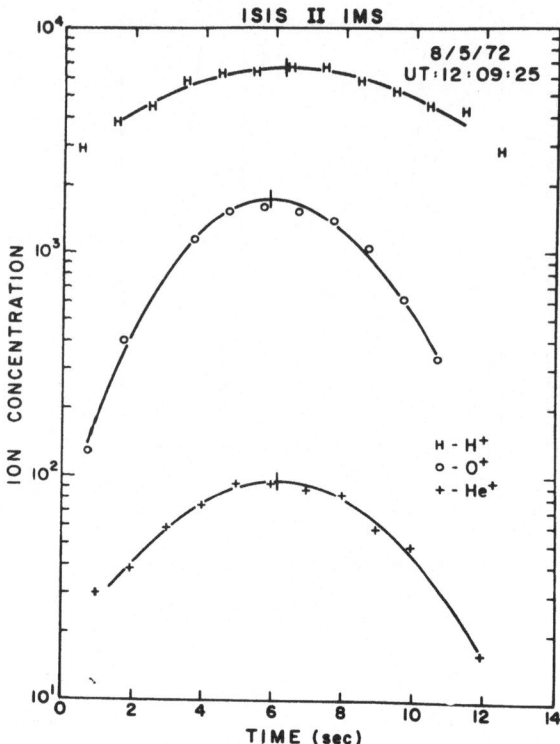

Fig. 18.   Ion concentrations as a function of time showing roll modulation and phase shifts of light ion species. A cubic equation is fitted to the top 10 points for each ion species (solid curve). The maximum of the curve (vertical dash) gives the ion concentration. Time difference between H$^+$ or He$^+$ and O$^+$ maxima relative to spacecraft spin period is a measure of light ion flow velocity (from HOFFMAN and DODSON (1980)).

(HOFFMAN and DODSON, 1980). Examples of the H$^+$ and He$^+$ outflow during winter solstice are shown in Fig. 19 (H$^+$) and Fig. 20 (He$^+$). In these figures, the correlation between the increase in outflow velocity and the decrease in light ion concentration can clearly be seen, as can the constant light ion outward flux, indicating flux limited flow conditions.

The measured winter solstice He$^+$ outward flux of 1.2×10^7 cm^{-2}sec^{-1} compares well with the values of 1.3×10^7 cm^{-2}sec^{-1} calculated by the model of RAITT et al. (1978a, 1978b) described earlier. For summer solstice conditions, the measurements show an outward flux of 10^6 cm^{-2}sec^{-1}, while the model value is 6×10^5 cm^{-2}sec^{-1}. The ratio of the winter to summer flux of 20, predicted by the model, is in close agreement with the density ratio of the neutral helium density of 20 to 30 due to the phenomenon known as the winter helium bulge. The similarity of these ratios is consistent with the source of He$^+$ ions being photoionization of neutral helium. The relative high summer flux of He$^+$ ions measured by HOFFMAN and DODSON (1980) is attributed to the measured velocity of He$^+$ being slightly higher in summer than in winter.

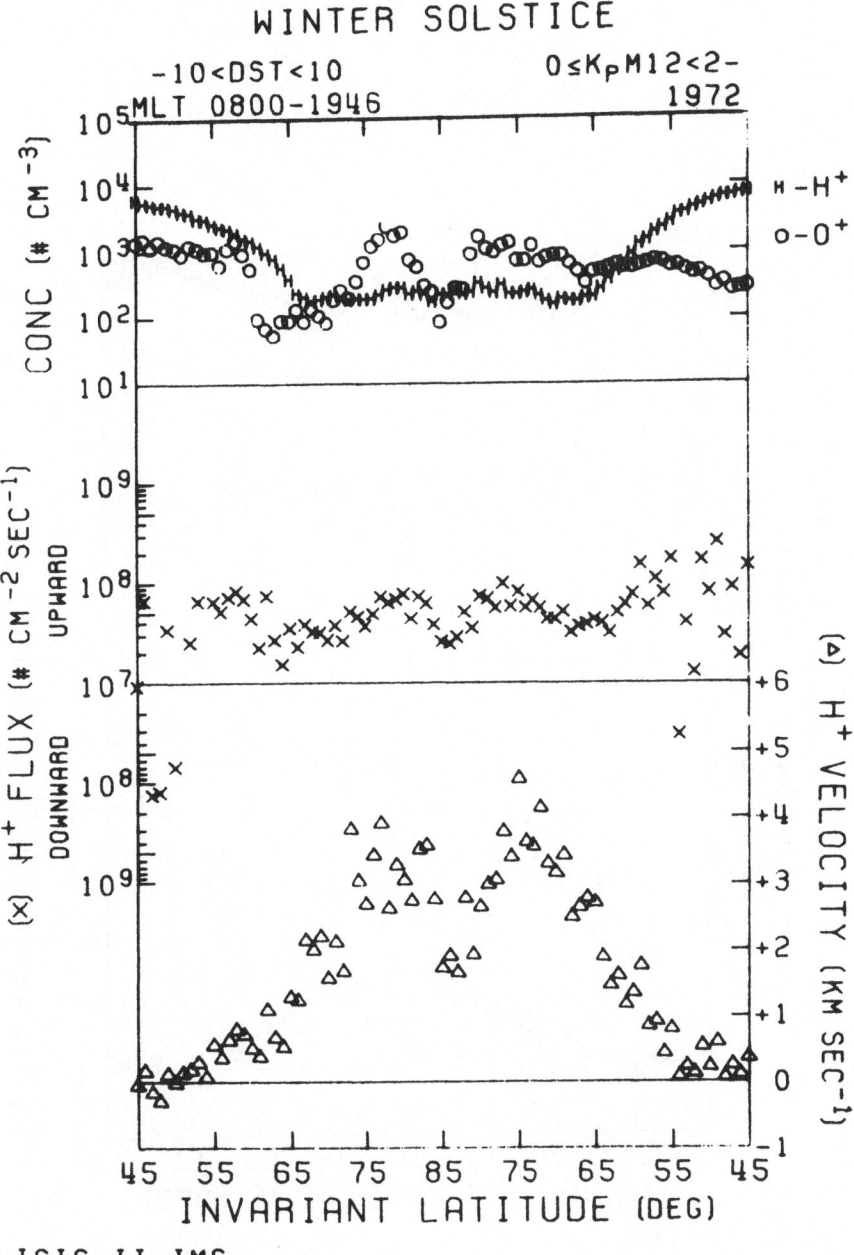

Fig. 19. H$^+$ and O$^+$ concentrations, H$^+$ flux, and H$^+$ flow velocity as a function of invariant latitude. Each point on the graph is an average of data from 5 to 30 satellite passes. Dusk is to the left, and dawn to the right. The time period over which the data were averaged is 4 weeks around winter solstice (from HOFFMAN and DODSON (1980)).

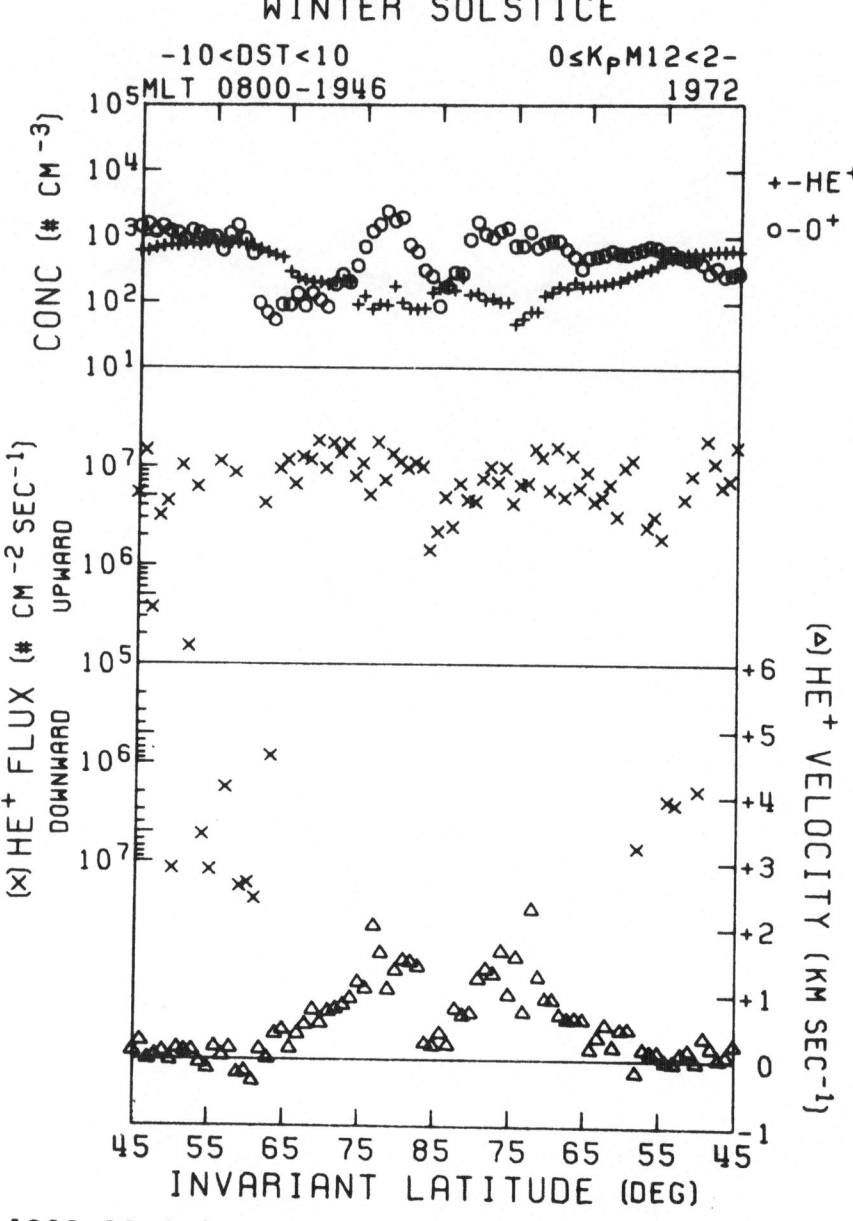

Fig. 20. He$^+$ and O$^+$ concentrations, He$^+$ flux and He$^+$ flow velocity as a function of invariant latitude. Each point on the graph is an average of data from 5 to 30 satellite passes. Dusk is to the left, and dawn to the right. The time period over which the data were averaged is 4 weeks around winter solstice (from HOFFMAN and DODSON (1980)).

HOFFMAN and DODSON (1980) have compared their measured H^+ fluxes with the model of BANKS and DOUPNIK (1974) and find they are a factor of 2 lower than the model values of $1-2 \times 10^8 \text{ cm}^{-2}\text{sec}^{-1}$. The measured value of $7 \times 10^7 \text{ cm}^{-2}\text{sec}^{-1}$ for the upward flux is even further from the typical values of around $3 \times 10^8 \text{ cm}^{-2}\text{sec}^{-1}$ for the model described by RAITT et al. (1975). However, as pointed out by HOFFMAN and DODSON (1980) and illustrated by BANKS (1972), the H^+ escape flux shows a proportional dependence on the neutral hydrogen density in the exosphere which is expected to show significant variations during a solar cycle.

The main indirect measurement used to show the existence of the polar wind has been the marked latitudinal variation of H^+ and He^+ number densities at a constant altitude. The light ion number density at altitudes above about 600 km shows a rapid decrease of about an order of magnitude from its low latitude value at magnetic latitudes of about 60°. This feature is often referred to as the light ion trough. An example of an OGO-2 measurement of the light ion density variation with latitude is shown in Fig. 21 (TAYLOR and WALSH, 1972) for a single pass and in Fig. 22 (TAYLOR and CORDIER, 1974) for a series of OGO-4 satellite passes. The average global morphology of light ions was studied over a six-month period in 1973 by RAITT and DORLING (1976) using data from an in-situ thermal ion probe flown on the ESRO-4 satellite. The averaged light ion number density observed clearly showed the existence of a light ion trough at all local times, but the slope of the density change with latitude varied greatly with local time. This feature was interpreted in terms of a continual

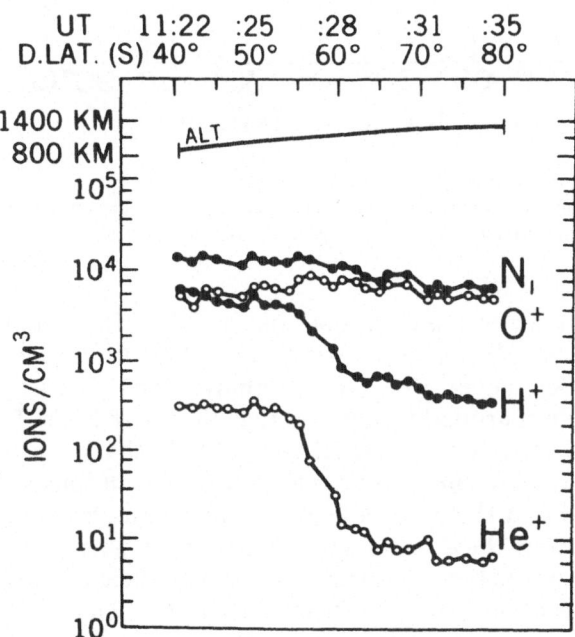

Fig. 21.   An example of the light ion trough observed in H^+ and He^+ ions, N_e shows no trough effect. Data from OGO-2 (from TAYLOR and WALSH (1972)).

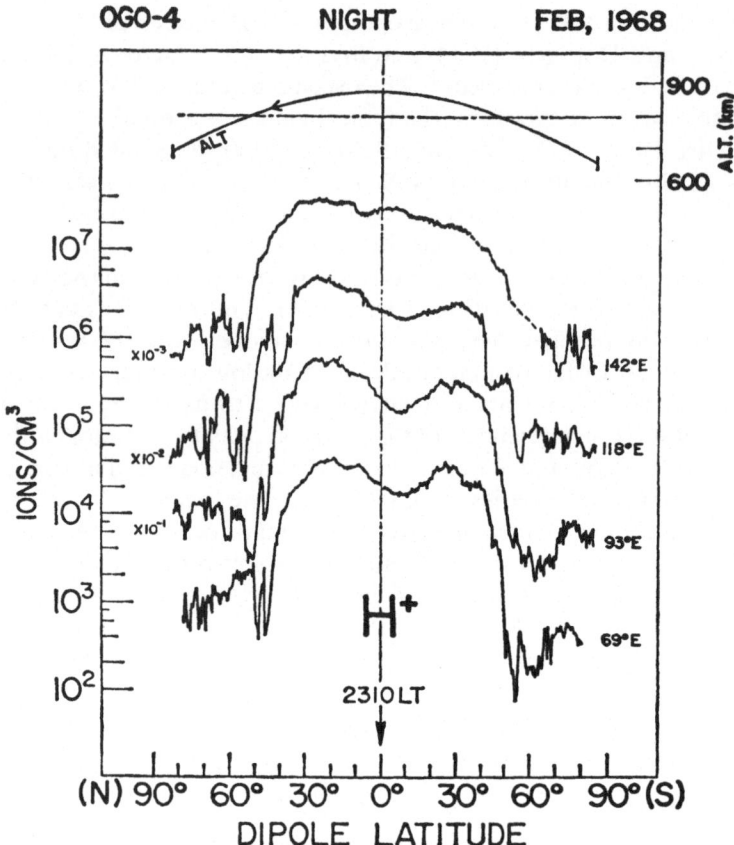

Fig. 22. A sequence of pole-to-pole profiles of H^+ density showing the light ion trough and the irregularities observed within the trough. The densities are scaled by the figure on the left-hand side of each plot to make the diagram more legible (from TAYLOR and CORDIER (1974)).

outflow over the polar regions and an inner plasmaspheric region of ebb and flow. Figure 23 shows contours of $\log_{10}[N(H^+)]$ in the 700–800 km altitude range averaged for the period March-September, 1973. The steepness of the H^+ density decline can be seen to be much greater at around 6 hours LT than that at 18 hours LT. However, at latitudes greater than about 60° magnetic, there is little change in the H^+ number density with local time, indicating continuous outflow of light ions at these latitudes.

A result of the reduced H^+ number density at high altitudes in light ion outflow regions is that O^+ becomes the dominant ion up to much higher altitudes than in the plasmasphere. This results in the ionospheric plasma scale height being typical of that of an O^+-electron plasma up to about 2,000 km altitude. The scale height can be measured remotely by using topside sounders to obtain plasma density-altitude profiles. This technique has been used with Alouette-II data by BANKS and DOUPNIK (1974). The data clearly show a transition from a light ion dominated topside

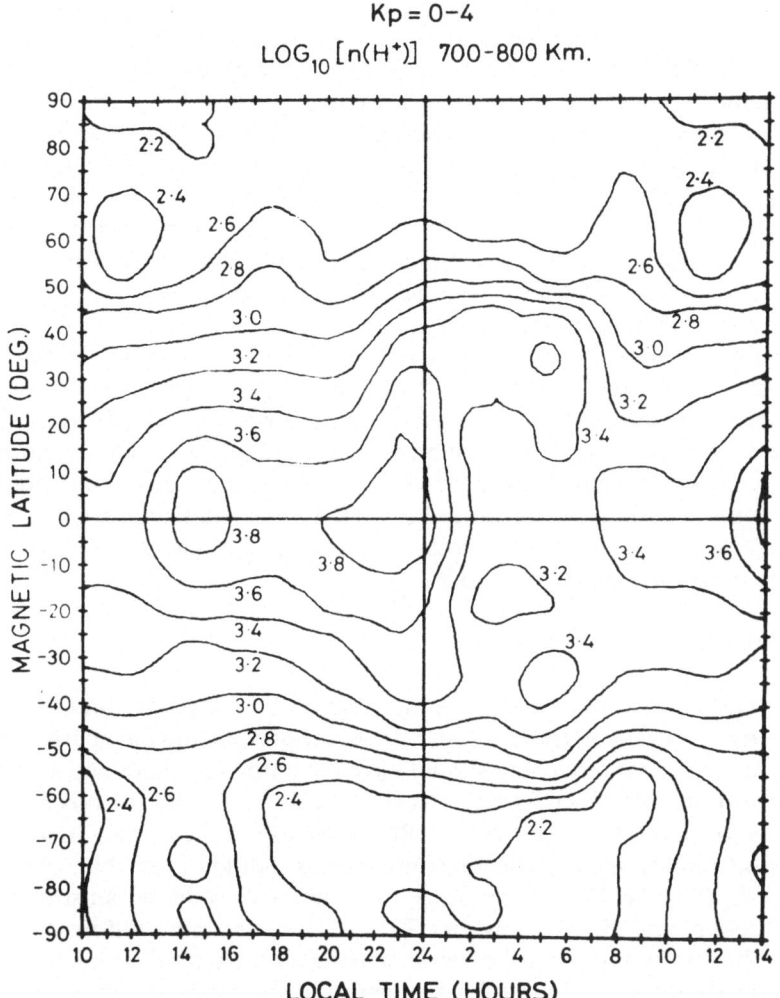

Fig. 23. Contours of constant values of $\log_{10} [N(H^+)]$ in the magnetic latitude/local time plane in the altitude band 700–800 km for values of Kp in the range 0–4. The data are averaged over the period March–September 1973 (from RAITT and DORLING (1976)).

ionosphere above 1,500 km at latitudes below 60° magnetic to an O^+ dominated ionosphere at higher latitudes in the midnight local time sector. In the morning sector, however, they find evidence of flow at magnetic latitudes as low as 48° in addition to outflow at higher latitudes. As with the other light ion trough measurements cited earlier, the lower latitude effects are associated with plasmaspheric, closed flux tube filling processes, while the high latitude effects are associated with flow of light ions from the ionosphere into the outer regions of the magnetosphere.

The light ion trough is an indirect indication of outflow since, as shown by Fig. 3a, the models show a marked decrease in H^+ number density from the diffusive equilibrium value at low latitudes to outflow conditions at higher latitudes. Averaged altitude profiles of H^+ number density have been made by BRINTON et al. (1971) and show the same general characteristics as those predicted by the models. The intermediate region between the high light ion density at low latitudes and the low light ion density at high latitudes shows the ebb and flow conditions corresponding to filling and emptying of closed flux tubes in the vicinity of the plasmapause, shown in more detail by RAITT and DORLING (1976). All of the light ion density measurements show that the light ion trough is seen at all local times and for both summer and winter hemispheres, demonstrating that there is a continual escape of light ions from the polar regions into the magnetospheric tail.

Another ionospheric parameter which is predicted by the models to be indirectly related to light ion outflow is the temperature increase of the light ions as their outflow velocity increases, resulting from frictional heating effects between H^+ and O^+ ions. MAIER and HOFFMAN (1974) have published some data from Isis-2 showing H^+ temperatures to be significantly elevated above O^+ temperatures when the phase shift in the roll modulation discussed earlier indicated that the satellite was in a region of light ion outflow. The ratio T_{H+}/T_{O+} was found to vary from near unity to about 10, with values typically in the range of 1.3 to 2.

4. Conclusion

At the present time, the bulk of the information on the polar wind derives from theoretical models. Based on these model studies, the plasma flowing up from the polar ionosphere into the magnetosphere should have the following characteristics: (1) The composition is dominated by the light ions H^+ and He^+; (2) The H^+ flux should vary from $10^7 - 5 \times 10^8$ $cm^{-2}s^{-1}$ and the He^+ flux from $10^5 - 10^7$ $cm^{-2}s^{-1}$, depending on the geophysical conditions; (3) The ion temperatures should be less than $8,000°K$, i.e., less than 1 eV; (4) The H^+ temperature distribution should be anisotropic, with $T_\parallel > T_\perp$ for supersonic H^+ outflow and $T_\parallel < T_\perp$ for subsonic outflow; (5) The H^+ velocity distribution should be asymmetric, with an elongated tail along the magnetic field in the upward direction; and (6) The electron temperature distribution should be anisotropic if the H^+ outflow is supersonic.

The polar wind characteristics listed above were determined from steady state models, and as yet, there have been no time-dependent model studies of the polar wind. Also, only a few of the theoretical predictions have been verified, owing to the lack of measurements. The composition, H^+ and He^+ fluxes, and ion temperatures have been measured at an altitude of 1,400 km by the ISIS spacecraft, and these measurements are in good general agreement with the theory. However, no definitive measurements are currently available on whether the light ion outflow is subsonic or supersonic, i.e., on whether it is a polar breeze or a polar wind. Furthermore, there are no measurements bearing on the anisotropy or asymmetry of the ion distribution functions nor are there any altitude profiles of the relevant polar wind properties, such as density, drift velocity and temperature. Hopefully, in the near future the Dynamics Explorer satellites will provide answers to some of the questions relating to the polar wind.

Looking further ahead to the proposed OPEN program of NASA, we hope that the instrumentation provided on the various satellites, and in particular, on the Polar Plasma Laboratory (PPL), will have modes of operation which will characterize the outflowing plasma to the extent that the theoretical models can be verified and refined. The polar wind is, after all, probably a significant origin of plasma in the earth's environment.

This research was supported by NASA grant NAGW-77, NSF grant ATM-8015497 and Air Force Contract F19628-79-C-0025 to Utah State University.

REFERENCES

Axford, W. I. and C. O. Hines, A unifying theory of high-latitude geophysical phenomena and geomagnetic storms, *Can. J. Phys.*, **39**, 1433–1464, 1961.

Bailey, G. J. and R. J. Moffett, Temperatures in the polar wind, *Planet. Space Sci.*, **22**, 1193–1199, 1974.

Banks, P. M., Behavior of thermal plasma in the magnetosphere and topside ionosphere, in *Critical Problems of Magnetospheric Physics*, edited by E. R. Dyer, pp. 157–178, Washington, D.C., Inter-Union Comm. Solar Terr. Phys., N.A.S., 1972.

Banks, P. M., Ion heating in thermal plasma flows, *J. Geophys. Res.*, **78**, 3186–3188, 1973.

Banks, P. M. and J. R. Doupnik, Thermal proton flow in the plasmasphere: The Morning Sector, *Planet. Space Sci.*, **22**, 79–94, 1974.

Banks, P. M. and T. E. Holzer, The polar wind, *J. Geophys. Res.*, **73**, 6846–6854, 1968.

Banks, P. M. and T. E. Holzer, Features of plasma transport in the upper atmosphere, *J. Geophys. Res.*, **74**, 6304–6316, 1969a.

Banks, P. M. and T. E. Holzer, High-latitude plasma transport: The polar wind, *J. Geophys. Res.*, **74**, 6317–6332, 1969b.

Banks, P. M. and G. Kockarts, *Aeronomy*, Part B, Academic Press, New York, 1973.

Banks, P. M., A. F. Nagy, and W. I. Axford, Dynamical behavior of thermal protons in the mid-latitude ionosphere and magnetosphere, *Planet. Space Sci.*, **19**, 1053–1067, 1971.

Banks, P. M., R. W. Schunk, and W. J. Raitt, Temperature and density structure of thermal proton flows, *J. Geophys. Res.*, **79**, 4691–4702, 1974.

Banks, P. M., R. W. Schunk, and W. J. Raitt, The topside ionosphere: A region of dynamic transition, *Annl. Rev. Earth Planet. Sci.*, **4**, 381–440, 1976.

Bauer, S. J., *Physics of Planetary Ionospheres*, Springer-Verlag, Berlin-Heidelberg, 1973.

Brinton, H. C., J. M. Grebowsky, and H. G. Mayr, Altitude variation of ion composition in the mid-latitude trough region: Evidence for upward plasma flow, *J. Geophys. Res.*, **76**, 3738–3745, 1971.

Conrad, J. R. and R. W. Schunk, Diffusion and heat flow equations with allowance for large temperature differences between interacting species, *J. Geophys. Res.*, **84**, 811–822, 1979.

Dessler, A. J. and P. A. Cloutier, Discussion of the letter by Peter M. Banks and Thomas E. Holzer, "The polar wind," *J. Geophys. Res.*, **74**, 3730–3733, 1969.

Dungey, J. W., Interplanetary magnetic field and the auroral zones, *Phys. Rev. Lett.*, **6**, 47–48, 1961.

Geisler, J. E., On the limiting daytime flux of ionization into the protonosphere, *J. Geophys. Res.*, **72**, 81–85, 1967.

Hanson, W. B. and T. N. L. Patterson, The maintenance of the night-time *F*-layer, *Planet. Space Sci.*, **12**, 979–997, 1964.

Hoffman, J. H., Studies of the composition of the ionosphere with a magnetic deflection mass spectrometer, *Int. J. Mass Spectrom. Ion Phys.*, **4**, 315–322, 1970.

Hoffman, J. H. and W. H. Dodson, Light ion concentrations and fluxes in the polar regions during magnetically quiet times, *J. Geophys. Res.*, **85**, 626–632, 1980.

HOFFMAN, J. H., W. H. DODSON, C. R. LIPPINCOTT, and H. D. HAMMACK, Initial ion composition results from the Isis-2 satellite, *J. Geophys. Res.*, **79**, 4246–4251, 1974.

HOLZER, T. E., J. A. FEDDER, and P. M. BANKS, A comparison of kinetic and hydrodynamic models of an expanding ion-exosphere, *J. Geophys. Res.*, **76**, 2453–2468, 1971.

LEMAIRE, J., O^+, H^+, and He^+ ion distributions in a new polar wind model, *J. Atmos. Terr. Phys.*, **34**, 1647–1658, 1972.

LEMAIRE, J. and M. SCHERER, Model of the polar ion-exosphere, *Planet. Space Sci.*, **18**, 103–120, 1970.

LEMAIRE, J. and M. SCHERER, Simple model for an ion-exosphere in an open magnetic field, *Phys. Fluids*, **14**, 1683–1694, 1971.

LEMAIRE, J. and M. SCHERER, Ion-exosphere with asymmetric velocity distribution, *Phys. Fluids*, **15**, 760–766, 1972.

LEMAIRE, J. and M. SCHERER, Kinetic models of the solar and polar winds, *Rev. Geophys. Space Phys.*, **11**, 427–468, 1973.

MAIER, E. J. and J. H. HOFFMAN, Observation of a two-temperature ion energy distribution in regions of polar wind flow, *J. Geophys. Res.*, **79**, 2444–2447, 1974.

MANGE, P., The distribution of minor ions in electrostatic equilibrium in the high atmosphere, *J. Geophys. Res.*, **65**, 3833–3834, 1960.

MARUBASHI, K., Escape of the polar-ionospheric plasma into the magnetospheric tail, *Rep. Ionos. Space Res. Japan*, **24**, 322–346, 1970.

OTTLEY, J. A. and R. W. SCHUNK, Density and temperature structure of helium ions in the topside polar ionosphere for subsonic outflows, *J. Geophys. Res.*, **85**, 4177–4190, 1980.

RAITT, W. J. and E. B. DORLING, The global morphology of light ions measured by the ESRO-4 satellite, *J. Atmos. Terr. Phys.*, **38**, 1077–1083, 1976.

RAITT, W. J., R. W. SCHUNK, and P. M. BANKS, A comparison of the temperature and density structure in high and low speed thermal proton flows, *Planet. Space Sci.*, **23**, 1103–1117, 1975.

RAITT, W. J., R. W. SCHUNK, and P. M. BANKS, The influence of convection electric fields on thermal proton outflow from the ionosphere, *Planet. Space Sci.*, **25**, 291–301, 1977.

RAITT, W. J., R. W. SCHUNK, and P. M. BANKS, Helium ion outflow from the terrestrial ionosphere, *Planet. Space Sci.*, **26**, 255–268, 1978a.

RAITT, W. J., R. W. SCHUNK, and P. M. BANKS, Quantitative calculations of helium ion escape fluxes from the polar ionospheres, *J. Geophys. Res.*, **83**, 5617–5624, 1978b.

SCHUNK, R. W., Mathematical structure of transport equations for multispecies flows, *Rev. Geophys. Space Phys.*, **15**, 429–445, 1977.

SCHUNK, R. W. and J. C. G. WALKER, Thermal diffusion in the topside ionosphere for mixtures which include multiply-charged ions, *Planet. Space Sci.*, **17**, 853–868, 1969.

SCHUNK, R. W. and J. C. G. WALKER, Thermal diffusion in the F2-region of the ionosphere, *Planet. Space Sci.*, **18**, 535–557, 1970.

SCHUNK, R. W. and D. S. WATKINS, Comparison of solutions to the thirteen-moment and standard transport equations for low speed thermal proton flows, *Planet. Space Sci.*, **27**, 433–444, 1979.

SCHUNK, R. W. and D. S. WATKINS, Electron temperature anisotropy in the polar wind, *J. Geophys. Res.*, **86**, 91–102, 1981.

SCHUNK, R. W. and D. S. WATKINS, Proton temperature anisotropy in the polar wind, *J. Geophys. Res.*, **87**, 171–180, 1982.

SCHUNK, R. W., W. J. RAITT, and P. M. BANKS, Effect of electric fields on the daytime high-latitude E and F regions, *J. Geophys. Res.*, **80**, 3121, 1975.

SCHUNK, R. W., W. J. RAITT, and A. F. NAGY, Effect of diffusion-thermal processes on the high-latitude topside ionosphere, *Planet. Space Sci.*, **26**, 189–191, 1978.

SHEPHERD, G. G., J. H. WHITTEKER, J. D. WINNINGHAM, J. H. HOFFMAN, E. J. MAIER, L. H. BRACE, J. R. BURROWS, and L. L. COGGER, The topside magnetospheric cleft ionosphere observed from the Isis 2 spacecraft, *J. Geophys. Res.*, **81**, 6092–6102, 1976.

SOJKA, J. J., W. J. RAITT, and R. W. SCHUNK, Effect of displaced geomagnetic and geographic poles on high latitude plasma convection and ionospheric depletions, *J. Geophys. Res.*, **84**, 5943–5951, 1979.

SOJKA, J. J., W. J. RAITT, and R. W. SCHUNK, A theoretical study of the high-latitude winter F-region at solar minimum for low magnetic activity, *J. Geophys. Res.*, **86**, 609–621, 1981a.

SOJKA, J. J., W. J. RAITT, and R. W. SCHUNK, Plasma density features associated with strong convection in the winter high-latitude F-region, *J. Geophys. Res.*, **86**, 6908–6916, 1981b.

ST.-MAURICE, J.-P. and R. W. SCHUNK, Diffusion and heat flow equations for the mid-latitude topside ionosphere, *Planet. Space Sci.*, **25**, 907–920, 1977.

STROBEL, D. F. and E. J. WEBER, Mathematical model of the polar wind, *J. Geophys. Res.*, **77**, 6864–6869, 1972.

TAYLOR, H. A., Jr. and G. R. CORDIER, In-situ observations of irregular ionospheric structure associated with the plasmapause, *Planet. Space Sci.*, **22**, 1289–1296, 1974.

TAYLOR, H. A., Jr. and W. J. WALSH, The light ion trough, the main trough and the plasmapause, *J. Geophys. Res.*, **77**, 6716–6723, 1972.

Energetic Ion Composition in the Earth's Magnetosphere, edited by R. G. Johnson, 143–165.

Low-Altitude Energetic Ion Composition Observations

B. A. WHALEN

*Herzberg Institute of Astrophysics,
National Research Council of Canada, Ottawa, Canada*

(Received February 26, 1982)

Energetic ion precipitation has been detected at low altitudes directly from sounding rockets and satellites, as well as indirectly from the ground using photometric techniques. Mass composition observations at energies up to 20 keV indicate that these particles have their origin in both the solar wind and the ionosphere. The spatial and temporal precipitation patterns of the major ion species (H^+, He^+, He^{++}, and O^+)· have been studied and it has been shown that some of these features are closely correlated with geomagnetic activity. None of the expected minor constituents have been observed at low altitudes. Upward flowing ions with mass distributions characteristic of the ionosphere, presumably the signature of the ionospheric source region, have been detected at high (auroral) latitudes. These data show that mechanisms capable of accelerating ions up to several keV exist at low altitudes in the auroral zone. Although the characteristics of the source are reasonably well known, the acceleration mechanism has yet to be definitively identified. Tracer techniques, using natural and artificial ion sources are being applied to magnetospheric problems and the use of these techniques along with improved ion mass spectrometers promises a rapid expansion in the understanding of ion transport, energization and loss processes.

1. Introduction

Recent energetic ion composition measurements have led to a rapid expansion in our knowledge of magnetospheric ion composition. The earliest observations made at low altitudes indicated that the magnetospheric plasma was dominated by ions of solar wind origin; however, this concept has been superseded by one in which either the ionosphere or the solar wind may be the dominant source region, depending on geophysical parameters such as magnetic activity, latitude and altitude.

These composition measurements reflect the combined effects of injection and loss mechanisms and both terms strongly influence the ambient energetic ion composition. Several reviews of high altitude magnetospheric ion composition observations and discussions of loss mechanisms are presented in this volume. In this report we concentrate on low altitude ion composition measurements in and near the atmospheric loss cone for energetic magnetospheric ions and in the source region for energetic ions of ionospheric origin.

Some of the more significant problems which can only be properly investigated at low altitudes are: i) atmospheric precipitation and loss rates, ii) ionospheric source characteristics, iii) low altitude ion acceleration, and iv) atmospheric accretion rates (AXFORD, 1970).

Low-altitude observations are necessitated by measurements which indicate that low-altitude adiabatic and non-adiabatic processes can occur which perturb low-altitude charged-particle trajectories. These processes make definitive high altitude measurements, relating to the above problems either difficult or, more likely, impossible.

2. Ground-Based Photometric Observations

The first direct evidence that energetic ions were bombarding the earth's atmosphere came from ground-based measurements of Doppler shifted neutral hydrogen (Balmer series) emissions from the aurora (MEINEL, 1951). The Doppler shift in wavelength scans of emissions near the zenith indicated that the emitting (precipitating) ions had energies in the keV energy range which implied that energetic protons were involved in the production of the optical aurora. Subsequent direct observations have confirmed that the auroral ion flux is generally dominated by protons. An excellent review of these early ground-based observations was given by EATHER (1967).

Attempts have also been made to observe emissions due to the precipitation of other ion species such as helium, which is expected at times to be the most intense flux. EATHER (1966) estimated that the most easily detected emissions would be from energetic He^+ ions, which were presumed in this case to result from charge exchange of the primary He^{++} beam with the neutral atmosphere. Subsequent searches for these emissions were made by EATHER (1968) and STOFFREGEN (1969). Eather reported negative results; however, STOFFREGEN observed transient bursts of helium I emissions at 5,876 Å with intensities ($\sim 120\ R$) well in excess of the anticipated levels. The latter observation has not been confirmed to this date.

HENDRICKSEN (1978) and SIVJEE et al. (1980) have recently reported observations of He^+ emissions in the polar cap which are more consistent with preconception. They observed He I emissions at 3,889 and 5,876 Å with average intensities of $\sim 2\ R$ and peak readings near 12 R . They attributed the presence of these emissions to a solar wind source for the ions.

Oxygen ions (O^+) are also known to be a major component of energetic ion precipitation (SHELLEY et al., 1972), particularly during large geomagnetic storms (see Section 4). An estimate of the optical emissions stimulated when these ions interact with the atmosphere was reported by TORR et al. (1974). They showed that 6,300 Å oxygen emissions could be as bright as 200 R for typical O^+ precipitation events; however, these emissions are difficult to unambiguously identify in auroral spectra since the ambient airglow is often of this order of magnitude or greater at high latitudes.

YEE and HAYS (1980) suggested that the O^+ energy flux could be monitored via the 7,320 Å emission of O^+ which they estimated could be enhanced by 10 to 20 R during oxygen precipitation events. High altitude emissions of the 7,320 and 7,330 Å O^+ doublets at mid-latitudes was reported by YEE et al. (1980). The emission profiles at altitudes above 550 km indicated a much higher ionospheric O^+ temperature than anticipated. In this case the increased temperature was attributed to fast atomic fragments created by dissociative recombination of molecular ions on the dayside rather than to energetic oxygen precipitation. The data, however, do suggest that optical signatures of energetic O^+ precipitation may yet be found in auroral emissions.

To summarize, useful information on the composition and spatial and temporal distribution of energetic ion precipitation can be derived from ground-based photometric observations; however, the techniques are not yet sufficiently well developed to provide a reliable quantitative monitor of even the major ion fluxes. In situ ion mass spectrometer measurements are required to give accurate energy and mass distribution information.

3.   Direct Ion Composition Measurements—Early Sounding Rocket Results

The first attempts at direct measurement of auroral ion mass composition were made using sounding rockets. In discussing these observations it is important to keep in mind how low-altitude measurements are affected by the presence of the neutral atmosphere. As charged energetic ions penetrate into the atmosphere they suffer charge exchange interactions with the ambient neutrals and may even be converted to energetic neutral atoms without significant energy loss. These neutrals are able to cross magnetic field lines before striking the dense atmosphere where their energy is lost and most of the optical signal is generated. Interaction of the beam with the atmosphere therefore spreads the beam spatially (DAVIDSON, 1965) and has a profound effect on the charge state of the ion beam.

Displayed in Fig. 1, which is from REASONER et al. (1968), is the variation in charge state composition as a function of altitude of a primary 200 keV He^{++} ion beam incident on the atmosphere. Shown here is the fraction F of the beam in three charge states, He^{++}, He^+, and neutral He. At approximately 400 km charge exchange interactions begin to have a significant effect on the charge state of the beam. By 200 km the beam has reached a charge exchange equilibrium condition where most of the beam is singly ionized or neutral.

It should be noted here that the high altitude atmospheric density is highly dependent on solar activity. Therefore the altitude above which charge exchange interactions may be ignored may vary from ~ 300 km to 500 km, the former referring to low and the latter to high levels of solar activity. Also, since charge exchange cross-sections are energy and species dependent (ALLISON, 1958), the profiles shown in Fig. 1 should only be considered as a guide, when charge exchange effects are a concern.

The first attempt at direct measurement of the auroral ion mass distribution was reported by ALBERT (1967). These observations were carried out at low altitudes (~ 250 km) in a region where charge exchange effects are expected to dominate the beam charge state. Albert reported a He^{++}/H^+ integral flux ratio of 7% at energies greater than 10 keV. These results were strongly criticized by WAX and BERNSTEIN (1968) on the basis that extrapolation of these results to high altitudes, using known charge exchange cross-sections, suggested that a ratio of 600 would be observed above the atmosphere which is grossly in excess of anticipated levels.

Soon after this report, two observations of energetic ion precipitation in the

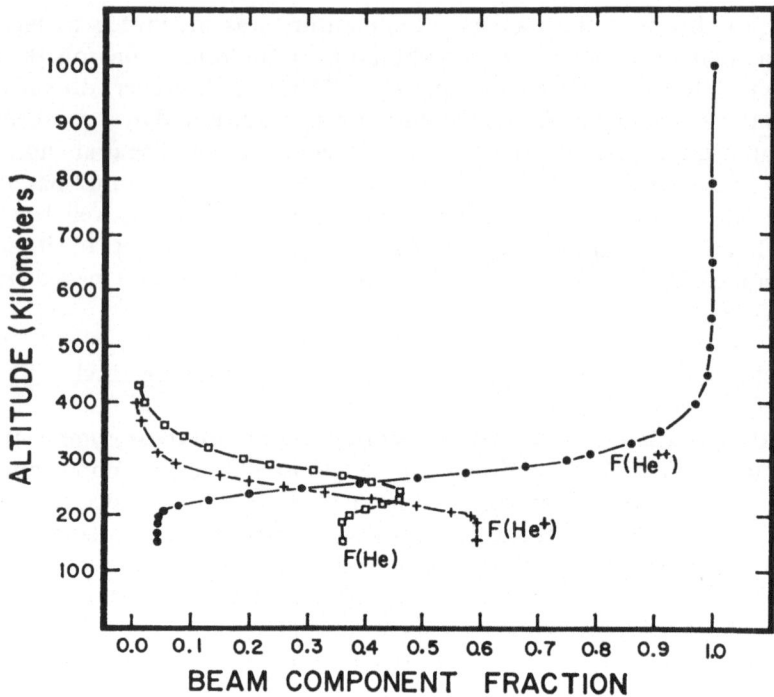

Fig. 1.   Charge-state composition of 200 keV doubly charged helium beam incident on the atmosphere.
Note that significant deviations of the beam from pure He^{++} occur at approximately 400 km (from
REASONER *et al.*, 1968).

aurora appeared in the literature. Both instruments relied on solid state detectors as the
primary energy sensing element and were thus limited to high energies. WHALEN and
MCDIARMID (1968) observed He^+ and H^+ ions in the energy per unit charge range
$50 \leq E/Q \leq 150$ kV at altitudes up to 360 km and, using the known equilibrium
fractions, estimated that the incident helium to proton ratio varied between 0.06 and
0.3 during the flight.

REASONER *et al.* (1968) reported ion charge state measurements of 100 keV singly
ionized and 200 keV doubly ionized ions (assumed to be H^+ and He^{++}, respectively) at
altitudes well above the atmosphere (750 km). The ratio of He^{++} to H^+ at 100 keV per
unit charge was found to be 1.6%. Both REASONER *et al.* and WHALEN and MCDIARMID
suggested that their observations were most consistent with a solar wind source for
energetic auroral ion precipitation.

The introduction of channel electron multipliers as a sensor for auroral ions led to
the application of more standard spectrometer designs. Mass spectrometers capable of
making mass per unit charge (m/Q) measurements in the energy range appropriate to
auroral events $(1-20$ keV$)$ were developed and flown into auroral displays on high
altitude sounding rockets. Electrostatic and magnetic deflection, where the magnetic
field was generated by a permanent magnet, was employed to determine the energy per

unit charge (E/Q) and mass per unit charge (m/Q) distribution of precipitating ions.

The first results from this type of mass spectrometer were reported by WHALEN *et al.* (1971). Their instrument operated in the 2 to 20 keV energy range and was capable of resolving ions in the mass per unit charge range $1 \leq m/Q \leq 6$, which includes H^+, He^{++}, and He^+. The objective of this experiment was to measure the charge state of the helium ion precipitation in an aurora to infer either a solar wind origin (if mostly He^{++}) or an ionospheric origin (if He^+) for the ions (AXFORD, 1970). Measurements were made up to an altitude of 775 km, well above the charge exchange region, and a He^{++}/H^+ ratio of ~4% was reported. No He^+ precipitation was detected.

Similar results from an improved version of the same instrument were reported by WHALEN and McDIARMID (1972). Figure 2 shows a mass spectrum in one of the four momentum channels of the instrument. Peaks due to H^+ and He^{++} ($m/Q = 1$ and 2) are clearly visible; however, no enhancement above noise is apparent near He^+ ($M/Q = 4$). A crude energy spectrum for both H^+ and He^{++} was derived from the four channels and is shown in Fig. 3. The authors commented that the spectral peaks at the same energy per unit charge for both ion species was suggestive of an electrostatic acceleration mechanism. Also, since the He^+ flux was estimated to be at least one order

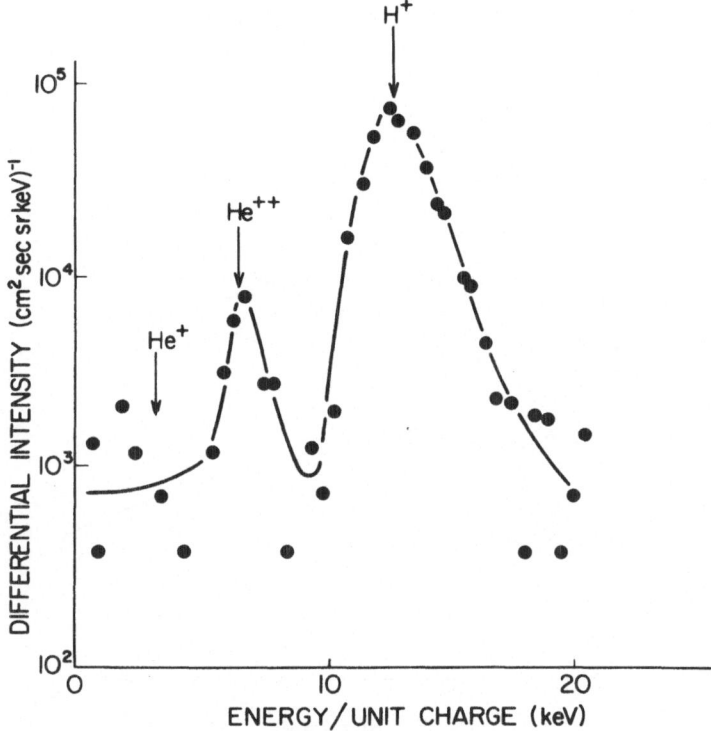

Fig. 2.   Mass per-unit-charge (m/Q) spectrum observed in a high altitude sounding rocket flight into an aurora. The arrows indicate the expected peak positions for various ion species. Only H^+ and He^{++} were observable above background (from WHALEN and McDIARMID, 1972).

of magnitude below the He^{++} flux, the results were consistent with a solar wind origin for auroral ions.

An entirely different method for measuring the integral flux of energetic noble gas isotopes (e.g., ^{3}He, ^{4}He, ^{20}Ne, ^{21}Ne, ^{22}Ne, and ^{34}Ar) was developed to measure the solar wind composition on the Apollo missions (GEISS et al., 1970). This technique involved the exposure of high purity aluminum and platinum foils to the ion bombardment. The incident energetic ions become embedded or trapped in the foil, the trapping probability being dependent on the ion species and energy. Generally ions with $E > 1$ keV are trapped. The foils are recovered after exposure and the embedded gases are driven off the foils and analysed in a laboratory mass spectrometer. This technique provides no information on the ion charge state and only a very limited indication of the incident energy. Some knowledge of the energy may be derived since the trapping efficiency is dependent on energy.

Initial results from two rocket flights, in which foils were exposed to ions precipitating in a bright auroral display, were published by AXFORD et al. (1972). Here, the authors reported only the ^{4}He flux measurements. In a subsequent publication of results from these flights, BUHLER et al. (1976) confirmed the previous

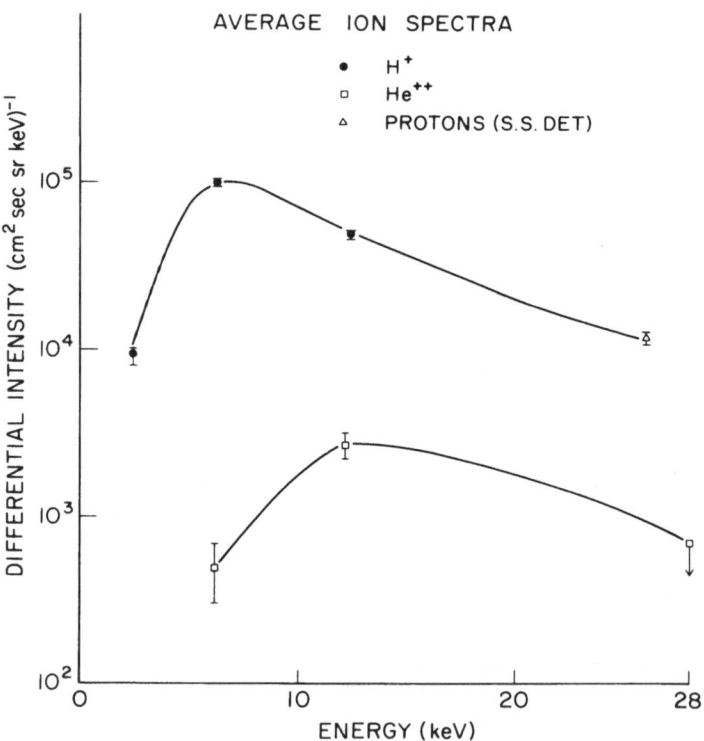

Fig. 3.   Proton and alpha particle energy spectra measured in the aurora (from WHALEN and McDIARMID, 1972).

^{4}He measurements and also reported ^{3}He flux estimates. The ^{4}He/^{3}He ratio was found to be 2,950 in the first flight which is similar to values found in the solar wind (2,350) using the same technique but grossly different from the nominal atmospheric composition ratio of $\sim 7.3 \times 10^5$. A solar wind origin for auroral ions was suggested.

An unsuccessful attempt was also made to measure the ^{20}Ne flux; however, an upperlimit on the ^{4}He/^{20}Ne ratio was set at ~ 100 which is well below the anticipated solar wind level. The reader is also referred to BUHLER et al. for an interesting discussion of the implications of these measurements with respect to the atmospheric helium budget.

Metal foil collectors were also installed on the Skylab structure to collect energetic magnetospheric helium, neon and argon ion precipitation. This spaceraft, which was in a 443 km circular orbit with a 50° inclination, carried an array of aluminum and platinum foils which were exposed to the ambient for approximately one year (from May 1973 to February 1974). The foils therefore collected ions at midlatitude although for short periods of time the spacecraft did pass through the low latitude edge of the auroral zone (the expected region of maximum ion flux). Laboratory analysis of the samples indicated that two classes of ions, corresponding to thermal particles swept up at the spacecraft velocity and energetic (~ 30 keV) ions of magnetospheric origin, had been entrapped in the foils (LIND et al., 1979).

As expected, gases of terrestrial origin, He^4 and neon and argon were found to dominate the low energy or thermal flux to the foils. At higher energies (that is, longer range particles) the averaged ^{4}He/^{3}He ratio was found to be 3100 ± 200 which is consistent with a solar wind source for the ^{3}He. A small admixture of terrestrial ^{4}He was suggested to account for the ratio being slightly larger than the nominal solar wind value of 2350 (GEISS et al., 1972).

Precipitating energetic (several keV) neon isotopes (^{20}Ne and ^{22}Ne) were also detected and comparisons with the helium fluxes led the authors to conclude that these particles also originated in the solar wind.

A similar energetic ion sampling technique was employed by WARASILA and SCHAEFFER (1975) to measure the low latitude, low altitude ion composition. They analysed the ^{3}He and ^{4}He embedded in a recovered section of the second stage of a Saturn test flight. This material, an aluminum antenna housing, had been in space and exposed to energetic ion bombardment for over two years.

A ^{4}He/^{3}He ratio near 145 was measured which is much lower than either the solar wind or atmospheric ratios. No entirely satisfactory explanation for this low ratio was given although solar cosmic rays were suggested as a possible source.

4.   Low-Altitude Satellite Observations of an Ionospheric Source

In all the reports discussed previously an ionospheric source for energetic ions was considered to be inconsistent with observations. The first indication that the ionosphere can make a significant contribution to the energetic magnetospheric ion population came from a low-altitude (800 km) polar-orbiting satellite (1971-089A). Using a mass spectrometer which combined electrostatic energy analysis with a crossed electric and magnetic field velocity selector (Wein filter), SHELLEY et al. (1972) observed large fluxes of heavy ($m/Q \cong 16$, assumed to be O^+) ions in the 0.7 to 12 keV energy range precipitating into the atmosphere. Mass spectra taken at a number of different energies are displayed in Fig. 4. The arrow on the right hand side indicates the expected position of the H^+ peak and the arrow on the left the O^+ peak position. Mass peaks are observed at both positions and at all energies from 0.7 to 12.1 keV.

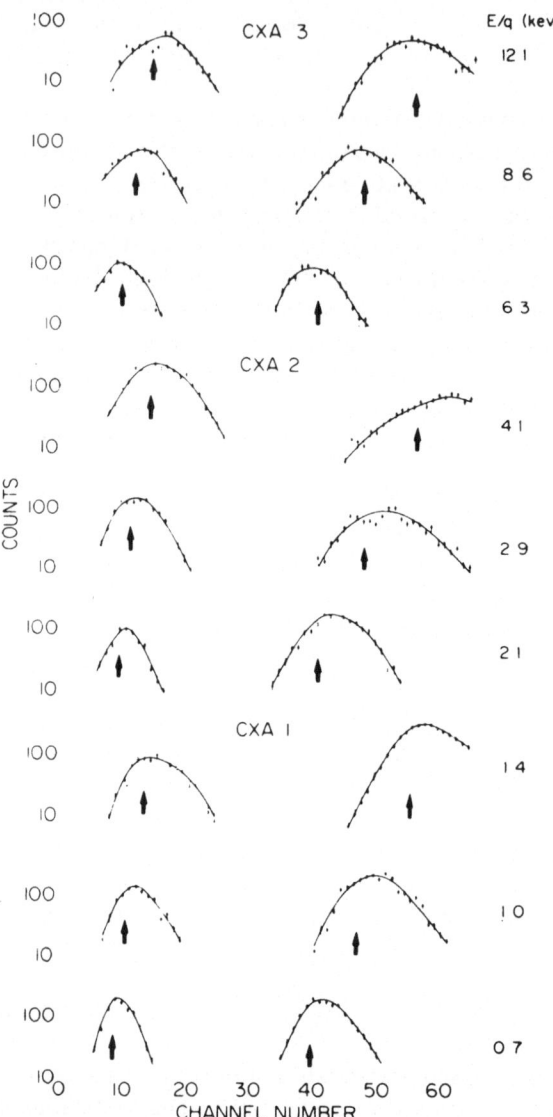

Fig. 4.   Mass (m/Q) spectra observed at nine energies on the low-altitude satellite 1971-089A during a geomagnetic storm. The arrows indicate the predicted centroid positions for $m/Q = 1$ and $m/Q = 16$, presumed to be H^+ and O^+ ions (from SHELLEY et al., 1972).

This heavy ion (O^+) precipitation was found to be strongly correlated with geomagnetic activity, the peak flux, which at times exceeded H^+, being observed during a large geomagnetic storm. The latitudinal range of the O^+ precipitation was found to coincide approximately with, but to extend equatorward of, the H^+ precipitation.

The spatial and temporal correlation of O^+ precipitation with the geomagnetic indices D_{st} and AE was investigated further by SHARP et al. (1976a). Figures 5 and 6, which are from this report, show the O^+ energy flux, integrated over latitude, as a function of time. Also plotted is the corresponding D_{st} index (Fig. 5) and AE index (Fig.

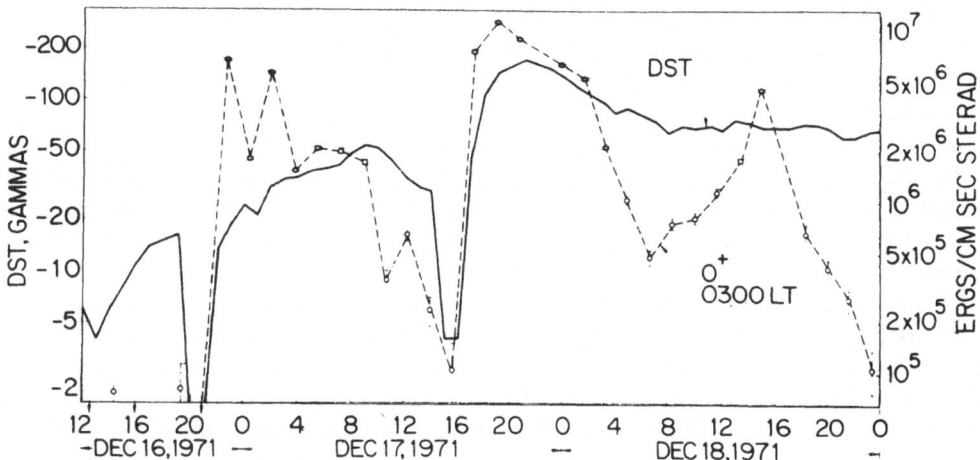

Fig. 5. Correlation of measured O^+ latitude integrated intensity with the geomagnetic disturbance index D_{st} (from SHARP et al., 1976a).

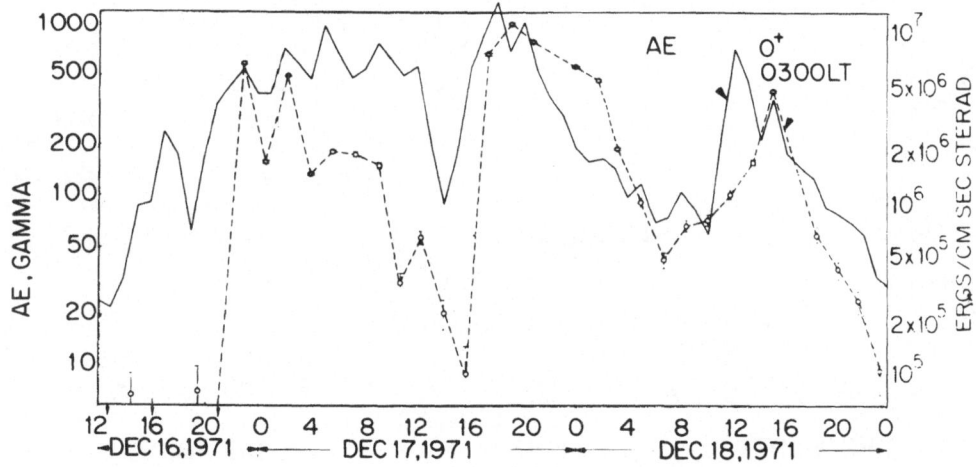

Fig. 6. Correlation of O^+ latitude integrated intensity with auroral magnetic activity index AE (from SHARP et al., 1976a).

6). These figures clearly demonstrate the close correlation between the O^+ precipitation and the level of geomagnetic activity. In particular, we note the close correspondence of the O^+ energy flux with D_{st} during the main phase of the storms and perhaps a better correlation with AE during the recovery phase. JOHNSON *et al.* (1975) reported a similar close correlation with the AE index during two isolated auroral substorms.

The spatial correlation of O^+ and H^+ precipitation during the same two storms depicted in Figs. 5 and 6 at two different local times is shown in Fig. 7 which is from SHARP *et al.* (1976a). Two sudden commencements (SSC) are indicated in the figure as dashed vertical lines. The latitudes of O^+ and H^+ precipitation zones are indicated by solid bars which span the 10% to 99% latitude integrated flux levels. The 50% level is marked on each bar. This figure demonstrates that the O^+ zone tends to be displaced several degrees equatorward of the H^+ zone and that both zones respond in concert to variations in geomagnetic activity.

In a related study of the same storm, SHARP *et al.* (1976b) found only small differences in the proton and oxygen ion energy spectra in the 0.7–12 keV energy range, with H^+ having slightly higher average energy. It was noted that O^+ precipitation occurred at $L < 2$ on one occasion associated with a large main phase decrease in D_{st}.

Since the ionosphere was identified as the source of the O^+ it was expected that He^+ would also be easily observed if the source region were at high altitudes where He^+ is the next most abundant ion after H^+. JOHNSON *et al.* (1974) observed these ions using similar (Wein filter type) instruments on the two satellites 1969-25B and 1971-089A. In two examples, H^+, O^+, and He^+ precipitation was observed simultaneously.

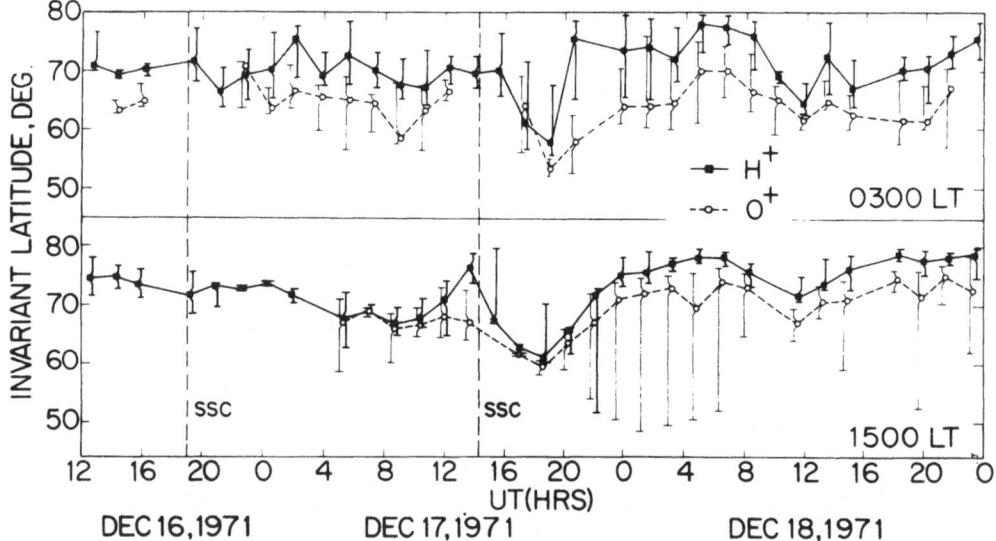

Fig. 7.   Locations of O^+ and H^+ precipitation zones during two geomagnetic storms in 1971. The dashed vertical lines indicate the times of sudden commencements (from SHARP *et al.*, 1976a).

Mass spectra recorded during these events from satellite 1971-089A, at three different energies, are shown in Fig. 8. Peaks due to He$^+$ and H$^+$ are clearly visible at all three energies; however, the O$^+$ precipitation, which was observed simultaneously at higher

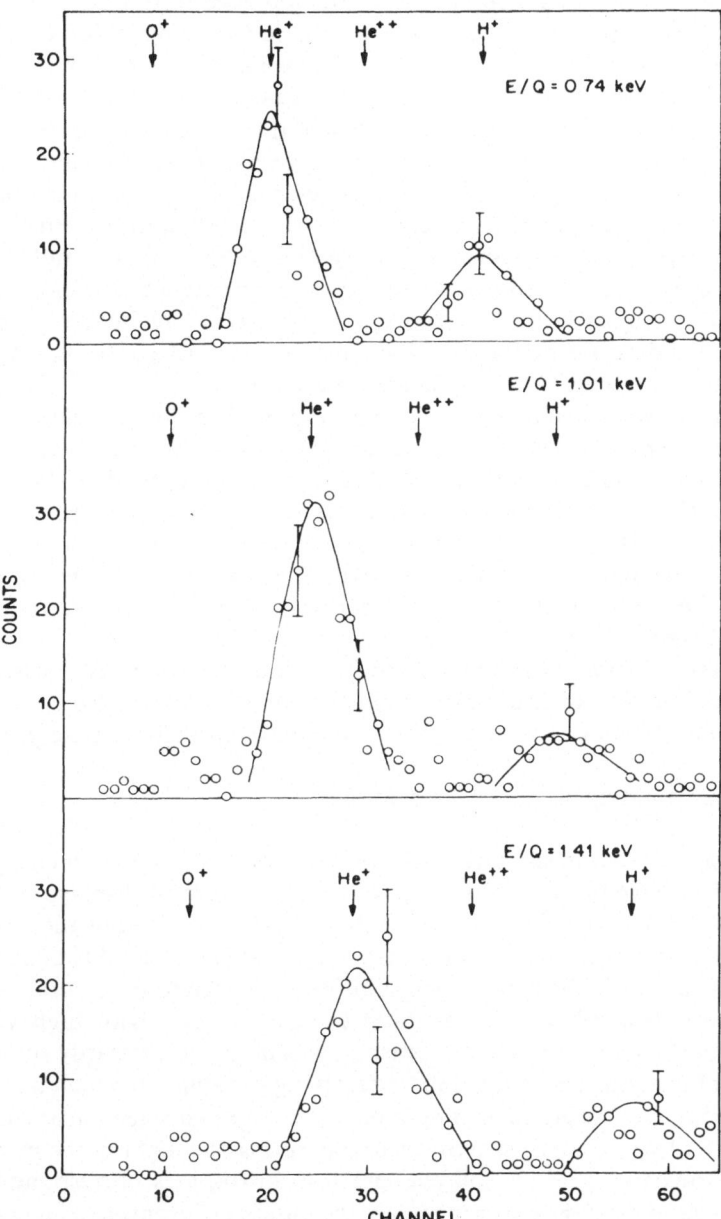

Fig. 8.   Mass (m/Q) spectra at three different energies from satellite 1971-089A. The arrows indicate the expected positions of various ion species (from JOHNSON *et al.*, 1974).

energies, does not appear above background in these mass spectra. No He^{++} ions, or He$^+$ with energies greater than 1.4 keV, were detected. Once again these authors considered their results to be consistent with an ionospheric source and comparisons of relative energy spectra led them to suggest that a mechanism which imparted equal velocities to all species was responsible for the ion acceleration.

As mentioned previously, the magnetospheric ion composition can be strongly influenced by loss as well as acceleration mechanisms. One of the major loss mechanisms for energetic ring current ions, charge-exchange, was considered by both TINSLEY (1976) and LYONS and EVANS (1976) and they concluded that He$^+$ may be the dominant ion in the ring current during the recovery phase of geomagnetic storms. SHARP et al. (1977a), using the mass spectrometer on the low altitude polar-orbiting satellite (1971-89A), searched for evidence of the He$^+$ precipitation during the recovery phase of a large geomagnetic storm with a negative result. Only H$^+$ and O$^+$ ions were observed. Recent measurements in the equatorial plane at high altitudes (e.g., GEISS et al., 1978) and at higher latitudes (JOHNSON et al., 1977) have tended to support the earlier low-altitude observations in that He$^+$ appears to contribute only a small fraction to the total density of ring current particles.

Ions of solar wind origin were also observed on these low-altitude satellites. For example, Fig. 9, which is from SHARP et al. (1974), shows mass spectra at three different energies which were taken over the nightside auroral zone. These data show that ions of ionospheric (O$^+$) as well as solar wind (He^{++}) origin may both be present in auroral precipitation. The He^{++} and H$^+$ energy per unit mass spectra were measured and found to have very similar shapes in this event. The integral He^{++}/He$^+$ intensity ratio was found to be approximately 4%, similar to solar wind values, and to increase with increasing latitude.

Thus, results from these low-altitude satellites corroborated earlier sounding rocket measurements which indicated a solar wind origin and also showed that the ionosphere can at times also be a major contributor to high latitude precipitation.

5.   Recent Sounding Rocket Observations

Second generation mass spectrometers with either an improved energy range or mass resolution have been developed and flown into auroral displays. For example, LYNCH et al. (1976) reported on a charge state analyser developed by their group which was similar in concept to that of REASONER et al. (1968) but capable of detecting ions with energies less than 20 keV, which is more appropriate to auroral events. This instrument, which combined electrostatic energy selection with high voltage pre-acceleration and solid state detector sensing, was designed primarily to measure the charge state of ions but had limited mass resolving capability for high mass particles.

The instrument was launched into a post-breakup hydrogen aurora and reached an apogee of 820 km. Charge spectra recorded during the flight in the most sensitive channel are shown in Fig. 10. Displayed here in six histograms, corresponding to three altitude intervals and two energy ranges, are the number of events as a function of pulse height in the solid state detector, where to a first order the pulse height is dependent on the incident ion charge state. The large peak marked H$^+$ corresponds to the sum of the

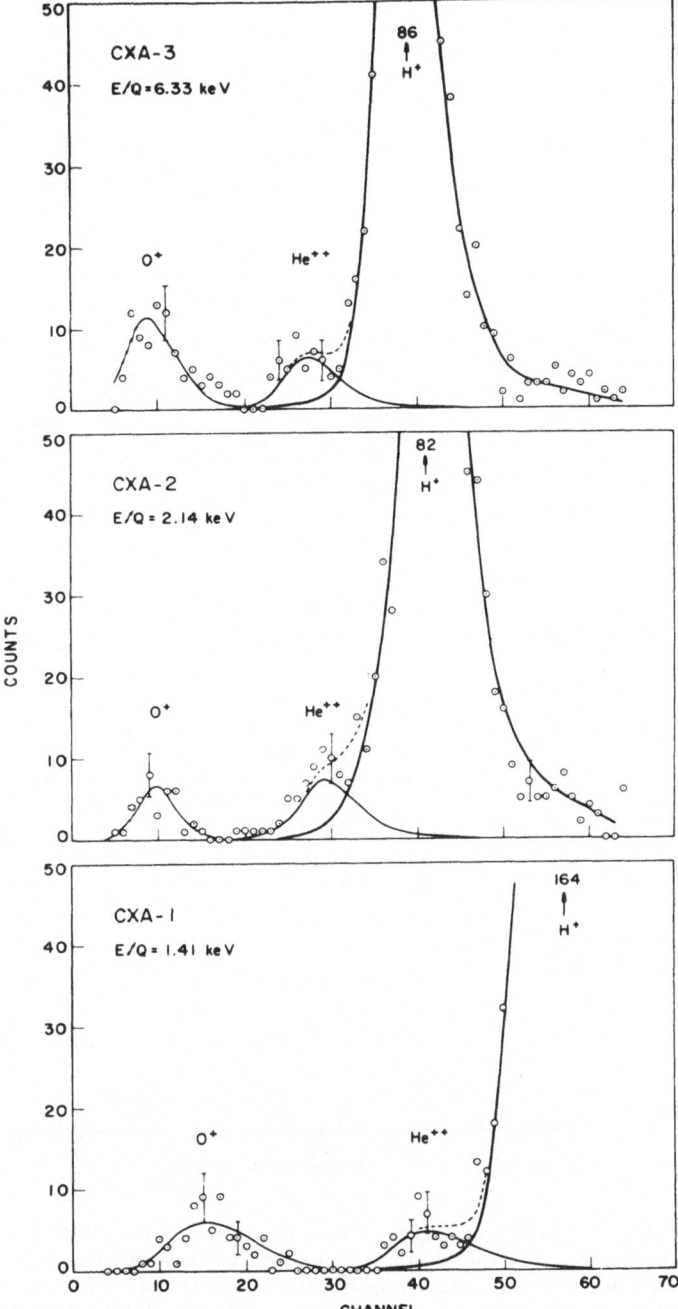

Fig. 9.   Mass (m/Q) spectra at three energies in the nightside auroral zone from the low-altitude spacecraft 1971-089A (from Sharp *et al.*, 1974).

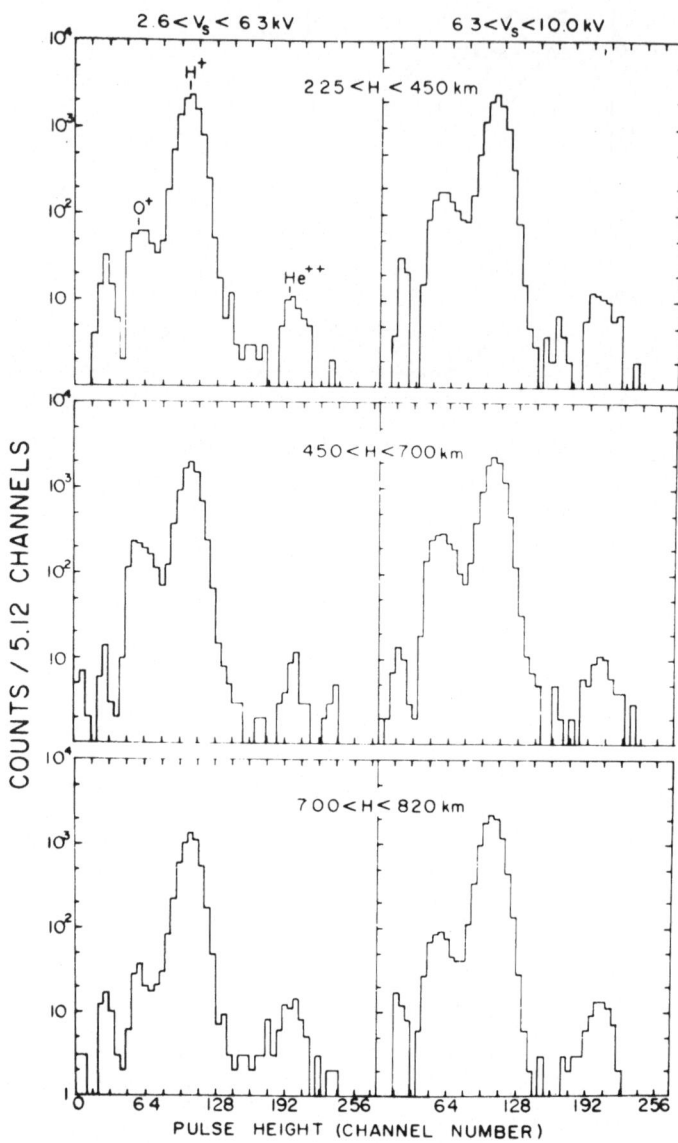

Fig. 10. Pulse height distribution from a charge-state analyser launched into a hydrogen aurora. The light ion pulse height is proportional to the incident ion charge state; for heavier ions the pulse height is reduced due to the pulse height defect. The histograms are divided into three altitude and two energy ranges. Three distinct pulse height groups are visible and are assumed to be due to H^+, He^{++} and O^+ (from LYNCH et al., 1976).

flux of all singly charged light ions and could include an admixture of He^+. The doubly charged ion peak is assumed to be He^{++}. The third peak at low pulse heights is marked O^+ but may be due to any energetic singly charged heavy ion species (e.g., C^+, N^+, or O^+).

The O^+/H^+ ratio for this flight was estimated to be approximately 10% and the He^{++}/H^+ ratio was found to have an average value of less than 1% and did not exceed 1.5% at its peak.

A search for highly charged ions such as O^{6+} or O^{7+}, which may be expected in auroral precipitation if a solar wind source is postulated (AXFORD, 1970), was also made with negative results. An upper limit on the flux of these highly charged ions was set at approximately 0.15% of the H^+ flux, which is slightly above the nominal solar wind value of 0.1%.

Data from this flight was further analysed by LYNCH et al. (1977). They observed correlated fluctuations in the intensity of various ion species and related these effects to variations in the source mechanism. In one instance simultaneous enhancements of H^+ and O^+ were observed at all energies. The lack of any measurable delay in arrival times was interpreted as an indication of a local (distance <1 R_e) source region. At other times during the flight velocity dispersion effects from a distant (~ 20 R_e) source were noted.

An excellent example of a similar event where time-of-flight velocity dispersion from a distant injection event separated the precipitating ions into distinct energy and mass groups was presented by CARLSON and TORBERT (1980). Using electrostatic (E/Q) analysers on two high altitude (550 km apogee) sounding rockets launched in the morning auroral oval, systematic variations in multiple peaked ion energy spectra were observed. Examples of two such energy spectra appear in Fig. 11. These peaks first

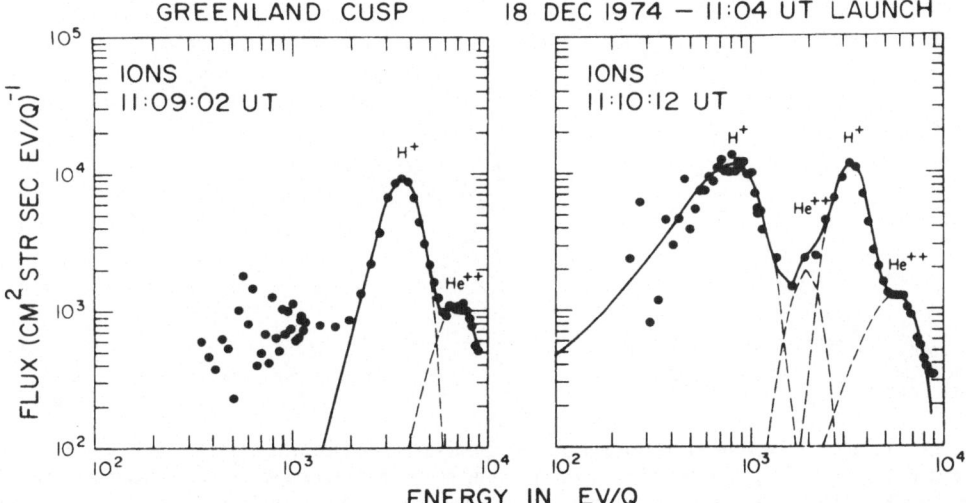

Fig. 11.   Energy spectra in the morning auroral oval. Peaks in the energy spectra are identified with different ion species (from CARLSON and TORBERT, 1980).

appeared at high energies and the peak energies decreased monotonically with time. By identifying each peak with a specific ion species and assuming a simultaneous injection of all ions at a distant source, the authors were able to reconstruct the temporal history of the event.

To test the model, dispersion curves were constructed where the abscissa was universal time and the ordinate was the inverse of the ion velocity at the peaks in the energy spectra. These results are displayed in Fig. 12. If the ions were injected simultaneously the locus of the points should be a straight line and the intercept should be the injection time. The H^+ line was derived by least squares fitting the data and the He^{++} and He^+ were predicted based on the H^+ result. As can be seen in this figure, reasonably good agreement with observation is found if it is assumed that the ion precipitation is composed of H^+, He^{++}, and He^+. The best fit source distance was found to be $\sim 12\ R_e$.

The authors concluded that the ion injection was associated with turbulent plasma entry into the magnetosphere at the magnetopause and that the presence of the large He^{++} flux was indicative of a solar wind source for most of the ions. A mixing of ionospheric and solar wind plasma was suggested as the source for He^+. The consistency of the time-of-flight model and the similarity of the observed energy spectra to typical magnetosheath spectra, implied that no significant acceleration (or deceleration) occurred between the injection point and the ionosphere.

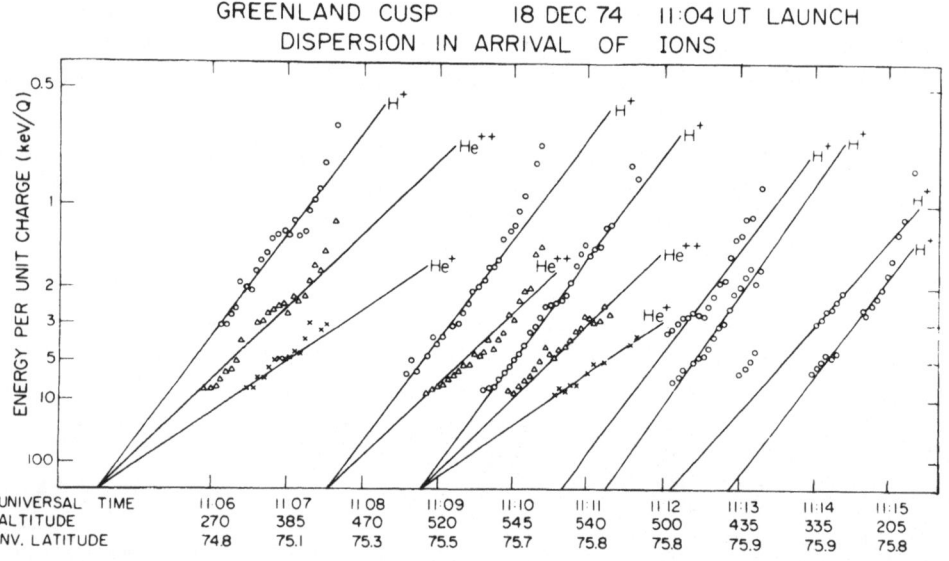

Fig. 12.   Velocity dispersion curves for various ion species identified in energy spectra (as in Fig. 11). The inverse of the ion velocity, labelled by the corresponding energy per-unit-carge, is plotted against universal time. The H^+ line was derived by least squares fitting the positions of the dominant peaks. The He^{++} and He^+ lines are predicted, that is, the slopes are $1/\sqrt{2}$ and $1/\sqrt{4}$ of the H^+ with the same intercept (from CARLSON and TORBERT, 1980).

An energetic ion mass spectrometer was recently developed by MOORE and EVANS (1979) and launched into a diffuse aurora near local midnight. This instrument combined electrostatic and magnetic analysis in a focusing geometry to give high mass resolution while still retaining a reasonably good geometric factor (sensitivity). A microchannel plate array was placed in the focal plane of the instrument to record the mass spectrum. Only H^+ and He^{++} were detected and the He^{++} flux was close to the noise level. An upper limit on the He^{++}/H^+ ratio was set at the 2 to 4% level. These results were once again indicative of a solar wind origin although it was pointed out that a component of the H^+ may have had its origin in the ionosphere.

6.   The Ionosphere as a Source for Magnetospheric Ions

Observations of the magnetospheric ion population have clearly shown that the ionosphere as well as the solar wind can be a major ion source. Low-altitude observations can only give indirect information on solar wind entry, as indicated previously; however, they are crucial to the understanding of the ionospheric source mechanism. In this section we review observations in the ionospheric source region and discuss some of the proposed injection mechanisms.

One possible mechanism for injecting ionospheric ions into the magnetosphere was suggested by BANKS and HOLZER (1968). They showed that the hydromagnetic expansion of the ionospheric plasma on open field lines could produce a supersonic flow of ions out of the polar cap ionosphere and named this the polar wind. Those ions would presumably populate the geomagnetic tail and then be convected into the magnetosphere to form a portion of the energetic magnetospheric ion population. Upward flowing ionospheric H^+ and He^+ have been detected in the polar cap by HOFFMAN et al. (1974); however, a corresponding O^+ flux was not detected which makes the polar wind an unlikely candidate for the major ionospheric source mechanism.

A source capable of producing the required energetic ion mass composition was first identified by SHELLEY et al. (1976a) using an ion mass spectrometer on the S3-3 satellite. They observed energetic H^+ and O^+ ions at approximately 5,500 km altitude flowing along field lines away from the ionosphere and suggested a parallel electric field acceleration mechanism for these ions.

A second type of upward flowing ion event was first observed by KLUMPAR (1975) in the electrostatic (E/Q) analyser on the ISIS 2 spacecraft. These events are characterized by ion pitch-angle distributions peaked between 90° and 180°, where 180° corresponds to ions moving away from the ionosphere, and have energy spectra similar to the field-aligned events reported at higher altitudes. Since the characteristic angular distribution is produced when ions are accelerated perpendicular to the local magnetic field lines, they are commonly referred to as Transversely Accelerated Ions (TAI). These events have also been called "conics" because of the conical shape of the angular distributions observed at high altitudes.

A clear example of a TAI event was reported by WHALEN et al. (1978). They flew electrostatic analysers on a high altitude sounding rocket which was launched into the expansive phase of an auroral substorm near local midnight. When the rocket reached

~400 km intense beams of energetic (up to 500 eV) ions at 90° pitch angles were encountered and as the attitude increased the beam appeared at progressively larger pitch-angles. An example of the pitch-angle distributions observed near apogee is shown in Fig. 13. At high energies (5.3 keV) an isotropic downcoming distribution with a normal atmospheric loss cone at large pitch angles is observed whereas at low energies an intense highly collimated beam-like structure is present at ~105°. The authors commented that the beam angular width was less than or of the order of the instrument resolution (10°). The angles of the peaks in the angular distributions for the five lowest energy channels of the spectrometer are shown in Fig. 14. The systematic change in angle with altitude, obvious in these data, was shown to be consistent with adiabatic expansion (i.e., conservation of magnetic moment) of an ion beam injected at 90° in the 400 to 500 km altitude range.

The mass distribution of TAI events has not been measured at low altitudes; however, the composition of the corresponding high altitude distributions has been reported, for example, by SHARP et al. (1977b). They observed large fluxes of upward flowing ion "conics" with energy spectra similar to the low altitude TAI events and composition measurements indicated that a substantial fraction of these ions were O^+. Thus these "conics", and the beams discussed earlier (SHELLEY et al., 1976a), seem to be signatures of the required ionospheric source mechanism.

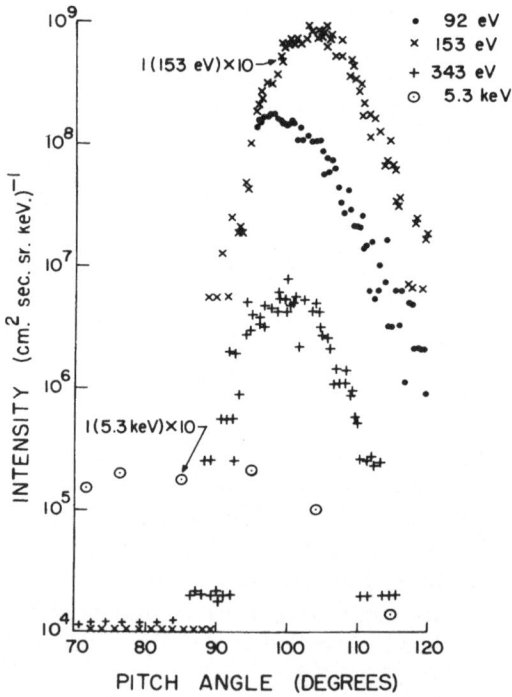

Fig. 13.   Ion pitch-angle distributions observed in a sounding rocket flight into an active aurora. The three low-energy ion distributions are beam-like, peaked near 105°, whereas the energetic (5.3 keV) ions are isotropic over the downcoming hemisphere (from WHALEN et al., 1978).

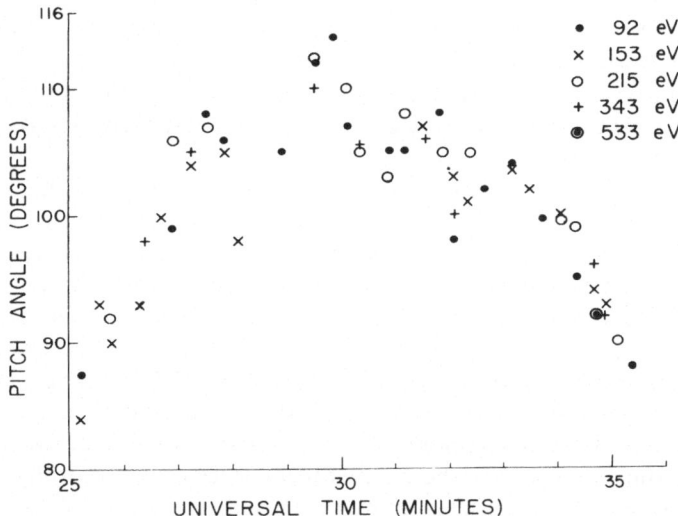

Fig. 14.   Angles of the ion pitch-angle distribution peaks as a function of time for five energy channels. This population of low energy ions was first observed at 90° near 400 km and shifted to 110° at apogee (from WHALEN *et al.*, 1978).

For summaries of the low altitude observations of TAI events the reader is referred to KLUMPAR (1979) and for high altitude observations reviews by SHELLEY (1979) and JOHNSON (1979) are recommended along with related papers in this volume.

Indications that low altitude ion acceleration may occur near the equator appeared in an article by PRANGÉ and CRIFO (1977). On a rocket flight through the equatorial ionosphere ($L \sim 1.3$) during a magnetic storm, large fluxes of suprathermal (E/Q up to 1 keV) ions were observed flowing along magnetic field lines out of the ionosphere. These observations were made at low altitudes (~ 350 km) where charge exchange effects are very important. Parallel acceleration of ionospheric ions at very low altitudes to high energies is therefore implied by these observations. Confirmation of these results and some indication of the frequency of occurrence of such events is required before concluding that such a mechanism plays a significant role in the production of energetic ionospheric ions.

7.   Discussion

High altitude energetic ion mass composition measurements present a picture of the magnetospheric population whose composition may be dominated by ions of solar wind or ionospheric origin with ionospheric ions apparently playing an increasingly more important role during periods of high geomagnetic activity. For reviews of these observations see SHELLEY (1979), JOHNSON (1979). Support for this picture may be found in some of the low-altitude observations presented here; however, it is not clear that a completely consistent picture has emerged for high latitude auroral events where ions of solar wind origin seem to dominate the relatively small sampling. This inconsistency may simply be a statistical fluctuation or may be a reflection of the geomagnetic and geo-electric field topology associated with the auroral acceleration mechanism. Improved sampling of the precipitating auroral ion composition should remove this apparent uncertainty and give a clearer picture of the relative contribution of the two competing sources.

One of the obvious applications for ion mass spectrometers is in the area of ion tracer experiments. Naturally occurring precipitation can be used as tracers of magnetospheric transport, energization and loss mechanisms as demonstrated in several of the reports presented here. For example, CARLSON and TORBERT (1980) used time-of-flight dispersion measurements of various ion species to infer the distance and characteristics of the ion source and also to investigate possible acceleration regions between the source and the ionosphere. Similar electric field tracer experiments were reported by SHELLEY et al. (1976b) where latitude profiles of energetic ion fluxes were used to infer the poleward component of the magnetospheric convection velocity. Improvements and extensions of these tracer techniques using naturally occurring ion fluxes will be attempted in the near future using multiple spacecraft on both the DE (Dynamics Explorer) and the proposed OPEN (Origin of Plasmas in Earth's Neighlorhoods) program.

A number of ion tracer experiments involving artificial releases (e.g., Lithium and Barium) which are easily distinguished from naturally occurring ions, are also planned. The AMPTE (Active Magnetospheric Particle Tracer Experiment) program will make use of the technique to study solar wind access to the magnetosphere and high latitude transport and energization processes. Plans are also under way to perform active ion tracer experiments at low altitudes on the Shuttle-Spacelab missions.

Intensive investigation of the low altitude ionospheric source mechanism(s) is expected to continue. The signatures of low altitude ion acceleration were summarized previously. Clearly the conical distributions resulting from TAI events and the field-aligned fluxes observed at high altitudes have the characteristics (i.e., composition, flux and energy distribution) required to explain the ionospheric component of the observed magnetospheric ion composition. Very little, however, is known of the mechanism responsible for the generation of these beams. Basic questions relating to the mechanism remain to be answered, such as: What is the free energy source for the acceleration? How is this energy transferred to the ionospheric ions? Is the mechanism mass or charge dependent? What ambient conditions are required?

Some preliminary attempts at modelling the source mechanism have been made. For example, UNGSTRUP et al. (1979) suggested that electrostatic ion cyclotron waves induced by a current driven instability (KINDEL and KENNEL, 1971) were responsible for the perpendicular heating and subsequent injection into the magnetosphere. WHALEN et al. (1978) also found their data to be consistent with such an interpretation. However, extensive plasma diagnostic measurements in the low-altitude source region itself along with an in-depth theoretical treatment are required to definitely identify the process.

It is clear from the preceeding that ion mass spectrometer measurements have produced a relatively complete, but mostly empirical, picture of the magnetospheric ion population, although some crucial holes in the picture do exist in the 20 to 200 keV energy range. Our knowledge or understanding of the plasma energization and transport processes which feed this population is, however, very limited. The challenge now is to develop techniques, both experimental and theoretical, which will address this more fundamental shortcoming.

REFERENCES

ALBERT, R. D., Detection of the presence of alpha particles in auroral fluxes (abstract), *EOS Trans. AGU*, **48**, 866, 1967.

ALLISON, S. K., Experimental results on charge-changing collisions of hydrogen and helium atoms and ions at kinetic energies above 0.2 keV, *Rev. Mod. Phys.*, **30**, 1137, 1958.

AXFORD, W. I., On the origin of radiation belt and auroral primary ions, in *Particles and Fields in the Magnetosphere*, edited by B. M. McCormac, p. 46, D. Reidel, New York, 1970.

AXFORD, W. I., F. BÜHLER, and H. J. A. CHIVERS, Auroral helium precipitation, *J. Geophys. Res.*, **77**, 6724, 1972.

BANKS, P. M. and T. E. HOLZER, The polar wind, *J. Geophys. Res.*, **73**, 6846, 1968.

BÜHLER, F., W. I. AXFORD, H. J. A. CHIVERS, and K. MARTI, Helium isotopes in an aurora, *J. Geophys. Res.*, **81**, 111, 1976.

CARLSON, C. W. and R. B. TORBERT, Solar wind ion injections in the morning auroral oval, *J. Geophys. Res.*, **85**, 2903, 1980.

DAVIDSON, G. T., Spatial distribution of low-energy protons precipitated in the auroral zone, *J. Geophys. Res.*, **70**, 1061, 1965.

EATHER, R. H., Radiation from positive particles penetrating the auroral atmosphere, *J. Geophys. Res.*, **71**, 4133, 1966.

EATHER, R. H., Auroral proton precipitation and hydrogen emissions, *Rev. Geophys. Space Phys.*, **5**, 207, 1967.

EATHER, R. H., A search for helium emissions in the auroral zone, *Space Res.*, **8**, 201, 1968.

GEISS, J., P. EBERHARDT, F. BUHLER, J. MEISTER, and P. SIGNER, Apollo 11 and 12 solar wind composition experiments: fluxes of He and Ne isotopes, *J. Geophys. Res.*, **75**, 5972, 1970.

GEISS, J., H. BALSIGER, P. EBERHARDT, H. P. WALKER, L. WEBER, and D. T. YOUNG, Dynamics of magnetospheric ion composition as observed by the Geos mass spectrometer, *Space Sci. Rev.*, **22**, 537, 1978.

HENDRICKSEN, K., Intensity enhancements on the wavelength of the He I 5,876 Å emission observed at high latitude, Proc. Esrange Symposium, Ajaccio, April (EAS-SP-135, June), 1978.

HOFFMAN, J. H., W. H. DODSON, C. R. LIPPINCOTT, and H. D. HAMMACK, Initial ion composition results from the ISIS 2 satellite, *J. Geophys. Res.*, **79**, 4246, 1974.

JOHNSON, R. G., Energetic ion composition in the earth's magnetosphere, *Rev. Geophys. Space Phys.*, **17**, 696, 1979.

JOHNSON, R. G., R. D. SHARP, and E. G. SHELLEY, The discovery of energetic He$^+$ ions in the magnetosphere, *J. Geophys. Res.*, **79**, 3135, 1974.

JOHNSON, R. G., R. D. SHARP, and E. G. SHELLEY, Composition of the hot plasmas in the magnetosphere, in *Physics of the Hot Plasma in the Magnetosphere*, edited by B. Hultqvist and L. Stenflo, p. 45, Plenum, New York, 1975.

JOHNSON, R. G., R. D. SHARP, and E. G. SHELLEY, Observation of ions of ionospheric origin in the storm-time ring current, *Geophys. Res. Lett.*, **4**, 403, 1977.

KINDEL, J. M. and C. F. KENNEL, Topside current instabilities, *J. Geophys. Res.*, **76**, 3055, 1971.

KLUMPAR, D.. M., Auroral zone non-thermal positive ions observed with ISIS 2 (Abstract) in *Program and Abstracts,*: XVII General Assembly, IAGA Bull. 36, International Association of Geomagnetism and Aeronomy, Grenoble, 1975.

KLUMPAR, D. M., Transversely accelerated ions: An ionospheric source of hot magnetospheric ions, *J. Geophys. Res.*, **84**, 4229, 1979.

LIND, D. L., J. GEISS, and W. STETTLER, solar and terrestrial noble gases in magnetospheric precipitation, *J. Geophys. Res.*, **84**, 6435, 1979.

LYNCH, J., D. PULLIAM, R. LEACH, and F. SCHERB, The charge spectrum of positive ions in a hydrogen aurora, *J. Geophys. Res.*, **81**, 1264, 1976.

LYNCH, J., R. LEACH, D. PULLIAM, and F. SCHERB, Composition and energy spectrum variations of auroral ions, *J. Geophys. Res.*, **82**, 1951, 1977.

LYONS, L. R. and D. S. EVANS, The inconsistency between proton charge exchange and the observed ring current decay, *J. Geophys. Res.*, **81**, 6197, 1976.

MEINEL, A. B., Doppler shifted auroral hydrogen emission, *Astrophys. J.*, **113**, 50, 1951.

MOORE, T. E. and D. S. EVANS, Distribution of energetic positive ion species above a diffuse midnight aurora, *J. Geophys. Res.*, **84**, 6443, 1979.

PRANGE, R. and J. CRIFO, Suprathermal particle observations in the nighttime ionosphere at low latitudes, *Geophys. Res. Lett.*, **4**, 141, 1977.

REASONER, D. L., Auroral helium precipitation, *Rev. Geophys. Space Phys.*, **11**, 169, 1973.

REASONER, D. L., R. H. EATHER, and B. J. O'BRIEN, Detection of alpha particles in auroral phenomena, *J. Geophys. Res.*, **73**, 4185, 1968.

SHARP, R. D., R. G. JOHNSON, and E. G. SHELLEY, Satellite measurements of auroral alpha particles, *J. Geophys. Res.*, **79**, 5167, 1974.

SHARP, R. D., R. G. JOHNSON, and E. G. SHELLEY, The morphology of energetic O^+ ions during two magnetic storms: Temporal variations, *J. Geophys. Res.*, **81**, 3283, 1976a.

SHARP, R. D., R. G. JOHNSON, and E. G. SHELLEY, The morphology of energetic O^+ ions during two magnetic storms, latitudinal variations, *J. Geophys. Res.*, **81**, 3292, 1976b.

SHARP, R. D., E. G. SHELLEY, and R. G. JOHNSON, A search for helium ions in the recovery phase of a magnetic storm, *J. Geophys. Res.*, **82**, 2361, 1977a.

SHARP, R.D., R. G. JOHNSON, and E. G. SHELLEY, Observation of an ionospheric acceleration mechanism producing energetic (keV) ions primarily normal to the geomagnetic field direction, *J. Geophys. Res.*, **82**, 3324, 1977b.

SHELLEY, E. G., Heavy ions in the magnetosphere, *Space Sci. Rev.*, **23**, 465, 1979.

SHELLEY, E. G., R. G. JOHNSON, and R. D. SHARP, Satellite observations of energetic heavy ions during a geomagnetic storm, *J. Geophys. Res.*, **77**, 6104, 1972.

SHELLEY, E. G., R. D. SHARP, and R. G. JOHNSON, Satellite observations of an ionospheric acceleration mechanism, *Geophys. Res. Lett.*, **3**, 654, 1976a.

SHELLEY, E. G., R. D. SHARP, and R. G. JOHNSON, He^{++} and H^+ flux measurements in the dayside cusp: Estimates of convection electric field, *J. Geophys. Res.*, **81**, 2363, 1976b.

SIVJEE, G. G., K. HENDRICKSEN, and C. S. DEEHR, Orthohelium emissions at 3,889 and 5,876 Å in the polar upper atmosphere, *J. Geophys. Res.*, **85**, 6043, 1980.

STOFFREGEN, W., Transient emissions on the wavelength of helium I, 5,876 Å, recorded during auroral break-up, *Planet. Space Sci.*, **17**, 1927, 1969.

TINSLEY, B. A., Evidence that the recovery phase ring current consists of helium ions, *J. Geophys. Res.*, **81**, 6193, 1976.

TORR, M. R., J. C. G. WALKER, and D. G. TORR, Escape of fast oxygen from the atmosphere during geomagnetic storms, *J. Geophys. Res.*, **79**, 5267, 1974.

UNGSTRUP, E., D. M. KLUMPAR, and W. J. HEIKKILA, Heating of ions to suprathermal energies in the topside ionosphere by electrostatic ion cyclotron waves, *J. Geophys. Res.*, **84**, 4289, 1979.

WARASILA, R. L. and O. A. SCHAEFFER, $^3He/^4He$ ratios in the lower radiation belt as measured by trapped particles in a recovered satellite, *Geophys. Res. Lett.*, **2**, 480, 1975.

WAX, R. L. and W. BERNSTEIN, Discussion of abstract by R. D. Albert, "Detection of the presence of alpha particles in auroral fluxes," *J. Geophys. Res.*, **73**, 4452, 1968.

WHALEN, B. A. and I. B. MCDIARMID, Direct measurement of auroral alpha particles, *J. Geophys. Res.*, **73**, 2307, 1968.

WHALEN, B. A. and I. B. MCDIARMID, Further low-energy auroral-ion composition measurements, *J. Geophys. Res.*, **77**, 1306, 1972.

WHALEN, B. A., J. R. MILLER, and I. B. MCDIARMID, Evidence for a solar wind origin of auroral ions from low-energy ion measurements, *J. Geophys. Res.*, **76**, 2406, 1971.

WHALEN, B. A., W. BERNSTEIN, and P. W. DALY, Low altitude acceleration of ionospheric ions, *Geophys. Res. Lett.*, **5**, 55, 1978.

YEE, J. H. and P. B. HAYS, The oxygen polar corona, *J. Geophys. Res.*, **85**, 1795, 1980.

YEE, J. H., J. W. MERIWETHER, Jr., and P. B. HAYS, Detection of a corona of fast oxygen atoms during solar maximum, *J. Geophys. Res.*, **85**, 3396, 1980.

Energetic Ion Composition in the Earth's Magnetosphere, edited by R. G. Johnson, 167–193.
Copyright © 1983 by Terra Scientific Publishing Company (TERRAPUB), Tokyo.

Hot Plasma Composition Results from the S3-3 Spacecraft

R. D. Sharp, A. G. Ghielmetti, R. G. Johnson, and E. G. Shelley

*Space Sciences Laboratory, Lockheed Palo Alto Research Laboratory,
3251 Hanover Street, Palo Alto, California, U.S.A.*

(Received March 26, 1982)

The S3-3 satellite discovered the principal auroral acceleration region at altitudes of about 1 R_E over the auroral zone. Intense fluxes of upward flowing O^+ and H^+ ions with keV energies were commonly observed in this region of the magnetosphere. The detailed morphology of these upflowing ions is described, including their latitude, local time, altitude, and magnetic activity dependences and their relationship to the trapped keV electron population. The first measurements of the composition of the trapped keV ions in the radiation belts are also described, showing the importance of the ionospheric source term to the storm time population of ions with energies ≤ 16 keV/e.

1. Introduction

Beginning in about 1970, certain characteristics of the distribution functions of the precipitating auroral electrons began to be interpreted as evidence that an electrostatic acceleration process was acting to energize them (Gurnett, 1972; Evans, 1975). Satellite measurements were unable to provide any direct confirmation of these inferences because the satellites were generally limited to low altitude polar or geostationary orbits, outside the region where the acceleration process occurs. In July 1976 the Air Force satellite S3-3 with a modest payload of energetic particle and field detectors was placed into a near polar orbit with an apogee of ~8,000 km. It immediately became clear that at that altitude it was encountering the region where the principal auroral acceleration mechanisms were operative (Shelley *et al.*, 1976; Sharp *et al.*, 1977b). Direct measurements of electric field intensities of up to several hundred millivolts/meter were provided by the U. C. Berkeley experiment (Mozer *et al.*, 1979). Signatures of in situ acceleration processes were found with the particle detectors (Mizera and Fennell, 1977; Sharp *et al.*, 1979). The directions of the inferred electrostatic potentials were such that they energized electrons in the downward direction and precipitated them into the atmosphere while simultaneously accelerating ambient ionospheric ions upward, injecting them into the radiation belts. The relative scarcity of observations of downward flowing ions in the altitude range of S3-3 suggested that this injection process was quite efficient (Ghielmetti *et al.*, 1979).

The Lockheed experiment on S3-3 contained an energetic ion mass spectrometer and provided the added dimension of ion composition information to the understanding of the complex plasma processes operative in this region. These measurements have provided useful information in three principal areas:

1. The study of the detailed characteristics of the several acceleration mechanisms found to be operative in the 4,000–8,000 km altitude region on auroral field lines.

2. The definition of the morphology of the upward flowing energetic ions which are the principal source term for that portion of the magnetospheric particle population which is of ionospheric origin.

3. The composition of the stormtime ring current. S3-3 provided the first direct measurements of the ion composition of this population of particles (JOHNSON et al., 1977).

In this review we shall examine the role of the ion composition data in understanding the magnetospheric processes operating in each of these three areas.

The S3-3 ion mass spectrometer experiment contained three individual sensors, each of which covered a separate portion of the energy per unit charge range from 0.5 to 16 keV/e. The sensors were mounted with their view directions perpendicular to the spacecraft spin axis so that they obtained a complete angular scan at three energies approximately every 20 sec. A 12 point energy spectrum of ions with mass per charge from 1 through 32 was acquired every 64 sec. A more complete description of the experiment is contained in SHARP et al. (1977b).

2. Ion Acceleration Mechanisms

These results have recently been reviewed (SHARP, 1981) and we will present here only a brief summary of the principal conclusions to this time.

Upward flowing H^+ and O^+ ions with energies in the keV range were frequently observed on auroral field lines by S3-3 at altitudes above 4,000 km. He^+ ions were also observed but substantially less frequently and with much lower intensity than H^+ and O^+. The upward flowing ions were often found in association with the signatures of parallel electric fields in the keV electron distributions. These signatures on occasions allowed a quantitative estimate to be made of an electrostatic potential difference below the spacecraft, and in a number of cases when direct comparisons were made this potential was found to correspond approximately to the energy at which the upflowing ions exhibited a peak in their energy spectrums (CLADIS and SHARP, 1979; MIZERA et al., 1981). This implies that a large fraction of the ion energy was acquired electrostatically. The spectral peaks were quite broad, however, and the angular distributions of the upflowing ion beams were typically wider than would be expected if the beams resulted from a pure electrostatic acceleration of the ambient thermal plasma. Also, upon occasions the upflowing ions exhibited a "conical" pitch angle distribution (with a local minimum in the distribution along the direction of the magnetic field). These characteristics implied that some transverse acceleration mechanism was also involved in the energization process.

Statistical results, which will be discussed below in more detail, showed that the energy of the O^+ beams was typically $1\frac{1}{2}$ to 2 times that of the H^+, and their angular widths were substantially broader. The median value of the half widths was 24° for the O^+ beams and 15° for the H^+ (COLLIN et al., 1981). If we further consider that the upflowing ion beams (with $E > 0.5$ keV) are only found at altitudes above about 4,000 km (GHIELMETTI et al., 1978), these parameters allow us to set limits on the relative strengths of the transverse and parallel ion acceleration mechanisms. It was concluded by COLLIN et al. (1981) that the H^+ ion beams could typically only receive a small fraction of their energy from a transverse acceleration mechanism acting above 4,000 km or they would have angular widths larger than those observed. The O^+ ions on the other hand could derive almost half of their energy from such a mechanism. These results were taken to imply that:

1. The H^+ ions were accelerated primarily by parallel electric fields.
2. The O^+ received on the average the same parallel acceleration as did the H^+.
3. The excess energy in the O^+ relative to the H^+ was provided by some transverse acceleration mechanism that preferentially acted on the O^+ constituent of the ion beams.

Some of the other results of these statistical studies which are relevant to our understanding of the properties of the ion acceleration mechanism as well as to the characterization of the ionospheric source term for the ring current are discussed in Section 4.

3. Morphology of Upward Flowing Ions

Some of the results of a study of the characteristics of the upward flowing ion events were reported by GHIELMETTI (1978), GHIELMETTI et al. (1978), and SHELLEY et al. (1980). The data which were utilized were acquired during the period July 13, 1976 to September 27, 1977, and included 925 orbits covering the complete range of local time. Because of the phasing of the orbit with season, almost all of the data were acquired over the summer hemisphere. The polar regions between 60 and 84 degrees invariant latitude (ILA) were subdivided into unit bins of size $2°$ ILA by three hours magnetic local time (MLT) by 1,000 km altitude. A bin was defined to have been sampled if at least one complete pitch angle scan was acquired while the satellite passed through it. An upward flowing ion (UFI) event was defined as the occurrence of at least one observation within a unit bin of an anisotropic pitch angle distribution with a maximum in the upward moving direction. Additionally it was required that the flux exceed the sensitivity threshold of $\sim 2 \times 10^6$ keV/(cm^2 sec ster keV) and also exceed the penetrating energetic electron background.

The probability of occurrence was defined as the ratio of the number of UFI events to the number of samples. For this analysis no distinction was made between field aligned distributions (beams) and conical events but the latter were relatively rare in this energy range and so the results are generally representative of field aligned events. For each event, in addition to the occurrence probability, the following parameters were recorded for the two most commonly observed ion species (H^+ and O^+): The peak differential energy flux, the energy at the peak flux, the maximum energy at which upflowing ions were observed, the magnetic local time, the invariant latitude, the altitude, and K_p. A total of about 19,397 samples were obtained in the period of this analysis and about 936 upward flowing ion events were observed.

Figure 1 shows a principal result of the study, a magnetic local time, invariant

PERIOD: 7-13-76 - 9-27-77   ALTITUDE 6-8000 km

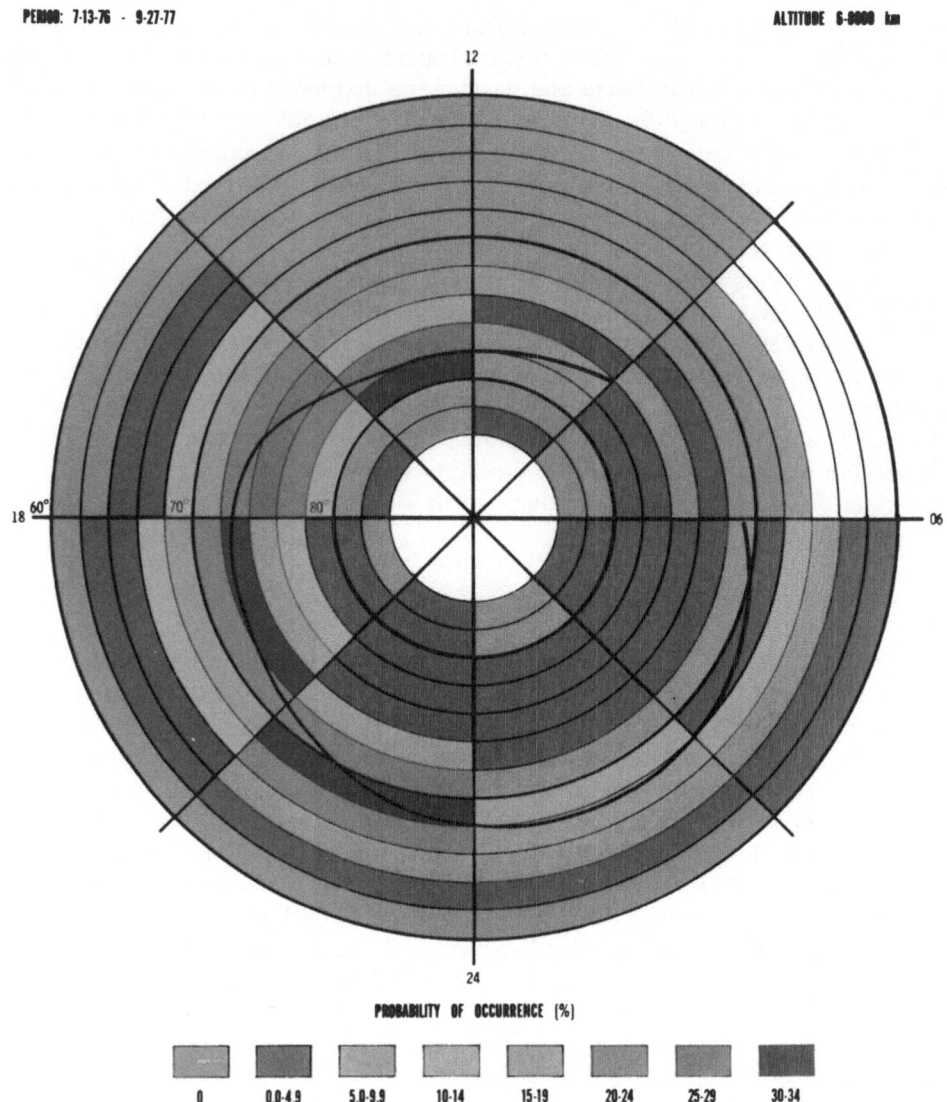

Fig. 1. The probability of occurrence of upward flowing ions as observed by the ion mass spectrometer on S3-3 between 6,000 and 8,000 km altitude. The solid black curve connects the values of the average latitude within each 3 hour local time sector.

latitude map of the probability of occurrence of upward flowing ions observed between 6,000 and 8,000 km altitude within the energy and sensitivity ranges defined above. The solid black curve connects the values of the average latitude within each local time sector. The white area in the 06–09 hour sector was deleted because it was not adequately sampled. The principal features of this distribution, the dramatic maximum in local evening and the general correspondence to the auroral oval, are in agreement with the previous less detailed results derived from the more limited data set published earlier (GHIELMETTI *et al.*, 1978).

3.1 Local time dependence

The local time asymmetry in the probability of occurrence is more strikingly evident in Fig. 2 which was obtained from the results in Fig. 1 by summing over the latitudinal distributions within each three hour local time sector. The error bars in this and subsequent figures represent the standard deviations of the means of the probability distributions. In order to investigate the dependence of this distribution on magnetic storm activity the data were sudivided into "disturbed" periods, during the inital phase or early recovery phase of magnetic storms, and "quiet" periods which had relatively low *Dst* and were at least several days after the recovery phase of the last previous magnetic storm. These results are shown in Figure 3. One sees that there is only a modest increase in occurrence probability during the storms and no qualitative difference in the pattern between disturbed and quiet times.

The local time dependence of the energy of the upflowing ions was investigated by forming the average of the energies at which the peak energy flux was recorded for each event. These results are illustrated in Figure 4. One sees a strong dependence on local time with the hardest events being observed in the premidnight sector. The O^+ is seen to be substantially harder than the H^+. A rough indication of the energy width of the ion distribution functions is also illustrated for the O^+ events by the dashed curve in Figure 4. This is the average of the maximum energies at which flux was observable on each event.

The ion composition was also observed to have a significant local time dependence. Figure 5 shows the ratio of the occurrence frequencies of O^+ and H^+ events

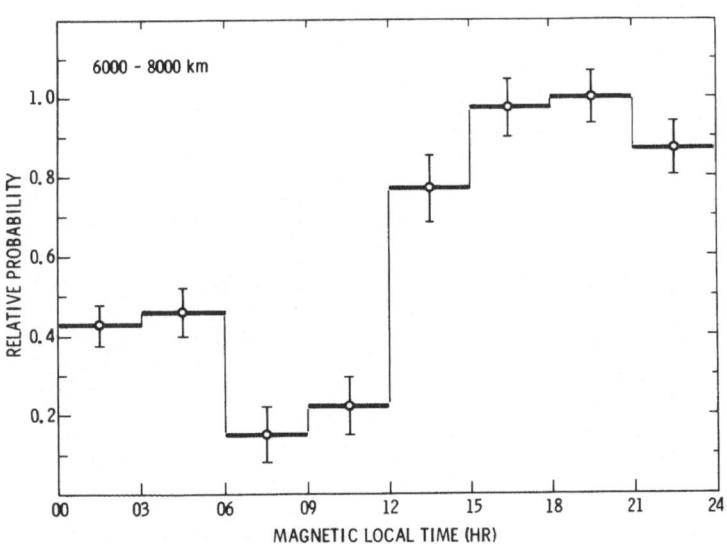

Fig. 2. Relative probability of occurrence of upward flowing ions as a function of magnetic local time.

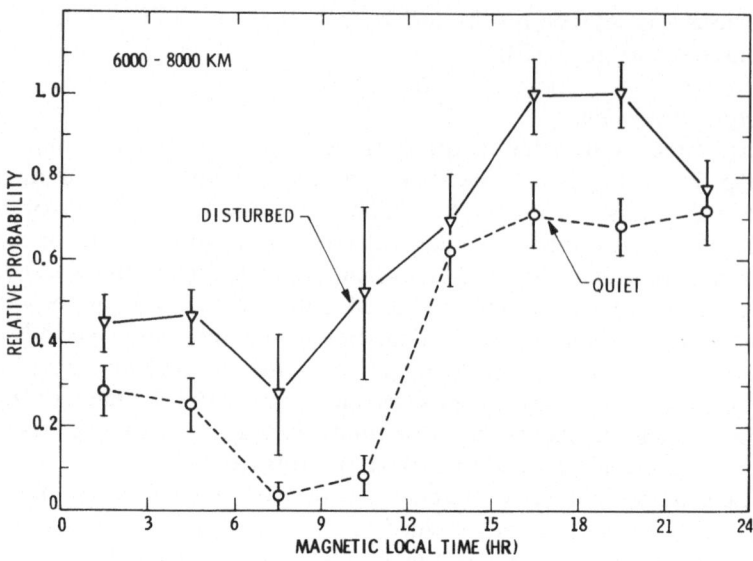

Fig. 3. A comparison of the probabilities of upward flowing ions between quiet and disturbed periods.

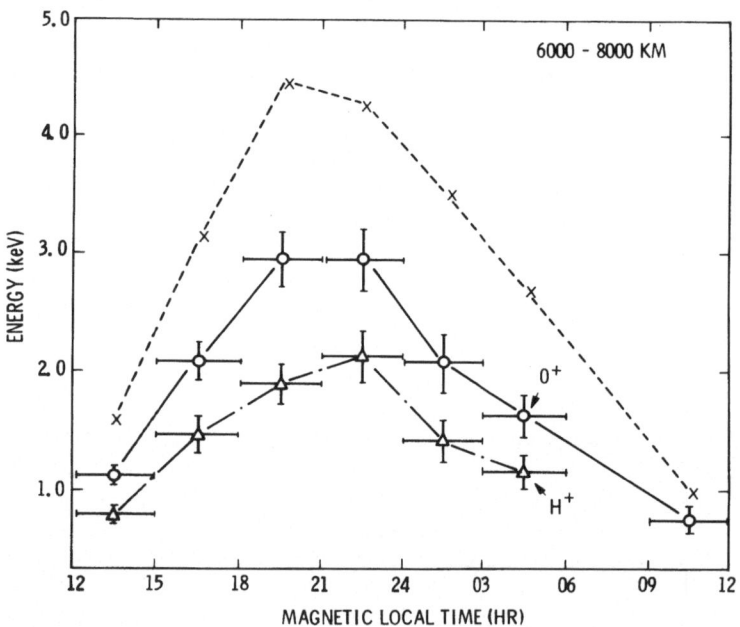

Fig. 4. Average of the energies at which at the peak energy flux was observed as a function of magnetic local time (circles represent O^+ and triangles represent H^+) and average of the maximum energies at which O^+ flux was observed (crosses).

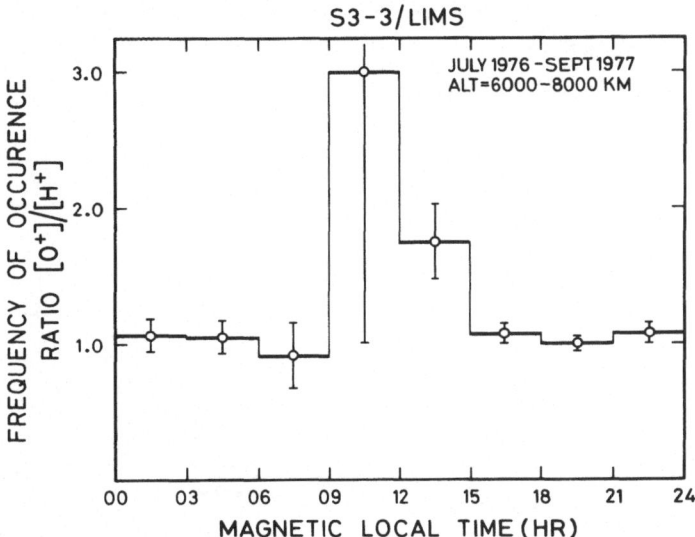

Fig. 5. Ratio of the occurrence frequencies of O^+ and H^+ events as a function of magnetic local time.

as a function of magnetic local time. One sees a dramatic change in composition in the vicinity of local noon associated with the acceleration of ionospheric O^+ ions in the dayside cusp (SHARP et al., 1977b; SHELLEY, 1979). As will be discussed below, this acceleration acts most frequently on the transverse component of the O^+ ion energy, leading to the observation of conical pitch angle distributions. The plot shown in Figure 5 should be considered only as a qualitative indicator of the composition in the region of the cusp since it is quite difficult to identify transversely accelerated ionospheric H^+ ions in the presence of the intense fluxes of precipitating and magnetically reflecting H^+ ions of solar origin that are present on the cusp field lines. It does serve to characterize the ionospheric acceleration region associated with the cusp as qualitatively different from that in other sectors, in agreement with the results of the morphological studies of GORNEY et al., (1981).

3.2   Altitude dependence

Since the local time distributions exhibited a broad flat maximum over the 15–24 hour sector this subset of the data were utilized to study the altitude dependence. For each three hour local time sector the averaged occurence probabilities in 1,000 km altitudinal bins were formed from sums over the latitude range within which 90% of the events occurred. This altitude distribution is shown in the left panel of Figure 6. In the absence of strong angular diffusion each upflowing beam should be observable at any altitude above its point of origin and the differential of this curve should characterize the location of the source region. This was obtained by subtracting probabilities in adjacent altitude bins and is shown in the right panel of Figure 6. These plots show that ion acceleration to energies above 500 eV occurs primarily at altitudes greater than 4,000 km and also suggests that we may in fact have gone over the peak of the

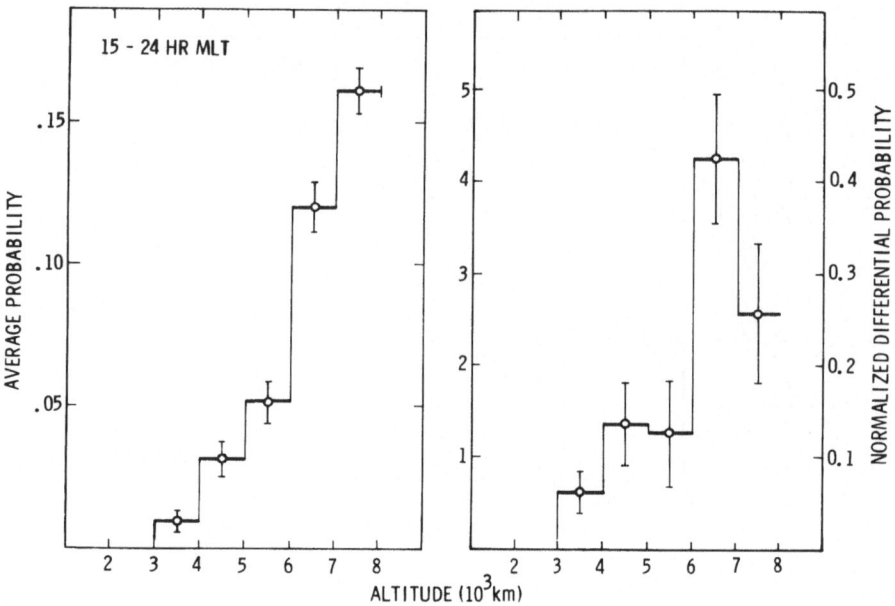

Fig. 6.   Left panel: Average occurrence probability per 1,000 km altitude bin in the 15–24 hour magnetic
local time sector. Right panel: Normalized differential probability obtained by subtracting probabilities in
adjacent altitude bins.

differential altitude distribution at $\sim 6{,}000$ km. In consideration of the statistical
uncertainties, however, the latter result cannot be considered conclusive.

The altitude dependence of the average of the energies at which the peak ion flux
was observed is shown by the solid curves in Figure 7. Since the ions receive a
substantial fraction of their energy from an electrostatic acceleration mechanism, this
quantity should be related to the average electrostatic potential difference in this
altitude and local time range. As discussed in Section 2 however, it is apparent from the
mass and angular dependencies of the accelerated ions that other mechanisms besides
electrostatic acceleration are operative. In view of the statistical uncertainties, no
significance is attached to the "peak" at 5,000–6,000 km in the O^+ distribution. The
dashed curves represent the averages of the maximum energies at which flux was
detectable in each event and again give a rough measure of the energy widths of the ion
distributions.

It is significant that at the lowest altitude of observation the ion energy at the peak
flux intensity is substantially above the 500 eV energy threshold of the spectrometer
and is comparable to the ion energy observed at the highest altitudes. This suggests that
on average the acceleration process is nonlinear in altitude and that a substantial
energization occurs in a relatively narrow range of altitude near 4,000 km.

The altitude dependence of the ratio of the O^+ to H^+ occurrence probabilities is
shown in Figure 8. Within the rather large statistical uncertainties, the results do not
show any evidence for an altitude dependence of this ratio and therefore support the

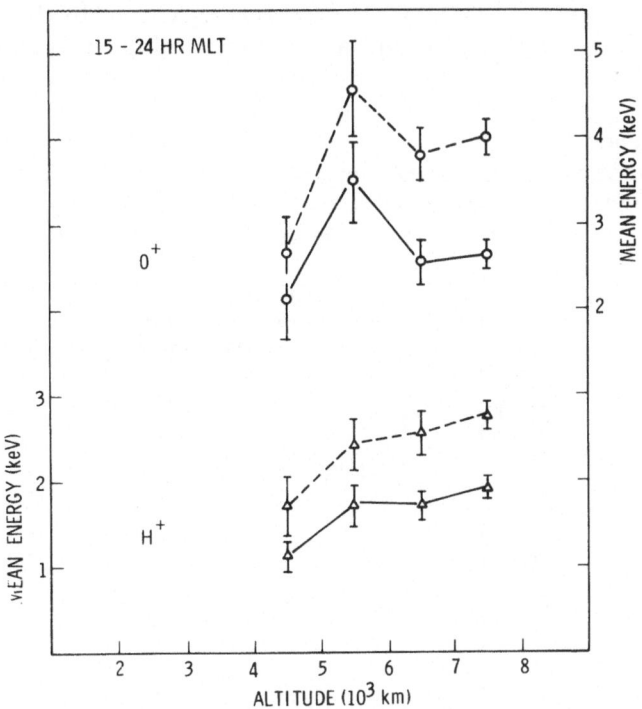

Fig. 7. Average of the energies at which peak flux was observed (solid curves) and average of the maximum energy at which flux was observed (dashed curves). Circles represent O^+ and triangles represent H^+.

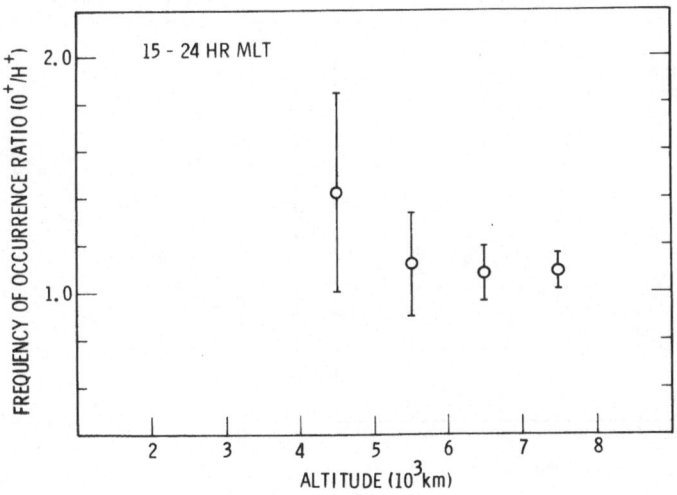

Fig. 8. Ratio of the occurrence frequencies of O^+ and H^+ events as a function of altitude.

inference by COLLIN *et al.* (1981) that the excess energy carried by the O^+ ions did not result from their having entered the acceleration region at systematically lower altitudes than did the H^+.

3.3 K_p dependence

The distribution of K_p for the data base utilized for this study is shown in Figure 9. Note that reasonable sampling was acquired for $K_p \lesssim 5$. The K_p dependence of the frequency of occurrence is illustrated in Figure 10, for the O^+ and H^+ separately, for

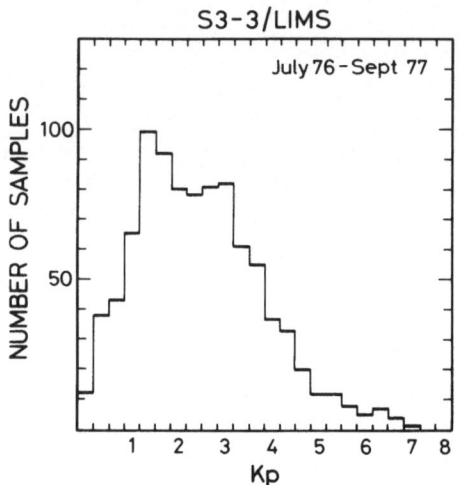

Fig. 9.   Distribution of K_p for the data base utilized for this study.

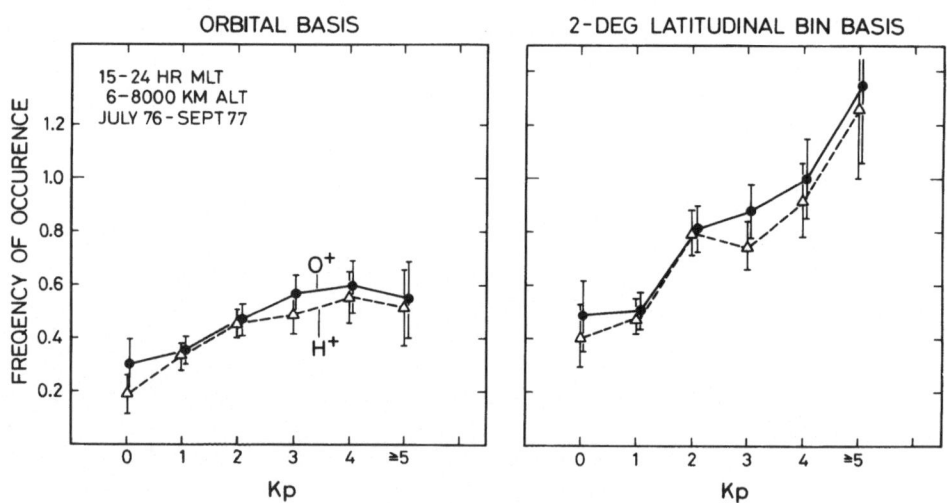

Fig. 10.   Frequence of occurrence of O^+ events (circles) and H^+ events (triangles) as a function of K_p. See text for definitions of the two types of frequencies.

events between 15 and 24 hours magnetic local time and with altitudes between 6,000 and 8,000 km. In the left panel the frequency of occurrence is defined on an orbital basis as the number of orbits in which at least one event was observed, divided by the total number of orbits in that K_p range. In the right panel it is defined as the number of 2 degree latitudinal bins in which at least one event was observed divided by the total number of orbits in that K_p range. Note that the latter definition allows frequency of occurrence values greater than 1. The ratio of these two occurrence frequencies gives a rough indication of the latitudinal extent of the UFI region as a function of magnetic activity. An examination of Figure 10 shows that the orbital frequency of occurrence increases from about 25% to about 50% as K_p increases from 0 to 5. The ratio of the two curves (two degree/orbital) ranges from about 1.5 to 2.3 with a small but not clearly significant increase with increasing K_p. An important result is that the H^+ and O^+ ions behave almost identically, with no apparent changes with magnetic activity in the composition as defined by the peak differential energy fluxes.

In order to investigate the relative flux intensities of O^+ and H^+ and their dependence on magnetic activity, the data were organized as follows. Within each range of K_p, the peak energy flux values were arranged according to the energy at which they occurred, and then averaged. This resulted in a set of curves of the type shown in Figure 11. These distributions have the units of a differential flux but should not be confused with such. The curve shown in Figure 11 is the grand average of all of the individual curves in the various K_p ranges and gives a synoptic picture of the energy and mass dependence of the peak flux intensity resulting from the still poorly understood acceleration mechanism responsible for the upward flowing ion fluxes. To illustrate the

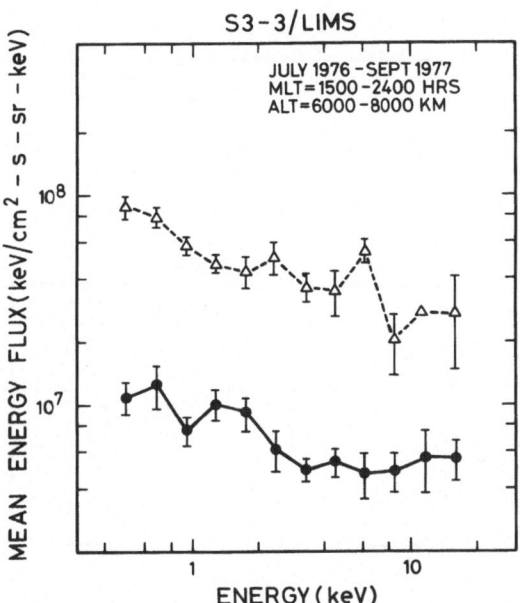

Fig. 11.   Averages of peak energy fluxes as a function of energy for O^+ events (circles) and H^+ events (triangles).

K_p dependence of the relative peak fluxes of O^+ and H^+, the individual curves of the type shown in Figure 11 at each K_p level were integrated over energy to provide a parameter with the units of energy flux. The variation of this parameter with K_p is shown in Figure 12. Again one should note that this is not a true energy flux but the integral of a peak flux distribution. We note that there is no substantial dependence of this parameter on magnetic activity. The corresponding number flux data are shown in Figure 13 where the ratios of the averages of the peak flux integrals for H^+ and O^+ are plotted.

Figure 14 shows two sets of curves computed in a similar manner illustrating the K_p dependence of the mean energy at which the peak flux intensity was observed (lower panel) and the mean of the maximum energies at which a significant flux was observed (upper panel). Here we see an increase with magnetic activity but no significant

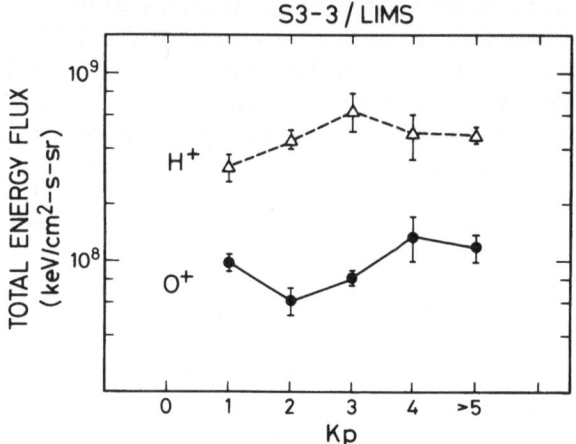

Fig. 12. Integrals of curves of the type shown in Figure 11 as a function of K_p for O^+ events (circles) and H^+ events (triangles).

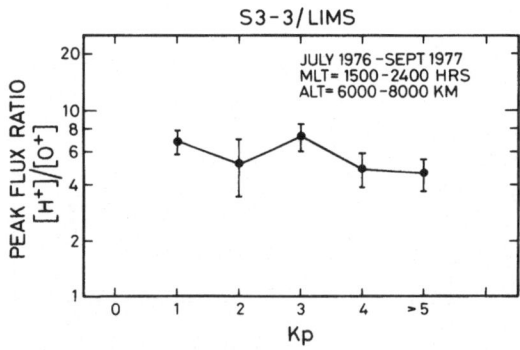

Fig. 13. Number flux data corresponding to the energy flux data presented in Figure 12, shown in the form of the ratios of the peak flux integrals for H^+ and O^+.

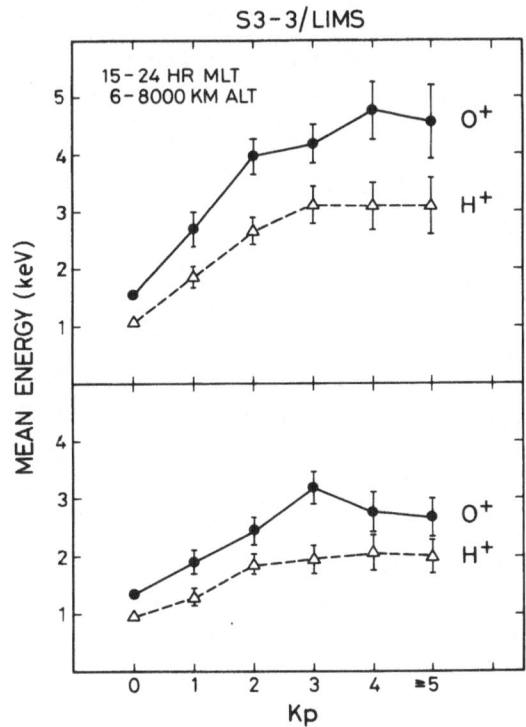

Fig. 14.   Mean value of the energy at peak flux intensity (lower panel) and mean value of the maximum energy at which flux was observed (upper panel) as a function of K_p.

difference in the K_p dependence of O^+ and H^+. This latter result is shown more explicitly in Figure 15 which gives the O^+ to H^+ average energy ratios as a function of K_p. The values represented by the solid dots were formed from the curves in the lower panel of Figure 14 while the values represented by the open triangles were formed from the curves in the upper panel.

3.4   Additional studies

A further study of some of the characteristics of the upward flowing ions beams and their relationship to auroral electrons was conducted by COLLIN *et al.* (1981). This study was based on 44 passes through the upflowing ion regions in the 1800–2400 hour local time sector at altitudes above 6,000 km. In order to focus on a single phenomena, a few wide "conics" (i.e. the events in Figure 19 with widths greater than 50°) were deleted from the remainder of the study. A possibly significant difference between this and the above described study was that COLLIN *et al.* computed the integrated number flux and flux weighted average energy from the peak spectrometer response at each of the 3 measured energies on each spin, rather than utilizing just the flux at the peak energy. Figure 16 shows the frequency of occurrence of upflowing H^+ and O^+ ions as a function of flux intensity. One sees that the maximum H^+ flux was

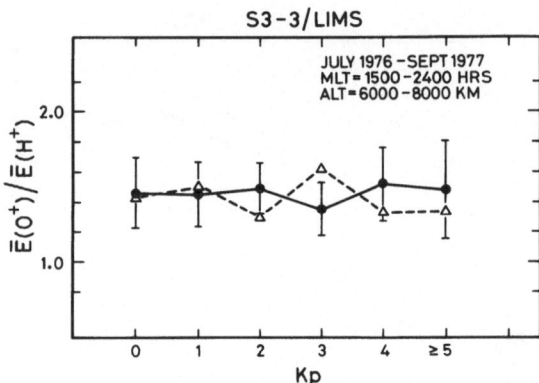

Fig. 15. Ratio of O^+ energy to H^+ energy for the curves in Figure 14. Solid circles represent energy at peak flux intensity and open triangles represent maximum energy at which flux was observed.

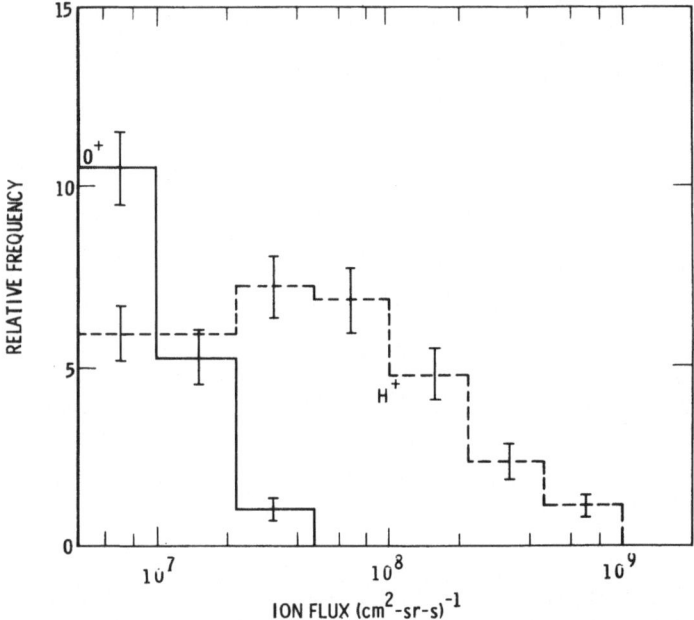

Fig. 16. Occurrence frequency distribution for upflowing O^+ and H^+ ions as a function of flux intensity. Error bars represent statistical uncertainties (Collin *et al*, 1981).

about 20 times as intense as the maximum O^+ flux. The median value of the ratio of H^+/O^+ intensity was about 7 and no correlation between the flux intensities of the two species was observed.

Both species of ions were observed with energies thoughout the range of the instrument. Figure 17 shows the distribution of frequency of occurrence of average energies. The oxygen ions are seen to be significantly more energetic than the protons in

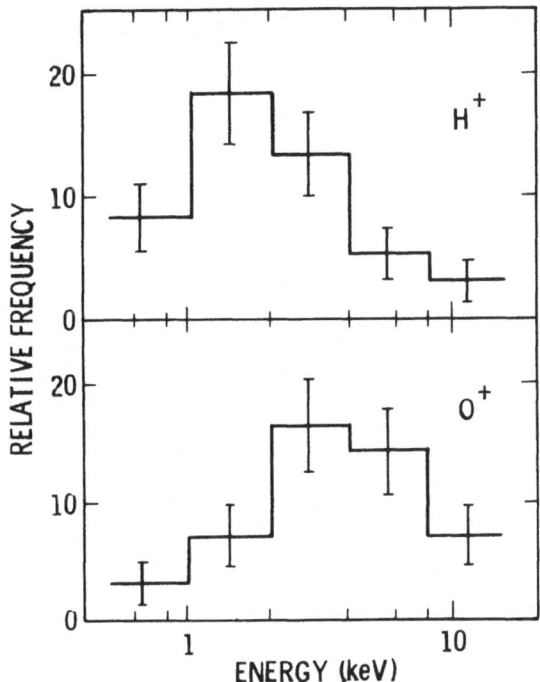

Fig. 17. Occurrence frequency distribution for upflowing O^+ and H^+ ions as a function of average energy (COLLIN *et al.*, 1981).

agreement with the results from the earlier study. The distributions imply that the energy range of the experiment was adequate to characterize the energy range of the ion acceleration mechanism.

Figure 18 shows that there is a clear systematic association between the energies of the two species. The oxygen energy was on the average 1.7 times as great as the proton energy, and the ratio between them ranged from about 0.8 to 3.

The events entering this study were nominally classified as ion "beams" since, as indicated above, the few broad "conics" which appeared to be qualitatively different in character from the majority of the events were deleted from the study. A close examination of the pitch angle distributions showed however that in about 45% of cases for O^+ and 20% of cases for H^+ the pitch angle distributions appeared to have a minimum at the angle of closest approach to the field line. In many of these cases the minimum was close to the limit of statistical significance. The remainder did not show a field-aligned minimum, but in many of these cases the possibility that such a minimum existed could not be ruled out because of the poor counting statistics and because the spectrometers did not always sample closer than 10° to the field line. An estimate of the half width of each distribution was made by finding the pitch angle at which the count rate had dropped to half its maximum observed value. Their occurrence probabilities (Figure 19) show that O^+ had significantly wider pitch angle distributions than H^+. As

AVERAGE ENERGY OF UPFLOWING O$^+$ IONS VERSUS UPFLOWING H$^+$ IONS

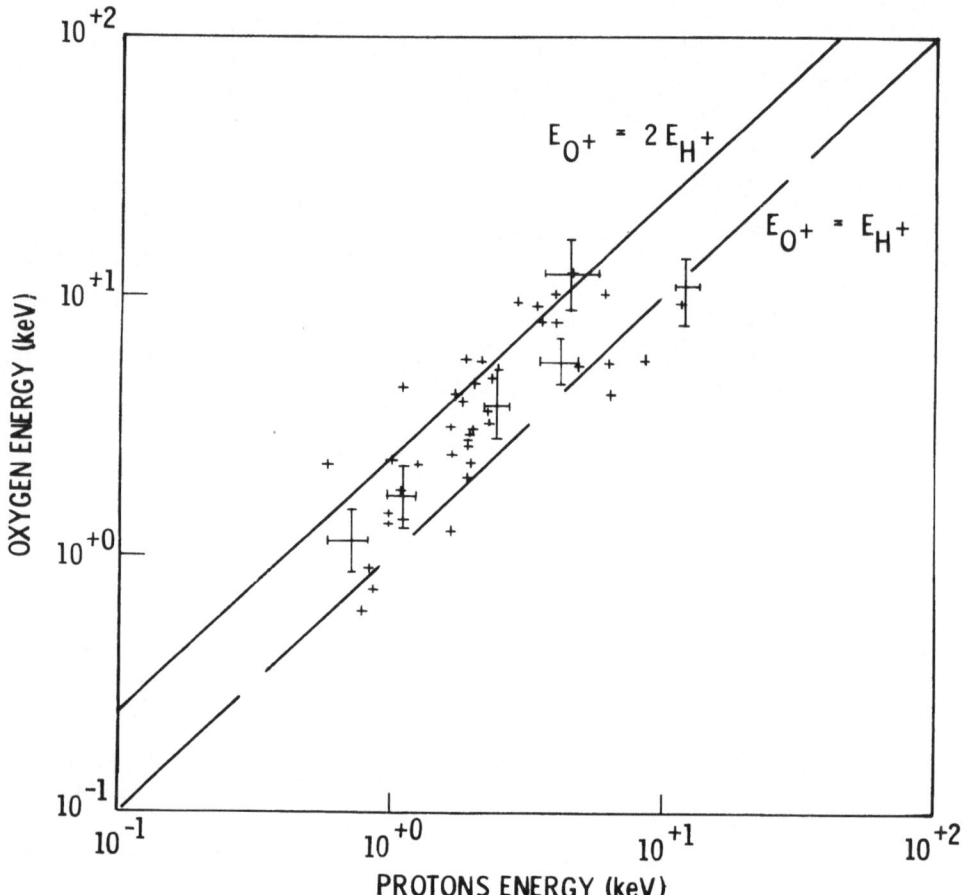

Fig. 18. Scatter plot of the average energies of simultaneously observed upflowing O$^+$ and H$^+$ ions. The dashed line represents equal ion energies and the solid line represents O$^+$ energies twice that of H$^+$. (Collin *et al.*, 1981).

discussed in Section 2 these results were taken to imply that the transverse acceleration mechanism responsible for the conical nature of the pitch angle distributions (the minimum along the field direction) preferentially acted on the O$^+$ ions and on average provided about half of their energy while the H$^+$ ions were primarily accelerated by a quasistatic parallel electric field.

In addition to the upflowing ion parameters, Collin *et al.* also recorded the characteristics of the trapped and precipitating energetic electrons (.07 $\leq E \leq$ 24 keV) on each spin in which upflowing ions were observed. Figure 20 shows a scatter plot of the oxygen ion energy versus the average energy of the trapped electrons (electron pitch angles between 80° and 100°). An equally good correlation was obtained with the

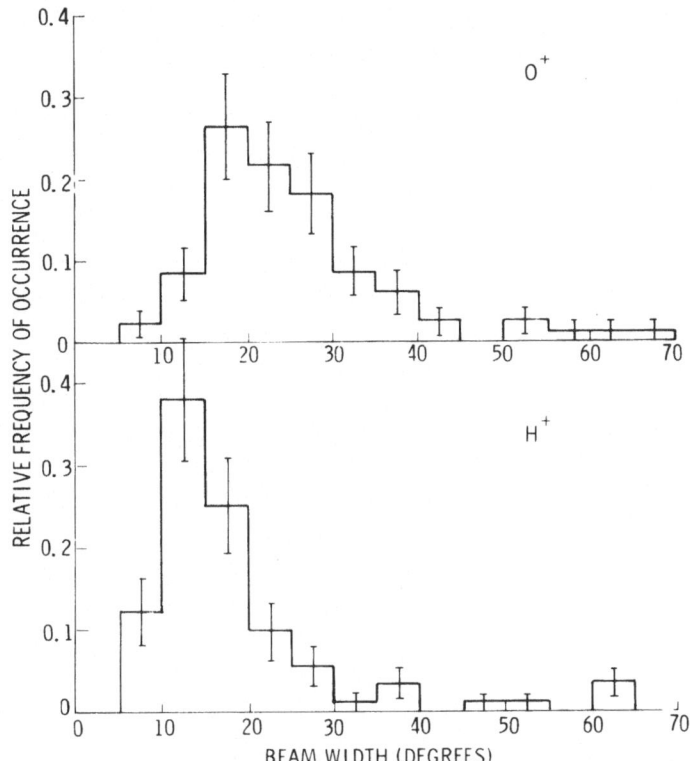

Fig. 19.   Occurrence frequency distributions of beam half width at half maximum for upflowing O^+ and H^+ ions (COLLIN *et al.* 1981).

protons. The average electron energy was generally found to lie between the average proton energy and the average oxygen energy. A more detailed correspondence between the ions and electrons was apparent when the trapped electron data were sorted into 5 groups according to the average energy of the oxygen ion beam with which they were associated. Figure 21 shows the average differential electron energy spectrum for each group. A shoulder in the spectrum is apparent which moves to higher energies and develops into a peak as the ion energy increases. These results suggest that the electrons are also substantially energized by the electrostatic field that accelerates the ions and that the magnitude of their electrostatic acceleration is comparable to that of the ions. Thus statistically the ∫ Edz above and below the 6,000 to 8,000 km altitude region of this study are approximately equal.

The results of a statistical study of some of the characteristics of the upward flowing O^+ ions accelerated in the low altitude dayside cusp were reported by SHELLEY (1979a). From 57 clearly identified cusp crossings on which S3-3 crossed 75° invariant latitude at an altitude above 5,000 km, upstreaming O^+ ion fluxes were detected in at least 58% of the cases. Conical distributions were approximately twice as frequently occurring as field aligned distributions. Because of energy and spatial limitations of the

AVERAGE TRAPPED ELECTRON ENERGY VERSUS AVERAGE O$^+$ ION
ENERGY IN UPFLOWING ION EVENTS

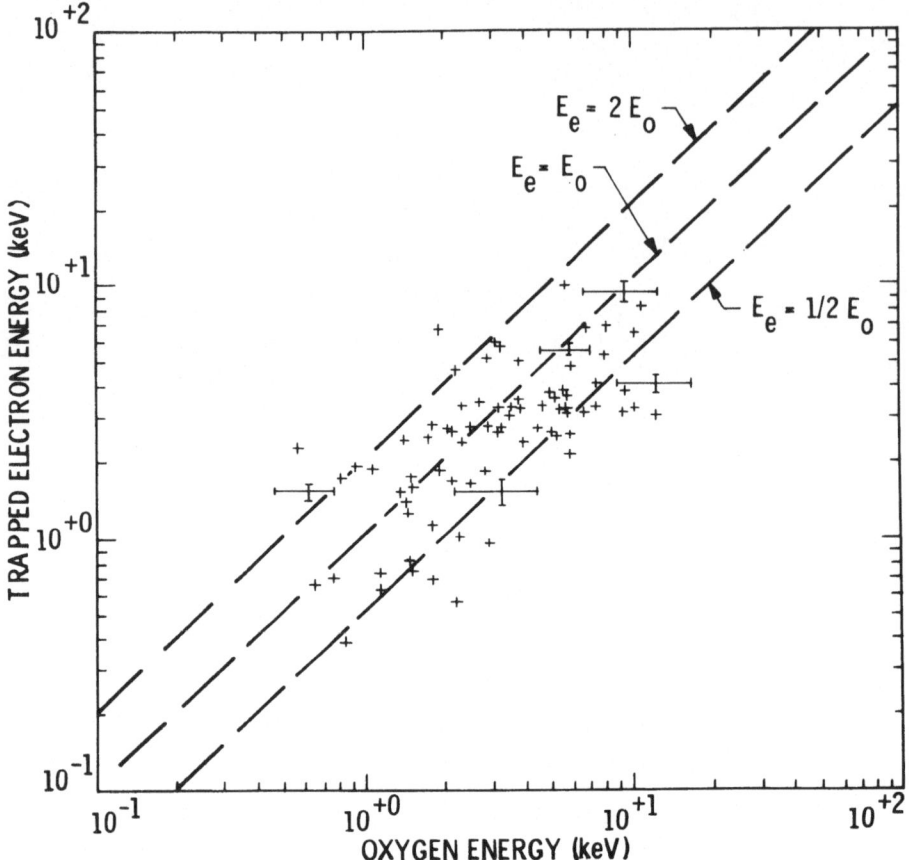

Fig. 20.   Scatter plot of upflowing O$^+$ ion energy and average energy of electrons with pitch angles of 90°
± 10°. The dashed lines represent O$^+$ ion energies of twice, equal, and half the electron energies. (Collin *et al.*, 1981).

S3-3 experiment, the O$^+$ fluxes could easily have been missed on these traversals and so
it was concluded that ionospheric ions are being accelerated in the cusp on a nearly
continuous basis. These ions are injected into the boundary layer along with the plasma
of solar wind origin which make up the bulk of the ion population in this region. They
have been detected in the boundary layer or plasma mantle both in the subsolar regions
(Peterson *et al.*, 1981) and in the magnetotail (Frank *et al*, 1977). From the latter
location they can be convected inward to the plasma sheet by the cross tail electric field
and thereby enter the magnetospheric circulation system where they contribute to the
ionospheric portion of this plasma population. (See Sharp *et al.* (1982) in this volume
for further details of the ISEE-1 results.)

AVERAGE ELECTRON ENERGY SPECTRA AS A FUNCTION OF AVERAGE
UPFLOWING O$^+$ ION ENERGY

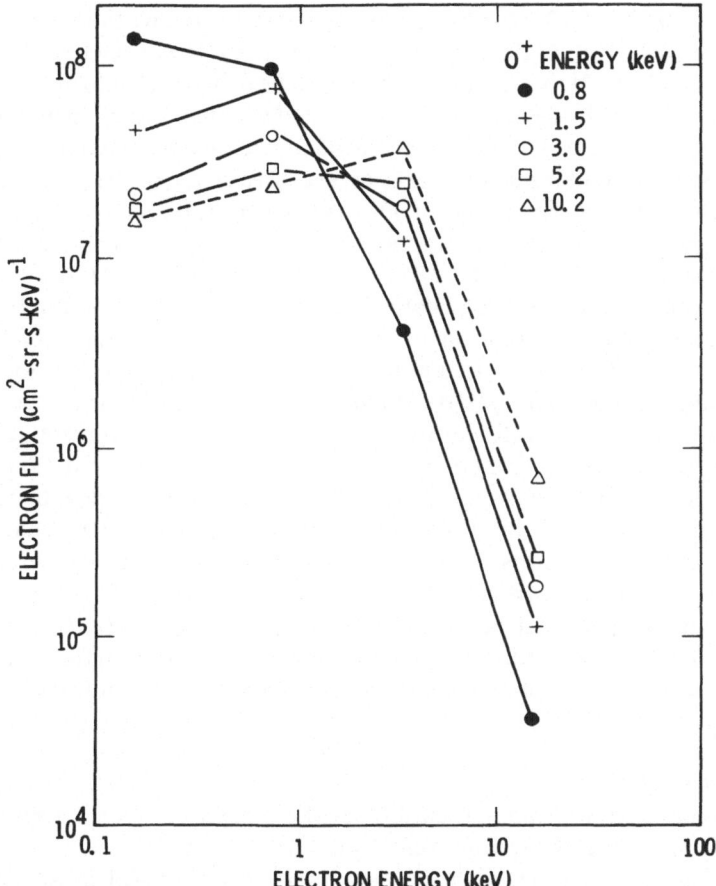

Fig. 21. Averaged differential electron energy spectrums for spins which were grouped according to the energy of the upflowing oxygen ions. (COLLIN *et al.*, 1981).

   These statistical studies of the characteristics of the upward flowing ions have been supplemented in an important way by the results from the Aerospace Corporation electrostatic analyzer experiment on S3-3 (GORNEY *et al.*, 1981). Although lacking mass discrimination capabilities, this analyzer had higher sensitivity and covered a lower energy range (90 eV to 3.9 keV) and thereby was able to detect an important component of the upflowing ion population that was not accessible to the ion mass spectrometer. The statistical study reported by GORNEY *et al.* focussed on the significant differences in morphology between the ion distributions that had peak fluxes along the magnetic field direction (beams), and those with local minimums in that direction (conics). During quiet conditions ($K_p \leq 3$) the conical distributions were found to show a local time distribution centered near noon, and a spatial association with the polar cusp, particularly for the events with $E < 400$ eV. The altitude distribution of these events was nearly uniform above an altitude of $\sim 2,000$ km. This contrasts with the results for the higher energy ions detected by the mass spectrometer and implies a low altitude generation region for these lower energy events. During disturbed times ($K_p > 3$) the conics were found to be relatively uniformly distributed in local time and their altitude distribution suggested a higher altitude generation region, particularly in the dusk sector. In contrast the ion beams observed during both quiet and disturbed times had a maximum occurrence frequency in the premidnight sector and were observed primarily in the 5,000–8,000 km altitude region in agreement with the characteristics of the more energetic ions detected by the mass spectrometer.

As discussed by GHIELMETTI *et al.* (1978), GORNEY *et al.* (1981), and SHARP (1981), the ability to distinguish between beams and conics depends on a number of factors including: The angular resolution of the instrument; the altitude of the satellite relative to the acceleration region; and the minimum angle between the local magnetic field vector and the spin plane of the satellite. The Aerospace ion spectrometer had a field of view of $10° \times 25°$ (full width) while the Lockheed instrument subtended $6° \times 5°$. As mentioned above, COLLIN *et al.* found that a careful examination of the pitch angle distributions of upflowing ion events which were nominally classified as "beams" (events remaining after the obviously wide conics were sorted out) revealed numerous cases of angular distributions with local minimums along the field direction. They concluded that even the "beams", particularly the O^+ beams, have typically experienced a substantial transverse acceleration. GORNEY *et al.* 's results show that there is a class of low energy events with an even larger relative contribution from the transverse acceleration mechanism that exhibit a substantially different morphology than the other events.

3.5   Summary

To summarize the principal features of the morphology of the more energetic ($E > .5$ keV) upward flowing ion events, we found:

1.   The average energies of the O^+ and H^+ beams were well correlated with each other and with the average energy of the electrons.

2.   The flux intensities of O^+ and H^+ beams and the associated electrons did not show any significant correlations.

3.   The O^+ beams were typically an order of magnitude less intense than the H^+ beams and about a factor of 1.5–2 more energetic.

4.   A dramatic evening sector maximum in the occurrence frequency of the upflowing ions was observed which was not qualitatively different in quiet and disturbed times. The average energy of the peak upflowing ions was also substantially higher in this local time sector.

5.   The composition of the upflowing ions was qualitatively different near local noon than in other local time sectors with O^+ conics dominating the observations.

6.   The principal energization of the upflowing ions was shown to occur above 4,000 km and some evidence for a possible peak in the frequency of occurrence near 6,000 km altitude was presented.

7.   The frequency of occurrence of the upflowing ions increased with increasing K_p but there was no evidence of a change in the relative occurrence of O^+ and H^+. The peak flux intensities of both H^+ and O^+ did not show a significant correlation with K_p. The mean and maximum energies of the ions increased with K_p but there was no change in the O^+ relative to H^+. These results do not exclude the possibility of a K_p dependence to the angular or energy widths of the ion distributions which might affect the total upward flow of ions. There is some limited evidence that the angular widths of the upflowing O^+ ions broadens with increasing ion energy while the H^+ cone widths narrow. [see figure 20 of SHELLEY (1979b).] Since the ion energy increases with increasing K_p, this suggests that there might in fact be an increase in the O^+/H^+ ratio during active times. More work is clearly required in this area.

4.   Trapped Particles

The S3-3 spacecraft with apogee at 2.3 R_e provided the first opportunity to investigate the composition of the trapped hot plasmas in the radiation belts and particularly the composition of the inner ring current during geomagnetic storms. Although the trapped ions observed with S3-3 at high latitudes corresponded to equatorial pitch angles near the loss cone, the pitch angle measurements in the ring current below $L = 4$ typically corresponded to equatorial pitch angle ranges from $15°$ to $40°$, and near the inner edge of the ring current, equatorial pitch angles as high as $50°$ were sampled (JOHNSON, et al., 1977).

Prior to the S3-3 launch, observations of large fluxes of O^+ ions precipitating from the magnetosphere with energies up to 12 keV during magnetic storms had been reported from low altitude satellite measurements (SHELLEY et al., 1972; SHARP, et al., 1974; and JOHNSON, et al., 1975). However, due to the spacecraft and spectrometer orientations, no measurements on the trapped component of the ion fluxes were obtained. Thus, direct injection and acceleration of the observed ions within or near the loss cone could not be excluded, although other considerations of the spatial and temporal distributions of the fluxes made it appear likely that significant fluxes of trapped ions were also present.

Using S3-3 data, JOHNSON et al (1977) investigated the composition of the storm-time ring current during the early main phases of magnetic storms on 29 December 1976, 6 April 1977, and 19 April 1977 which had peak intensities near $-100 \, \gamma$. From the mass spectrums shown for selected energies in Figure 22, it can be seen that O^+ and H^+ are the dominant ions observed although measurable amounts of He^+ are also observed. These data are from the L-shell range from about 3 to 4 and from the equatorial pitch angle range from $15°$ to $41°$.

Energy spectrums of the O^+ and H^+ ions, corresponding to the data intervals of Figure 22, are shown in Figure 23. The high O^+ fluxes are evident, and the resulting O^+ number densities for these intervals in temporal order were 2.9, 3.3, and 7.1 with O^+/H^+ density ratios of 2.1, 3.0, and 1.5, respectively. Based on these high density ratios and in consideration of the low oxygen ion abundance in the solar wind (see BAME et al., in this volume), it was concluded that the ionosphere was a major contributor to the storm-time ring current in the measured energy range.

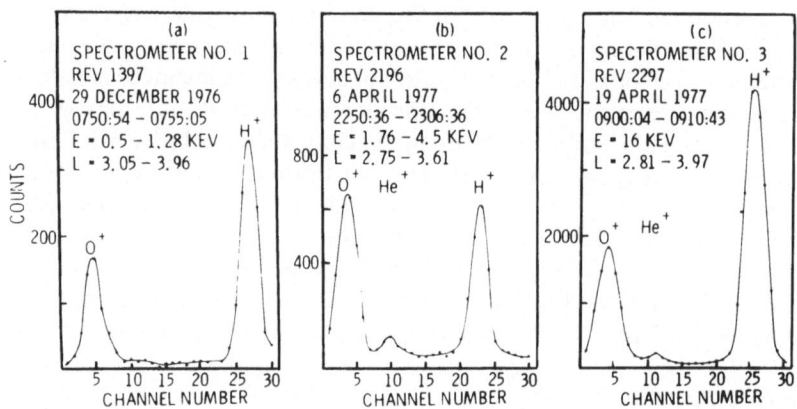

Fig. 22. Examples of mass spectrums from the three ion spectrometers during the main phases of the 29 December 1976, 6 April 1977, and 19 April 1977 magnetic storms. (JOHNSON *et al.*, 1977).

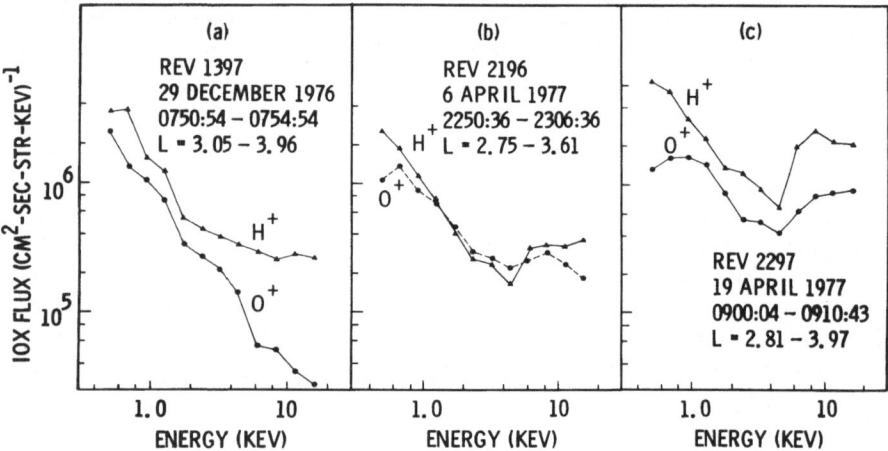

Fig. 23. Energy distributions for O^+ and H^+ during the main phases of the 29 December 1976, 6 April 1977, and 19 April 1977 magnetic storms. (JOHNSON *et al.*, 1977).

During the late recovery phase of the 29 December 1976 storm, the observed composition at low L-values was significantly different from that found during the main phase of the storm. As seen from the mass spectrum in Figure 24 (from JOHNSON, *et al.*, 1977) in comparison with the mass spectrums in Figure 22, there is an enhancement of He^+ and a depletion of H^+ relative to the O^+. This temporal evolution in the ring current composition at low L-shells is *qualitatively* consistent with expectations based on the charge exchange loss process for the ion constituents (Tinsley [1976]; Lyons and Evans [1976]).

In addition to the charge exchange loss of ions from the ring current, as discussed above, evidence for a pitch angle scattering loss process for ring current ions has also

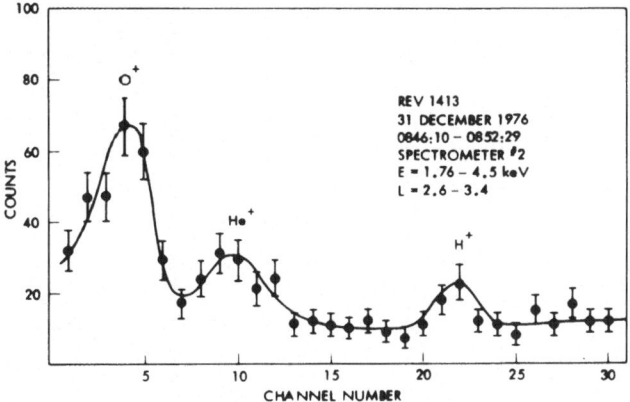

Fig. 24. An example of the mass spectrum during the recovery phase of the 29 December 1976 magnetic storm. (JOHNSON *et al.*, 1977).

been presented by JOHNSON *et al.* (1979). A segment of S3-3 data acquired during the main phase of the 11 December 1977 magnetic storm is shown in Figure 25. The abscissa shows universal time (SYST), longitude, latitude, altitude in kilometers, invariant latitude, and magnetic local time. The four lowest panels show electron spectrometer data (see JOHNSON *et al.*, 1979, for details of this format that are not discussed here). The panel labeled PITCH shows the pitch angle of the instrument look direction. The next four panels show the logarithm of the counts from ions with M/Q (mass/charge) $= 1, 2, 4$, and 16 summed once per second from the temporally selected output channels from each of the three mass spectrometers, giving an approximate measure of the relative flux of the relevant species. The next three panels (CXA-3, CXA-2, and CXA-1) display the mass spectrums accumulated at the selected energy step of each spectrometer. The selected energy step is given by the top panel with each spectrometer at its lowest energy for Step 1, etc. . The energy-per-unit-change values are 0.5, 0.68, 0.94, and 1.28 keV for spectrometer #1; 1.76, 2.4, 3.3, and 4.5 keV for spectrometer #2; and 6.2, 8.5, 11.6, and 16.0 keV for spectrometer #3. The evidence for mass dependent pitch angle scattering can be seen by comparing the $M/Q = 1$ response, the $M/Q = 16$ response, and the pitch angle (PITCH) of the instrument look direction in the time period from 1200:22 to 1205:30. Loss cones for the H^+ ions in this region are generally well developed in both directions along the magnetic field; whereas, for the O^+ ions, the loss cone region near $0°$ pitch angle (the instrument pointed upward) contains high fluxes, and in some regions, nearly isotropic fluxes. From this event and other similar events, it is concluded that strong pitch angle scattering is occurring above the satellite for the O^+ ions, but not for H^+ ions. Thus, this loss process can contribute directly to the decay of the ring current by moving ions into the single transit loss cone. Possibly more important, even a slower rate of pitch angle scattering will continuously lower the mirror altitudes of the ions and thereby transport the ions into the denser region of the atmosphere where the charge exchange losses are more rapid. These observations of the mass dependence of the ion pitch angle

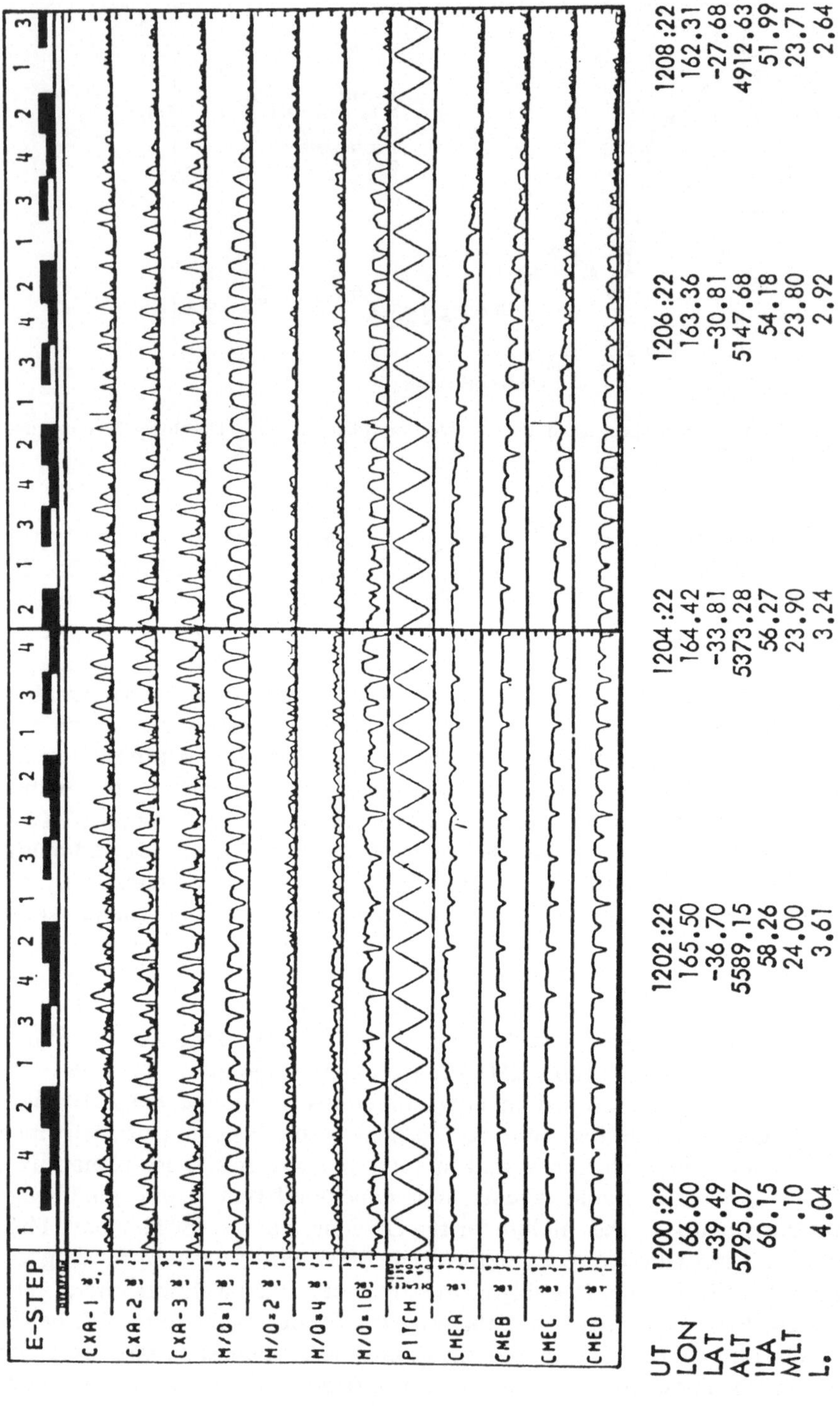

R. D. Sharp et al.

Fig. 25. Survey plot of S3-3 ion and electron data during the main phase of the 11 December 1977 magnetic storm. Large fluxes of precipitating O⁺ ions are observed from 1200:22 to 1204:30 U.T. (Johnson et al., 1979).

scattering rates, coupled with the spatial and/or temporal variations in these rates, suggested by the data in Figure 25, provide a further indication that quantitative modeling of the ring current decay by using only charge exchange loss processes may be applicable to only very limited regions of the magnetosphere and/or for limited geophysical conditions even in the equatorial regions of the magnetosphere.

5. Discussion

Since the launch of S3-3, a number of satellites carrying energetic ion mass spectrometers have been placed into high altitude orbits where they can sample the near equatorial trapped ion populations in the ring current and the distant plasma sheet (see reviews in this volume). As we see in these articles the results are complex and we do not yet have a clear understanding of the relative importance of the ion sources and the detailed operation of the dominant transport and loss mechanisms.

An important element of the puzzle is the morphology of the principal ionospheric source term, i.e. the composition and intensity, and the latitude, local time and magnetic activity dependence of the upward flowing energetic ions. (Other suggested ionospheric sources such as the plasmasphere and the polar wind have been shown to be of minor significance because of the general dominance of O^+ over He^+ in the trapped particle populations). The initial morphological results from S3-3 presented here are the best available information on the ionospheric source at this time.

As discussed by SHARP et al. (this volume) some of these results, e.g., the implied lack of dependence of the O^+/H^+ ratio on K_p, help us to understand the substorm associated changes in the composition of the distant plasma sheet. The principal process for removal of ions from that region is inward convection of the plasma with a velocity which is mass independent. One, therefore, would expect changes in the relative importance of the ionosphere and solar wind as source terms to be directly reflected by an increase in O^+ density as He^{++} decreases and vice versa. This is indeed the case. The situation is apparently more complicated in the $L = 6$ to 11 region, where both O^+ and He^{++} increase in the trapped population relative to H^+ during the early phase of large magnetic storms (LENNARTSSON, 1981). Possible explanations could be an increase in the O^+/H^+ ratio of the ionospheric source term, mass dependent losses, or both.

Figure 1 shows that almost all of the upward flowing ion events occurred at invariant latitudes above $66°$ ($L > 6$). These are the events for which the above described magnetic activity variations are pertinent. As we have seen, these results suggest that there was not an increase in the O^+/H^+ ratio with increasing K_p. However, the comparison of statistical results taken during different periods is subject to many uncertainities. There may be seasonal or long term (solar cycle) variations. Also, as discussed above, the study was based on peak flux intensities rather than the integral upflowing ion flux. Furthermore, K_p is not the ideal parameter to represent the early phase of magnetic storms. It is probably more relevant to substorm activity and therefore to the ISEE results in the distant plasma sheet where there is no apparent inconsistency.

As indicated above, another possibility is mass dependent loss terms. One expects that charge exchange would be more important in the inner magnetosphere than in the distant plasma sheet but this gives the reverse effect to that which is observed. H^+ ions have the largest cross section for charge exchange in this energy range and should be depleted relative to the heavy ions in the trapped population. Thus during quiet times one would expect an increase in O^+ and He^{++} relative to H^+, opposite to the observed trend in the $L > 6$ range. The expected trend is in fact observed at lower altitudes where the charge exchange mechanism is more important (JOHNSON *et al.*, 1977; LUNDIN *et al.*, 1980; LENNARTSSON, 1981).

Pitch angle diffusion into the loss cone is another significant loss process for ring current ions (SHARP *et al.*, 1977a) and there is evidence that at least on some occasions and in some locations in the magnetosphere this loss is enchanced for O^+ relative to H^+. As we have seen in Figure 25, for invariant latitudes below about 60° the H^+ fluxes exhibit nearly empty loss cones both parallel and antiparallel to the magnetic field direction and appear to be relatively stably trapped while the O^+ pitch angle distributions show evidence for substantial pitch angle diffusion and subsequent precipitation. Unfortunately in the relevant range of $L > 6$ during quiet times the flux intensities are generally too low to see if this effect persists.

We would like to thank T. C. SANDERS, E. HERTZBERG, J. D. MATHEWS, J. D. MCDANIEL, L. HOOKER, and D. L. CARR for their important contributions to the S3-3 ion composition experiment and thus to the advances in space science resulting therefrom.

We also thank W. LENNARTSSON for his comments on the manuscript and R. WRIGHT for his assistance with the figures. This work has been supported by the Office of Naval Research under contract N00014-78-C-0479, the Atmospheric Sciences Section of the National Science Foundation under grant ATM-7911174, and NASA under contract NASW-3395.

REFERENCES

CLADIS, J. B. and R. D. SHARP, Scale of electric field along magnetic field in an inverted V event, *J. Geophys. Res.*, **84**, 6564, 1979.

COLLIN, H. L., R. D. SHARP, E. G. SHELLEY, and R. G. JOHNSON, Some general characteristics of upflowing ion beams over the auroral zone and their relationship to auroral electrons, *J. Geophys. Res.*, **86**, 6820, 1981.

EVANS, D. S., Evidence for the low altitude acceleration of auroral particles in *Physics of the Hot Plasma in the Magnetosphere*, edited by B. HULTQVIST and L. STENFLO, Plenum Publishing Co., New York, 1975.

FRANK, L. A., K. L. ACKERSON, and D. M. YEAGER, Observations of atomic oxygen (O^+) in the earth's magnetotail, *J. Geophys. Res.*, **82**, 129, 1977.

GHIELMETTI, A. G., Upward flowing ion characteristics in the high latitude ionospheric acceleration regions, (abstract), *EOS*, **59**, 1155, 1978.

GHIELMETTI, A. G., R. G. JOHNSON, R. D. SHARP, and E. G. SHELLEY, The latitudinal, diurnal, and altitudinal distributions of upward flowing energetic ions of ionospheric origin, *Geophys. Res. Lett.*, **5**, 59, 1978.

GHIELMETTI, A. G., R. D. SHARP, E. G. SHELLEY, and R. G. JOHNSON, Downward flowing ions and evidence for injection of ionospheric ions into the plasma sheet, *J. Geophys. Res.*, **84**, 5781, 1979.

GORNEY, D. J., A. CLARKE, D. CROLEY, J. F. FENNELL, J. LUHMAN, and P. F. MIZERA, The distribution of ion beams and conics below 8,000 km, *J. Geophys. Res.*, **86**, 83, 1981.

GURNETT, D. A., Electric field and plasma observations in the magnetosphere, in *Critical Problems of Magnetospheric Physics*, edited by E. R. DYER, National Academy of Sciences, Washington, D.C., 1972.

JOHNSON, R. G., R. D. SHARP, and E. G. SHELLEY, Study of stormtime fluxes of heavy ions, Final report, NASA contract NASW-3112, LMSC/D673774, April 1979.

JOHNSON, R. G., R. D. SHARP, and E. G. SHELLEY, Observations of ions of ionospheric origin in the stormtime ring current, *Geophys. Res. Lett.*, **4**, 403, 1977.

JOHNSON, R. G., R. D. SHARP, and E. G. SHELLEY, Composition of the hot plasma in the magnetosphere, in *Physics of the hot plasma in the magnetosphere*, edited by B. HULTQVIST and L. STENFLO, Plenum, New York, 1975.

LENNARTSSON, W., A comparison of the near equatorial ion composition between quiet and disturbed conditions, *J. Geophys. Res.*, **87**, 6109, 1982.

LUNDIN, R., L. R. LYONS, and N. PISSARENKO, Observations of ring current composition at $L < 4$, *Geophys. Res. Lett.*, **7**, 425, 1980.

LYONS, L. R. and D. S. EVANS, The inconsistency between proton charge exchange and the observed ring current decay, *J. Geophys. Res.*, **81**, 6197, 1976.

MIZERA, P. F. and J. F. FENNELL, Signatures of electric fields from high and low altitude particle distributions, *Geophys. Res. Lett.*, **4**, 311, 1977.

MIZERA, P. F., J. F. FENNELL, D. R. CROLEY, A. L. VAMPOLA, F. S. MOZER, R. B. TORBERT, M. TEMERIN, R. LYSAK, M. HUDSON, C. A. CATTELL, R. G. JOHNSON, R. D. SHARP, A. GHIELMETTI, and P. M. KINTNER, The aurora inferred from S3-3 particles and fields, *J. Geophys., Res.*, **86**, 2329, 1981.

MOZER, F. S., C. A. CATTELL, M. TEMERIN, R. B. TORBERT, S. VONGLINSKI, M. WOLDORFF, and J. WYGANT, The DC and AC electric field, plasma density, plasma temperature, and field-aligned current experiments on the S3-3 satellite, *J. Geophys. Res.*, **84**, 5875, 1979.

PETERSON, W. K., E. G. SHELLEY, G. HAERENDEL, G. PASCHMANN, Energetic ion composition in the subsolar magnetopause and boundary layer, *J. Geophys. Res.*, **87**, 2139, 1982.

SHARP, R. D., R. G. JOHNSON, E. G. SHELLEY, and K. K. HARRIS, Energetic O^+ ions in the magnetosphere, *J. Geophys. Res.*, **79**, 1844, 1974.

SHARP, R. D., E. G. SHELLEY, and R. G. JOHNSON, A search for Helium ions in the recovery phase of a magnetic storm, *J. Geophys. Res.*, **82**, 2361, 1977a.

SHARP, R. D., R. G. JOHNSON, and E. G. SHELLEY, Observations of an ionospheric acceleration mechanism producing energetic (keV) ions primarily normal to the geomagnetic field direction, *J. Geophys. Res.*, **82**, 3324, 1977b.

SHARP, R. D., R. G. JOHNSON, and E. G. SHELLEY, Energetic particle measurements from within ionospheric structures responsible for auroral acceleration processes, *J. Geophys. Res.*, **84**, 480, 1979.

SHARP, R. D., Positive ion acceleration in the 1 R_E altitude range, in *The Physics of Auroral Arc Formation*, edited by S. I. AKASOFU and J. R. KAN, Geophysical Monograph 25, American Geophysical Union, Washington, D.C., 1981, p. 112.

SHELLEY, E. G., R. G. JOHNSON, and R. D. SHARP, Satellite observations of energetic heavy ions during a geomagnetic storm, *J. Geophys. Res.*, **77**, 6104, 1972.

SHELLEY, E. G., R. D. SHARP, and R. G. JOHNSON, Satellite observations of an ionospheric acceleration mechanism, *Geophys. Res. Lett.*, **3**, 654, 1976.

SHELLEY, E. G., Ion Composition in the dayside cusp: Injection of ionospheric ions into the high latitude boundary layer, Proceedings of Magnetospheric Boundary Layers Conference, Alpbach, 11–15 June, 1979 (ESA SP-148, August 1979a).

SHELLEY, E. G., Heavy ions in the magnetosphere, Space Science Reviews 23, **465**, 1979b.

SHELLEY, E. G., R. G. JOHNSON, and A. GHIELMETTI, Relationship of upward flowing energetic ions to geomagnetic activity, (abstract), *EOS*, **61**, 1080, 1980.

TINSLEY, B. A., Evidence that the recovery phase ring current consists of helium ions, *J. Geophys. Res*, **81**, 6193, 1976.

Energetic Ion Composition in the Earth's Magnetosphere, edited by R. G. Johnson, 195–230.
Copyright © 1983 by Terra Scientific Publishing Company (TERRAPUB), Tokyo.

The Composition of Thermal and Hot Ions Observed by the GEOS-1 and -2 Spacecraft

H. Balsiger, J. Geiss, and D. T. Young*

Physikalisches Institut, University of Bern, 3012 Bern, Switzerland

(Received March 24, 1982)

The GEOS-1 and -2 ion composition experiments have surveyed the plasma composition of the near equatorial inner magnetosphere in the energy per charge range below 16 keV/e since May 1977. In this paper we review contributions of these studies to plasmasphere and ring current composition, to wave particle interactions, to the questions of large scale plasma circulation and to origins of magnetospheric ions.

1. Introduction

The launch of the European Spacecraft GEOS-1 on April 20, 1977 represented the start of a quite important contribution to magnetospheric research not only because the entire satellite was dedicated to science, but also because it was electromagnetically clean and it had quite advanced and novel experiments on board. One of these experiments was the first good-resolution ion composition experiment (ICE), especially developed for magnetospheric research (Balsiger *et al.*, 1976). Among its special properties were wide mass range to make it suited for Ba^+ tracer experiments, good mass and energy resolution, high sensitivity, large energy per charge range (from thermal to 16 keV/e) and a large suppression factor of $> 10^4$ against internally scattered ions. Furthermore the electronics, in particular the power supplies, have proven to be extremely stable, which allowed for the accumulation of data during long periods and for direct data comparison over several years (see synoptic studies, Section 6). The orbit of GEOS-1, due to a malfunction of the Delta launcher, was elliptical (apogee 38,000 km, perigee 2,050 km) rather than the planned geostationary and good data were obtained in an L range between 3 and 8. This orbit gave GEOS-1 the unforeseen opportunity to measure the storm time ring current down to $L \simeq 3$ as well as the plasmasphere. On July 14, 1978 GEOS-2, with identical equipment, was placed into the originally planned geostationary orbit which is ideally suited for synoptic studies of magnetospheric particle populations.

The main contributions of the GEOS ion mass spectrometers were: (a) to establish the importance of hot plasma of terrestrial origin at high altitudes, (b) first

---

*Present adress: Mail stop 438, Los Alamos National Laboratory, Los Alamos, New Mexico 87545, USA.

composition measurements of the storm time and recovery phase ring current at the equator ($\lesssim 16\,\mathrm{keV/e}$) at $L \gtrsim 3$, (c) discovery of doubly charged thermal ions (O^{++}, He^{++}) in the plasmasphere, (d) measurement of the composition of thermal ions inside and outside the plasmaphere, in particular establishing the (relatively high) abundance of helium and oxygen ions, (e) survey of composition of magnetospheric ion populations over several years by means of synoptic studies, in particular the dependences of composition at geostationary orbit on local time, magnetic and solar activity, (f) to establish the increase of the average oxygen content while approaching solar maximum, and (g) to demonstrate in a cooperative study among several GEOS experiments that cold He^+ ions, in conjunction with an enhanced anisotropy of energetic protons, control processes of ULF wave propagation and generation.

The goal of plasma composition measurements is, of course, to assess the sources of magnetospheric plasmas in the different regions; to study entry, acceleration, transport, and loss mechanisms; and to monitor large scale circulation by using specific ions as tracers. In Fig. 1 we present three possible sources (cf. BALSIGER *et al.*,

LARGE SCALE MAGNETOSPHERIC CIRCULATION

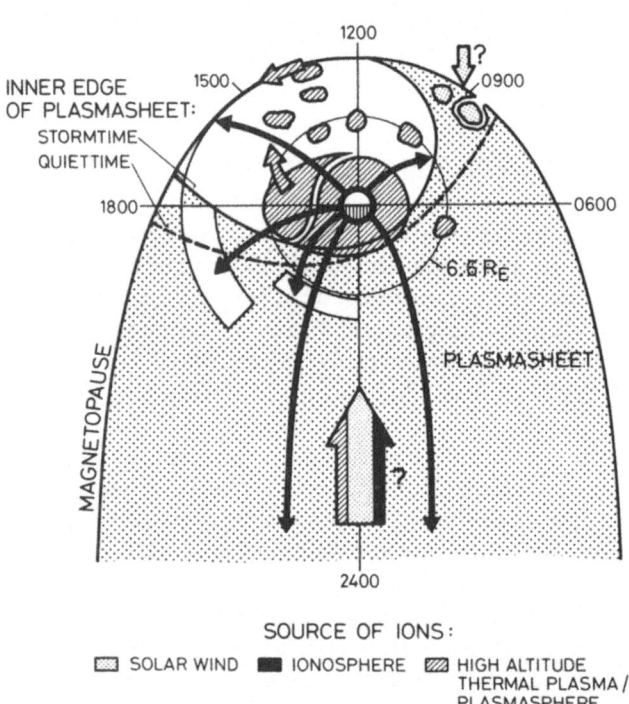

SOURCE OF IONS:

☐ SOLAR WIND   ■ IONOSPHERE   ▨ HIGH ALTITUDE THERMAL PLASMA / PLASMASPHERE

Fig. 1.   Schematic of large scale magnetospheric circulation. Indicated is the possibility that plasmas from three sources are mixed in the plasmasheet and injected during magnetic substorms and storms from local midnight into the inner magnetosphere and the ring current. Shown are also observed field aligned beams out of the ionosphere and the (asymmetric) inward motion of the inner edge of the plasmasheet during magnetic storms, as inferred from observations of the "mixed region". For further discussion see text. From BALSIGER (1981).

1980; BALSIGER, 1981) for hot magnetospheric plasma—the solar wind, the ionosphere and the thermal plasma at high altitudes (for simplicity called "plasmaspheric")—and we indicate some suggested large scale circulation patterns. The GEOS mass spectrometers have surveyed the inner magnetosphere since 1977 and were able, in cooperative studies with other mass spectrometers on polar orbiting satellites and in the magnetotail, to gather important (although not conclusive) data towards solving the question of large scale ion circulation.

2.   Thermal and Suprathermal Ions

Prior to GEOS-1 the abundance of heavy ions in the plasmasphere was believed to be quite low ($\sim 1\%$ for He^+ abundance) based on OGO measurements (TAYLOR et al., 1965; HARRIS et al., 1970). First surveys with the GEOS-1 mass spectrometer showed that the He^+ concentration is more typically $\sim 10\%$ in the plasmasphere (YOUNG et al., 1977; GEISS et al., 1978), and that there exists an important thermal and suprathermal ($< 50\,eV$) plasma population outside the plasmasphere with both O^+ and He^+ as important constituents (YOUNG et al., 1977; YOUNG, 1979). These observations have since been confirmed by the ISEE mass spectrometer (cf. HORWITZ et al., 1981).

Composition data for thermal and suprathermal ions were obtained in the so called thermal (T) mode by means of a retarding potential analyzer (RPA) in the entrance section of the mass spectrometer (BALSIGER et al., 1976). In order to increase time resolution, and because fluxes were quite low, the doubly charged ions He^{++} and O^{++} were usually only measured with the RPA set at 0 Volts. Rough density and temperatures were obtained by means of a fitting program applied to counting rate versus spin-angle plots, assuming Maxwellian ion distributions rammed by the moving satellite (FARRUGIA, 1982). Comparison with densities and temperatures derived from RPA scans for the singly charged ions gives fair agreement (within a factor of $\lesssim 2$) with the values obtained with the fitting method.

Figures 2 and 3 show examples of this procedure for a plasmapause crossing on December 11, 1977. Data points are plotted against spin phase angle with the ram direction being at $0°$. Figure 2 shows typical plasmasphere composition. The satellite potential is assumed to be 0 Volts in this case, which is consistent with results from the DC E-field experiment inside the plasmasphere (A. Pedersen, private communication, 1981). In this figure there is no pitch angle (symbol P) information available because the GEOS magnetometer saturates at low altitudes. The spin angle distributions of the ions are peaked in the ram direction, hence we infer an isotropic, isothermal plasma of $1-2\,eV$ (the spread in temperature between the different species is believed to be not significant in this case and to reflect the expected accuracy of the fitting method). Density and temperature for H^+ are given as lower and upper limits, respectively, because saturation effects occurring at these counting rates cannot be accurately corrected. In Fig. 3 we show data typical for the plasmapause and the thermal plasma outside the plasmasphere. Here, in addition to the isotropic plasma, ions ordered by the magnetic field direction are seen (cf. BAUGHER et al., 1980). Data with the RPA at $+ 6$ Volts are shown for the singly charged ions in order to emphasize the non-isotropic components. These are mainly at $90°$ pitch angle for H^+ and He^+, and along

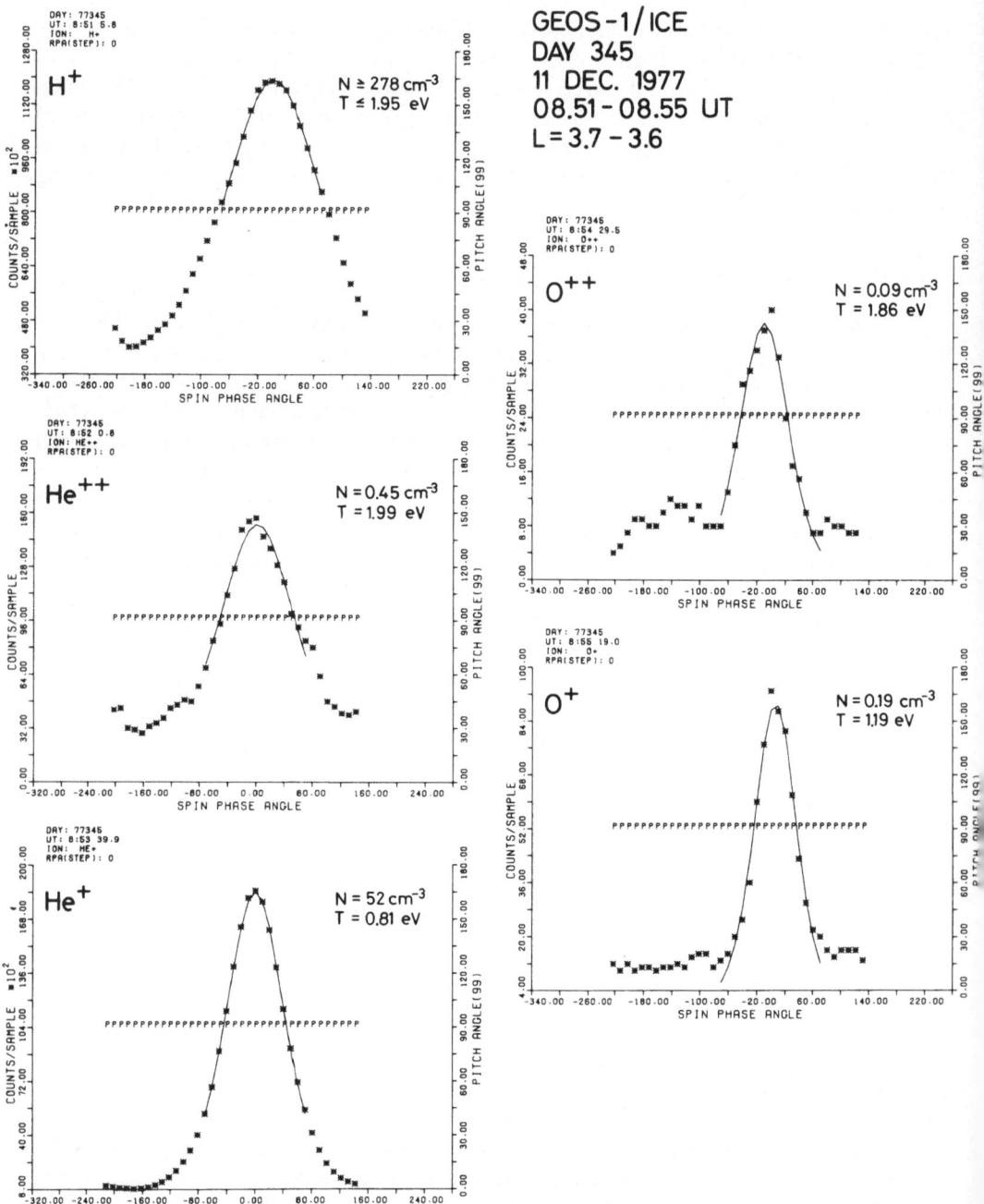

Fig. 2.   Typical thermal ion composition inside the plasmasphere measured after a plasmapause crossing on December 11, 1977. Counting rate is plotted versus spin phase angle. Densities and temperatures are obtained by means of a fit program, assuming a satellite potential of 0 Volts (see text). Values given for H^+ are lower and upper limits, respectively, because of detector saturation. No pitch angle information (P) is obtained because the magnetometer saturates at low altitudes. Data are taken with the retarding potential analyzer (RPA) at 0 Volts.

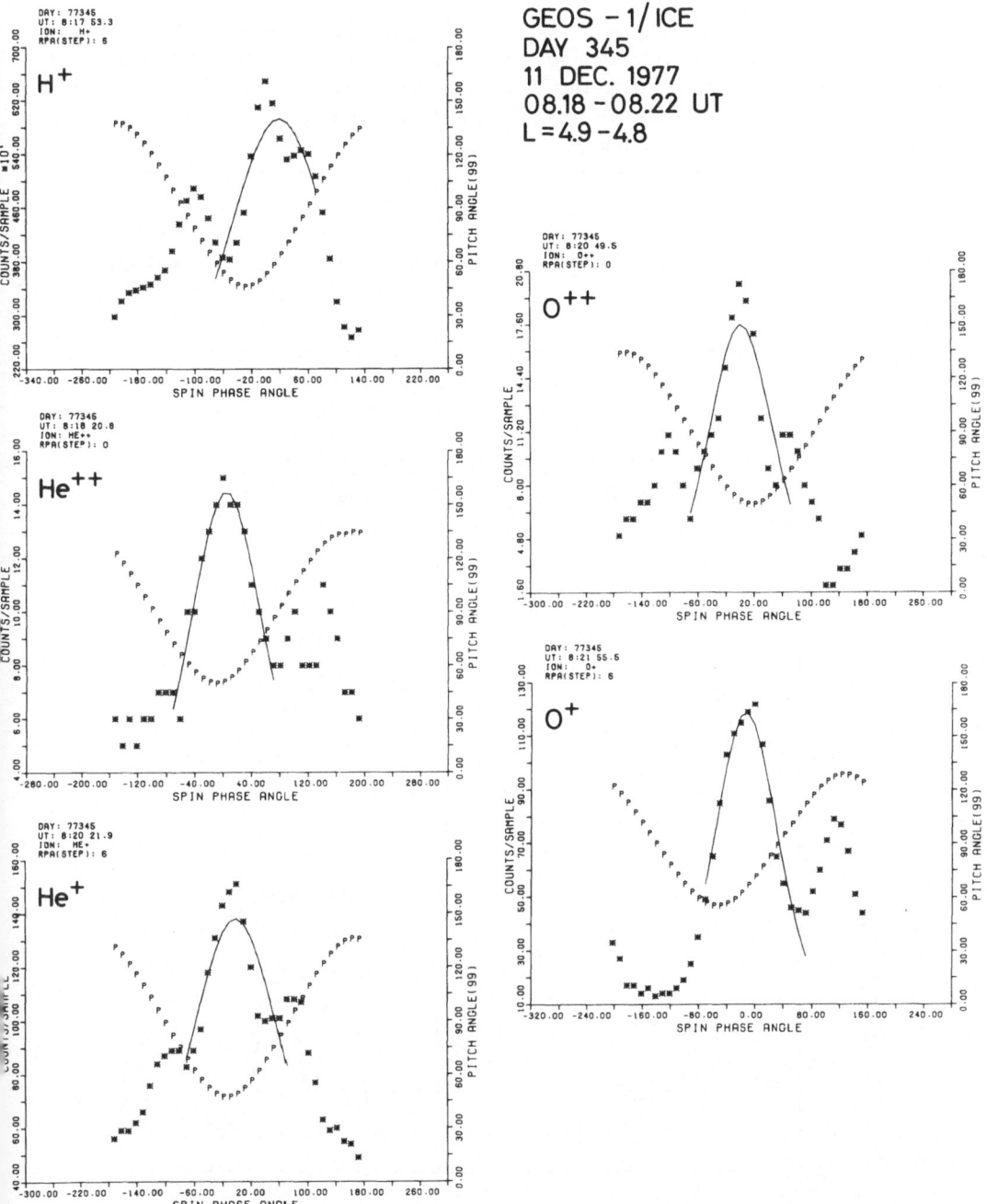

Fig. 3. Typical composition for thermal and suprathermal ions outside and at the plasmapause with the RPA at 6 Volts (see text) for the singly charged ions, in order to better demonstrate the non-thermal components. The fit program cannot be applied to such distributions and relative composition must be estimated by comparing peak heights. Pitch angles are indicated by the symbol P.

the magnetic field line for O^+. Of course the fitting program cannot be applied to such distributions and the relative composition must be estimated by comparing peak heights, whereas the temperatures can be estimated from RPA scans. Typical densities of H^+ and He^+ are lower, and those of O^+ are higher than inside the plasmasphere; temperatures are several to a few tens of eV.

In the previous paragraph we have not discussed the doubly charged thermal and suprathermal ions of helium and oxygen which are also shown in Figs. 2 and 3. These ions were first observed in the magnetosphere by the GEOS-1 mass spectrometer (YOUNG *et al.*, 1977), and it has since been confirmed by GEOS-2 and ISEE-1 that they are consistently seen at thermal and suprathermal energies within the magnetosphere. But, as discussed in the next chapter, they are also found as part of the hot (100 eV to several keV) population, although usually as minor constituents. (Of course the main part of the observed hot He^{++} ($\gtrsim 1$ keV) is of solar wind rather than of plasmaspheric origin; cf. BALSIGER *et al.*, 1980.) Figure 4, from GEISS *et al.*, 1978, shows a mass

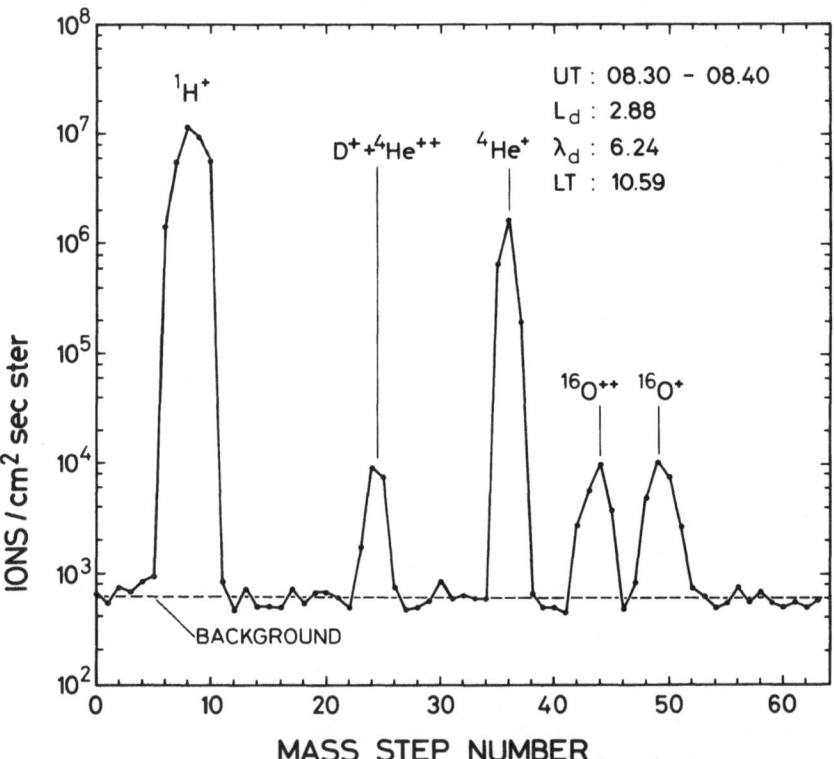

Fig. 4. Integral thermal (T) mode mass spectrum taken with a low resolution angular scan and the RPA fixed at the 2 eV/e step. The top of the $^1H^+$ peak is distorted due to saturation of detector electronics. From GEISS *et al.* (1978).

spectrum of these ions integrated over the range $2 \lesssim E/Q \lesssim 110$ eV/e. From the mass spectrum alone we could not distinguish between deuterium (D^+) and He^{++} both of which appear at $M/Q = 2$. However, Geiss et al. (1978) have shown that, based on a temperature measurement (by means of the RPA in front of the mass spectrometer) and assuming thermal equilibrium, the two species can be distinguished. For the case shown here the ions were mainly D^+: For D^+ a temperature of 0.8 eV was computed, similar to the one for O^+ and O^{++} of 0.7 ± 0.1 eV, whereas for He^{++} the temperature would have had to be 1.7 eV. In general we feel that in cases where the abundance ratio between masses per charge one and two is larger than $\sim 1,000$, one has to consider the possibility of a D^+ admixture.

The observation that the abundance ratio of O^{++}/O^+ could approach unity in the equatorial plasmasphere at $L \sim 3$, which is two to three orders of magnitude greater than the relative O^{++} abundance at corresponding latitudes in the topside ionosphere, led Geiss and Young (1981) to study theoretically the production and transport of O^{++} in the ionosphere and plasmasphere. They numerically solved the time dependent diffusion equation for a gas with H^+ and O^+ as major constituents and O^{++} as a minor ion, and concluded that thermal diffusion of O^{++} driven by the temperature gradient between the ionosphere and equatorial plasmasphere is responsible for the observed enrichment of doubly charged ions. This is mainly due to the fact that the thermal diffusion coefficients are roughly proportional to the square of the charge. They also showed that it is hard to overcome this diffusive separation by convection. Figures 5 to 7 are taken from Geiss and Young, 1981. Figure 5 shows the time integrated

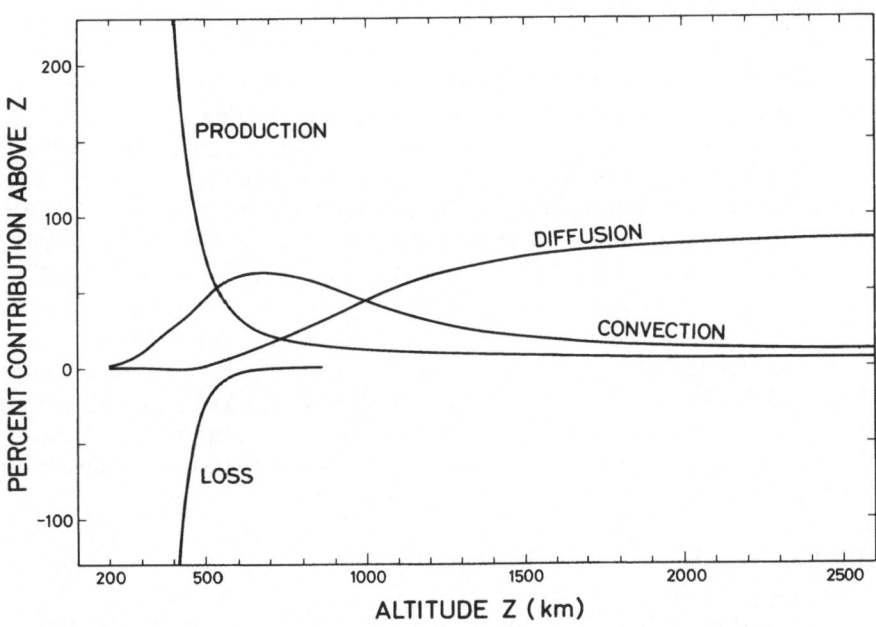

Fig. 5.   The time-integrated contributions of production, loss, diffusion and convection to the O^{++} content in the $L = 3$ flux tube above altitude Z is given for day 5, 1200 hours LT. From Geiss and Young (1981).

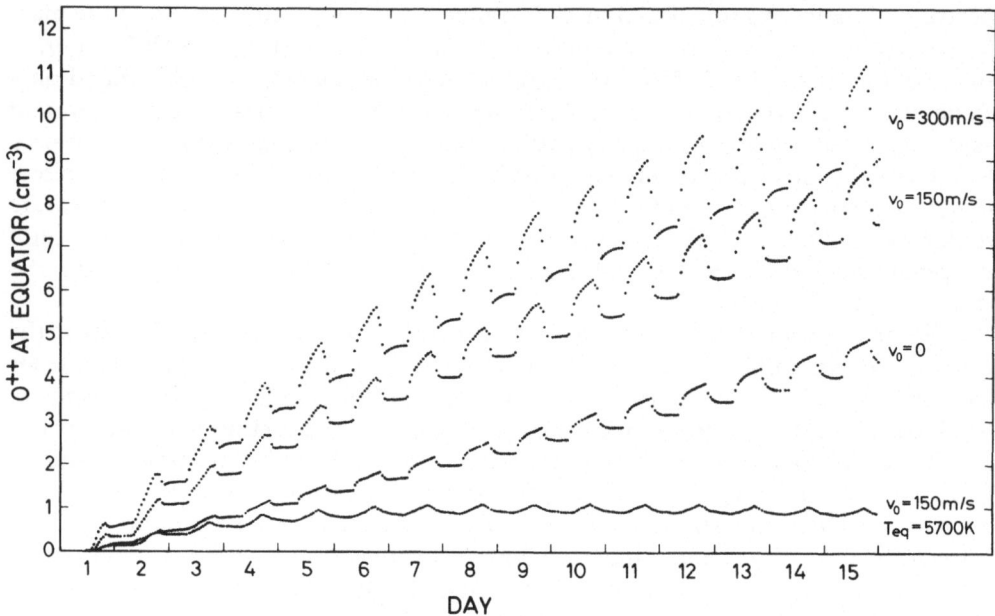

Fig. 6.   Buildup of the O^{++} density at high altitude in the $L = 3$ flux tube for the first 15 days. The influence of the convection velocity is shown in the upper 3 curves. The daily variations are the combined results of the daily variations of production and convection. The lowest curve shows the effect of choosing a lower equatorial temperature. From Geiss and Young (1981).

contribution of different processes to the O^{++} content in the $L = 3$ flux tube above a certain altitude Z for day 5 after the start of the flux tube refilling at noon. Diffusion is the dominant process above 1,000 km and contributes more than 90% to the O^{++} content above 2,000 km. Figure 6 shows how the O^{++} grows with time at the equator (V_0 is the assumed maximum velocity in the diurnal convection pattern). The lowest curve shows the dramatic effect of choosing a lower equatorial temperature. This reduction in the average temperature gradient results in a virtual switch-off of the thermal diffusion effect. Figure 7 gives the O^{++} content for days 2 and 7, with and without ($\alpha = 0$) thermal diffusion, as well as its local time dependence. The lower dotted curve shows the O^{++}/O^{+} ratio for day 7 as a function of altitude. Unity is reached above 6,600 km. Hence, the calculations of Geiss and Young (1981) show that under the right conditions in the plasmasphere the observed high O^{++}/O^{+} ratios can be obtained in several days. He^{++}/He^{+} should also grow with time more slowly due to the lower relative production rate (Geiss *et al.*, 1978).

3.   Hot Ions Accelerated at High Altitudes

The observation of hot O^{++} (up to several keV) with densities in the range 10%–100% of the O^{+} density led to the suggestion of a third source of hot magnetospheric ions, namely the high altitude thermal or suprathermal plasma

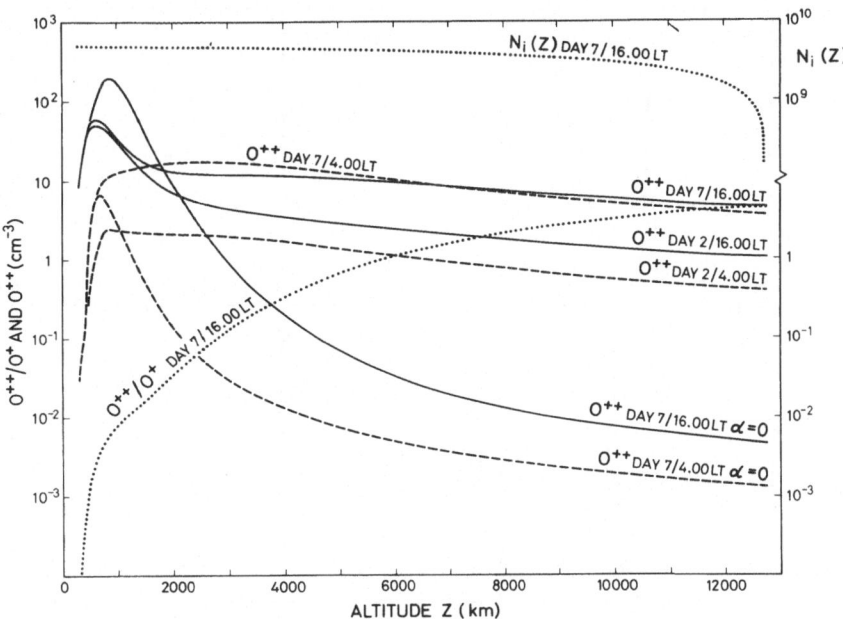

Fig. 7.   Density O^{++} as a function of altitude for selected times at $L = 3$. The upper four O^{++} curves give the results for the reference model (with convection velocity $V_0 = 150$ m/s). In the hypothetical case of no thermal diffusion ($\alpha = 0$, two lower curves) the lack of upward motion leads to buildup of O^{++} around 700km. The two dotted lines give the O^{++}/O^{+} ratio and the total O^{++} content in the flux tube between altitude Z and the equator. From GEISS and YOUNG (1981).

(BALSIGER *et al.*, 1980) as described in the introductory chapter. For simplicity this source is called "plasmaspheric" although it is not clear whether these ions are accelerated from the cold plasmasphere population or from the warm (suprathermal) ions which are found typically in the vicinity of the plasmapause (the latter could, of course, also come from the plasmasphere). Figure 8 shows four examples where hot multiply charged ions have been clearly identified. On the left hand side averaged mass spectra are given in the energy/charge range $0.1–0.9$ keV/e. Note, however, that O^{++} can also be accelerated to higher energies as seen in the energy spectra on the right hand side of Fig. 8 (e.g. for December 12, 1978). The O^{3+} ion also appears in this figure. It was reported for the first time by BALSIGER (1981), and has since been identified in a few other cases although its production mechanism is not yet fully understood. At the same time that O^{3+} was seen, there was an exceptionally high He^{++}/He^{+} ratio of 0.3; the He^{++} in this case was of terrestrial rather than solar origin as judged by its soft energy spectrum.

The hot plasmaspheric ion population has been studied for 50 GEOS-2 periods of between 1 and 21 hours duration (BALSIGER *et al.*, 1981) in order to establish (1) what conditions are required for observing multiply charged hot ions in substantial amounts in the magnetosphere (at geostationary orbit) and (2) what mechanism is accelerating these ions.

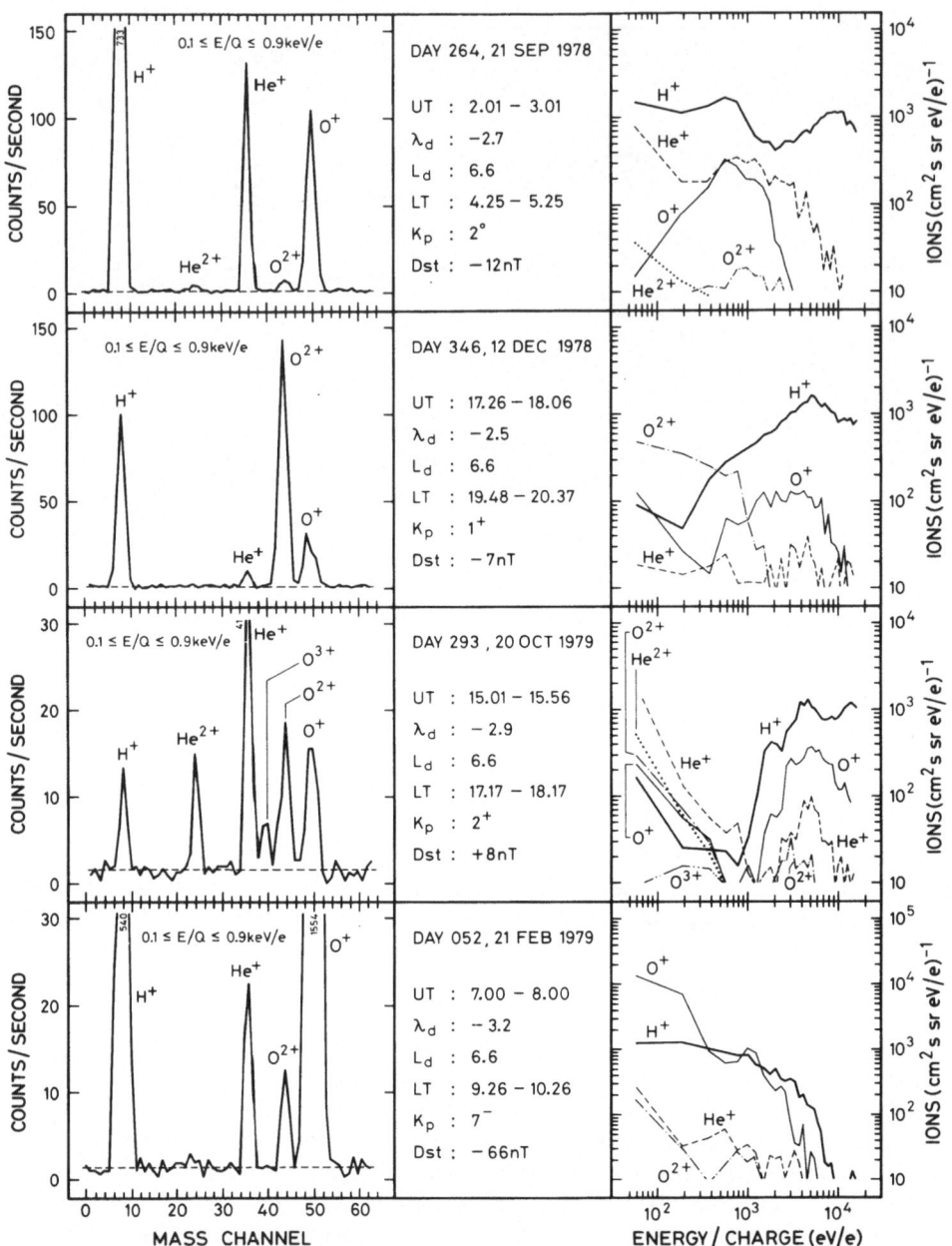

Fig. 8. Ions observed from the "plasmaspheric" source (see text). Mass and energy spectra measured at geostationary orbit by the GEOS-2 mass spectrometer. Mass spectra (left) are averaged over the energy per charge range 0.1–0.9 keV/e. Counts per second are proportional to differential flux. Where mass peaks exceed the scale, the according counting rates are given. Energy spectra (right) cover the full range of the mass spectrometer. In all four examples the "plasmaspheric" source is clearly recognizable by the source specific ions He$^+$ and/or O^{++}. For the first time O^{3+}, resulting from further ionization of O^{++} is clearly recognizable on October 20, 1979; at the same time "plasmaspheric" He^{++} resulting from further ionization of He$^+$ is exceptionally high (He^{++}/He$^+ \simeq 0.3$). From BALSIGER (1981).

The theoretical paper of GEISS and YOUNG (1981) shows that multiply charged thermal ions are enriched at high altitudes during long quiet periods if the temperature gradient is sufficiently large—a result which explained previous observations (YOUNG *et al.*, 1978). Prior to the period shown in Fig. 8, the magnetosphere was quiet, and in particular before days 346 and 293 it was quiet for more than one week. Our survey of hot plasmaspheric ions confirms that high O^{++}/O^+ ratios ($\gtrsim 0.5$) are observed after long periods ($\gtrsim 4$ days) of magnetic quiet ($K_p \leq 3$) in agreement with GEISS *et al.*, 1978.

It seems that the acceleration of plasmaspheric ions from thermal or suprathermal temperatures does not necessarily require strong magnetic activity. Hot doubly charged ions were found both in the middle of very quiet, and during magnetically active periods. On September 14, 1979 GEOS encountered such ions over a quiet period of 21 hours (Fig. 9). On the other hand the spectra in Fig. 8 for December 12, 1978 and October 20, 1979 were taken during substorm activity. Hot plasmaspheric ions were observed in many other substorms (preferentially during the expansion phase) as well as during initial phases of magnetic storms (when *Dst* was increasing). No local time dependence was seen in the cases studied.

The composition of the plasmaspheric hot ions is quite variable. The general feature is that abundances of He^+ and O^{++} are relatively high. A preliminary analysis gives similar abundances for O^+, He^+, H^+ and average ratios of the doubly charged ions of ~ 0.6 and ~ 0.1 for O^{++}/O^+ and He^{++}/He^+, respectively. (An admixture of

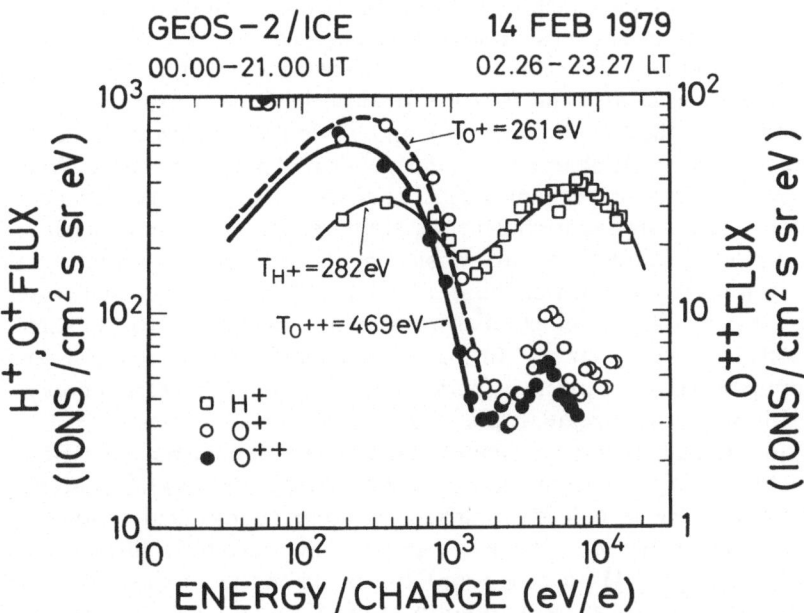

Fig. 9.   Ion flux for H^+, O^+, and O^{++} plotted versus energy/charge demonstrating the similarity of the lower energy ions ($\lesssim 2$ keV/e). Also given are curves from a fit program where maxwellians are fitted to the datapoints (the lowest datapoint is excluded in the fit because these ions are not isotropic). The fit also shows similar temperatures for the low energy part of singly charged ions, whereas O^{++} has roughly twice the energy of O^+, suggesting a charge and not a mass dependent energization mechanism.

D^+ to the He^{++} is unlikely in these cases of relatively low H^+ abundances, but cannot be excluded without more detailed analysis.) Similar composition is found for the inner edge of the storm time ring current (see Section 5).

The GEOS-2 spin axis is oriented perpendicular to the equatorial plane and hence the instrument covers only the population near 90° pitch angle (note that the acceptance angle depends somewhat on energy; BALSIGER et al., 1976). However we have compared one of our suprathermal ion cases with data from the UCSD plasma instrument on SCATHA (E. C. Whipple, private communication, 1982) and in this case our observation matches with the equatorially trapped plasma population as described by OLSEN (1981). We infer from this that hot plasmaspheric ions are trapped with pitch angles close to 90°, but further comparison with other experiments is needed to confirm this. OLSEN (1981) has shown that the equatorially trapped populations are associated in 90% of their occurrences with near-equatorial electrostatic noise in the 20–200 Hz range (GURNETT, 1976). A more complete study of gyroresonant wave-particle interactions has been carried out with GEOS data (YOUNG et al., 1981a) and is reported in the next section. At least in the case of He^+, they find energization perpendicular to the magnetic field direction ($\boldsymbol{B}$) under circumstances similar to those reported here.

In a discussion of acceleration mechanisms this population is particularly interesting, because it not only contains a wide variety of ion masses, but two of the elemental species are both singly and doubly ionized. This gives one the possibility of investigating the mass- and charge-dependence of the acceleration mechanisms. In order to separate the plasmaspheric ions from the other hot population, the energy spectrum of each species was fit with two maxwellian velocity distributions and temperature and density were computed for each of the two populations. By comparing ion temperatures a clear trend is seen: for each individual case studied, all singly charged ions have the same temperature and all doubly charged ions have *twice* the temperature of the singly charged ions, independent of mass per charge. We infer from this that the energization mechanism must be charge-, and not mass-dependent. As the fitting method is not always applicable, we compared the energy spectra of the different species by plotting differential flux versus energy per charge. In Fig. 9 we give an example demonstrating how well the spectra match. Ninety percent of all cases studied gave the same result; i.e. singly charged ions are accelerated to the same, doubly charged ions to twice the energy. In Fig. 10 a case is shown where the second, higher energy population has also plasmaspheric composition and where the energy per charge spectra, although quite irregular, are strikingly similar.

The question whether the plasmaspheric source is an important contributor to the hot magnetospheric population is still open. Our survey has shown that acceleration of O^{++} to several keV is indeed occurring for some geophysical conditions. If one considers in Fig. 10 the ions above ~ 1 keV/e as evidence for a further energization stage (with preferential H^+ acceleration), this could be the missing H^+ source which replenishes the magnetosphere with hot protons during quiet times (BALSIGER et al., 1980).

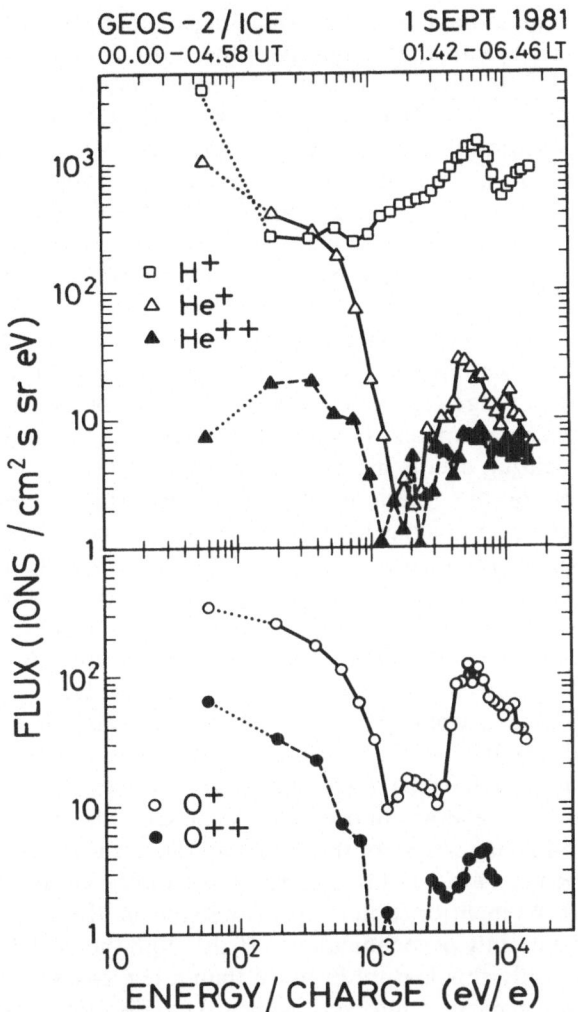

Fig. 10. Ion fluxes for five major ions plotted versus energy/charge, demonstrating the similarity of the spectra for the lower energy ions ($\lesssim$ 1.2 keV/e). But also the shape of higher energy ion spectra are very similar in this case (see text for further discussion).

4. Observations of Heavy Ions in Wave-Particle Interactions

It has been known theoretically for some time that the presence of even small amounts of heavy ions in a hydrogenic plasma can cause new resonances and cutoffs to appear and can couple the different propagation modes at the so-called cross-over frequency. These characteristic frequencies depend both on the heavy ion mass and on its abundance. Furthermore the presence of heavy ions will alter the wave group velocity and can enhance the growth rate of ULF waves in the vicinity of the heavy ion gyrofrequency. These ideas have been confirmed in a cooperative study of ULF waves involving the ion mass spectrometer and five other GEOS experiments.

On GEOS-1 and -2, the S-300 experiment often observes intense ULF waves near the local He^+ gyrofrequency (roughly pc1) and a detailed study was undertaken on the generation and propagation of these waves (YOUNG et al., 1981a; ROUX et al., 1982). The role of cold He^+ in the generation of electrostatic ion cyclotron waves has been reported separately by JONES et al. (1981). Briefly, the GEOS observations show that intense (up to 1 γ^2/Hz) harmonically structured quasi-monochromatic ULF emissions are commonly observed on both spacecraft. These are identified as ion cyclotron waves which are strongly tied to the presence of He^+ and in turn have a profound effect on the cold He^+ population observed by the mass spectrometer. The waves occur frequently ($\simeq 8\%$ of the time over a 4-month period) and are concentrated in the local dusk sector near the plasmaspheric bulge region for wave frequencies below Ω_{He+} (ROUX et al., 1982). The ion mass spectrometer shows that both the He^+ concentration and the occurrence of He^+ heating events are also most frequent there.

An investigation of the relationship among the ULF waves, the cold He^+ population and the hot protons has led to the following conclusions. ULF waves are generated by a temperature anisotropy ($T_\perp > T_\parallel$) in hot magnetospheric protons of ~ 10–100 keV energy. The waves are driven unstable convectively below Ω_{He+} in the presence of a He^+ concentration of $\gtrsim 5\%$. Above Ω_{He+}, the presence of more He^+ stabilizes the waves, however we find that the proton anisotropy and plasma β are near the threshold conditions for absolute instability. In this case He^+ plays an interesting role because it alters wave propagation to the extent that a laser-like effect takes place along field lines and amplification can still occur (ROUX et al., 1982).

We have also found that the waves act back preferentially on the cold He^+ population and heat it to several hundred eV's (YOUNG et al., 1981a). This seems to occur through a non-linear process of trapping and diffusion of He^+ ions in the very strong wave fields (ROUX et al., 1982). Figure 11 shows one of the events studied by YOUNG et al. (1981a). As mentioned in the previous section, this interaction with He^+ produces acceleration mainly perpendicular to B although the GEOS pitch angle scan limits our ability to study this feature in more than a few cases.

The picture of wave-particle interactions near Ω_{He+} is now reasonably complete and offers interesting implications for the exchange of energy in the collisionless magnetospheric plasma. We find that it is a fairly common phenomenon in which energetic protons give energy to cold He^+ which itself is a catalyst for the interaction. The medium through which this exchange takes place is the ion cyclotron wave whose propagation characteristics are strongly controlled by He^+. Magnetospheric sub-storms provide the ultimate energy source since they create the hot proton anisotropy in the first place (ROUX et al., 1982).

We would also like to investigate whether or not a similar phenomenon occurs with O^+ near Ω_{0+}. Although large amounts of hot O^+ are seen in the magnetosphere, the cold ion abundance is usually $\sim 1\%$ i.e. about 10 times less than that of He^+. Unfortunately the GEOS spin frequency (.17 Hz) is very near the nominal O^+ gyrofrequency at geostationary orbit (.10–.15 Hz) and makes any search for ULF waves near Ω_{0+} very difficult.

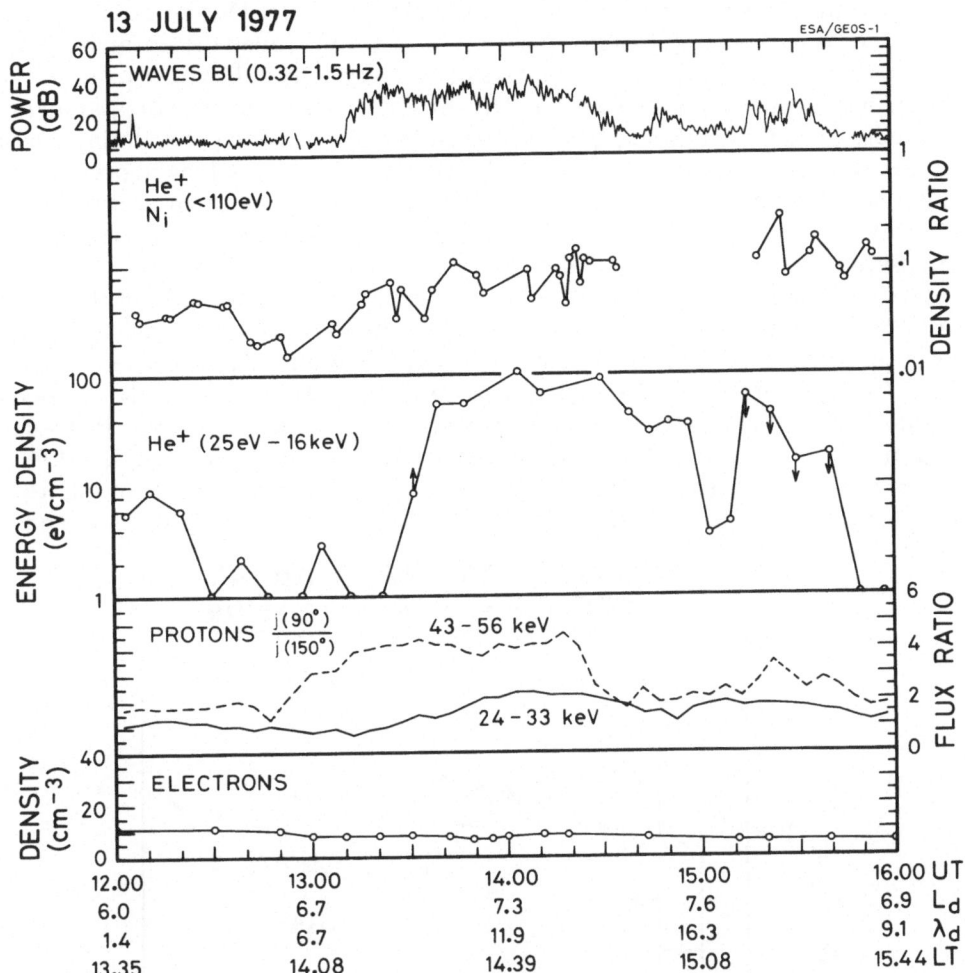

Fig. 11.   Composite data for July 13, 1977. Top panel gives the wave power in the L mode referenced to $10^{-5}\gamma^2/Hz$. The second panel is the He$^+$ relative abundance below 110 eV. The third panel is the He$^+$ energy density above 25 eV, indicative of He$^+$ heating by the waves. The fourth panel gives the energetic proton flux anisotropy (defined as the ratio of 90° to 150° pitch angle fluxes) in the three lowest energy ranges. The bottom panel gives the local cold electron density determined by S301 and S304 experiments. Error bars on He$^+$ data refer to statistical uncertainty of 1 standard deviation. Upper limits result from measurements near background levels. Data are displayed as a function of UT, dipole L value, magnetic latitude and local time. From YOUNG *et al.* (1981a).

5.   Storm Time and Recovery Phase Ring Current

Under conditions of the magnetic storm main phase and also during the recovery phase up to the point where the ion energy density has again decreased to prestorm values, it is possible to speak of a distinct ring current population. The inferred center of the ion ring current moves earthward to $L \sim 3$–4 and its inner edge approximately coincides with the shrunken boundary of the plasmasphere. During several storms GEOS-1 measured (at $E \lesssim 16$ keV/e) the ion composition around the storm maximum and also during the following several days of recovery. The only earlier composition measurements had been those performed by the Lockheed group on precipitating ions at altitudes $\lesssim 1,000$ km (Sharp *et al.*, 1977) and on trapped and precipitating ions at altitudes $\lesssim 8,000$ km (Johnson *et al.*, 1977). Although the GEOS-1 mass spectrometer did not cover the energy range around the maximum of ring current energy density (~ 50 keV), it performed the first measurements close to the equator and also provided composition as a function of radial distance.

A coordinated study with the mass spectrometers on S3-3 and ISEE-1 (Balsiger *et al.*, 1979) during the recovery phase of the December 11, 1977 storm gave very good agreement between O^+ densities measured by S3-3 at low altitude ($\sim 6,000$ km) and by GEOS-1 at high altitude ($\sim 25,000$ km) near local noon. However, the H^+ density at S3-3 was an order of magnitude lower than on GEOS-1, thus demonstrating in this case the probable loss of H^+ due to charge exchange. On the other hand, GEOS-1 measurements confirmed that O^+ (not He^+) was the dominant heavy ion during storm main phase and recovery (Geiss *et al.*, 1978; Balsiger *et al.*, 1980). Figures 12 and 13

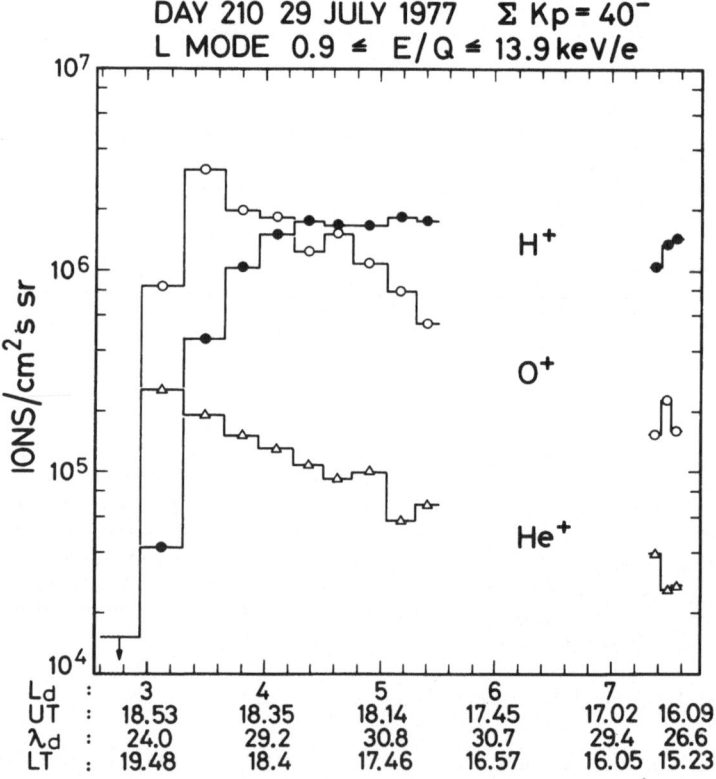

Fig.12.   Radial profile of composition during the July 29, 1977 magnetic storm (maximum $D_{st} = 100$ nT). Each set of data at a given L_d is obtained from L mode scans (see Balsiger *et al.*, 1976 for a description of modes) by summing over the 25 energy steps between 0.9 and 13.9 keV/e. From Balsiger *et al.* (1980).

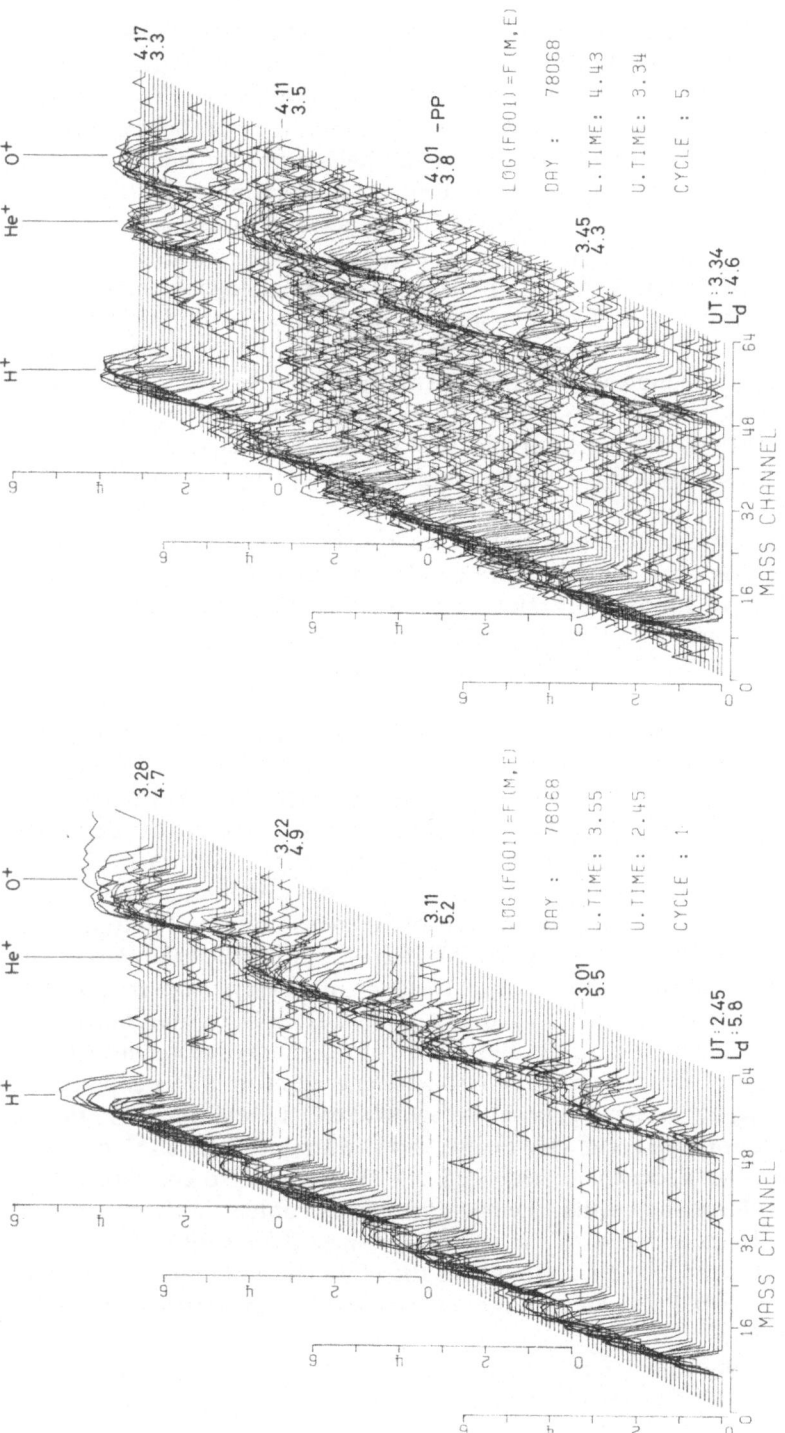

Fig. 13.   Sequence of three-dimensional L mode spectra for the geomagnetic storm on March 8–9, 1978 (maximum $D_{st} = -111$ nT). Dashed lines separate complete mass-energy scans, each lasting 5.9 min. Zero and 1 count per sample (0.15 s) are plotted as zero, otherwise data are shown as they appear from telemetry. The abscissa gives mass per charge (M/Q), the ordinate is logarithm of counts per sample and the third axis gives time and energy/charge. Energy/charge is scanned between two dashed lines from 60 eV/e to 15.7 keV/e in 32 logarithmic steps. In between L modes, other scans were made which are not shown here. The plasmapause (PP) on this day is clearly localized by a steep gradient in electron density. From BALSIGER *et al.* (1980).

summarize our major findings on the storm time ring current. Figure 12 shows a typical radial composition profile. The H^+ flux is nearly constant in the range $L \simeq 4.5-7$ and drops off at the inner edge of the ring current, whereas both O^+ and He^+ increase towards the inner edge where O^+ becomes the dominant ion. A similar behavior can be seen in Fig. 13 showing a sequence of three-dimensional spectra for the March 8–9, 1978 storm. Additional information in this figure are: 1) The plasma pause (PP) which could be clearly distinguished by the S-300 experiment on GEOS-1 as a steep gradient in electron density; note that the inner edge of the ring current reaches beyond this plasmapause. 2) The ions (mainly the O^+ and He^+) at the inner edge are consistently softer than at higher altitudes and have a similar composition (including He^+ and O^{++}) as the local thermal or suprathermal plasma. This led us in the previous chapter to suggest that, at times, the high altitude thermal plasma is accelerated to several keV and thus becomes a source to the hot population and in particular to the inner edge ring current population. The general characteristics of storm time ring current composition have since been confirmed by ISEE-1 (LENNARTSSON et al., 1979) and PROGNOZ-7 (LUNDIN et al., 1980).

In an attempt to estimate the contribution of terrestrial ions over a wider energy range, the 0–16 keV composition data were combined with the ion energy spectra obtained from the MPAE Lindau solid state telescopes S-321 on GEOS-1. Figure 14 shows how well the ion spectrum matches the H^+ spectrum of the mass spectrometer for the moderate storm of April 30, 1978. It was concluded that in this case most of the energetic ions were protons. We have fitted these data to Maxwellian distributions (Fig. 15) starting at the high energy end and, assuming that the ions measured by S-321 are all protons, their distribution was extrapolated to low energies. Subtracting this distribution from the mass spectrometer data, a low energy proton Maxwellian was computed, and also the O^+ data were fitted with a Maxwellian (for further detail see BALSIGER et al., 1980). Assuming that O^+ and the low energy portion of the H^+ spectrum are of terrestrial origin and that the high energy ions are of solar wind origin, the following limits for the relative contributions of the two sources to the total local energy density were computed for this case: $\geq 20\%$ terrestrial, $\leq 80\%$ solar. However, other cases are more complicated as seen in Fig. 16, taken in the early recovery phase of the July 21, 1977 storm near the center of the ring current. Here O^+ alone contributes 23% of the total energy density (BALSIGER, 1981) and from its energy spectrum it cannot be excluded that oxygen has a second peak at higher energies (the S-321 experiment would not register singly charged oxygen ions below 100 keV). In reviewing GEOS and ISEE-1 composition data, WILLIAMS (1981) also shows examples where O^+ could in principle be the dominant ion at energies > 25 keV. We must also emphasize that YOUNG et al. (1981b) have shown that the average oxygen content in the magnetosphere has dramatically increased with the solar cycle since these GEOS-1 measurements were made (see next chapter). We would therefore assume that during solar maximum oxygen ions could play an even more important role as contributors to the storm time ring current.

BALSIGER et al. (1980) have analyzed four storms up to four days after storm maximum in order to study the composition of the recovery phase ring current. In no case was He^+ the dominant ion either during or after the storm; however, O^+ could

DAY 120, 30 APRIL 1978, UT = 15.51−15.57

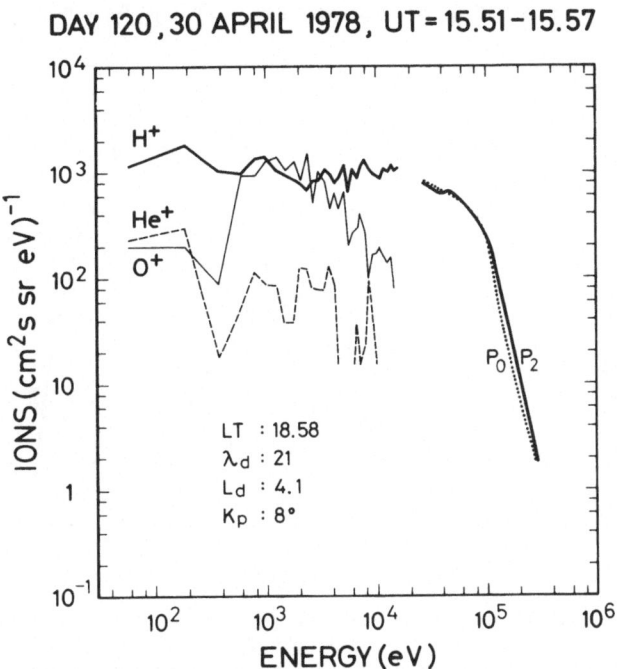

Fig. 14. Combined energy spectra of GEOS-1 experiments S-303 (for H^+, O^+, and He^+, below 16 keV) and S-321 (total ions above 24 keV) for the center of the storm time ring current on April 30, 1978. Data from S-321 have been provided by the Max-Planck-Institut, Lindau. The solid line above 24 keV represents the detector P_2 looking nearly parallel to S-303. Detector P_0 (dotted line) looks almost parallel to the spin axis. The difference between the two curves is a measure of the pitch angle distribution (nearly isotropic in these cases). The proton curve of S-303 matches fairly well the total ion curve of S-321, implying that ring current ions above 20 keV are mainly of solar wind origin in this example. From BALSIGER *et al.*, 1980.

dominate over H^+ (in terms of number density) within the GEOS mass spectrometer energy range. If, as has been suggested (LYONS and EVANS, 1976; TINSLEY, 1976), charge exchange with neutral hydrogen were the main loss mechanism for the ring current ions and if there were not a continuous H^+ source, protons should diminish relative to O^+ with progressing time. In order to test this the H^+/O^+ ratios in the region $3 \leq L \leq 5$ for several days following the storms have been computed and are given in Table 1. In three out of four cases the H^+/O^+ ratio remained rather constant or even increased with time; for the fourth case (November 26, 1977), it was argued in BALSIGER *et al.* (1980) that 0^+ had again been injected by a substorm one day after the main phase, when K_p rose to 5°. One day later the H^+/O^+ ratio was rising as in the other cases. As a possible explanation BALSIGER *et al.* (1980) suggested that the losses of H^+ due to charge exchange are compensated for by a proton-rich source, continuously supplying hot H^+ even during the recovery phase.

DAY 120, 30 APRIL 1978, UT = 15.51 – 15.57

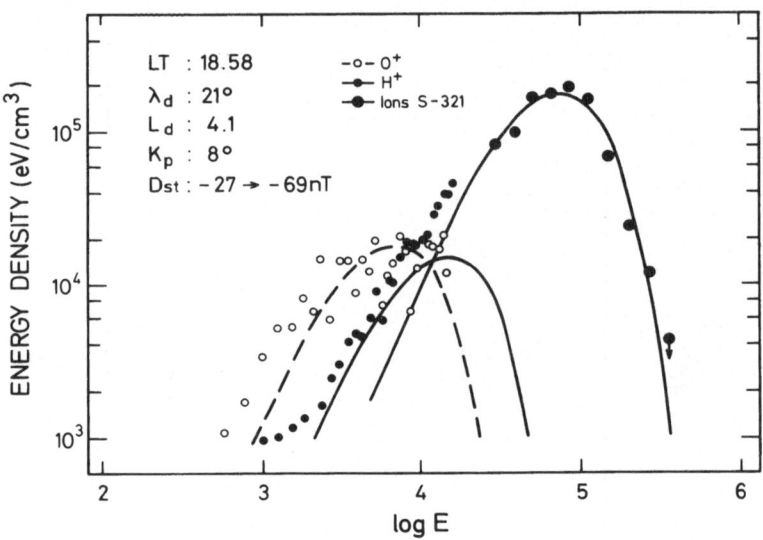

Fig. 15.   Differential spectrum of energy density ($N \cdot E$) versus the dimensional parameters $I = \log E$ ($\mathrm{d}(NE)/\mathrm{d}I = f(I)$) where E is measured in electron volts. The curves are Maxwellian fits for O^+ (dashed line) and low-energy and high-energy protons (solid lines); for an explanation see BALSIGER et al. (1980). From BALSIGER et al. (1980).

6.   Synoptic Studies near Geostationary Orbit

The geostationary orbit provides a unique possibility for long time studies of magnetospheric particles because spatial variations are reduced to the single dimension of local time. Furthermore a high data recovery rate is possible ($\simeq 98\%$ for GEOS-2 up to August 1, 1980 and about 75% in 1981 because of limited GEOS-2 operation) which, in combination with an excellent longtime stability of the electronics, allows for the accumulation of large, complete data sets with good statistics so that rare ions (e.g. O^{++}, He^{++}) may also be observed routinely. We have taken advantage of these features of the GEOS-2 mission and are reducing all data from the mass spectrometer to the form of 3-hour averages which we term synoptic data. The 3-hour intervals have been arranged such that each interval corresponds to one of eight local time sectors: 0–3, 3–6, ..., 21–24 hours. The energy range of the GEOS mass spectrometer has been divided in two, 0–860 eV/e and 0.9–15.9 keV/e, respectively, based on our observation that energy spectra tend to have a minimum around 0.9–1 keV/e which separates the "warm" from the "hot" ions (cf. Fig. 10). These synoptic GEOS-2 data plus selected intervals of GEOS-1 data will be discussed in this section with respect to composition variations due to 1) geomagnetic activity and 2) solar activity as measured by the 10.7 cm solar radio flux. We shall concentrate our discussion mainly on the hot ions, i.e. on the energy range 0.9–15.9 keV/e. Ion composition at geostationary orbit shows also a certain local time dependence as seen in Fig. 17 from YOUNG (1980). In particular

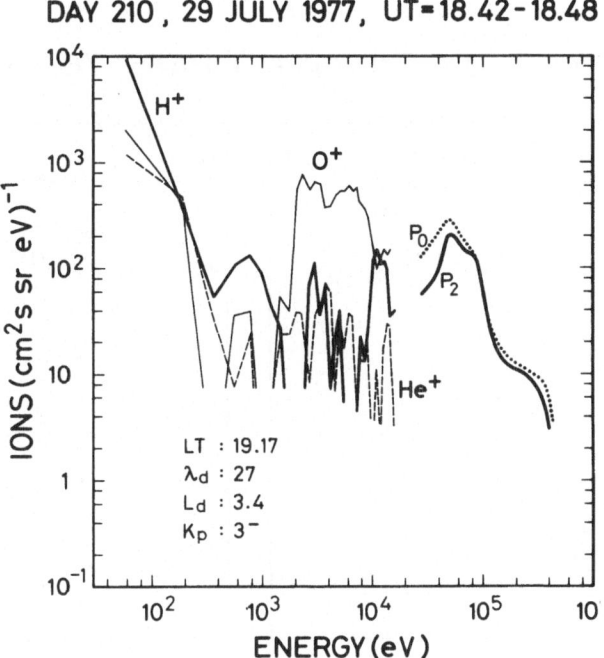

Fig. 16.   Combined spectra of GEOS-1 experiments S-303 (for H^+, O^+, and He^+, below 16 keV) and S-321 (total ions above 24 keV) for the center of the ring current in the early recovery phase on July 29, 1977. For description see Fig. 14 and text. From BALSIGER (1981).

Table 1.   Composition of ring current ($3 \lesssim L_d \lesssim 5$) during storm recovery phase.

Date	Day of storm	Days after storm			
		$+1$	$+2$	$+3$	$+4$
29 July 1977, Day 210	0.16	0.12	0.29	1.83	
26 Nov 1977, Day 330	0.31	0.06	0.11	0.17	
04 Jan 1978, Day 004	0.49	0.61	0.64	(0.44)	(1.25)
09 March 1978, Day 068	0.54	0.49	0.66	0.49	0.42

The number density ratio H^+/O^+ is given as a function of time. Due to large background corrections (penetrating electrons) the errors may be quite high: $\sim 10\%$ for storm days and ~ 30–50% for days after the storm. Numbers in parentheses have an error of $\sim 100\%$. From BALSIGER et al. (1980).

O^+ shows a maximum around noon whereas H^+ and He^{++} peak around local midnight. This can be explained for the oxygen by the GEOS-1 observation that the O^+ density increases with decreasing L for a fixed local time (cf. Section 5) and at the geostationary orbit GEOS-2 is probing lower L values around noon than at the other local times. For H^+, which showed rather constant density in the ring current profiles over several L (see Fig. 12), we explain the increase in density by the fact that the

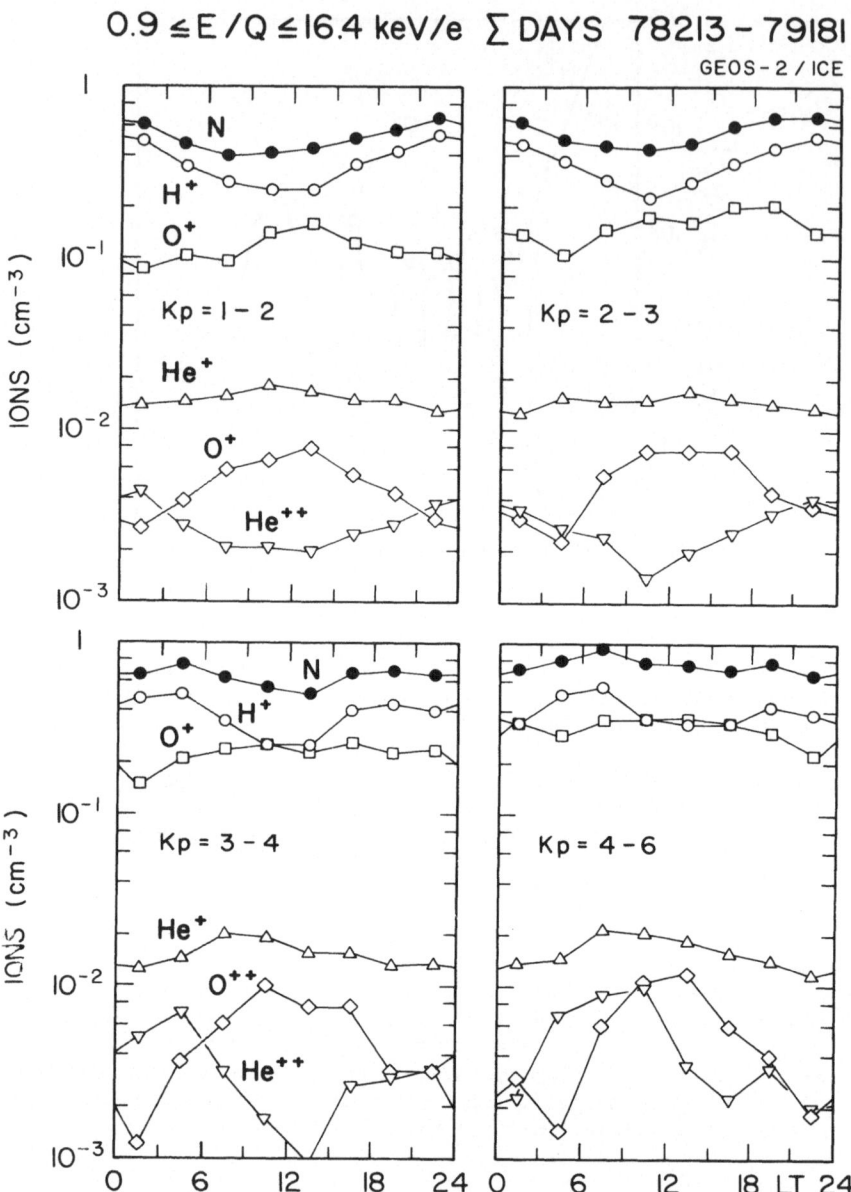

Fig. 17. Synoptic data for hot ions ($0.9 \leq E/Q \leq 16.4$ keV/e) averaged over 11 months. Density for five major ions and total ion density N are plotted versus local time (LT) in four different K_p ranges. From YOUNG (1980).

satellite crosses the inner edge of the plasmasheet in the midnight sector; this also explains the simultaneous increase in the He^{++} density. The fact that composition data often show a quite distinct boundary between plasmasheet and ring current region will be discussed in the last chapter.

Ion composition and magnetic activity: Figure 18 gives ion densities averaged over almost one year as a function of K_p (YOUNG, 1980). Both H^+ and He^{++} are rising with K_p by about a factor of 2 with a roughly constant He^{++}/H^+ ratio of $\approx 1\%$ over the whole K_p range. Much more dramatic is the increase of the O^+ density, namely one decade, whereas He^+ and O^{++} are relatively constant, $\approx 2 \times 10^{-2}\,cm^{-3}$ and $\approx 8 \times 10^{-3}\,cm^{-3}$, respectively, over the whole energy range. We shall return to further discussion of the oxygen ions in the following paragraph.

Solar cycle effects on ion composition: By combining GEOS-1 and -2 results, YOUNG *et al.* (1981b) have analyzed nearly four years of data for long term trends in average ion composition. The data have been correlated with the monthly averaged

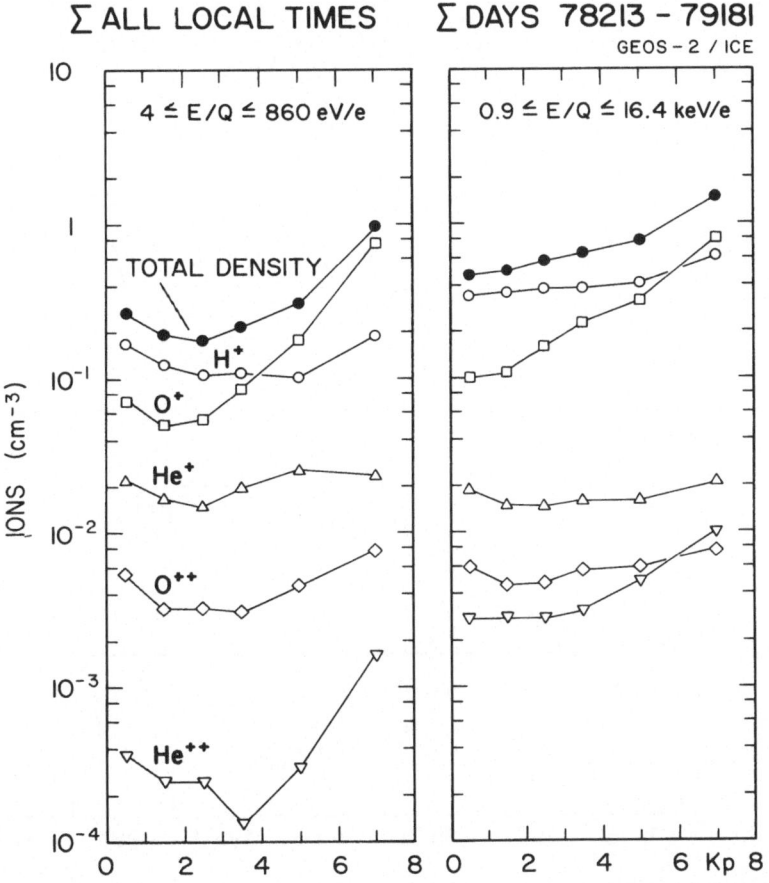

Fig. 18. Synoptic data averaged over 11 months. Individual ion densities averaged over all local times, as well as total ion density N are plotted versus K_p in two energy ranges. From YOUNG (1980).

10.7 cm solar radio flux ($\bar{F}_{10.7}$) and the Zurich sunspot number ($\bar{R}_z$) as measures of solar activity. The $\bar{F}_{10.7}$ index is commonly used as a measure of the solar EUV variations and is easily available, whereas actual EUV spectra are not. In Fig. 19 O^+ density from magnetically quiet periods ($K_p < 2_0$) is plotted along with the monthly averaged solar indices $\bar{F}_{10.7}$ and $\bar{R}_z$ and the geomagnetic activity index $\bar{A}_p$. It is evident that the increase in O^+ is related to the solar cycle. At the same time, there is a similar increase in He^+ and O^{++} whereas H^+ and He^{++} show hardly any long term change. For magnetic quiet the average H^+ density was 0.3 cm^{-3} and the He^{++}/H^+ ratio about 9×10^{-3} over a period of three years (Young *et al.*, 1981b). The O^{++} variation is closely linked to that of O^+. As a matter of fact, the O^{++}/O^+ ratio for $K_p < 2_0$ is

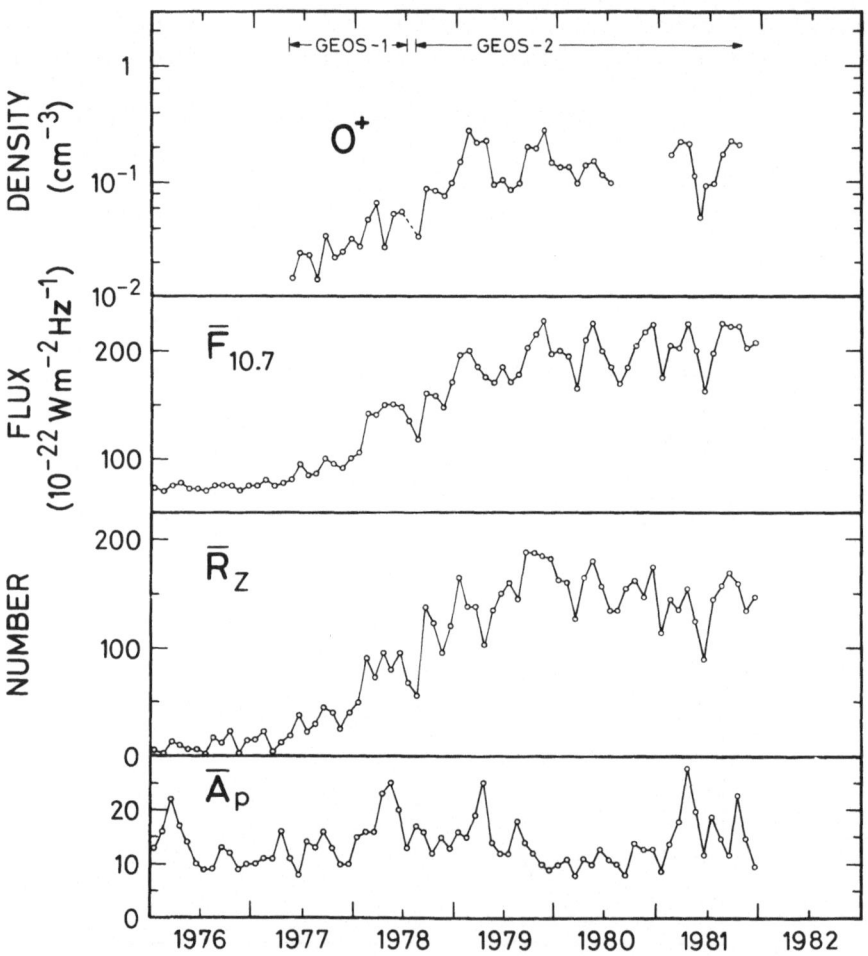

Fig. 19.   Effect of solar cycle on hot ion composition: O^+ density ($1 \leq E \leq 14$ keV) for magnetically quiet periods ($K_p < 2_0$) is plotted along with the monthly averaged solar indices $F_{10.7}$ and R_z and the magnetospheric magnetic activity index $\bar{A}_p$. The gap in GEOS-2 data from 7.80 to 2.81 corresponds to a period when the satellite was switched off.

nearly constant ($\sim 6\%$). This is also shown in Fig. 20 where O^{++} density is plotted versus O^+ density for 3 K_p ranges. For the two lower ranges O^{++}/O^+ is nearly constant, independent of solar cycle. The $K_p = 6-9$ data possibly show some solar cycle effect, however the data set is too limited to draw any conclusions. As expected from GEISS and YOUNG (1981), the O^{++} is lower relative to O^+ during magnetically disturbed times (compare discussion in Section 2). As reason for the increase of O^+ and O^{++} YOUNG et al. (1981b) suggest the rising of the scale height of neutral oxygen due to the heating of the upper atmosphere by the increased solar UV combined with increased ionization rates due to higher EUV fluxes (GEISS and YOUNG, 1981). A more thorough discussion of this data set is given in YOUNG et al. (1982).

7.   Coordinated Studies with Other Spacecraft

In order to separate spatial from temporal variations and to investigate large scale magnetospheric circulation, one needs to study simultaneous mass spectrometer data taken at different locations in the magnetosphere. Preliminary results of an extensive coordinated hot plasma study have been given by BALSIGER (1981) and by JOHNSON (1981). For the 21–22 February 1979 storms, simultaneous composition measurements were obtained on seven satellites: S3-3, GEOS-1, -2, ISEE-1, -3, PROGNOZ-7, and SCATHA. Their orbits within the magnetosphere are shown projected onto the geomagnetic equatorial plane in Fig. 21. Universal time is indicated along the orbits beginning on February 21. The magnetopause locations were not measured but are

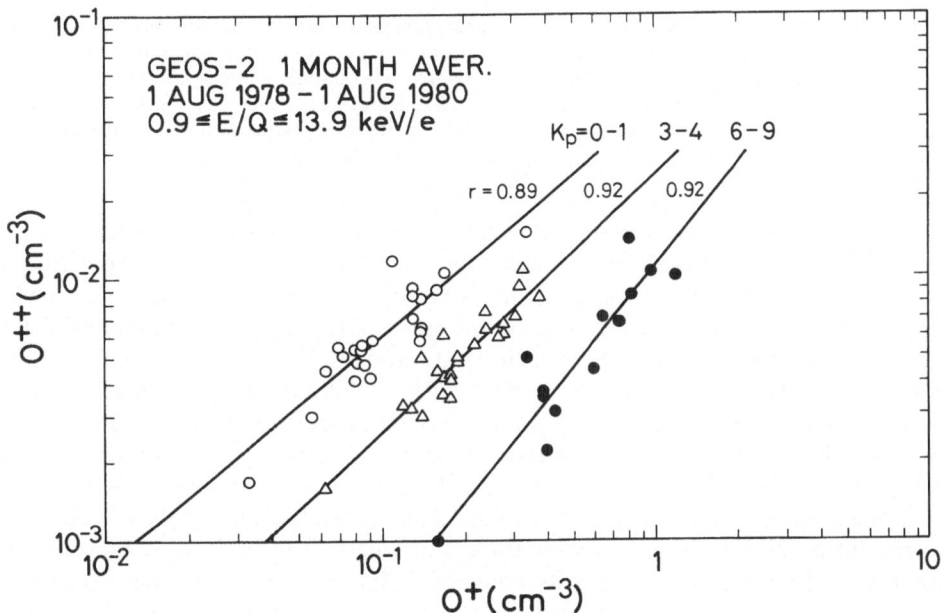

Fig. 20.   Monthly averages of O^{++} and O^+ densities in the period August 1, 1978–August 1, 1980 for three different K_p ranges. Note that the O^{++} variation is closely linked to that of O^+.

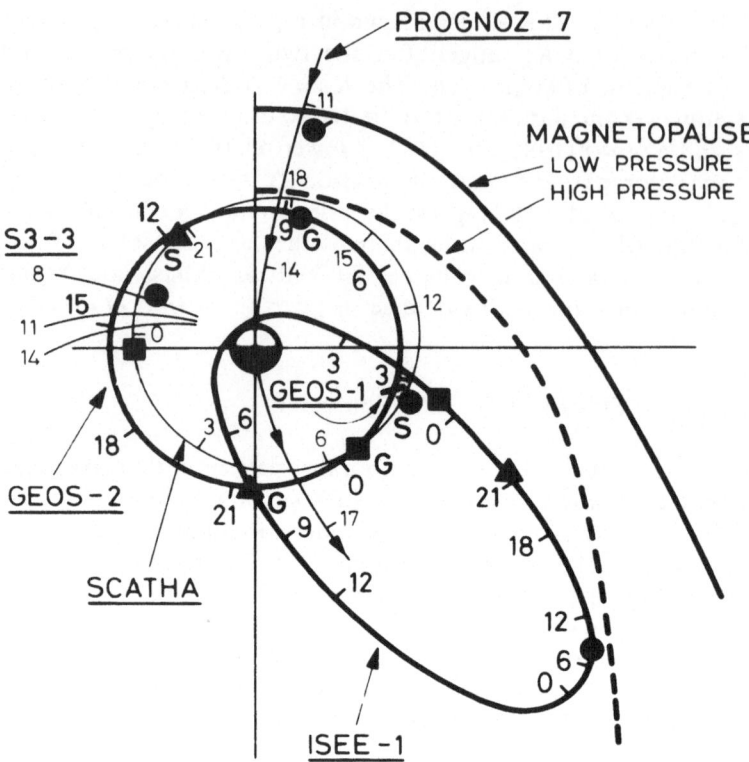

Fig. 21. Equatorial projections of orbits of satellites carrying mass spectrometers during the composition-coordination-study of the February 21–22, 1979 magnetic storms. Round, triangle and square symbols indicate the relative positions of the satellites during 3 consecutive minima in Dst (see also Figs. 22 and 23). During these three periods O^+ was the dominant ion at all satellite locations indicated with the exception of ISEE-1 around 22.00 UT, where O^+ contributed about 40% to the total plasmasheet number density. Orbits are only shown for times of data coverage. From Balsiger (1981).

indicated schematically for measured low and high solar wind pressures. ISEE-3 was at the sun-earth libration point and provided information on solar wind composition. In Fig. 22 GEOS-2 and ISEE-1 ion densities are plotted on the same time scale together with geomagnetic indices. At this stage in the study only qualitative statements can be made: (1) Around 16.45 UT on February 21 and 01.30 UT on February 23, the plasmasheet proton density as measured by ISEE-1 at about 20 R_E in the tail indicated an admixture of "new" plasma with a relatively soft spectrum. Peaks in the GEOS-2 proton density, with soft energy spectrums as well, appear in the hourly averages around 17.30 and 02.30 UT on the respective days; i.e. with an approximate time delay of only one hour. Comparison with ISEE-3 and more detailed data analysis with higher time resolution should lead to better understanding of the entry of magnetosheath plasma into the plasmasheet and its convection into the inner magnetosphere. (2) During storm maxima (as measured by D_{st}) O^+ is the dominant ion in the inner magnetosphere, and even in the plasmasheet it is quite abundant (cf. ISEE-1 before

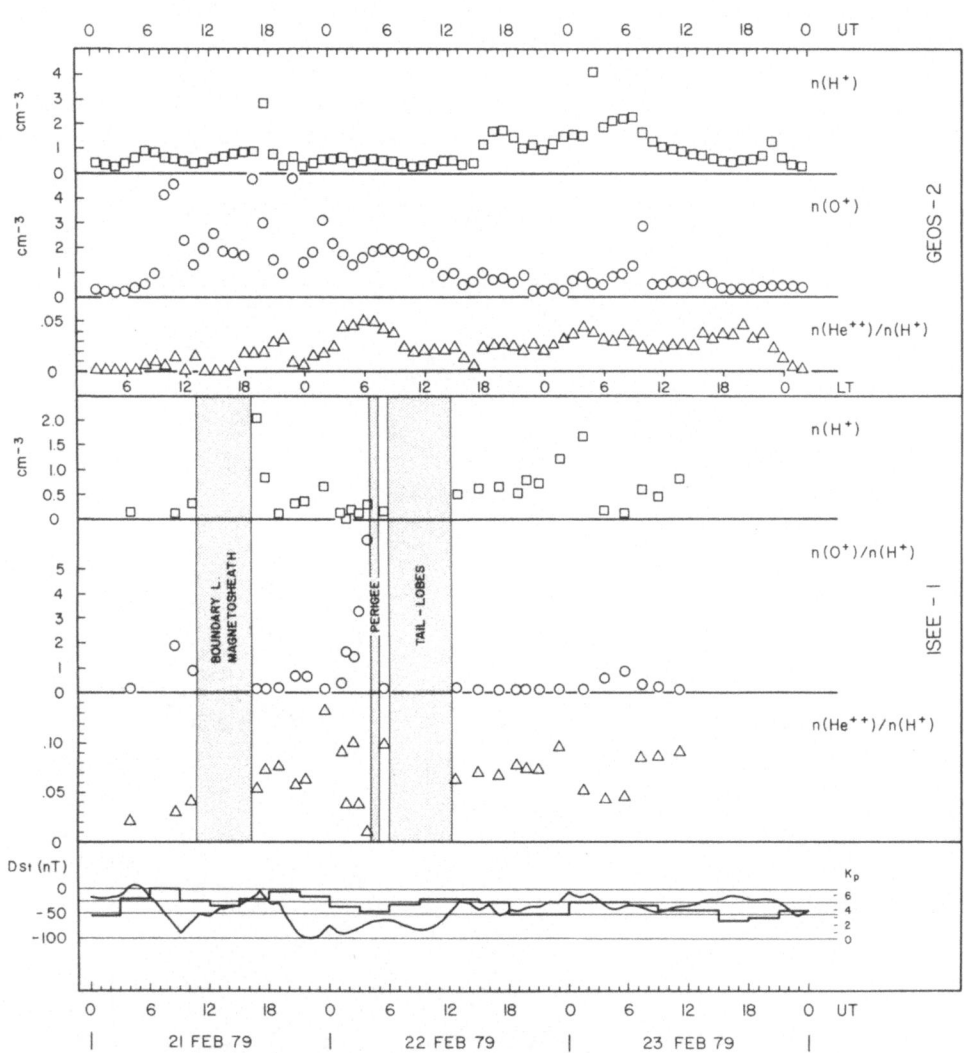

MAGNETIC STORM FEB 21/22 1979

Fig. 22.   Simultaneous ion composition measurements made by two almost identical instruments, one on GEOS-2 at 6.6 R_E and the other on ISEE-1 with an orbit as shown in Figure 21. Note the increase in H^+ density around 17.00 UT on February 21 and around 22.00 UT on February 23 at both spacecraft, being more than 13 R_E apart, and also the substantial increase of O^+ during the main phases of the storms. From BALSIGER (1981).

01.30 UT on February 22). Preliminary analysis shows O^+ to be the dominant ion at the location of SCATHA, S3-3 and PROGNOZ-7 as well during and near the storm main phase. The round, triangular, and square symbols in Fig. 21 indicate the locations of the satellites at or near the maximum storm intensity as indicated by the dotted lines in Fig. 23. Figure 23 from JOHNSON (1981) demonstrates that even at 32 keV the energy density of the O^+ ions can become equal to or larger than that of H^+ near these peaks. Hence the preliminary conclusions from this coordinated study are that during

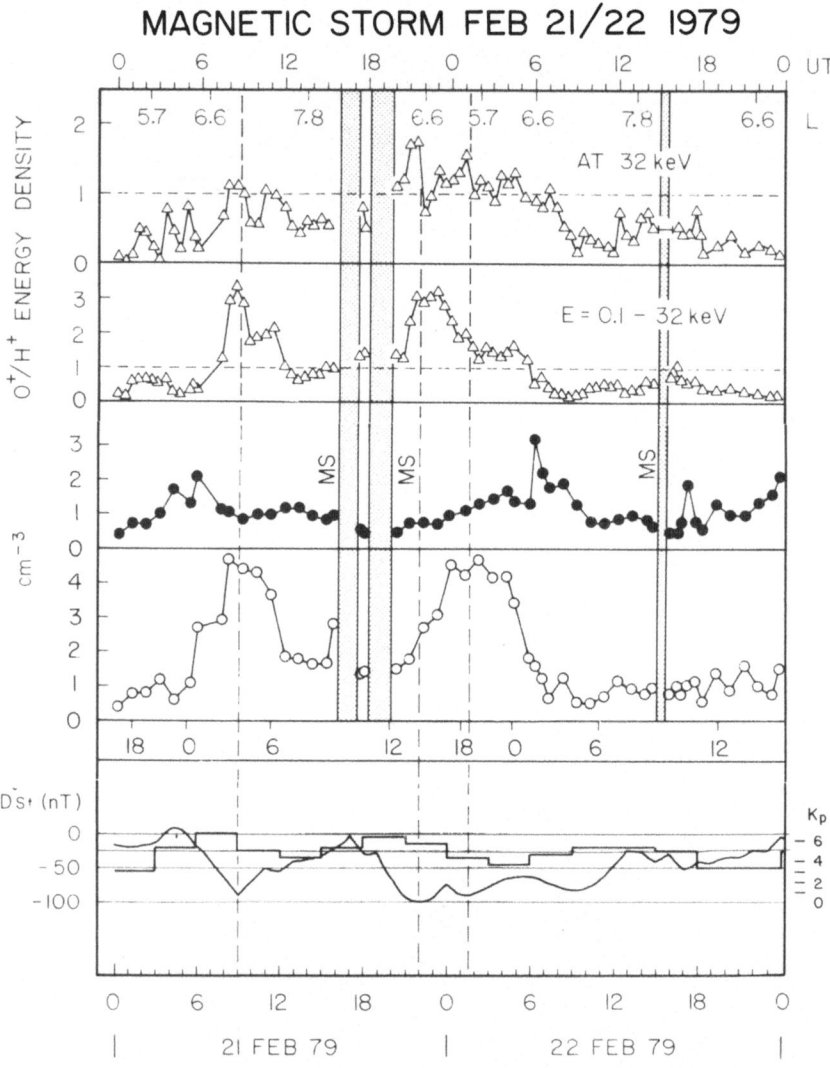

Fig. 23. SCATHA hot plasma composition data during the magnetic storms on 21–22 February 1979. O^+/H^+ energy density ratio reaches values of unity and above during storm maxima, even at 32 keV. From JOHNSON (1981).

magnetic storms the ionospheric ion injections occur nearly simultaneously over a wide range of local times and are distributed over a very large volume of the magnetosphere. The observed contribution of terrestrial ions to the total energy density becomes at times at least 50%, but could be much higher, considering the fact that at least a certain (unknown) amount of the H^+ is also of terrestrial origin. If the O^+ injection takes place along field lines, very efficient pitch angle scattering would thus be required. As an alternative it has been suggested that part of the hot plasma might be accelerated from local thermal plasma at high altitudes (BALSIGER *et al.*, 1980; see also Sections 2 and 3); as a matter of fact, throughout this storm, O^{++} ions were observed as part of the hot plasma at GEOS-2 (BALSIGER, 1981).

8. Summary and Discussion

The two GEOS spacecraft have provided a very large amount of composition data for the near equatorial inner magnetosphere. These data are of course still not fully digested. One of our main goals up to now was to study systematically as large a body of data as possible in order to establish the gross features of composition. From the synoptic studies (Section 6) perhaps the most interesting result obtained is the increase of terrestrial oxygen ions during increase of the solar cycle (YOUNG *et al.*, 1981b, 1982).

Coordinated composition measurements with several spacecraft have been performed (Section 7) in order to study large scale circulation. On the one hand we frequently observe substorms with characteristic energy dispersion signatures as shown in Figure 24 for H^+, O^+, and He^+ (STOKHOLM, 1981); they are indicative for a classical convection model (with ion injection and subsequent magnetic field drifts) as sketched qualitatively in Fig. 1. On the other hand there are several indications that a simple convection model with one source region cannot explain all the observations and that injection regions other than the inner boundary of the plasmsheet around local midnight must be considered: (1) The composition of the ions in Fig. 24 is not the same as that observed in the plasmasheet (PETERSON *et al.*, 1981) because the He^{++} abundance is too low. (2) During onsets of magnetic storms, the magnetosphere is sometimes compressed and the magnetopause moves inside geostationary orbit. On such occasions we have observed (Fig. 25) a quite pronounced boundary between ring current and plasmasheet (in this figure called "mixed region"), apparently separating two populations with distinct differences in composition (BALSIGER *et al.*, 1980). The ring current ions have a composition which is more typical for upflowing ionospheric ions observed mainly in the dusk to midnight sector by GHIELMETTI *et al.* (1978). (3) The nearly simultaneous appearance of O^+ over very wide regions of the magnetosphere during large magnetic storms as demonstrated in coordinated composition studies (Section 7) suggests a simultaneous local injection over a large local time region either along field lines (in combination with very effective pitch angle scattering) and/or through local acceleration of high altitude thermal plasma as discussed in Section 3. (4) The acceleration of plasmaspheric ions (characterized by high He^+ and large O^{++}/O^+ ratios) during the initial phase of substorms, as discussed in Section 3, is further evidence for an additional source other than the plasmasheet.

The question of what portion of magnetospheric plasma is of terrestrial origin (i.e.

H. Balsiger *et al.*

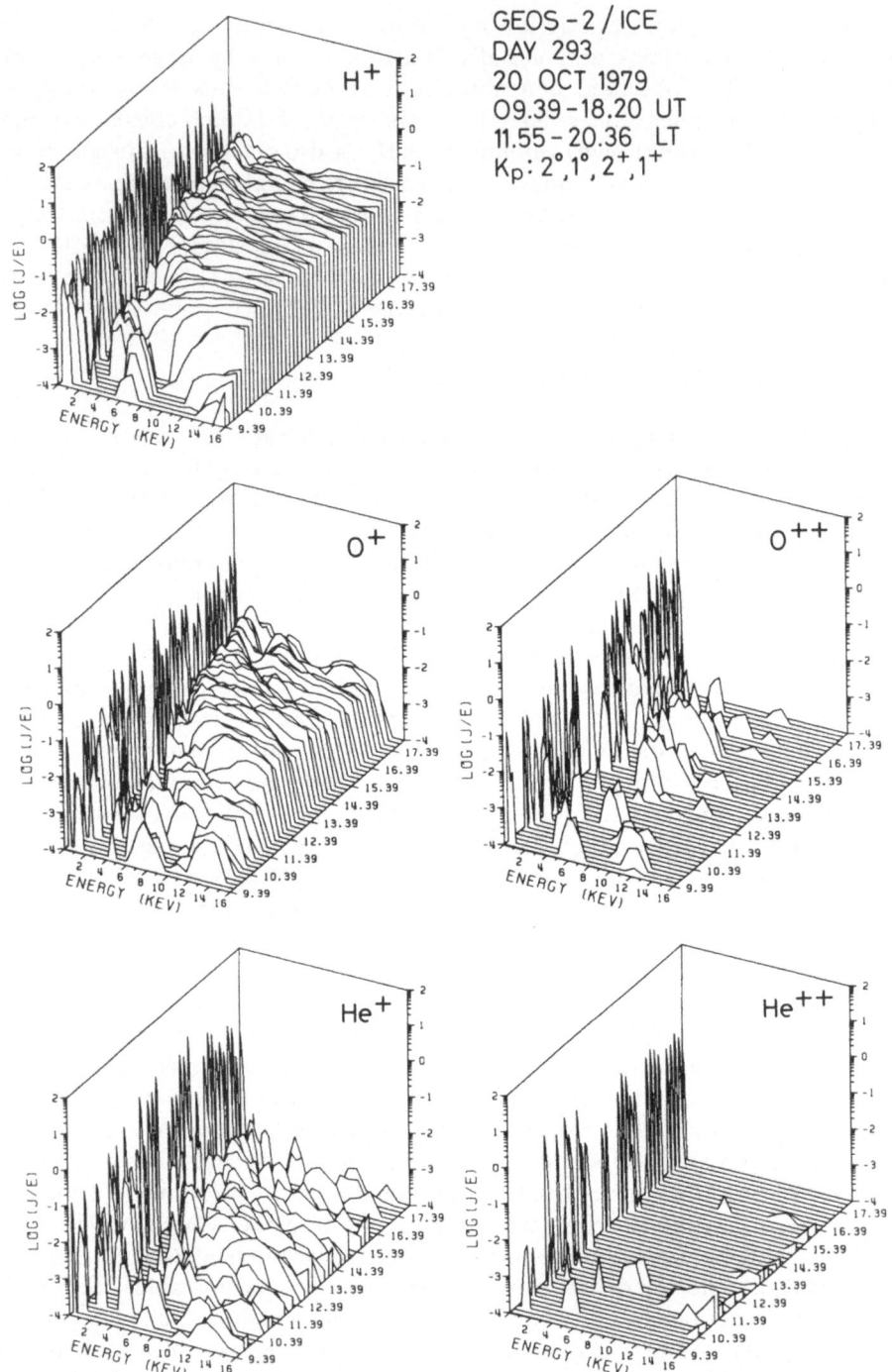

GEOS-2/ICE
DAY 293
20 OCT 1979
09.39-18.20 UT
11.55-20.36 LT
K_p: $2°,1°,2^+,1^+$

Fig. 24. Three-dimensional phase space density plots for five major ions during a substorm on 20 October 1979 showing similar characteristic energy dispersion signatures for H^+, O^+, and He^+ (STOKHOLM, 1981).

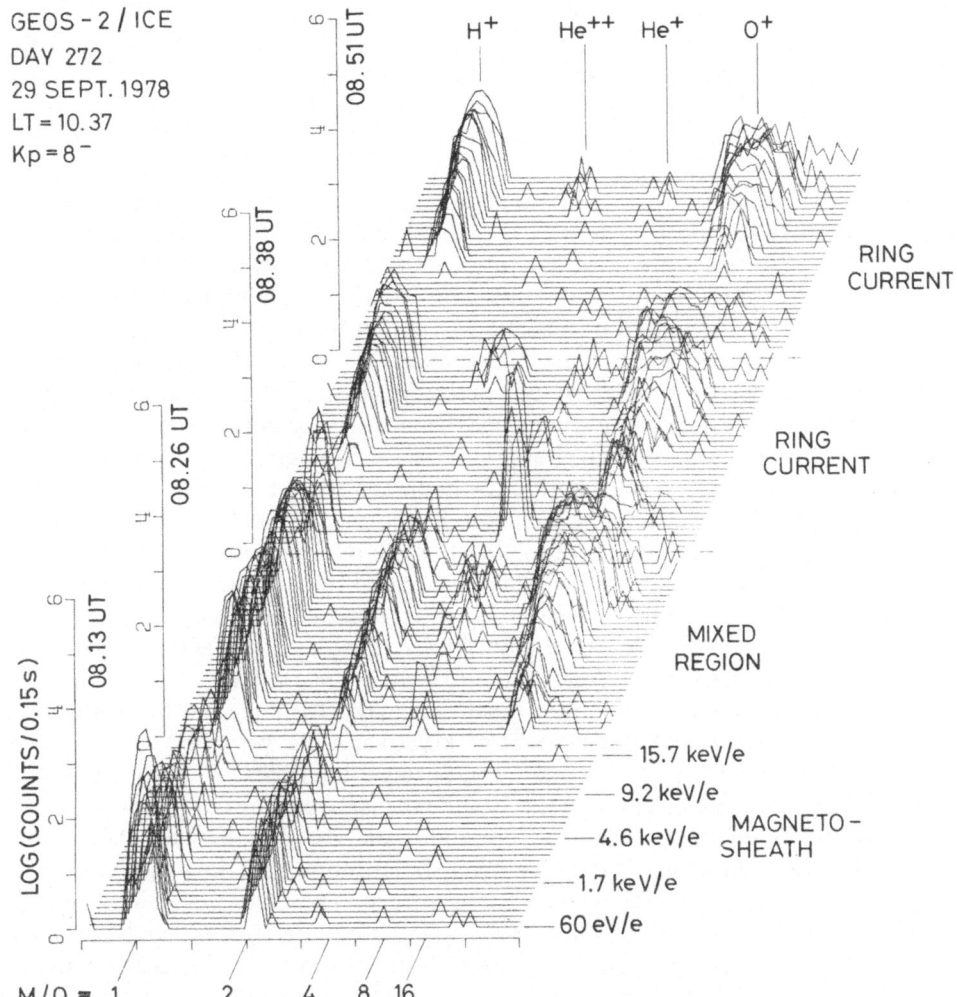

Fig. 25.   Sequence of four consecutive three-dimensional plots (for a description see Fig. 13)of the GEOS-2 mass spectrometer for the magnetic storm on September 29, 1978. Each plot covers 5.9 min and is separated by a different 6-min scan not shown here. Data are taken during and after an excursion of the magnetopause past geostationary orbit. In the first plot GEOS-2 is still in the magnetosheath, characterized by high H^+ and He^{++} at energies around 1–2 keV/e. The second plot demonstrates a "mixed region" with high fluxes of H^+, He^{++} and O^+ (and some He^+) at much higher energies than the sheath plasma; this region is clearly inside the magnetopause. Ions in this region have similar composition and temperatures as observed for the tail plasmasheet (PETERSON et al., 1981); hence we infer that this is an observation of the plasmasheet at geostationary orbit, as indicated in Fig. 1. The third and fourth plots show typical storm time ring current composition and energy spectra with mainly protons and O^+ and much less He^{++} than was observed in the "mixed region". Note the quite distinct boundary between plasmasheet and ring current. From BALSIGER et al. (1980).

ionosphere and plasmasphere) and what portion is of solar origin (i.e. solar wind) has yet to be answered. Rough estimates by GEISS et al. (1978) and JOHNSON (1981) come to about the same conclusion, i.e. that both sources contribute similar amounts. In Table 2 we add another estimate based on long time average composition data from five synoptic study periods. The very simple assumptions are that some unknown mixture of plasma is added during magnetically active times ($K_p = 4$–6) to the quiet time ($K_p = 0$–1) average plasma composition, and that the solar wind H^+ can be computed from the He^{++} (assuming a H^+/He^{++} solar wind ratio of 22). With these assumptions the difference between proton density at active times and that at quiet times was divided up in a solar and terrestrial component. Then He^{++} was added to the solar component, He^+ and O^+ to the terrestrial component, and the relative contributions of the two sources to the added-on plasma was computed (the terrestrial component includes both ionospheric and plasmaspheric ions). The result of this computation in Table 2 shows that the ratio between ions of terrestrial and solar origin varied during the last few years: with increasing solar cycle the terrestrial component increased with the exception of period 4 (a one month subset of period 3) where the ratio between solar and terrestrial component was exceptionally high, almost as high as during period 1. Even changing the assumptions would not change the result dramatically (e.g. by doubling the assumed H^+/He^{++} ratio, the relative amount of injected solar wind would be roughly doubled) and we would still come to the same general conclusions: the terrestrial source is an important one for the plasma of the inner magnetosphere and its importance may well increase with the rise in solar EUV activity due to the latter's influence on the topside ionosphere (YOUNG et al., 1982).

One disturbing fact concerning the quiet time magnetosphere has been addressed before by YOUNG (1979) and BALSIGER et al. (1980) and has yet to be solved: In the inner magnetosphere H^+ is the dominant ion during quiet times ($K_p \leqq 1$) and even at solar maximum as shown in Table 3; this despite the large amount of heavy ions being injected during magnetically disturbed periods, and despite the fact that charge exchange should, even at geostationary orbit, be an important loss mechanism for protons. Hence, a quiet time proton source is needed. The three possibilities are the solar wind (with discrimination against He^{++} at entry or acceleration; c.f. BALSIGER et al., 1980, Fig. 17), the high altitude thermal or suprathermal plasma (with an acceleration process favoring protons; e.g. resonant wave particle interaction) or the polar wind, all of which qualify.

During more than three years of quite careful composition studies it has become more and more clear that the question of origin of magnetospheric ions is more subtle than originally anticipated. For one thing, the main ion, H^+, cannot be easily divided up into two portions representative of the two parent populations. Also most of the other ions are often not totally unambiguous tracers and additional information like energy and pitch angle distributions etc. must be taken into account (c.f. BALSIGER et al., 1980 for a more detailed discussion of this problem). Although the present generation mass spectrometers have increased our knowledge by a big step, they have also opened new questions. For a further step, the time resolution as well as energy and pitch angle coverage of mass spectrometers needs to be improved, and extension of the energy range is clearly desirable. A future generation instrument must have at least as

Table 2. Average number densities of hot (0.9–16 keV/e) ions in the inner magnetosphere.

Period	Number density ($10^{-2}/cm^3$)																				Total ions (%)				
	H^+					He^{++}					He^+					O^+									
	1	2	3	4	5	1	2	3	4	5	1	2	3	4	5	1	2	3	4	5	1	2	3	4	5
K_p 0–1	15	35	35	39	31	0.10	0.27	0.26	0.27	0.22	0.4	1.8	1.9	1.6	1.9	1	7	14	9	11					
K_p 4–6	58	41	47	67	48	1.05	0.40	0.57	1.03	0.36	0.8	1.2	1.9	1.5	2.0	13	23	34	19	37					
DIFF	43	6	12	28	17	0.95	0.13	0.31	0.76	0.14	0.4	–	–	–	0.1	12	16	20	10	26					
SW*	21	3	7	17	3	0.95	0.13	0.31	0.76	0.14	–	–	–	–	–	–	–	–	–	–	39	14	23	46	7
TERR	22	3	5	11	14	–	–	–	–	–	0.4	–	–	–	0.1	12	16	20	10	26	61	86	77	54	93

Differences between densities at quiet times and magnetically active times are divided up in a solar wind (SW) and a terrestrial (TERR) portion for five periods.
Relative contributions of the two sources for these periods are computed in the last columns.
Period 1, May 77–June 78; 2, Aug–Dec 78; 3, Jan–Dec 79; 4, July 79; 5, Jan–July 80.
*Assuming $He^{++} = He^+_{sw}$, $H^+_{sw} = 22 He^{++}_{sw}$.

H. BALSIGER et al.

Table 3.   Average relative abundance of hot (0.9–16 keV/e) ions in the inner magnetosphere.

		RELATIVE NUMBER DENSITY (%)									
		Magnetic quiet ($K_p = 0$–1)					Magnetic active ($K_p = 4$–6)				
		H^+	He^{++}	He^+	O^{++}	O^+	H^x	He^{++}	He^{++}	O^{++}	O^+
GEOS-1	May 77–June 78	91.5	0.6	2.4		5.5	79.7	1.5	1.0		17.8
$L = 6.9$–8.2											
GEOS-2	Aug.–Dec. 78	78.8	0.6	4.2	1.0	15.4	63.1	0.6	1.8	0.6	33.9
$L = 6.6$	Jan.–Dec. 79	68.9	0.5	3.8	1.3	25.5	56.9	0.7	2.3	0.7	39.4
	Jan.–July 80	69.2	0.5	4.4	1.8	24.1	54.8	0.4	2.3	0.7	41.8

good a mass resolution as the GEOS/ISEE/DE type but background due to penetrating radiation and reflected ions must be further reduced (in order to measure rare ions as O^{++} and the $^3He/^4He$ ratios) and time and angular resolution must be increased by introducing two dimensional detectors.

We wish to acknowledge the contributions of many people responsible for design, development, calibration and data analysis during the GEOS project. In particular we are grateful to H. Rosenbauer of the Max-Planck-Institut, Lindau, and H. Loidl of the Max-Planck-Institut, Garching; to Mrs. R. Bänninger, Messrs. P. Eberhardt, Ch. Farrugia, J. Fischer, A. Ghielmetti, U. Rettenmund, M. Stokholm, and L. Weber of the University of Bern; and K. Knott of ESTEC. We thank A. Korth, A. Pedersen, and E. Whipple for providing data in advance of publication; Ch. Farrugia, A. Ghielmetti, R. G. Johnson, and M. Stokholm for valuable discussions; K. Bratschi and R. Wright for graphics; and Miss L. Reichert for typing. Finally we wish to thank the Swiss National Science Foundation (grant 2.705.72 and others) and also the State of Bern for supporting the design and construction of the sensor unit, and the German Bundesministerium für Forschung und Technologie (grant RV 14-B 44/73-SF20) as well as the Max-Planck-Society for supporting the design and construction of the electronics unit of the GEOS mass spectrometers. Data analysis was supported by the Swiss National Science Foundation (grants 2.467.79 and 2.881.80).

REFERENCES

BALSIGER, H., Composition of hot ions (0.1–16 keV/e) as observed by the GEOS and ISEE mass spectrometers and inferences for the origin and circulation of magnetospheric plasmas, Adv. Space Res., 1, 289, 1981.
BALSIGER, H., P. EBERHARDT, J. GEISS, A. GHIELMETTI, H. P. WALKER, D. T. YOUNG, H. LOIDL, and H. ROSENBAUER, A satellite-borne ion mass spectrometer for the energy range 0 to 16 keV, Space Sci. Inst., 2, 499, 1976.
BALSIGER, H., P. EBERHARDT, J. GEISS, D. T. YOUNG, R. G. JOHNSON, W. LENNARTSSON, R. D. SHARP, and E. G. SHELLEY, The composition of the hot magnetospheric plasma during the 11 December 1977 geomagnetic storm as observed by three satellites, Trans. Am. Geophys. Union, 60, 928, 1979.
BALSIGER, H., P. EBERHARDT, J. GEISS, and D. T. YOUNG, Magnetic storm injection of 0.9 to 16 keV/e solar and terrestrial ions into the high-altitude magnetosphere, J. Geophys. Res., 85, 1645, 1980.
BALSIGER, H., J. GEISS, and D. T. YOUNG, Evidence for hot magnetospheric ions accelerated of high altitudes, Trans. Am. Geophys. Union, 62, 995, 1981.
BAUGHER, C. R., C. R. CHAPPELL, J. L. HORWITZ, E. G. SHELLEY, and D. T. YOUNG, Initial thermal plasma observations from ISEE-1, Geophys. Res. Lett., 7, 657, 1980.
FARRUGIA, Ch., PhD thesis, University of Bern, 1982.

GEISS, J. and D. T. YOUNG, Production and transport of O^{++} in the ionosphere and plasmasphere, *J. Geophys. Res.*, **86**, 4739, 1981.

GEISS, J., H. BALSIGER, P. EBERHARDT, H. P. WALKER, L. WEBER, D. T. YOUNG, and H. ROSENBAUER, Dynamics of magnetospheric ion composition as observed by the GEOS mass spectrometer, *Space Sci. Rev.*, **22**, 537, 1978.

GHIELMETTI, A. G., R. G. JOHNSON, R. D. SHARP, and E. G. SHELLEY, The latitudinal diurnal, and altitudinal distributions of upward flowing energetic ions of ionospheric origin, *Geophys. Res. Lett.*, **5**, 59, 1978.

GURNETT, D. A., Plasma wave interactions with energetic ions near the magnetic equator, *J. Geophys. Res.*, **81**, 2765, 1976.

HARRIS, K. K., G. W. SHARP, and C. R. CHAPPELL, Observations of the plasmapause from OGO-5, *J. Geophys. Res.*, **75**, 219, 1970.

HORWITZ, J. L., C. R. BAUGHER, C. R. CHAPPELL, E. G. SHELLEY, D. T. YOUNG, and R. R. ANDERSON, ISEE-1 observations of thermal plasma in the vicinity of the plasmasphere during periods of quieting magnetic activity, *J. Geophys. Res.*, **86**, 9989, 1981.

JOHNSON, R. G., A review of the hot plasma composition near geosynchronous altitude, in *Spacecraft Charging Technology 1980*, Conference proceedings, edited by N. John Stevens and C. P. Pike, p. 412, NASA CP-2182, 1981.

JOHNSON, R. G., R. D. SHARP, and E. G. SHELLEY, Observations of ions of ionospheric origin in the storm time ring current, *Geophys. Res. Lett.*, **4**, 403, 1977.

JONES, D., A. KORTH, K. RONNMARK, and D. T. YOUNG, Helium cyclotron harmonic waves in the magnetospheric plasma, *Adv. Space Res.*, **1**, 319, 1981.

LENNARTSSON, W., E. G. SHELLEY, R. D. SHARP, R. G. JOHNSON, and H. BALSIGER, Some initial ISEE-1 results on the ring current composition and dynamics during the magnetic storm of December 11, 1977, *Geophys. Res. Lett.*, **6**, 483, 1979.

LUNDIN, R., L. R. LYONS, and N. PISSARENKO, Observations of the ring current composition at $L < 4$, *Geophys. Res. Lett.*, **7**, 425, 1980.

LYONS, L. R. and D. S. EVANS, The inconsistency between proton charge exchange and the observed ring current decay, *J. Geophys. Res.*, **81**, 6197, 1976.

OLSEN, R. C., Equatorially trapped plasma populations, *J. Geophys. Res.*, **86**, 11,235, 1981.

PETERSON, W. K., R. D. SHARP, E. G. SHELLEY, and R. G. JOHNSON, Energetic ion composition of the plasma sheet, *J. Geophys. Res.*, **86**, 761, 1981.

ROUX, A., S. PERRAUT, C. DEVILLEDARY, R. GENDRIN, G. KREMSER, A. KORTH, and D. T. YOUNG, Wave-particle interaction near Ω_{He+} observed in GEOS 1 and 2, 2. Generation of ion cyclotron waves and heating of He^+, submitted to *J. Geophys. Res.*, 1982.

SHARP, R. D., E. G. SHELLEY, and R. G. JOHNSON, A search for helium ions in the recovery phase of a magnetic storm, *J. Geophys. Res.*, **82**, 2361, 1977.

STOKHOLM, M., Das Vordringen geladener Teilchen in das Innere der irdischen Magnetosphäre, Master thesis, University of Bern, 1981.

TAYLOR, H. A., H. C. BRINTON, and C. R. SMITH, Positive ion composition in the magnetoionosphere obtained from the OGO-A satellite, *J. Geophys. Res.*, **70**, 5769, 1965.

TINSLEY, B. A., Evidence that the recovery phase ring current consists of helium ions, *J. Geophys. Res.*, **81**, 6193, 1976.

WILLIAMS, D. J., Ring current composition and sources: An update, *Planet. Space Sci.*, **29**, 1195, 1981.

YOUNG, D. T., Ion composition measurements in magnetospheric modeling, in *Quantitative Modeling of Magnetospheric Processes*, Geophys. Monogr. Ser., **21**, edited by W. P. Olson, p. 340, AGU, Washington, D. C., 1979.

YOUNG, D. T., Synoptic studies of magnetospheric composition, Habilitationsschrift, University of Bern, 1980.

YOUNG, D. T., J. GEISS, H. BALSIGER, P. EBERHARDT, A. GHIELMETTI, and H. ROSENBAUER, Discovery of He^{2+} and O^{2+} ions of terrestrial origin in the outer magnetosphere, *Geophys. Res. Lett.*, **4**, 561, 1977.

YOUNG, D. T., S. PERRAUT, A. ROUX, C. DEVILLEDARY, R. GENDRIN, A. KORTH, G. KREMSER, and D. JONES, Wave-particle interactions near Ω_{He+} observed on GEOS 1 and 2, 1. Propagation of ion cyclotron waves in He^+-rich plasma, *J. Geophys. Res.*, **86**, 6755, 1981a.

YOUNG, D. T., H. BALSIGER, and J. GEISS, Observed increase in the abundance of kilovolt O^+ in the magnetosphere due to solar cycle effects, *Adv. Space Res.*, **1**, 309, 1981b.

Young, D. T., H. Balsiger, and J. Geiss, Monthly averages of 1–16 keV ion composition near geostationary orbit: Correlations with geomagnetic and solar activity, submitted to *J. Geophys. Res.*, 1982.

Energetic Ion Composition in the Earth's Magnetosphere, edited by R. G. Johnson, 231–261.
Copyright © 1983 by Terra Scientific Publishing Company (TERRAPUB), Tokyo.

Hot Plasma Composition Results from the ISEE-1 Spacecraft

R. D. Sharp, R. G. Johnson, W. Lennartsson,

W. K. Peterson, and E. G. Shelley

*Space Sciences Laboratory, Lockheed Palo Alto Research Laboratory,
3251 Hanover Street, Palo Alto, California 94304, U.S.A.*

(Received March 13, 1982)

ISEE-1 carries the first energetic ion mass spectrometer to the distant magnetotail and some remarkable results have emerged from the experiment in that region. Within the plasma sheet a correspondence between the He^{++} and H^+ spectrums in a number of instances has provided new information on the mechanisms involved in the entry and thermalization of the solar wind. A statistical study has shown a high fractional O^+ content with increasing substorm activity and a simultaneously decreasing fraction of He^{++}. From these results one can infer that the ionosphere provides a comparable amount of the plasma sheet to the solar wind during active times. A separate and distinct population of low intensity streaming ions of recent ionospheric origin exists in both the distant plasma sheet and the lobes of the magnetotail. Because of the limitations of earlier instruments they have not previously been identified. They form the dominant plasma constituent in the tail lobes and we therefore come to the unexpected conclusion that the hot plasma in this distant region of the magnetosphere is predominately of ionospheric origin. In the inner magnetosphere ISEE has provided radial scans of the equatorially trapped plasma composition and given us new insight into the composition and dynamics of the ring current. A commonly observed spectral feature allows us to infer the typical gross plasma circulation pattern during major magnetic storms and thereby make the first indirect determination of the composition and relative source strengths for the principal ring current ion population which is above the energy range of any currently operating ion mass spectrometer. We conclude that the solar wind is relatively more important in this higher energy portion of the spectrum than it is at energies per charge below 17 keV/e but that the ionosphere is still a significant contributor to the ring current even at high energies. A recently completed statistical study has provided a comparison of the ion composition in the near equatorial magnetosphere during quiet and disturbed conditions and allowed a test of the theoretical models for the decay of the storm time ring current. We conclude that current models based on the charge exchange decay of a static trapped population are inadequate to explain the observations. The experimental results suggest the need to add a continuous injection mechanism to the models, even during periods of extended quiet conditions, and to include transport effects with time constants which are not long compared to charge exchange lifetimes as is currently assumed.

1. Introduction

The energetic ion mass spectrometer on the International Sun Earth Explorer

(ISEE-1) is one of a family of instruments utilizing the same basic ion optics that have been, or are being, constructed for a number of projects including GEOS-1, GEOS-2, ISEE, Dynamics Explorer, and AMPTE (Active Magnetospheric Particle Tracer Experiment). The basic instrument has been described by BALSIGER et al. (1976) and SHELLEY et al. (1978, 1981). It covers the energy per charge range from 0 to ~ 17 keV/e and the mass range from 1 to ~ 150 amu. It has high sensitivity (~ 1 cm^2 sr eV) and a resolution of $M/\Delta M \sim 10$ at focus.

Adapting the instrument to the requirements of the ISEE mission was a particular challenge because of the diverse nature of the plasmas encountered by ISEE-1 in its highly elliptical orbit. They include the cold, dense plasmas of the plasmasphere, the highly energetic ring current ions, the strongly flowing (high mach number) magnetosheath and boundary layer plasmas, the very tenuous tail lobe plasmas, the quasi-isotropic (low mach number) plasma of the plasma sheet, and the solar wind. The experiment has provided significant results in each of these regimes, in several of them for the first time with an ion mass spectrometer.

ISEE-1 was launched on 22 October 1977 into a geocentric elliptical orbit with apogee at an altitude of 138,120 km, perigee at 281 km, and an inclination of 28.7°. It has an orbital period of 2.4 days. It spins at about 20 RPM with its spin axis nearly normal to the ecliptic. The experiment view direction is approximately in the satellite spin plane.

The experiment is controlled through an onboard random access memory which can be programmed from the ground. Specialized modes which are frequently modified to satisfy evolving scientific objectives are developed for each plasma regime. A typical magnetospheric mode might consist of a sequence of measurements of a few selected ion species with each element of the sequence consisting of a 4 second (480° of rotation) measurement at one of 16 energy steps. Other modes include the acquisition of more detailed energy and mass spectrums with up to 64 steps being available in each parameter.

As of this writing the experiment is operating successfully and research is progressing in a number of areas. This review is in the nature of a progress report on the energetic magnetospheric plasma results only. The cold plasma results (BAUGHER et al., 1980; HORWITZ et al., 1981) and solar wind results (SCHMIDT et al., 1980) are described elsewhere.

2. Magnetotail

ISEE-1 carries the first energetic ion mass spectrometer to be used in the distant magnetotail and some remarkable results have emerged from the initial studies of the composition of the plasma in that region. The experiment samples three principal magnetotail plasma regimes: the plasma sheet, the tail lobes, and the magnetospheric boundary layer. The plasma sheet is characterized by a high temperature plasma with kT of the order of a few keV and density in the range of a few tenths per cm^{-3}. The bulk flow energies are generally small compared to the thermal energies. The composition is generally dominated by H$^+$ ions during quiet times, but as we shall see below, a strong O$^+$ component appears with increasing substorm activity. Within this quasi-isotropic population we find a separate population of ion streams with low thermal energies compared to their bulk flow energies. The streams apparently originate in the auroral acceleration region at altitudes of ~ 1 RE and consist primarily of O^+ and H^+ ions. They have been studied extensively with the ion mass spectrometer experiment on the low altitude S3-3 satellite (see the S3-3 review paper in this volume). The H^+ component of the streams is often difficult to resolve from the low energy tail of the H^+ component of the quasi-isotropic population but the O^+ component is generally readily distinguishable.

The streams also exist in the tail lobes with properties similar to those of the streams in the plasma sheet. The streams are generally the dominant constituent of the tail lobe plasma in this energy range. We therefore come to the unexpected finding that the hot plasma in this distant region of the magnetosphere is predominately of ionospheric origin and is energized by processes different from those producing the polar wind (see paper by Raitt and Schunk in this volume).

The third magnetotail plasma regime, the magnetospheric boundary layer, has been extensively studied with electrostatic analyzer experiments on previous satellites (HONES et al, 1972; ROSENBAUER et al., 1975; HARDY et al., 1975; FRANK et al., 1977). The different ionic constituents are typically flowing at the same bulk velocity and have thermal energies which are low compared to their bulk flow energies so that even a non-mass discriminating instrument can often be used to infer the presence of O^+ and H^+ components. We find that these boundary layer streams generally consist of H^+ and He^{++} ions in typical solar wind ratios with an occasional small O^+ component arising from an ionospheric acceleration process which has been observed to be operating in the polar cusp (SHELLEY, 1979).

In the following paragraphs we shall illustrate some of the characteristics of the magnetotail plasmas with examples and initial statistical results from the ISEE ion composition experiment.

2.1 Plasma sheet

PETERSON et al. (1981) presented a detailed analysis of plasma sheet parameters for six specific intervals of ~ 1 to 2 hours in duration when ISEE was at radial distances between 15 and 21 RE during the spring of 1978. The intervals were selected to illustrate the range and variability of the energetic ion composition.

In all the intervals except one, H^+ was the dominant constituent in the energy per charge range ≤ 17 keV/e. O^+ varied from 2.0% to 71% of the number density. He^{++} varied from 0.3% to 4.4% and He^+ was always less than 0.5%. In several instances a remarkable correspondence was found between the detailed shapes of the He^{++} and H^+ energy spectrums at high energies. This is illustrated for two of the intervals in Fig. 1 and 2. In Fig. 1 the data from the two ion species have been plotted on the same energy per charge scale, while in Fig. 2 the abscissas have been shifted relative to each other by a factor of 2 so that they are on the same energy per nucleon scale. Other examples were found where the best correspondence was achieved if the abscissas were shifted by some value intermediate between those in Fig. 1 and 2.

H^+ and He^{++} spectrums which are identical on an energy per nucleon scale are suggestive of a situation in which the solar wind plasma is injected into the

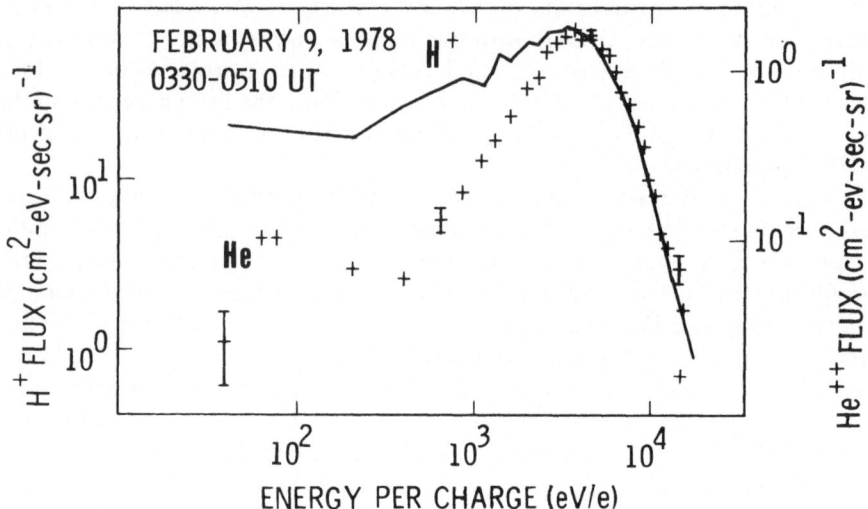

Fig. 1.   Comparative H$^+$ and He^{++} spectrums in the plasma sheet at 15 R_E geocentric radial distance. The data have been averaged over 360° and plotted versus energy per charge (PETERSON *et al.*, 1981).

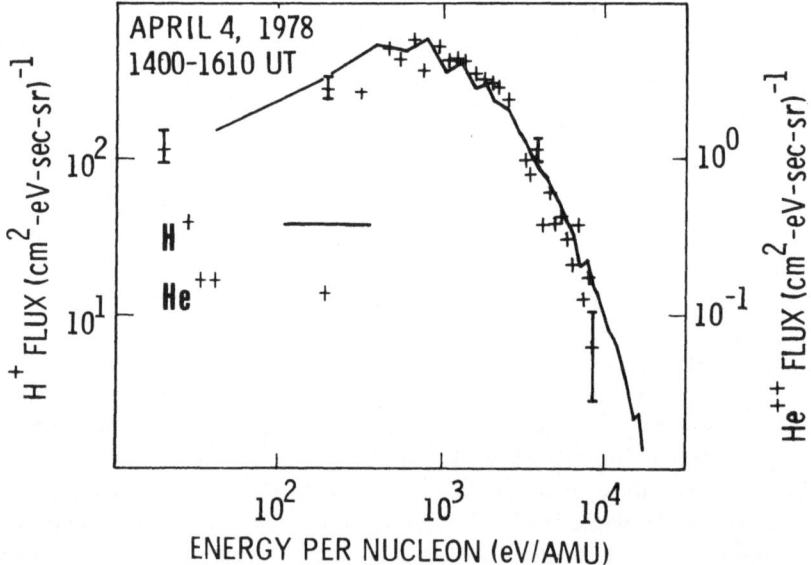

Fig. 2.   Comparative H$^+$ and He^{++} spectrums in the plasma sheet at 21 R_E geocentric radial distance. The data have been averaged over 360° and plotted versus energy per nucleon (PETERSON *et al.*, 1981).

magnetosphere by mass and charge independent processes preserving the relationship between the energy distributions which exists in the solar wind. He^+ and He^{++} spectrums which match well on an energy per charge scale suggest that mechanisms involving an electrostatic acceleration were involved in the injection process. It was concluded from these results that no single mechanism of solar wind entry and subsequent thermalization and acceleration was dominant at all times in the plasma sheet.

PETERSON et al., (1981) also estimated the fraction of the plasma sheet number density of terrestrial (ionospheric) origin. The problem in making such an estimate is equivalent to estimating the fraction of H^+ of terrestrial origin, since it is reasonable to assume that all observed He^{++} is of solar origin and all observed O^+ is of terrestrial origin. The basis for estimating the fraction of H^+ of terrestrial origin is illustrated in Fig. 1. In most of the intervals studied the distribution function of the He^{++} ions could be reasonably well approximated by a Maxwellian. The H^+ distribution function however often had an excess of ions at low energies in a region of the spectrum where O^+ ions of ionospheric origin made a substantial contribution.

Since the upflowing ion beams from the auroral acceleration region are known to contain both O^+ and H^+ ions it was inferred from these relationships that the low energy portion of the H^+ spectrum in excess of that predicted by a Maxwellian distribution was probably of ionospheric origin. An attempt was made to estimate the fractional ion density originating in the ionosphere (f_I) by summing these low energy protons with the O^+ and He^+ constituents. The values of f_I obtained for five of the intervals were: 0.10, 0.10, 0.25, 0.40, and 0.65.

A more extensive, statistical study of the plasma sheet composition is being performed utilizing data from the period March through May 1978 (SHARP et al., 1981a). The relationship between the observed changes in composition and substorm activity are being studied by use of auroral electrojet indices (AE) provided by KAMEI and MAEDA (1981). Some interesting preliminary results are available at this time. Intervals of approximately 1 hour in duration when the spacecraft was continuously in the plasma sheet were identified from survey plots, and averages of the number densities and average energies of various ionic constituents were calculated. The data were integrated over pitch angle and over the energy per charge range from 0.1 to 16 keV/e. Periods when the satellite was in the lobes or magnetospheric boundary layer were excluded. The intervals are characterized as "active" if the AE index exceeds approximately $500\,\gamma$ during the interval itself or the 1 hour period preceeding it. Similarily they are characterized as "quiet" if the AE index does not exceed approximately $100\,\gamma$ during the interval or the 1 hour period preceeding it.

The initial data set consists of 40 quiet and 31 active intervals obtained on 16 different days when the spacecraft was between about $10\,R_E$ and $23\,R_E$ geocentric radial distance. The plasma in this energy range generally consisted almost entirely of H^+, O^+, He^{++}, and He^+ and occasional small contributions of other species (e.g. O^{++}) were not included in the analysis. Histograms showing distributions with respect to the ratio of the number densities of He^{++}/H^+ and O^+/H^+ are shown in Figs. 3 and 4. The He^{++}/H^+ ratio shows a small but significant decrease with increasing activity. The mean value of the ratio varies from $.020 \pm .002$ (quiet) to $0.011 \pm .002$

He⁺⁺/H⁺

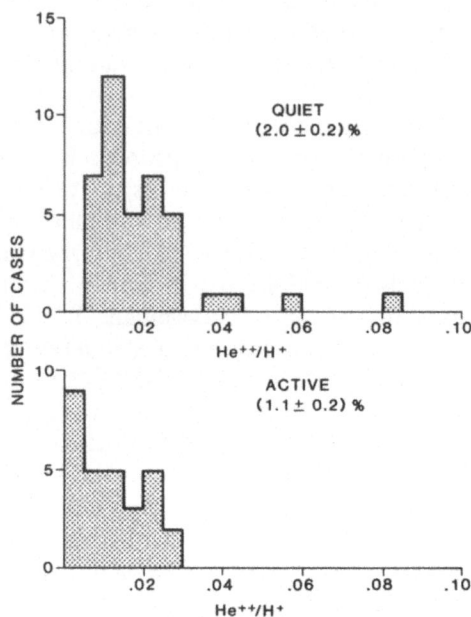

Fig. 3. Distributions of ratios of He⁺⁺/H⁺ densities in the plasma sheet during quiet and active times. The data cover the energy per charge range from 0.1 to 16 keV/e. The mean values of the distributions and their standard deviations are shown in parenthesis (Sharp *et al.*, 1981a).

(active). One should also note that the "active" distribution peaks in the lowest bin, whereas this bin contains no cases in the "quiet" distribution. As seen in Fig. 4, the effect for O^+/H^+ is much more dramatic and in the opposite direction. The mean value of the ratio varies from $0.022 \pm .003$ (quiet) to $0.39 \pm .06$ (active). The He^+/H^+ ratio (not shown) also increases significantly with increasing activity. The mean value varies from $0.004 \pm .001$ (quiet) to $0.008 \pm .001$ (active).

As has been indicated previously the principal source regions for the plasma in the plasma sheet are thought to be the solar wind, which provides primarily H^+ and He^{++} ions, and the ionosphere, which provides H^+, O^+, and He^+ ions. The above results can be interpreted qualitatively as resulting from an increased contribution of the ionosphere relative to the solar wind as substorm activity increases. It would be of great interest if one could make this interpretation more quantitative and estimate the fraction of the number density which can be attributed to each source. As outlined below, this is in fact possible if one is willing to make two principal assumptions.

We have 8 measured quantities, the number densities of the four principal species during active and quiet times: $H^+(A)$, $O^+(A)$, $He^{++}(A)$, $He^+(A)$, $H^+(Q)$, $O^+(Q)$, $He^{++}(Q)$, and $He^+(Q)$. As noted above, all the ion species except H^+ are essentially source specific so the problem reduces to determining the H^+ density originating in the

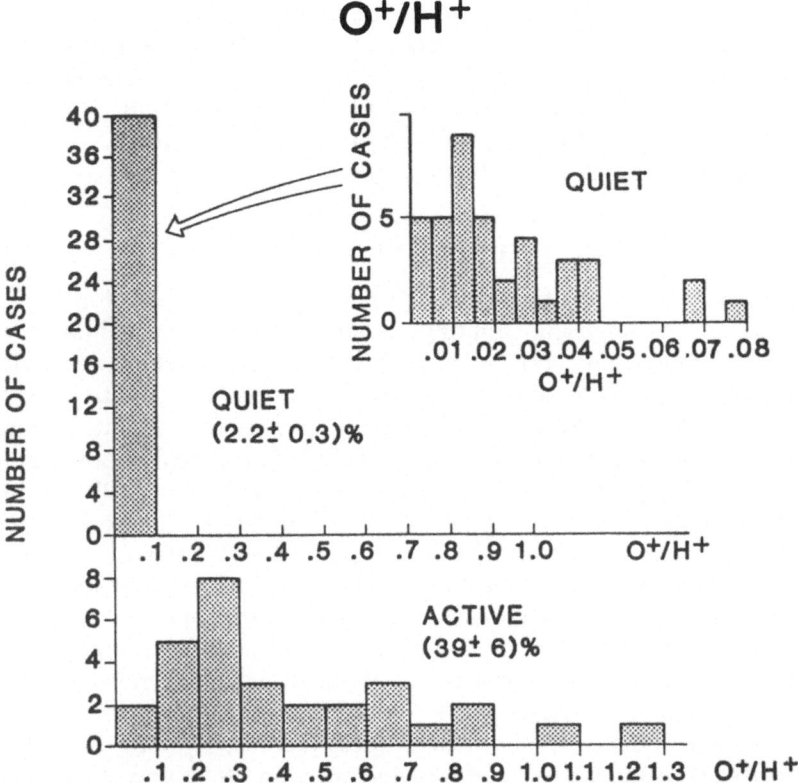

Fig. 4. Distributions of ratios of O^+/H^+ densities in the plasma sheet during quiet and active times. The data cover the energy range from 0.1 to 16 keV. The mean values of the distributions and their standard deviations are shown in parenthesis. The inset shows an expanded plot of the results for density ratios < 0.1 during quiet times (SHARP et al., 1981a).

ionosphere, $H_i^+(A)$, $H_i^+(Q)$; and in the solar wind, $H_{sw}^+(A)$, $H_{sw}^+(Q)$. We can solve directly for these four unknowns using the following 4 equations

$$H_i^+(A) + H_{sw}^+(A) = H^+(A) \tag{1}$$

$$H_i^+(Q) + H_{sw}^+(Q) = H^+(Q) \tag{2}$$

$$\frac{O^+(Q)}{H_i^+(Q)} = \frac{O^+(A)}{H_i^+(A)} = X_1 \tag{3}$$

$$\frac{He^{++}(Q)}{H_{sw}^+(Q)} = \frac{He^{++}(A)}{H_{sw}^+(A)} = X_2. \tag{4}$$

Equations (3) and (4) represent the principal assumption, i.e. that the ion composition of the two source terms does not vary systematically with substorm activity. Evidence in support of Eq. (3) has been presented by Ghielmetti and coworkers (see S3-3 review in this volume) and while not completely definitive it does strongly suggest that the O^+/H^+ ratio of the upflowing ion events does not vary appreciably with K_p. No published evidence exists either for or against the assumption in Eq. (4). It seems plausible however that short term fluctuations in AE are not dependent on the density of a minor constituent in the solar wind.

Simultaneously solving Eq. (1) through (4) we can compute X_1, X_2, and f_1, the fractional number density of ionospheric origin.

$$f_1 = \frac{(O^+ + H_i^+ + He^+)}{(O^+ + H_i^+ + He^+) + (H_{sw}^+ + He^{++})}. \tag{5}$$

The input parameters and the results of the calculation are presented in Table 1. The indicated uncertainties represent counting statistics only. The results indicate that the plasma sheet was almost entirely of solar wind origin during quiet intervals but had an important, perhaps dominant, ionospheric contribution during the active periods. The plausibility of the solution can be examined by comparing the estimated values of X_1 and X_2 with previous measurements. COLLIN *et al.* (1981) studied upflowing energetic ions in the auroral zone with the S3-3 satellite over the period July 1976 to July 1977. Their results lead to an estimate of $X_1 = 0.25$, a factor of 3.5 lower than the value obtained here. Their results were for a somewhat different energy range $(0.5 \leq E/q \leq 16 \text{ keV})$ and there is some evidence that the magnetospheric O^+/H^+ ratio has been increasing during the increasing phase of the solar cycle (YOUNG *et al.*, 1981) so this discrepancy does not appear to be unreasonable. Similarly the long term average value of X_2 in the solar wind is about 4% but the ratio varies widely (HUNDHAUSEN, 1972) and a value of 2.1% is not unreasonable.

As has been indicated, the results described apply only to the range $0.1 \leq E/q \leq 16 \text{ keV/e}$ and we must address the question of the uncertainty caused by those portions of the distributions outside this range. The plasma at lower energy was monitored by the instrument during these periods. The data come from a single wide-energy-band measurement $(.01 \leq E/q \leq .1 \text{ keV/e})$ and have larger uncertainties than the higher energy data so they are not routinely included in the averages. The only significant contribution to the plasma density was for the quiet time O^+. An approximate correction for this contribution yields 0.027 for the O^+/H^+ ratio and raises $f_1(Q)$ to 0.059 with neglible change in the other results. The contribution at higher energies is more likely to be significant, particularly for the He^{++} which has the highest average energy per charge of the several species. To assess the magnitude of this uncertainty we extrapolated the number densities to higher energies by assuming a Maxwellian distribution which would produce the measured overall average energy for each species in the energy range of the measurements. The principal effect on the

Table 1. Measured ion density ratios, $0.1 \leq E/q \leq 16 \text{ keV/e}$.

	He^+/H^+	He^{++}/H^+	O^+/H^+
Active	.008 ± .001	.011 ± .002	.39 ± .06
Quiet	.004 ± .001	.020 ± .002	.022 ± .003

Results of calculation; $X_1 = 0.87$, $X_2 = 0.021$, $f_1(A) = 0.60$, and $f_1(Q) = 0.049$.

density ratios was for He^{++}/H^+ during active periods which increased to .017. Including these corrections gave $X_1 = 2.0$, $X_2 = .021$, $f_i(A) = .41$, and $f_i(Q) = .042$. The discrepancy with the Collin *et al.* results is seen to increase substantially but the estimates of relative source strengths are not greatly affected.

Quantitative estimates and limits on the ratio of hot plasma ions of terrestrial origin to those of solar wind origin near geosynchronous altitude have also been made by BALSIGER *et al.* (1980) and JOHNSON (1981) using data from the ion composition experiment on the GEOS satellite in late 1978 and early 1979. Johnson also concludes that the ionospheric component of the plasma increases during magnetically disturbed periods. For $K_p \gtrsim 3$, using conservative assumptions for the ionospheric H^+ component, he infers that the ionospheric component is comparable to or exceeds the solar wind component, in agreement with the results for the distant plasma sheet reported here.

2.2 Ion streams

As indicated above, a population of streaming ions of ionospheric origin, usually with low densities and temperatures, are observed throughout the magnetotail, both in the plasma sheet and the lobes. These streams usually consist either of O^+ or H^+ ions. They differ in their properties from both the primary plasma sheet plasmas and the boundary layer plasmas. The former are characterized by a quasi-isotropic pitch angle distribution and average energies of a few keV. The latter are generally more intense than the streams from the ionosphere and have an He^{++} component moving at the same bulk velocity as the dominant proton component. The boundary layers are usually confined to the vicinity of the magnetopause in the region of the magnetotail accessible to ISEE.

Figure 5 shows a data segment containing two O^+ streams in the plasma sheet observed at a geocentric radial distance of 22 R_E (SHARP *et al.*, 1981b). The 0.63 keV ions are flowing approximately tailward and the 1.6 keV ions are flowing approximately earthward. The spectrometer response is shown during three satellite spins when it was set on energy steps 1, 3, and 7 as indicated. The upper curve gives the pitch angle of the measured ions. The number density and temperature of the streams are estimated to be: $n = 4 \times 10^{-3}$ cm^{-3}, $kT = 25$ eV for the 0.63 keV stream and $n = 3 \times 10^{-3}$ cm^{-3}, $kT = 96$ eV for the 1.6 keV stream. These are well below the sensitivity threshold of previous experiments that have operated in this region of the magnetosphere. The quasi-isotropic plasma sheet population at this time consisted primarily of H^+ ions and had a number density of approximately 0.1 cm^{-3} and a temperature of approximately 5 keV. As indicated previously this population tended to obscure the weaker H^+ streams in the plasma sheet and for this reason the initial statistical studies dealt only with the O^+ streams in this region (SHARP *et al.*, 1981b).

These studies were based on approximately 134 hours of data during the period February–May 1978 when ISEE was in the expected location of the central plasma sheet and lobes. (Geocentric radial distance $\geq 11\ R_E$, $|GSMY| \leq 12\ R_E$, $|dZ| \leq 12\ R_E$.) The principal results are illustrated in Table 2 and Fig. 6, 7, and 8. Table 2 gives the frequency of occurrence based on the number of times the experiment sampled the given ionic constituent in the energy range ($110\ eV \leq E/Q \leq 17\ keV$). A

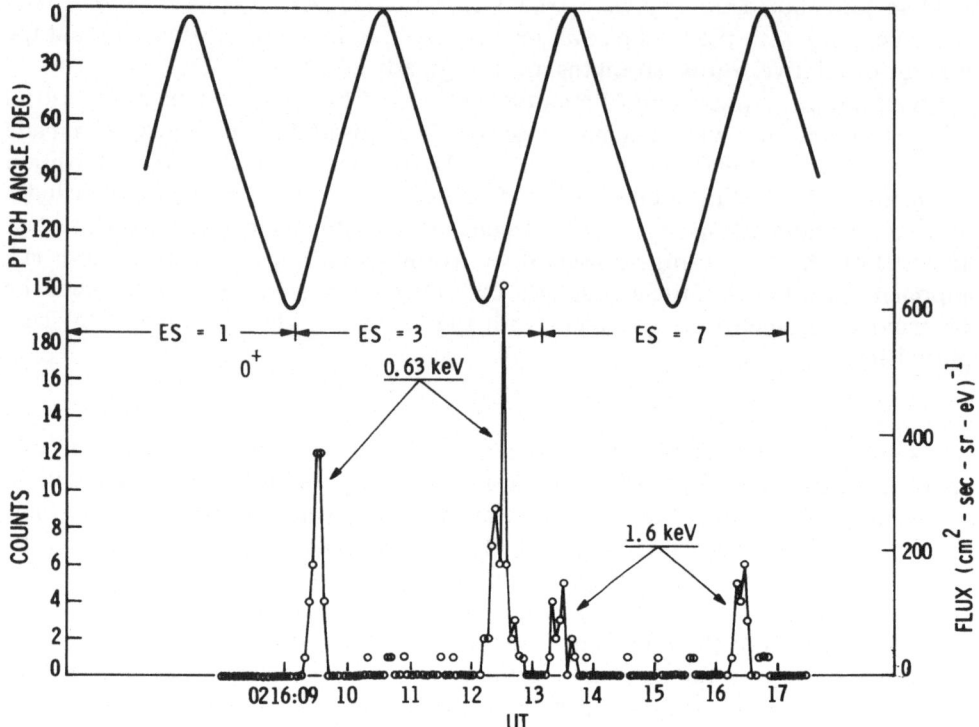

Fig. 5. 0^+ streams in the plasma sheet observed on March 2, 1978 (Sharp *et al.*, 1981b).

Table 2. Statistical study of ion beams in the magnetotail.

Ion	Location	K_p[†]	Samples	Frequency, %
H^+	lobes	lo	294	9.9 ± 1.8
		hi	249	11.6 ± 2.2
		all	543	10.7 ± 1.4
0^+	lobes	lo	294	9.9 ± 1.8
		hi	249	24.5 ± 3.1
		all	543	16.6 ± 1.7
0^+	Plasma sheet	lo	250	10.4 ± 2.0
		hi	285	45.6 ± 4.0
		all	535	29.2 ± 2.3

[†]lo ≤ 3^+, hi ≥ 4^-.

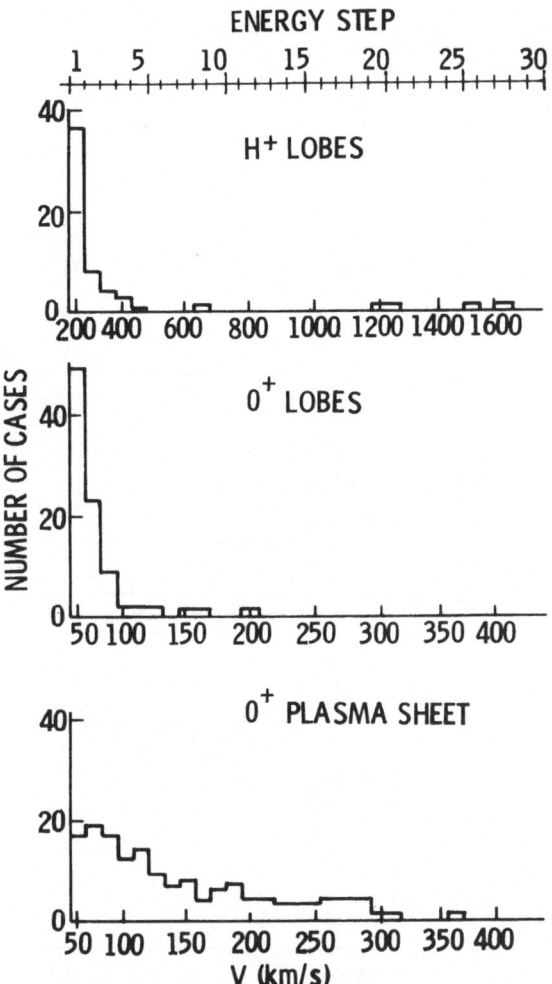

Fig. 6.   Distributions of ion streams with respect to the ion velocity at the peak flux intensity (SHARP *et al.*, 1981b).

division was made with respect to magnetic activity based on the K_p index and is indicated in the table. One sees that the overall frequency of occurrence varied between $\sim 10\%$ and 30%. O^+ was the most common constituent in the lobes. Increasing magnetic activity was related to a substantial increase in the plasma sheet streams and had a smaller effect in the lobes. For the H^+ streams in fact there was no discernable effect.

Figure 6 shows the distributions of the observed ion streams with respect to the energy step (top scale) at which there was maximum flux. The equivalent ion velocity is shown in the lower scales. One sees that the O^+ streams in the plasma sheet extend to significantly higher energies than in the lobes. The O^+ and H^+ distributions in the

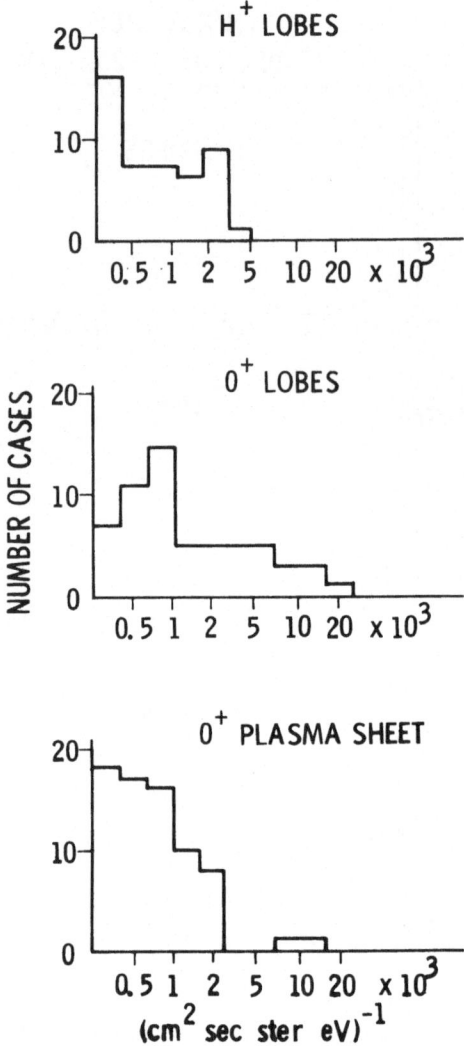

Fig. 7.   Distributions of ion streams with respect to peak flux intensity (Sharp *et al.*, 1981b).

lobes are quite similar. The streams occur in the same general range of energies and not in overlapping ranges of velocity. This is one of the characteristics of the tail lobe streams that distinguishes them from the boundary layer streams.

Figures 7 and 8 show the distributions of peak flux intensities and angular widths of the observed streams. One sees a general similarity of the properties of the O^+ and H^+ in the lobes and of the O^+ streams in the lobes and plasma sheet. Although the statistics are poor there is some evidence in Fig. 8 for a group of wide proton events in the lobes that are not evident in the O^+. The 112.5 to 120° bin represents a lower limit in that it includes some even wider examples, some of which had conical pitch angle

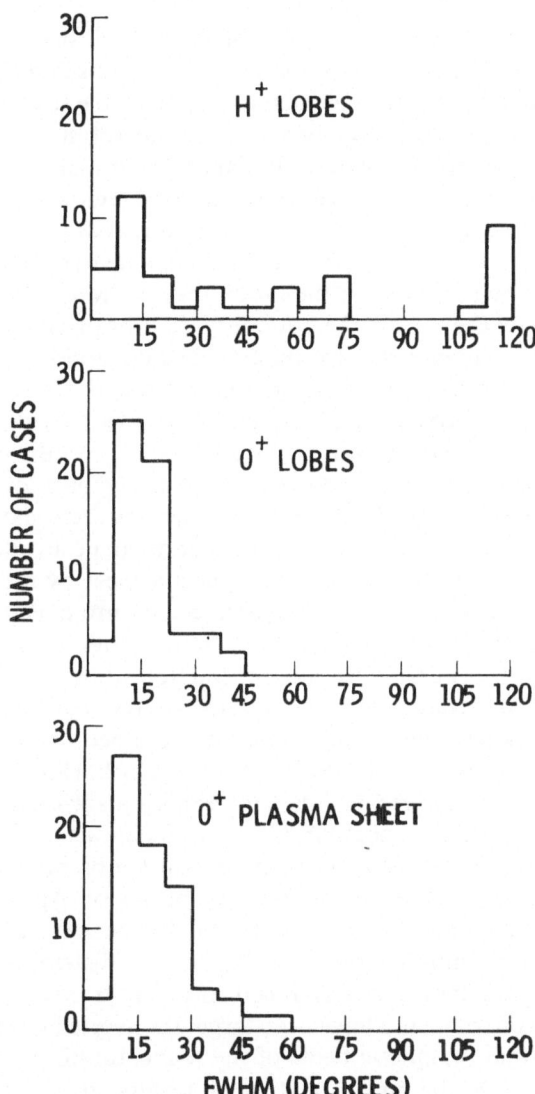

Fig. 8. Distributions of ion streams with respect to the observed full widths at half maximum of their angular distributions (SHARP *et al.*, 1981b).

distributions. As will be discussed below this may be the initial manifestation of a beam plasma instability which eventually thermalizes the tailward flowing lobe streams and injects them into the quasisotropic plasma sheet population.

The O^+ distribution in the lobes in Fig. 7 shows a significant peak indicating that these streams have a somewhat higher average intensity than the H^+. The combination of higher intensity and higher frequency of occurrence (Table 2) means that the O^+ was clearly the dominant constituent of the tail lobe plasmas with $E > 110$ eV during the period of this study.

We know that the O^+ streams must be of ionospheric origin and we infer from the similarity of the gross properties of the O^+ and H^+ streams that they originate from the same source. We infer from these results that it is probable that both the O^+ and most of the H^+ streams in the central regions of the magnetotail are of ionospheric origin. Streams with the characteristics of the boundary layer or mantle, which are primarily of solar wind origin, were not found within the limited region and period of this study. This inference results in some modification of recent ideas about the formation of the plasma sheet. HARDY et al. (1975, 1976, 1979) were able to detect the tail lobe streams with their electrostatic analyzer on the moon, but without mass discrimination they could not differentiate them from the boundary layer or plasma mantle streams. Their results were taken as evidence that the mantle plasmas could convect inward from the magnetopause to the plasma sheet at geocentric radial distances as low as 60 R_E, although as noted by HARDY et al. (1976) and CROOKER (1979), the magnitude of the typical crosstail field did not appear to be adequate to produce this effect.

PILIPP and MORFILL (1978) used these results as the basis of their model in which the entire plasma sheet was produced by the inward convection of mantle plasma. The present results suggest that the polar cap ionosphere can be an additional source of ion streams which are eventually carried to the plasma sheet by the crosstail convection field. Other ionospheric streams are directly injected onto plasma sheet field lines. As seen in the previous section, especially during active times, these ionospheric ions can form an important, occasionally dominant, fraction of the plasma sheet.

Some additional insight into the nature of the mechanism that eventually thermalizes the magnetotail streams can be obtained from the apparent mass dependence of this process. On those occasions when both O^+ and H^+ streams are observed simultaneously the H^+ beams often exhibit a substantially broadened pitch angle distribution or a conical angular distribution characteristic of a high transverse to parallel temperature ratio while the simultaneously observed O^+ is in the form of a narrow, nearly field-aligned beam. An example of this on April 19, 1978 is shown in Fig. 9. A similar example was presented in Fig. 4 of SHARP et al. (1981b). The high transverse temperature implied by these broad H^+ distributions must have been imparted by a local acceleration mechanism since any transverse energy acquired by the ions in the ionosphere would have been largely converted to parallel energy as they streamed into the weak magnetic fields of the magnetotail.

As seen in Fig. 8 the initial statistical results also indicated that the H^+ distributions in the tail lobes were more likely than the O^+ to exhibit high transverse temperatures. These results are somewhat puzzling since the opposite effect was found by the S3-3 satellite in the auroral acceleration region. In the upflowing ion beams with keV energies observed at altitudes greater than $\sim 5,000$ km, the O^+ angular distributions were found to be broader than the H^+ distributions and their energies were substantially higher (COLLIN et al., 1981). This was taken to be the result of a mass dependent transverse acceleration mechanism which preferentially acted upon the 0^+ component of the beams, both components of which had also been energized by quasistatic electric fields.

A possible explanation for this discrepancy may lie in the recent theoretical work of ASHOUR-ABDALLA et al. (1981). They have extended the analysis of KINDEL and

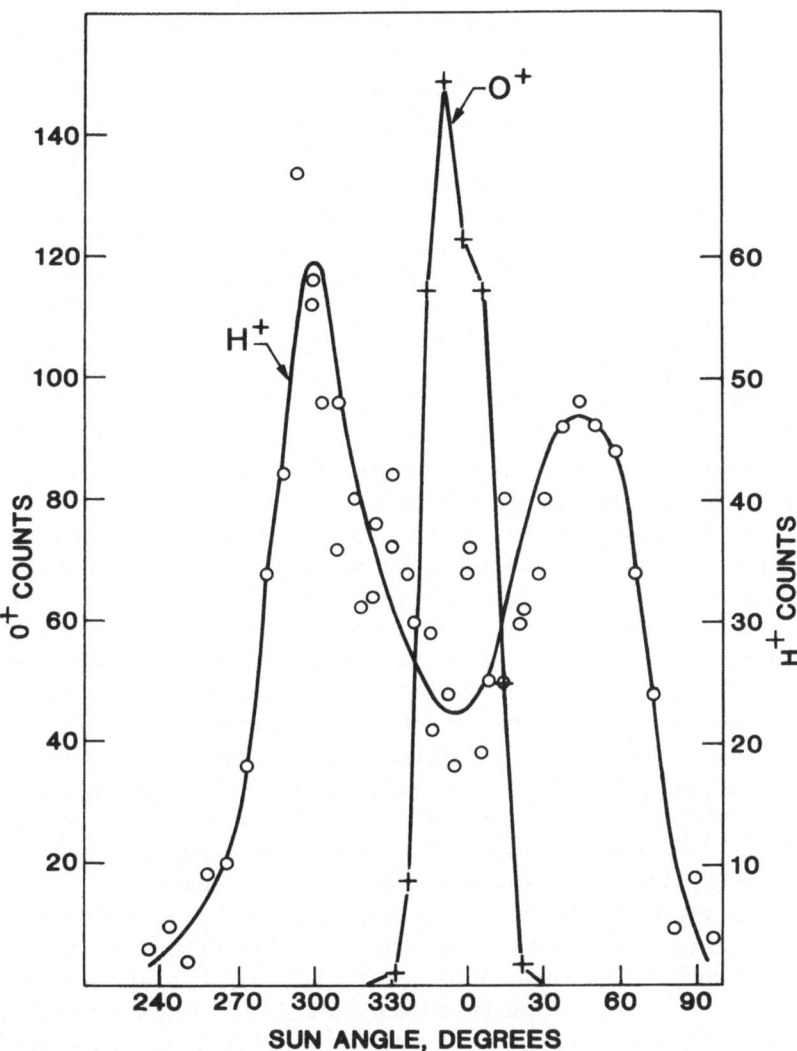

Fig. 9.   Angular distributions of 628 eV O^+ ions and 212 eV H^+ ions in a tail lobe stream on April 19, 1978 at 0500 UT. These are the energies corresponding to the peak response for the respective species.

KENNEL (1971) for the situation applicable to the auroral ionosphere in which the cold ionospheric electron population drifts relative to the ambient ionospheric ions, exciting ion cyclotron waves and transversely accelerating the ions. In a simulation study with an initial drifting electron Maxwellian, ASHOUR-ABDALLA *et al.* (1981) showed that the oxygen waves grew to only small amplitudes because the hydrogen waves extracted most of the free energy from the electrons before the oxygen waves could grow. This was contrary to the above-mentioned energetic ion observations in the auroral region and they suggested that the dilemma could be resolved if the electron velocity distribution was maintained by the constant injection of fresh electrons, inhibiting the above-described saturation effects. In the absence of this saturation they found that oxygen wave growth and transverse oxygen ion acceleration dominated over that of hydrogen in agreement with the S3-3 observations. In the tail lobes, however, the situation is reversed. Here we have ion streams flowing through a cold ambient electron plasma. The same electron population may be in contact with the ion stream for an extended period and some type of saturation effect may occur, leading to the preferential transverse energization of the H^+ component. A detailed simulation study using the appropriate plasma parameters for the tail lobes will be required to resolve this question but it is an interesting possibility for the resolution of the apparent conflict between the low and high altitude ion observations.

3.   Inner Magnetosphere

An important problem in magnetospheric research for which ion mass spectrometer data are particularly relevant is the origin of the storm time ring current. The first direct measurements of the composition of the trapped ring current ions were made with the S3-3 satellite and showed that in the energy range from 0.5 to 16 keV/e O^+ was often of comparable intensity to H^+ (JOHNSON et al., 1977). This confirmed earlier inferences of the importance of the ionosphere as a source term for this plasma population (SHELLEY et al., 1972). Later, on the basis of more extensive measurements from the GEOS satellites, it was inferred that in the regions of the magnetosphere accessible from the GEOS orbits, the solar wind (characterized by He^{++}), and the ionosphere (characterized by O^+) gave, on average, comparable contributions to the storm time plasma in the energy range from 0.9 to 16 keV/e (BALSIGER et al., 1980). Because of its unique orbital characteristics, the ISEE satellite can provide important complementary information to these earlier data sets, allowing direct measurements of the radial dependence of the near equatorial plasma population.

From the first 16 months of the ISEE data, LENNARTSSON et al. (1981) selected ten magnetic storms with peak $Dst \lesssim -100\,\gamma$ for which data in the inner magnetosphere were acquired during the early main phase. The data set provided relatively complete coverage of the inner magnetosphere between about 2 and 15 R_E at all local times. The average densities of the various ionic constituents were computed over time intervals corresponding to a spacecraft motion of about 1 R_E in radial distance. This amounted to about 20 minutes at $R \sim 2\,R_E$ and to more than 1 hour at $R \sim 15\,R_E$. The calculations assumed isotropic angular distributions and included energies in the range from 0.1 to 17 keV/e. The overall results are summarized in Fig. 10. These histograms give the frequency of occurrence of different ion concentrations in all of the intervals studied without any discrimination with respect to radial distance or local time. The remaining fraction in each case consisted essentially entirely of H^+. In some cases only upper limits on the densities could be established because of background due to penetrating radiation or poor counting statistics. If each upper limit is treated as a real density then the solid histograms apply. If it is neglected then the dashed histograms apply. The actual frequency of occurrence lies somewhere between the two.

The most striking feature of Fig. 10 is the apparent large and variable ionospheric

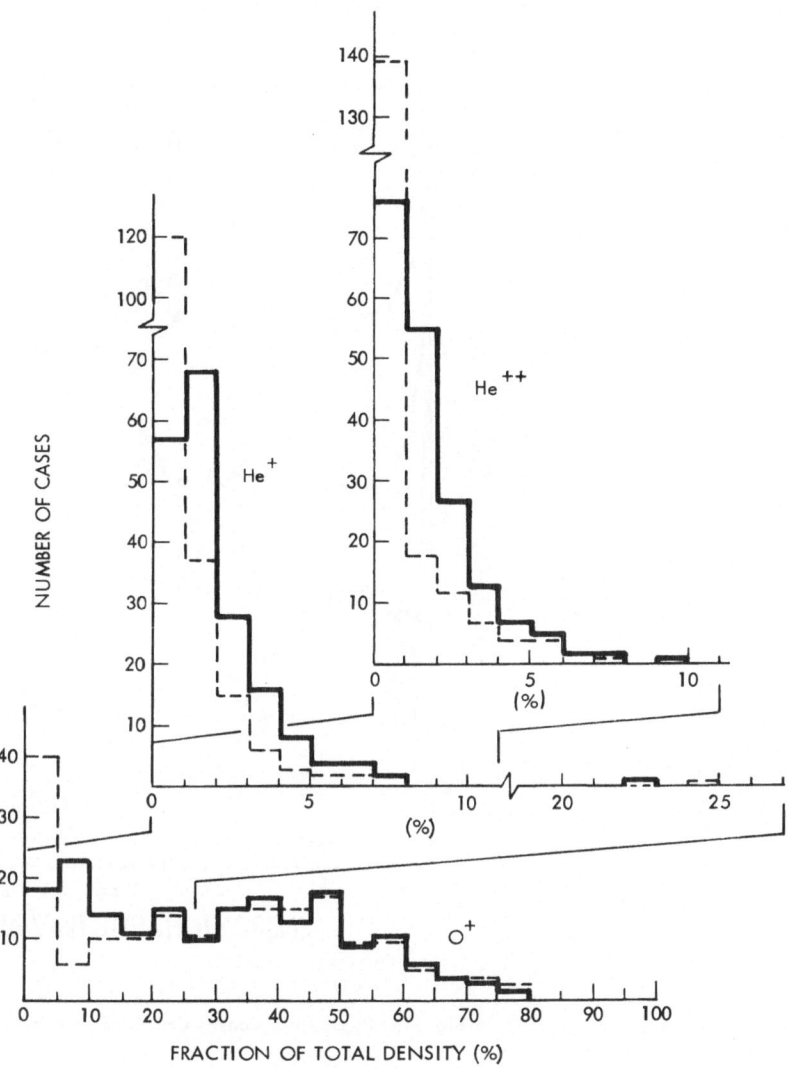

Fig. 10.   Distribution of fractional densities of O^+, He^+, and He^{++} in the inner magnetosphere $(2-15\ R_E$ geocentric distance) during magnetic storms (LENNARTSSON *et al.*, 1981).

component of the measured plasma. Since all of the O^+, almost all of the He^+, and some unknown fraction of the H^+ is of ionospheric origin it is clear that the ionosphere is an important, perhaps dominant contributor to the plasma in the energy range of this study. On the other hand, if one considers that the solar wind typically contains ~ 25 times as much H^+ as He^{++}, the high frequency of measurable He^{++} densities suggest that this source term is also important, in agreement with the GEOS result.

    Some insight into the particle dynamics during these magnetic storms can be obtained by an examination of the detailed distribution functions. A commonly

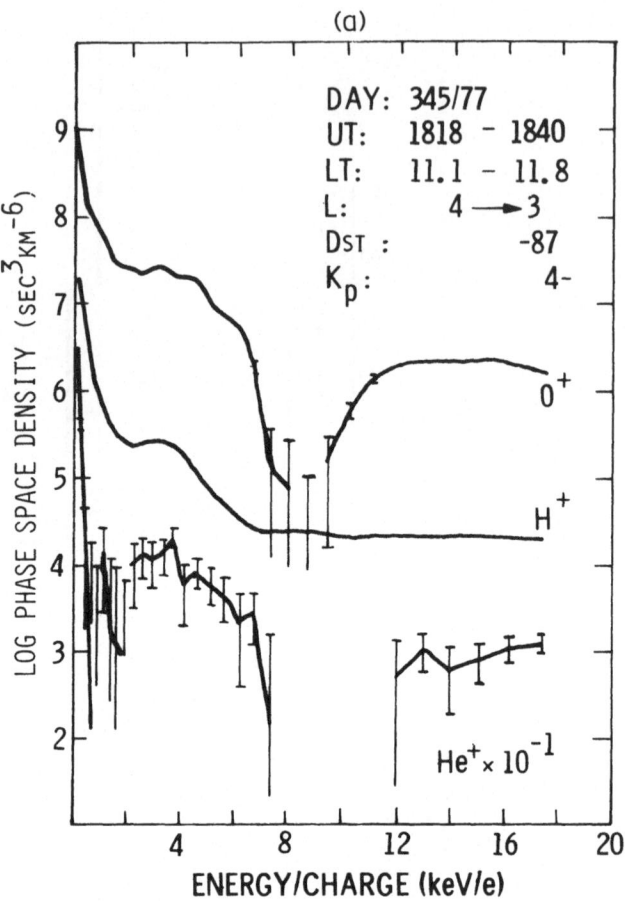

Fig. 11.   Phase space density distributions exhibiting a "hole"

observed feature of these distributions was a "hole" in the few keV range. Some examples are given in Fig. 11. This "hole" was most clearly seen in the dayside magnetosphere and occurred at an energy that varied with local time and radial distance and also varied to some extent from event to event. This is apparently the same feature as the "deep minimum" discussed by MCILWAIN (1972) in the energy flux spectrums taken by the geosynchronous satellite ATS-5. As reported by MCILWAIN, this flux minimum seems to be at a demarcation energy between ions (with low energy) that have drifted through the dawnside magnetosphere and ions (with higher energy) that have drifted through the dusk region. That is, the minimum corresponds to ions not yet arrived (due to slow drift close to the earth) or ions already lost (due to high loss rates close to the earth). The frequent occurrence of this spectral feature suggests that much of the time the large scale drift motion of the particles can be described by almost

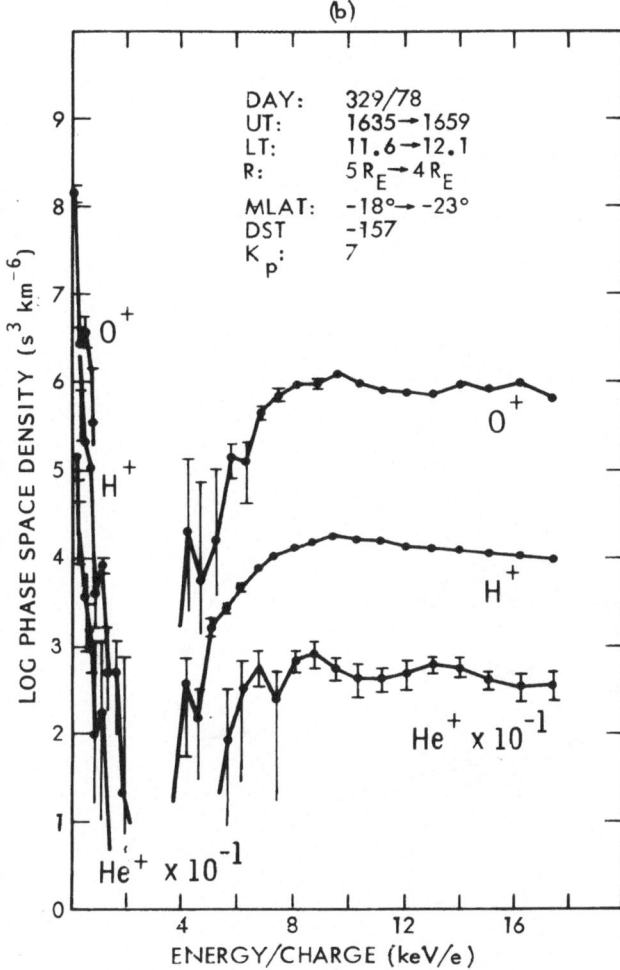

(b)

DAY: 329/78
UT: 1635→1659
LT: 11.6→12.1
R: $5R_E$→$4R_E$
MLAT: $-18°$→$-23°$
DST -157
K_p: 7

in the few keV range (LENNARTSSON et al., 1979, 1981).

constant magnetic moments, resulting in a relationship between the measured energy and spatial origin of the ions such as is illustrated in Fig. 12. In this schematic diagram the phase space density is assumed to be measured at a fixed pitch angle (e.g. 90°) and the ions are assumed to be injected at a fairly well defined radial distance and to preserve their magnetic moments while drifting sunward under the influence of a dawn-dusk electric field.

Some corroboration of this interpretation for the O^+ fluxes measured by ISEE during the December 11, 1977, magnetic storm was presented by LENNARTSSON et al., (1979). Data from this event taken between 11.1 and 11.8 hours local time and at L values between 3 and 4 have been shown in Fig. 11(a). From similar spectrums taken during this entire dayside magnetospheric traversal LENNARTSSON et al. have plotted the location in energy of the deepest minimum in the O^+ distribution as a function of

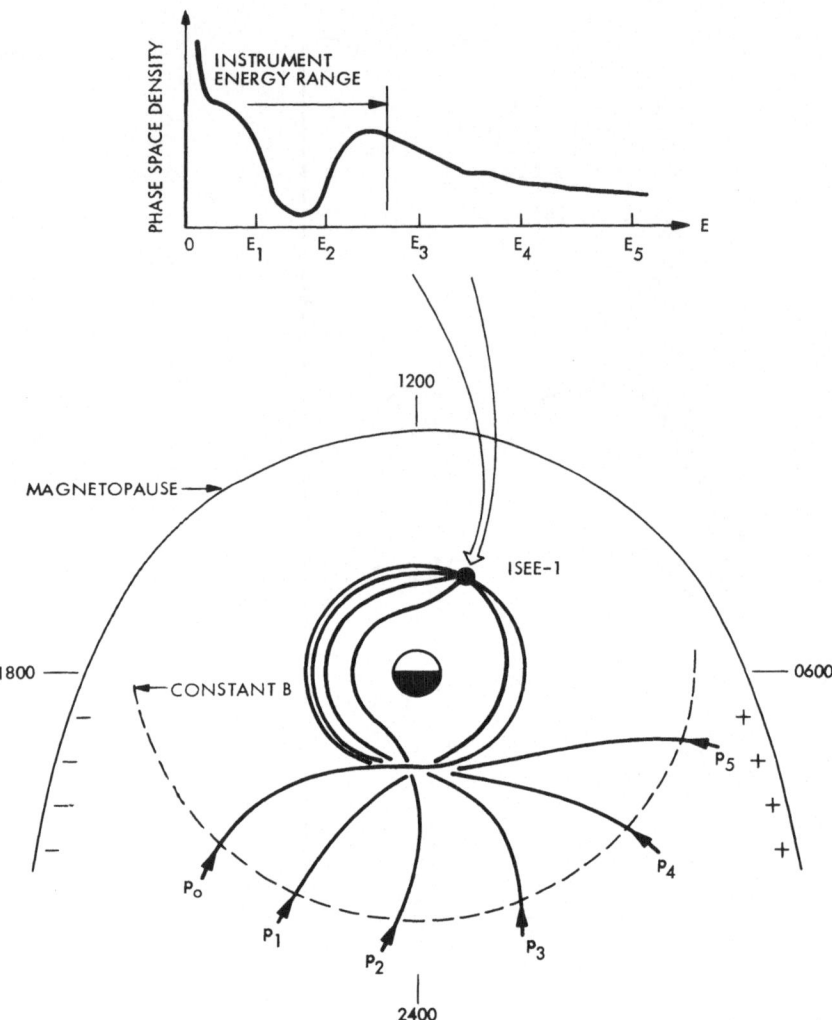

Fig. 12. Schematic illustration of a relationship between measured energy and spatial origin of ions (Lennartsson *et al.*, 1981).

geocentric distance (See Fig. 13). The dashed curve shows the energy at which westward gradient *B* drift in the equatorial plane is exactly cancelled by eastward co-rotation convection (for 90° pitch angle particles in a dipole field). The remarkable agreement provides strong support for the conclusion that the "hole" in the distribution function is the demarcation between low energy ions circling the earth in the eastward direction and high energy ions circling the earth westward.

The detailed motion of the ions in any individual event will of course depend on the spatial structure and time dependence of the electric field, but to the extent that the model represented by Fig. 12 is qualitatively valid, in some average sense, for the 10

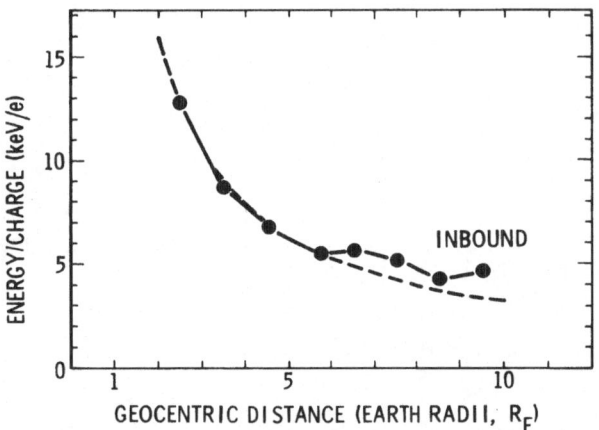

Fig. 13.   Energy at deepest minimum in phase space density distribution as a function of geocentric distance, for O^+ ions on December 11, 1977 (solid line) and energy at which co-rotation cancels gradient B drift (dashed line).

magnetic storms which contributed to the data set discussed above, we have the opportunity to use this data to investigate the important portion of the ring current distribution function that is above the energy range accessible to the present generation of ion mass spectrometers. As shown by SMITH and HOFFMAN (1973) and WILLIAMS (1980) the portion of the spectrum in the 50–100 keV range dominates the energy density of the trapped particles in the $L \sim 3$–4 region of the magnetosphere where most of the ring current energy is concentrated. By studying these ions at high altitudes in the predawn local time sector (P4 and P5 in Fig. 12) we can investigate their composition at a location in the magnetosphere where they are still within the energy range of the ISEE spectrometer.

This has been done by dividing the data represented by the histograms in Fig. 10 into two subgroups depending on their local time, and plotting them as a function of geocentric radial distance. Figure 14 shows the results for the O^+ to H^+ number density ratios. Samples taken in the 0100 to 0600 LT sector are indicated as triangles and the remaining data are represented by circles. Upper limits are marked by open symbols. The solid and dashed lines are the medians of triangular and circular data symbols respectively. One sees in Fig. 14 that the largest O^+/H^+ ratios (near unity) are found at low geocentric distances. Despite the large variation in individual ratios, one can see a significant difference in composition between the high altitude, predawn data and the high altitude data in the remaining local time region. This is shown more clearly by the two histograms on the right hand ordinate. These show the total number of points (solid and open) in certain vertical bins for $R > 7\ R_E$. The solid histograms corresponds to the 0100–0600 LT sector and the dashed histogram corresponds to 0600–0100 LT. These results suggest that the plasma which will later become the high energy portion of the ring current spectrum has a significantly reduced oxygen to hydrogen ratio relative to the plasma in other sectors.

A plot of the He^{++}/H^+ density ratios in the same format as Fig. 14 is shown in

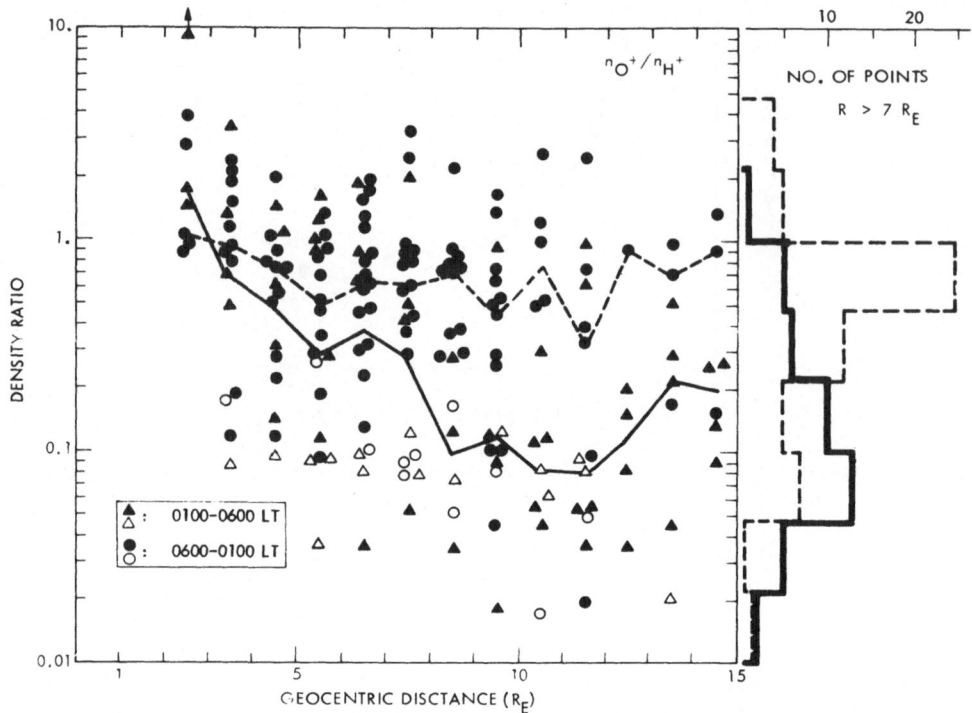

Fig. 14. $0^+/H^+$ density ratios during storms as a function of geocentric distance. Upper limits are indicated by open symbols. Triangles represent data in the 0100 to 0600 LT sector and circles represent data in the 0600 to 0100 LT sector. The two histograms to the right show the total number of points in the indicated vertical bins for $R > 7 R_E$. The solid histrogram corresponds to the 0100 to 0600 LT data and the dashed histogram to the 0600 to 0100 LT data (Lennartsson *et al.*, 1981).

Fig. 15. The two medians refer only to the solid data points in this figure. These ratios do not show a clearly significant radial dependence, but only limiting values are available at low geocentric distances because of background effects on the He^{++}. Although these results are less definitive than those in Fig. 14 it is probable that the data at large radial distances ($R > 7 R_E$) in the 0100 to 0600 LT sector is richer in He^{++} than the corresponding data at other local times. Taken together these results suggest that the solar wind is a more important contributor to the ring current energy than the available data in the energy range below ~ 17 keV/e would indicate. However, as is apparent in Fig. 14 and 15 the ratios are highly variable and the differences between their average values in the two local time sectors are not dramatic and do not suggest that the solar wind is always the dominant source of ions in the high energy ring current.

A recently completed study has allowed for a comparison of the ion composition in the near equatorial magnetosphere during quiet and disturbed conditions. Data acquired during some 24 traversals of the magnetosphere at different local times and $R \leq 15 R_E$ during very quiet periods were compared with the above described data

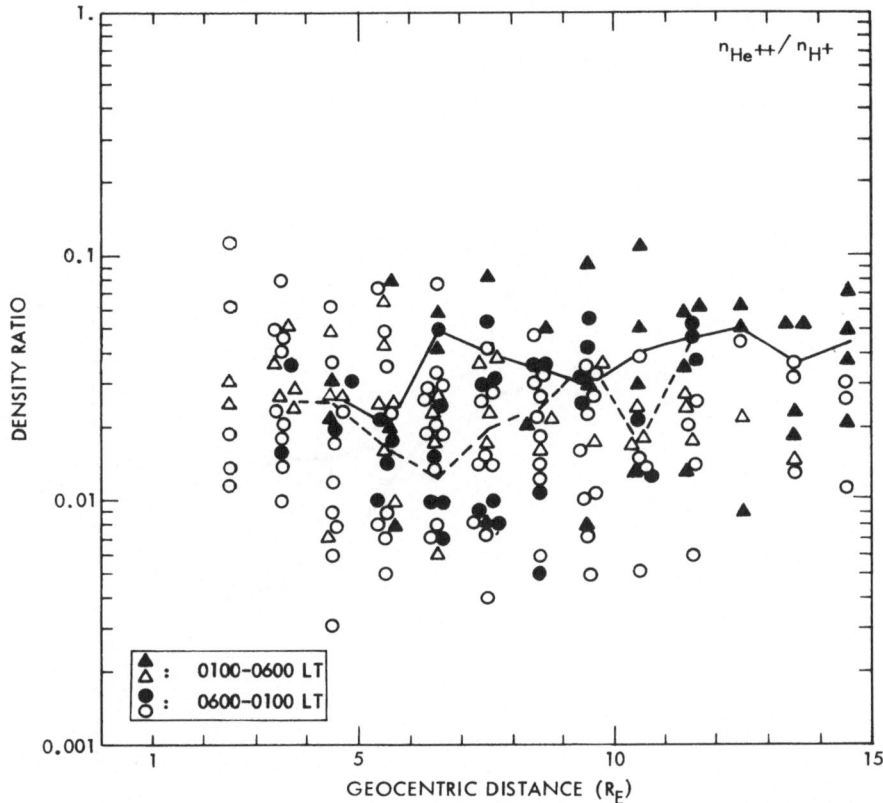

Fig. 15. He^{++}/H$^+$ density ratios in the same format as Fig. 14 (LENNARTSSON *et al.*, 1981).

taken primarily during the early main phase of magnetic storms (LENNARTSSON and SHARP, 1982). Figures 16, 17, and 18 show the average density ratios of the various ionic constituents in the energy range 0.1 to 17 keV/e as a function of L for the two data sets.

Looking first at the quiet time results, the most striking feature is the rapid rise in the He$^+$/H$^+$ and 0$^+$/H$^+$ ratios at low L values. This is qualitatively consistent with expectations from charge exchange since H$^+$ has by far the shortest lifetime of the three species. At 5 keV for example $\tau(\text{H}^+):\tau(0^+):\tau(\text{He}^+) \simeq 1:7:50$ (SMITH *et al.*, 1981). However, as discussed by LENNARTSSON and SHARP (1982), the results are not quantitatively consistent with these estimated charge exchange lifetimes and the expected ion drift paths. These lead to the prediction of much higher $0^+/H^+$ and He$^+$/H$^+$ ratios than those observed, particularly at $L \lesssim 4$. The dominance of H$^+$ at $L > 5$ in the quiet time data shows that charge exchange is not the primary element in establishing the ionic composition in that range. Additionally, the similarity in the shapes of the quiet time $0^+/H^+$ and He$^+$/H$^+$ curves in Fig. 16 and 17 indicates that the $0^+/\text{He}^+$ ratio remains at an almost constant value (~ 5) over the altitude range of the study. This generalization can even be extended to include the magnetotail data

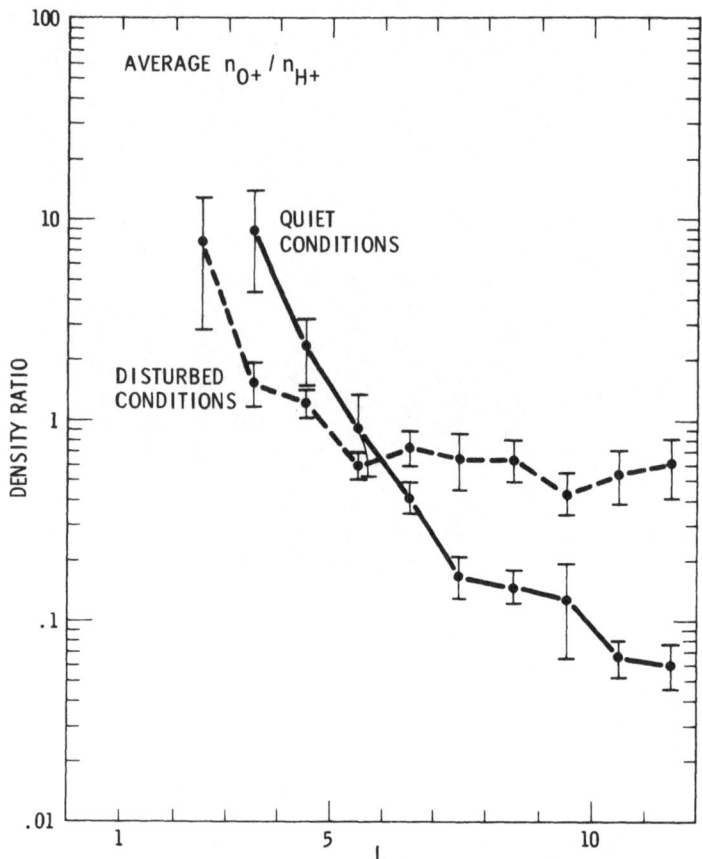

Fig. 16.   Average O^+/H^+ density ratios as a function of L for quiet and disturbed conditions (Lennartsson and Sharp, 1982).

discussed earlier. The quiet time data in Table 1 were acquired at a mean geocentric radial distance of 16.3 R_E (variance = 3.0 R_E) and indicate an O^+/He^+ density ratio of 5.6 ± 1.1. The approximate constancy of this ratio over such an extended altitude range suggests that mass dependent loss processes such as charge exchange are not substantially affecting these two ion species as they convect inward.

Taken together, these results suggest that models based on the charge exchange decay of static trapped populations are not likely to be successful in explaining the detailed evolution of the ring current (Lyons and Evans, 1976; Tinsley, 1976; Smith *et al.*, 1981). Instead, the results seem to require continuous injection, even during periods of extended quiet, and transport times which are not long compared to the charge exchange lifetimes of the heavier ion species.

If we now also examine the results under disturbed conditions we see some notable features. There is a crossover in the O^+/H^+ ratios at $L \sim 6$. Inside this location the O^+/H^+ and He^+/H^+ ratios decrease as magnetic activity increases. There is some

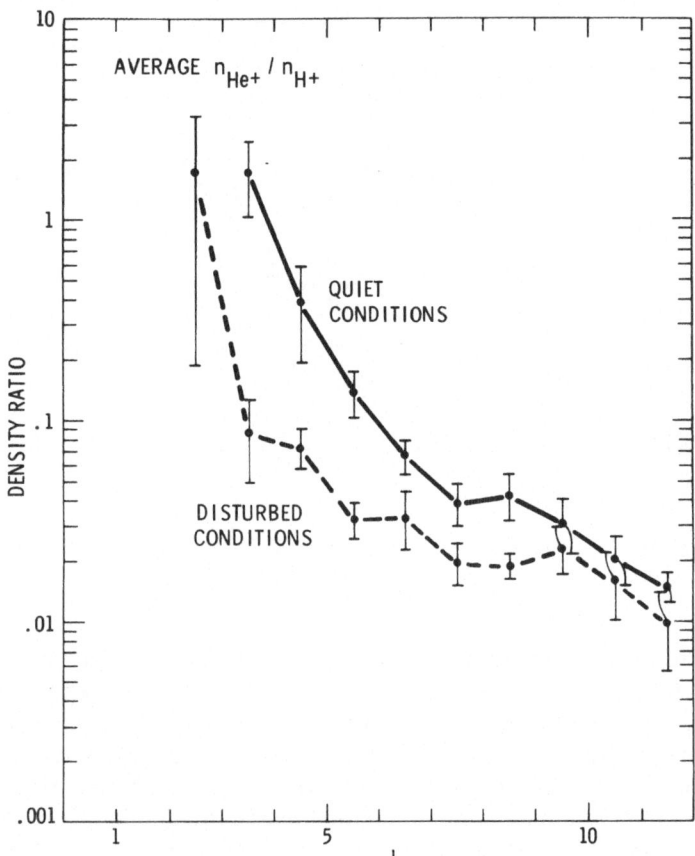

Fig. 17. Average He$^+$/H$^+$ density ratios as a function of L for quiet and disturbed conditions (LENNARTSSON and SHARP, 1982).

indication that the He^{++}/H$^+$ ratio may also be increasing in the $L = 5-7$ range but the data are complicated by relatively high background levels and are not completely definitive. The results do suggest, however, that the solar wind may be a more important source in this region relative to the ionosphere during storms than it is during quiet times. This is opposite to the effect we found for substorm activity in the deep tail.

Outside of $L \sim 7$ we find a more complicated situation. Both the He^{++}/H$^+$ and the O$^+$/H$^+$ ratios apparently increase during storms. BALSIGER et al. (1980) reported a similar result from the GEOS data. This cannot be understood simply as the result of a change in the relative source strengths and suggests that the ionospheric source might have had a higher O$^+$/H$^+$ ratio during storms than during quiet times. We also note that there is about a factor of 6–7 decrease in He$^+$ with respect to O$^+$ over almost the entire L range during disturbed conditions. This is probably also a source effect.

The only information we have on changes in the relative composition of the

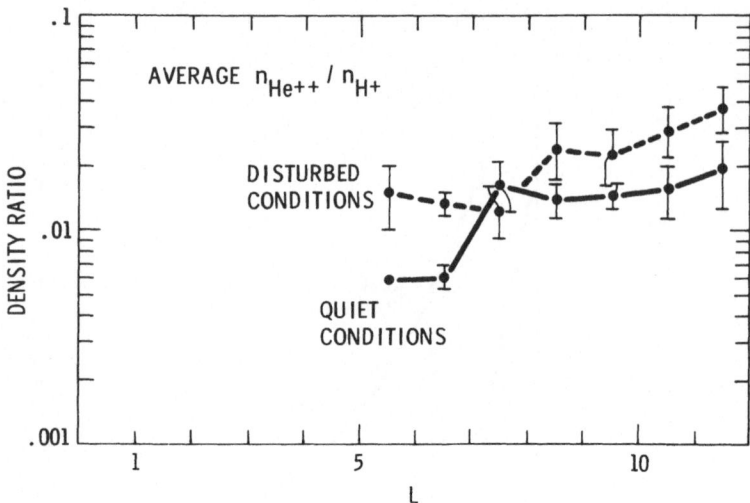

Fig. 18.   Average He^{++}/H^+ density ratios as a function of L for quiet and disturbed conditions (LENNARTSSON and SHARP, 1982).

ionospheric source with magnetic activity is the statistical study based on the K_p dependence of the S3-3 data which is summarized in the review paper in this volume. This study did not include the He^+ constituent and had other limitations which are discussed in that paper. The principal conclusion was that there was no significant difference in the O^+/H^+ ratio between quiet and active times. K_p is probably not a very good parameter to characterize the early main phase of magnetic storms, however, and so the applicability of that data set to the understanding of this problem is questionable. It is probably more relevant to the results in the deep tail which exhibited the substorm dependence discussed earlier. Other possible explanations for the apparent discrepancy with the high altitude results are discussed in the S3-3 review paper.

Another area of magnetospheric research in which composition measurements are of importance is the study of geomagnetic micropulsations. Magnetic pulsations in the 10 to 600 second period range (Pc 3-5) are a common phenomena in the dayside magnetosphere. They are generally thought to be the result of standing wave resonant field line oscillations. To compare the observed periods with theory it is necessary to know the mass density and thus the composition of the ambient plasma. Composition data have not generally been available and theoretically predicted resonance periods based on the assumption of a pure proton-electron plasma have often been in disagreement with observations.

A detailed analysis of a set of long period ULF pulsations which were observed by the ISEE magnetometers on December 11, 1977 has been conducted by SINGER *et al.* (1979). During the period of interest there was a substantial fraction of O^+ in the measured plasma resulting in a mass density five to nine times greater than would have been computed on the basis of a pure proton-electron plasma. The expected wave periods under these circumstances are two to three times longer than they would have been if H^+ were the only ion present. The observed periods were in good agreement with theoretical predictions for a second harmonic standing wave oscillation. (The fundamental magnetic field oscillation is thought to have a node at the magnetic equator near where the data were acquired.) It is concluded that composition data are a requirement for any serious quantitative intercomparisons of measurements with theory.

4. Magnetospheric Boundary Regions

Measurements of the relative H^+ and He^{++} distributions in the magnetosheath can provide insight into possible mass and charge dependent process that may occur in the bow shock and magnetosheath. Such distributions were examined by PETERSON et al. (1979) on eight magnetosheath traversals in the 0800–1100 local time range. Moments of the distributions functions were calculated and showed that bulk velocities of the two species, averaged over several minute intervals, were equal within experimental uncertainties. Both species exhibited temperature anisotropies with $T_\perp > T_\parallel$ and both had high energy tails, i.e. excess flux at high energies relative to a best fit Maxwellian. There was a remarkable difference in the distribution functions of the two species at low thermal speeds in the form of a lack of He^{++} ions relative to H^+ ions in this portion of the spectrum. An example of this effect is shown in Fig. 19. A similar effect was observed on all 8 traversals. A possible mechanism which could lead to such a phenomena was described by PETERSON et al. (1979) as follows:

Electric potentials in the bow shock are determined by the energy of the dominant solar wind ion (H^+). Minor heavy ions species, with high mass to charge ratios and high energies, pass through the shock front with only limited changes in kinetic energy. Behind the shock front irregular magnetic fields randomize velocity directions of the minor heavy ions but have little effect on kinetic energy in the center of mass coordinate frame. The result would be a heavy ion distribution function having the form of a hollow shell moving with the same bulk speed as the primary component (H^+).

PETERSON et al. (1982) have examined the composition data in nine cases when the ISEE satellite traversed the dayside magnetospheric boundary regions in the vicinity of the subsolar point for evidence of ions of ionospheric origin. The analysis was hampered by the relatively long cycle time of the spectrometer but the overall results provided convincing evidence that He^+ and O^+ ions with low flux intensities and keV energies can be found in the boundary layer, the magnetopause and the magnetosheath.

These data confirm earlier inferences based on the energy spectrums that magnetospheric particles can penetrate the magnetopause and contribute to the magnetosheath population (SCHOLER et al., 1981; SPEISER et al., 1981). The mass spectrometer results also suggested that the terrestrial contribution to the boundary layer ion fluxes might have had a different composition in the subsolar region than was typical of the boundary layer plasmas on the flanks of the magnetosphere. Preliminary results indicated that He^+ is most often dominant over O^+ in the subsolar region while O^+ is typically dominant on the flanks. This may be the signature of the process

R. D. Sharp *et al.*

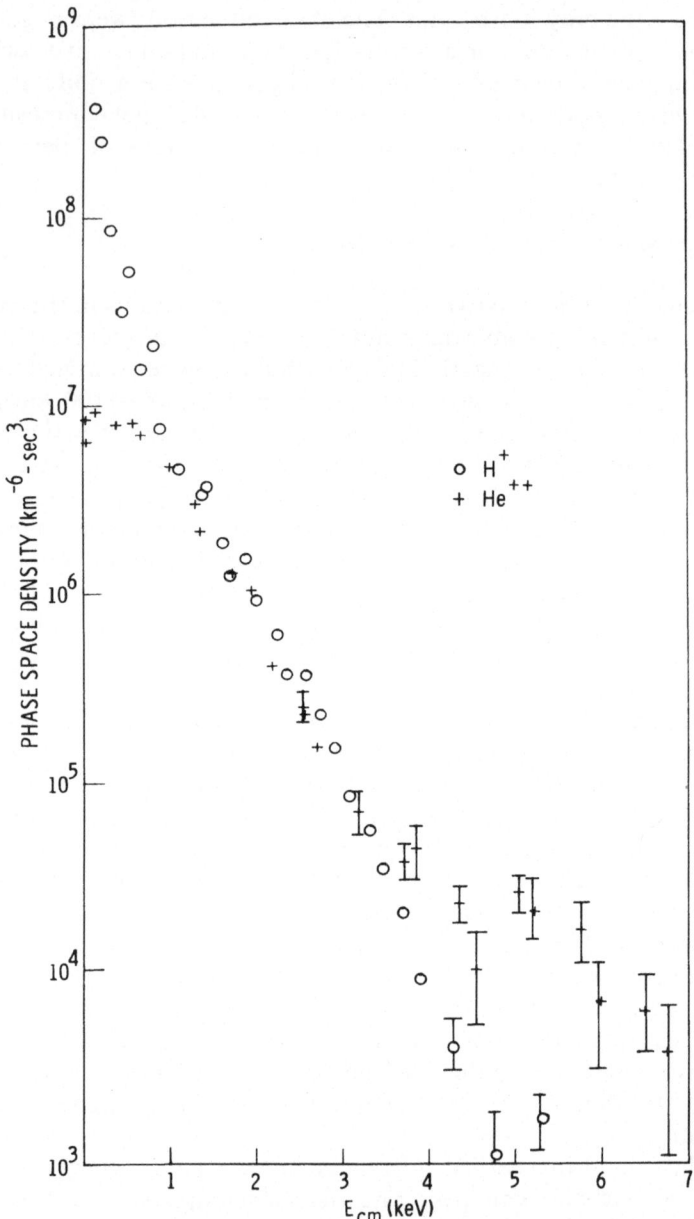

Fig. 19.   Phase space densities in the magnetosheath averaged over a 60° interval centered on the bulk flow direction and plotted against total energy in the center of mass frame. The data were acquired on 19 November, 1977 between 2330–2359 UT when ISEE was located between 13.7 and 13.2 R_E at ~0955 local time (Peterson *et al.*, 1979).

proposed by FREEMAN *et al.* (1977) in which plasmaspheric plasma, which typically is rich in He$^+$, is detached from the plasmapause and convected to the dayside magnetopause by the magnetospheric electric field. The He$^+$ ions could then become accelerated, mixed with the magnetosheath plasma, and injected into the plasma sheet in the magnetotail as discussed in the previous section.

The predominance of He$^+$ over O$^+$ is a relatively rare finding in the energetic plasma observations with the ion mass spectrometers on the several satellites that have probed throughout the magnetosphere. Except perhaps for the inner ring current during the late recovery phase of magnetic storms when charge exchange losses become important (LUNDIN *et al.*, 1980) O$^+$ is generally dominant over He$^+$. Therefore, as discussed by YOUNG (1979) the Freeman *et al.* mechanism is not likely to be a major source of energetic magnetospheric plasma. The ionospheric processes ocurring in the auroral acceleration region in which O$^+$ is strongly favored over He$^+$, seem to be the principal contributor of terrestrial ions to the magnetosphere.

We thank C. T. Russell for use of the data from the onboard magnetometer and our several co-investigators (H. Balsiger, J. Geiss, P. Eberhardt, G. Haerendel, and H. Rosenbauer) for their valuable contributions to the experiment. This work was principally supported by NASA under contract NAS 5-25773.

REFERENCES

ASHOUR-ABDALLA, M., H. OKUDA, and C. Z. CHENG, Acceleration of heavy ions on auroral field lines, *Geophys. Res. Lett.*, **8**, 795, 1981.

BALSIGER, H., P. EBERHARDT, J. GEISS, A. GHIELMETTI, H. P. WALKER, D. T. YOUNG, H. LOIDL, and H. ROSENBAUER, A satellite-borne ion mass spectrometer for the energy range 0 to 16 keV, *Space Sci. Inst.*, **2**, 499, 1976.

BALSIGER, H., P. EBERHARDT, J. GEISS, and D. T. YOUNG, Magnetic storm injection of 0.9–16 keV/e solar and terrestrial ions into the high altitude magnetosphere, *J. Geophys. Res.*, **85**, 1645, 1980.

BAUGHER, C. R., C. R. CHAPPELL, J. L. HORWITZ, E. G. SHELLEY, and D. T. YOUNG, Initial thermal plasma observations from ISEE-1, *Geophys. Res. Lett.*, **7**, 657, 1980.

COLLIN, H. L., R. D. SHARP, E. G. SHELLEY, and R. G. JOHNSON, Some general characteristics of upflowing ion beams over the auroral zone and their relationship to auroral electrons, *J. Geophys. Res.*, **86**, 6820, 1981.

CROOKER, N. V., Dayside merging and cusp geometry, *J. Geophys. Res.*, **84**, 951, 1979.

FRANK, L. A., K. L. ACKERSON, and D. M. YEAGER, Observations of atomic oxygen (O$^+$) in the earth's magnetotail, *J. Geophys. Res.*, **82**, 129, 1977.

FREEMAN, J. W., H. K. HILLS, T. W. HILL, P. H. REIFF, and D. A. HARDY, Heavy ion circulation in the earth's magnetosphere, *Geophys. Res. Lett.*, **4**, 195, 1977.

HARDY, D. A., H. K. HILLS, and J. W. FREEMAN, Jr., A new plasma regime in the distant geomagnetic tail, *Geophys. Res. Lett.*, **2**, 169, 1975.

HARDY, D. A., J. W. FREEMAN, and H. K. HILLS, Plasma observations in the magnetotail, in *Magnetospheric Particles and Fields*, edited by B. M. McCormac, p. 89, D. Reidel, Dordrecht, Holland, 1976.

HARDY, D. A., H. K. HILLS, and J. W. FREEMAN, Occurrence of lobe plasma at lunar distance, *J. Geophys. Res.*, **84**, 72, 1979.

HONES, E. W., Jr., J. R. ASBRIDGE, S. J. BAME, M. D. MONTGOMERY, S. SINGER, and S. I. AKASOFU, Measurements of magnetotail plasma flow made with VELA 4B, *J. Geophys. Res.*, **77**, 5503, 1972.

HORWITZ, J. L., C. R. BAUGHER, C. R. CHAPPELL, E. G. SHELLEY, and D. T. YOUNG, Pancake pitch angle distributions in warms ions observed with ISEE-1, *J. Geophys. Res.*, **86**, 3311, 1981.

HUNDHAUSEN, A. J., *Coronal Expansion and Solar Wind*, 99 pp., Springer-Verlag Berlin-Heidelberg, 1972.

JOHNSON, R. G., A review of the hot plasma composition near geosynchronous altitude, in *Spacecraft Charging Technology 1980*, Conference proceedings edited by N. J. Stevens and C. P. Pike, p. 412, NASA-CP-2182, 1981.

JOHNSON, R. G., R. D. SHARP, and E. G. SHELLEY, Observations of ions of ionospheric origin in the storm time ring current, *Geophys. Res. Lett.*, **4**, 403, 1977.

KAMEI, T. and H. MAEDA, Auroral electrojet indices (*AE*) for January–June 1978, World Data Center for Geomagnetism, Data Book No. 3, Kyoto University, April, 1981.

KINDEL, J. M. and C. F. KENNEL, Topside Current Instabilities, *J. Geophys. Res.*, **76**, 3055, 1971.

LENNARTSSON, W. and R. D. SHARP, A comparison of the near equatorial ion composition between quiet and disturbed conditions, *J. Geophys. Res.*, **87**, 6109, 1982.

LENNARTSSON, W., E. G. SHELLEY, R. D. SHARP, R. G. JOHNSON, and H. BALSIGER, Some initial ISEE-1 results on the ring current composition and dynamics during the magnetic storm of December 11, 1977, *Geophys. Res. Lett.*, **6**, 483, 1979.

LENNARTSSON, W., R. D. SHARP, E. G. SHELLEY, R. G. JOHNSON, and H. BALSIGER, Ion composition and energy distribution during 10 magnetic storms, *J. Geophys. Res.*, **86**, 4678, 1981.

LUNDIN, R., L. R. LYONS, and N. PISSARENKO, Observations of ring current composition at $L < 4$, *Geophys. Res. Lett.*, **7**, 425, 1980.

LYONS, L. R. and D. S. EVANS, The inconsistency between proton charge exchange and the observed ring current decay, *J. Geophys. Res.*, **81**, 6197, 1976.

MCILWAIN, C. E., Plasma convection in the vicinity of the geosynchronous orbit, in *Earth's Magnetospheric Processes*, edited by B. M. McCormac, p. 268, D. Reidel, Dordrecht, Holland, 1972.

PETERSON, W. K., E. G. SHELLEY, R. D. SHARP, R. G. JOHNSON, J. GEISS, and H. ROSENBAUER, H^+ and He^{++} in the dawnside magnetosheath, *Geophys. Res. Lett.*, **6**, 667, 1979.

PETERSON, W. K., R. D. SHARP, E. G. SHELLEY, and R. G. JOHNSON, Energetic ion composition of the plasma sheet, *J. Geophys. Res.*, **86**, 761, 1981.

PETERSON, W. K., E. G. SHELLEY, G. HAERENDEL, and G. PASCHMANN, Energetic ion composition in the subsolar magnetopause and boundary layer, *J. Geophys. Res.*, **87**, 2139, 1982.

PILLIP, W. G. and G. MORFILL, The formation of the plasma sheet resulting from plasma mantle dynamics, *J. Geophys. Res.*, **83**, 5670, 1978.

ROSENBAUER, H., H. GRUNWALDT, M. D. MONTGOMERY, G. PASCHMANN, and N. SCHOPKE, HEOS-2 plasma observations in the distant polar magnetosphere: The plasma mantle, *J. Geophys. Res.*, **80**, 2723, 1975.

SCHMIDT, W. K. H., H. ROSENBAUER, E. G. SHELLEY, and J. GEISS, On temperature and speed of He^{++} and O^{6+} ions in the solar wind, *Geophys. Res. Lett.*, **7**, 697, 1980.

SCHOLER, M. F., M. IPAVICH, G. GLOECKER, D. HOVESTADT, and B. KLECKER, Leakage of magnetospheric ions into the magnetosheath along reconnected field lines at the dayside magnetopause, *J. Geophys. Res.*, **86**, 1299, 1981.

SHARP, R. D., W. LENNARTSSON, E. G. SHELLEY, and W. K. PETERSON, ISEE measurements of plasma sheet ion composition, *Trans. Am. Geophys. Union* (abstract), **62**, 995, 1981a.

SHARP, R. D., D. L. CARR, W. K. PETERSON, and E. G. SHELLEY, Ion streams in the magnetotail, *J. Geophys. Res.*, **86**, 4639, 1981b.

SHELLEY, E. G., Ion composition in the dayside cusp: Injection of magnetospheric ions into the high latitude boundary layer, Proceedings of Magnetospheric Boundary Layers Conference, Alpbach, 11–15 June, 1979 (ESA SP-148, August 1979).

SHELLEY, E. G., R. G. JOHNSON, and R. D. SHARP, Satellite observation of energetic heavy ions during a geomagnetic storm, *J. Geophys. Res.*, **77**, 6104, 1972.

SHELLEY, E. G., R. D. SHARP, R. G. JOHNSON, J. GEISS, P. EBERHARDT, H. BALSIGER, G. HAERENDEL, and H. ROSENBAUER, Plasma composition experiment on ISEE-A, *IEEE Trans. Geosci. Electr.*, **GE-16**, 266, 1978.

SHELLEY, E. G., D. A. SIMPSON, T. C. SANDERS, E. HERTZBERG, H. BALSIGER, and A. GHIELMETTI, The energetic ion composition spectrometer (EICS) for the Dynamics Explorer-A, *Space Sci. Inst.*, **5**, 443, 1981.

SINGER, H. J., C. T. RUSSELL, M. G. KIVELSON, T. A. FRITZ, and W. LENNARTSSON, *Geophys. Res. Lett.*, **6**, 889, 1979.

SMITH, P. H. and R. A. HOFFMAN, Ring current particle distributions during the magnetic storms of

December 16–18, 1971, *J. Geophys. Res.*, **78**, 4731, 1973.

SMITH, P. H., N. K. BEWTRA, and R. A. HOFFMAN, Inference of the ring current ion composition by means of charge exchange decay, *J. Geophys. Res.*, **86**, 3470, 1981.

SPEISER, T. W., D. J. WILLIAMS, and H. A. GARCIA, Magnetospherically trapped ions as a source of magnetosheath energetic ions, *J. Geophys. Res.*, **86**, 723, 1981.

TINSLEY, B. A., Evidence that the recovery phase ring current consists of helium ions, *J. Geophys. Res.*, **81**, 6193, 1976.

WILLIAMS, D. J., Ring current composition and sources, in *Dynamics of the Magnetosphere*, edited by S.-I. Akasofu, Reidel, 1980.

YOUNG, D. T., Ion composition in magnetospheric modeling, in *Quantitative Magnetospheric Models*, edited by R. Olsen, Am. Geophys. Union, Washington, D. C., 1979.

YOUNG, D. T., H. BALSIGER, and J. GEISS, Observed increase in the abundance of kilovolt O^+ in the magnetosphere due to solar cycle effects, Advances in Space Research, COSPAR, 1981.

Energetic Ion Composition in the Earth's Magnetosphere, edited by R. G. Johnson, 263–286.
Copyright © 1983 by Terra Scientific Publishing Company (TERRAPUB), Tokyo.

Observations of Low-Energy Plasma Composition from the ISEE-1 and SCATHA Satellites

J. L. Horwitz,* C. R. Chappell,** D. L. Reasoner,**

P. D. Craven,** J. L. Green,** and C. R. Baugher**

*Department of Physics, The University of Alabama in Huntsville,
Huntsville, Alabama 35899, U.S.A.*
**Space Sciences Laboratory, Marshall Space Flight Center,
MSFC, Alabama 35812, U.S.A.*

(Received February 26, 1982)

Observations by the SCATHA and ISEE-1 satellites of the energy, pitch angle and compositional structure of low-energy ($E \lesssim 100$ electron volts) ions are reviewed. From these observations it appears that normally the dominant ion populations in the plasmasphere are cold, isotropic, quasi-Maxwellian distributions, with H^+ being the dominant species followed by He^+ and then O^+, with smaller amounts of O^{++} and He^{++} and/or D^+. Outside the plasmapause the ion populations are most often field-aligned with energies of 10's of electron volts ('warm'), with H^+ and O^+ the dominant constituents, with He^+ fluxes normally in lesser amounts and occasional traces of O^{++} and He^{++}/D^+. Within the plasmasphere, sometimes mixed with the cold population, are pancake distributions of warm ions, particularly in He^+ and H^+. At times these pancake distributions are the principal distributions observed in the vicinity of the outer part of the plasmasphere. Evidence for energization perpendicular to the magnetic field is seen in pancake and conical (peak fluxes between 0 and 90° to the magnetic field line direction) distributions. Field-aligned flows of very low-energy (1–2 electron volts) H^+ and He^+ have been observed in the outer plasmasphere during periods of magnetic quiet; these flows appear to be associated with the refilling of the plasmasphere. Very high fractional concentrations of He^+ are occasionally observed in the plasmasphere, including regions of apparently virtually pure He^+. The cause of these high He^+ concentrations is as yet unknown. Sunward convection flows of plasmaspheric plasma in the afternoon-dusk local time sector have been observed to reach speeds of 50 km/s during moderately high magnetic activity; such convection is likely to play the major role in the detachment and transport of plasmaspheric plasma from the dusk bulge region to the dayside magnetopause. Low-energy plasma of ionospheric origin has also been observed adjacent to the dayside magnetopause and in anti-sunward streams within the magnetotail lobes, both situations being likely to be associated with the loss of this plasma from the magnetosphere.

1. Introduction

One of the most prominent themes of recent experimental and theoretical research work in magnetospheric physics is the stress on understanding the role low-energy plasmas play in many general magnetospheric problems. The topic of wave-particle interactions within the radiation belts, ring current, and plasma sheet is one such area. ASHOUR-ABDULLAH and THORNE (1978) and many others have discussed how cyclotron instabilities may be quenched or enhanced depending upon the amount of cold plasma present and its temperature, while CORNWALL (1976) and others have considered the effects of cold ion composition on such instabilities.

A second area is simply the origin of the energetic plasmas in the magnetosphere. The presence of large amounts of energetic (keV) O^+ and He^+ found in the ring current, plasma sheet, tail lobes and elsewhere by the Lockheed group (e.g., SHELLEY et al., 1972; SHELLEY, 1979; SHARP, 1980; JOHNSON, 1979; PETERSON et al., 1981; LENNARTSSON et al., 1979), the GEOS-1 and -2 group (GEISS et al., 1978; BALSIGER et al., 1980, 1982), and the Prognoz-1 group (LUNDIN et al., 1981) has established that the ionosphere is a significant source of energetic particles to the magnetosphere in addition to the solar wind. The Lockheed group has also measured directly the energetic ions of ionospheric origin at ~ 1 R_E altitude flowing up the magnetic field lines into the magnetosphere (SHELLEY et al., 1976; SHARP et al., 1977, 1982a). YOUNG (1980) and BALSIGER et al. (1980) have also indicated that some regions of energetic plasmas with especially high He^+ content may have been freshly accelerated plasma-spheric plasma.

A third area where low-energy plasmas are of particular significance is in the measurement of convection within the magnetosphere. Electric dipole antennas suffer from photo-electron sheath obscuration in the low-density, long Debye length regions outside the plasmasphere and have thus far been inadequate for measurement of DC electric fields. Measurements of convection electric fields in the magnetosphere have, however, been achieved through bulk parameter fitting to three-dimensional medium-energy ($E \gtrsim 200$ eV) ion distributions to obtain the convection flow velocity (FRANK et al., 1978). Since detection of small flows is enhanced when the ratio of the flow velocity to the thermal velocity of the measured ions (Mach number) is large, similar techniques applied to the measurement of ions at energies $\lesssim 100$ eV can be helpful in this regard.

In this brief review, we shall describe some of the initial measurements of low-energy ion properties by ion composition detectors aboard the ISEE-1 and SCATHA satellites. ISEE-1 was launched in October, 1977 into a highly elliptical orbit with ~ 22.5 R_E apogee and $30°$ inclination to the equator. The Plasma Composition Experiment (PCE) (cf. SHELLEY et al., 1978) aboard ISEE-1 consists of a sequential combination of Retarding Potential Analyzer (RPA), Electrostatic Analyzer (ESA) and Ion Mass Spectrometer (IMS), and is capable of measuring ions in the energy per charge range 0–17 keV/e with discrimination among ion species according to the mass to charge ratio M/Z. Work pertaining more to the energetic (keV) component of the ion distributions is described elsewhere in this collection by SHARP et al. (1982b). This review will be limited to ions with energy per charge $\lesssim 100$ eV/e. SCATHA was launched in January 1979 into a near geosynchronous orbit with an apogee of 7.8 R_E, perigee of 5.3 R_E and a line of apsides which precesses in local time at the rate of about 20 min. per day. This spacecraft carried the Light Ion Mass Spectrometer (LIMS), somewhat similar in concept to the ISEE-1 PCE but without the intervening electrostatic analyzer. The LIMS measures the ions H^+, He^+, and O^+ with energies less than 100 eV. Due to an unfortunate failure in the electronics package shortly after initial turn-on, only about 10 days of data have been available for analysis. The review by JOHNSON et al. (1982) elsewhere in this collection describes ion composition results from a separate instrument on SCATHA measuring ions in the energy per charge range 0.1–32 keV/e.

The discussion is organized into four categories: (1) origin, (2) energization, (3) transport, and (4) loss of magnetospheric low-energy plasma. Low-energy plasma composition measurements from GEOS-1 and -2 are described elsewhere in this collection by BALSIGER et al. (1982).

2.   Origin of Low-Energy Plasma

Anticipating an ionospheric origin for the major portion of the low-energy magnetospheric plasma, it is useful to consider observations of the ionospheric ion composition. Figure 1 displays latitudinal profiles of concentrations for various ion species observed at a nearly constant altitude of 1,400 km by the Magnetic Ion Mass Spectrometer (MIMS) on board the polar orbiting satellite ISIS-2 (HOFFMAN et al., 1974). The upper panel shows typical concentration profiles for H^+, He^+, O^+, O^{++}, and N^+ in the normal daytime ionosphere observed near equinox. It is seen that at low-to-mid-latitudes H^+ is normally the dominant ion present, followed by either He^+ toward the winter hemisphere or O^+ toward the summer hemisphere. N^+ concentrations are typically a factor of ten below O^+ and vary similarly to O^+ in this latitude range. At the higher latitudes, owing to the outflow of the light ions H^+ and He^+ associated with the polar wind (BANKS and HOLZER, 1968; HOFFMAN and DODSON, 1980), O^+ becomes the dominant ion near both poles in the topside ionosphere.

The unusually strong and extended period of magnetic storm activity during August 1972 led to the abnormal ion concentrations displayed in the bottom panel of Fig. 1. In this case N^+ was actually the dominant ion at times (slightly exceeding concentrations of O^+) and there were significant concentrations of the molecular ions N_2^+, O_2^+, NO^+ in the range 10^2–10^3 ions/cm^3. Such high concentrations of nitrogen related species were probably caused by thermospheric heating (due to Joule heating and particle precipitation effects) causing upwelling of N_2 and O_2 from the lower thermosphere to higher altitudes.

Typical observations from ISEE-1 are presented in Fig. 2, which shows peak ion number fluxes versus setting on the mass analyzer sweep for data corresponding to the plasmasphere (upper panel) and the plasma trough region outside the plasmasphere (lower panel). The plasmaspheric ion distributions are usually cold ($kT \lesssim 1$ eV) Maxwellian distributions and we have converted these plasmaspheric number fluxes to estimates of ion concentrations in the upper panel. It is evident that the plasmaspheric composition is rather similar to the composition of its mid-latitude ionospheric magnetic flux tube feet (cf. Fig. 1), with order of ion dominance H^+, He^+, O^+, He^{++}, and O^{++}. The doubly ionized ions He^{++} and O^{++} were first reported by YOUNG et al. (1977) and the O^{++} has been seperately discussed by HORWITZ (1981); it has been found by GEISS

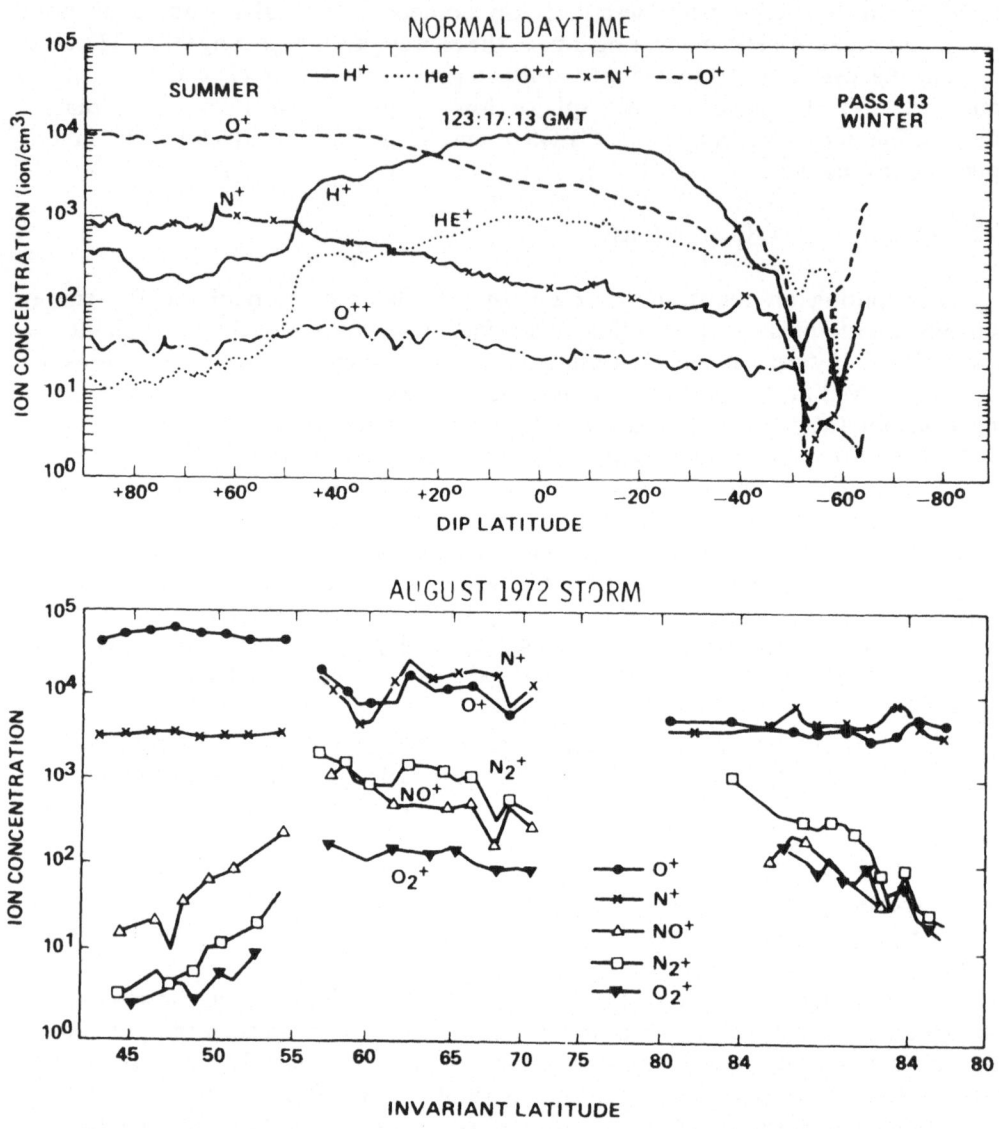

Fig. 1. Latitudinal profiles of ion concentrations at 1,400 km observed by ISIS-2 (adapted from HOFFMAN *et al.*, 1974). Upper panel displays profiles for normal equinoctial daytime ionosphere, while the lower panel shows abnormally high concentrations of N^+ and molecular ions during the August 1972 storm.

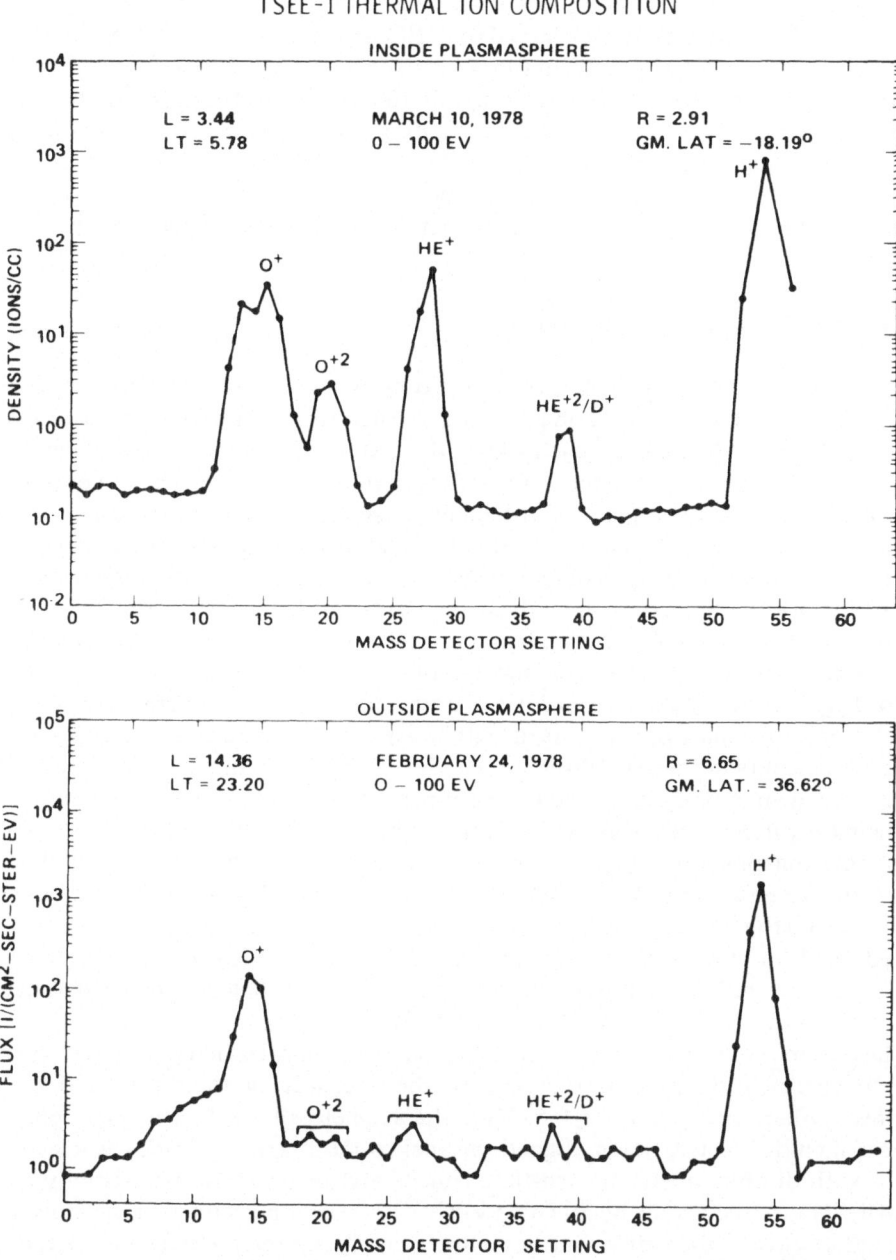

Fig. 2. Examples of ISEE-1 mass sweep observations for the plasmasphere (upper panel) and outside the plasmasphere (lower panel).

and YOUNG (1981) that upward thermal diffusion from the ionosphere is much stronger for O^{++} than O^+ and can explain plasmaspheric O^{++}/O^+ density ratios occasionally approaching unity as have been observed on GEOS-1 by GEISS et al. (1980) and ISEE-1 (HORWITZ, 1981). The lower panel of Fig. 1 illustrates that typically H^+ and O^+ are the principal ion constituents in the energy range 0–100 eV outside the plasmapause, with a lesser component of He^+. The increased percentage of O^+ in the low-energy ion fluxes beyond the plasmapause seen with ISEE-1 appears to be consistent with the increased O^+ at the feet of the corresponding high latitude ionospheric magnetic flux tubes, while the presence of H^+ and He^+ may be at least partially the polar wind perhaps observed at high altitudes.

3.   Energization of Ionospheric Ions

Evidence that important energization processes have acted on these low-energy ions is contained in their energy and pitch angle structure, illustrated in Fig. 3 for ISEE-1 and Fig. 4 for SCATHA. In each figure the insets are contour plots of observed counting rates as functions of energy and either spin or pitch angle. The upper panel of Fig. 3 displays, for data typical of the inner plasmasphere, contours which fall off rapidly with energy at $\lesssim 2$ eV for H^+ and He^+ and at significantly larger energies for O^+, while peaking near $0°$ spin phase angle which corresponds to the direction of the satellite ram motion. Such data are indicative of isotropic Maxwellian ion distributions with low temperatures ($kT \lesssim 1$ eV), being rammed due to the satellite orbital motion and to some extent the corotational motion of the plasmaspheric plasma. The lower panel of Fig. 3 shows data typical for the outer plasmasphere, where lower density, cold, isotropic plasma appears mixed with warm ion distributions (since there are significant contours in the 10–100 eV range) displaying pitch angle anisotropy. In this instance the warm ions are field-aligned, as indicated by the peaks near the minima and maxima in the pitch angle curve shown in the adjacent panel, although it is common in the outer plasmasphere to find pancake distributions (peaks at $90°$ pitch angle) of warm ions as well, mixed with the plasmaspheric cold components. Typical distributions seen by SCATHA (Fig. 4) are also the pancake distribution, occurring here near the expected location of the dusk plasmaspheric bulge region (it is not clear whether this distribution occurred inside the plasmasphere or not), while field-aligned distributions are shown for the regions beyond the plasmapause.

The presence of these warm ion distributions with energies of tens of electron volts points to energization processes acting on the ions during their transit from the ionosphere (where energies are $\lesssim 0.5$ eV) to the magnetosphere. Possibly energization processes include electric fields aligned parallel to the magnetic field, associated for instance with double layers, electrostatic shocks and/or anomalous resistivity along auroral field lines (see review by SHAWHAN et al., 1978) or ion cyclotron acceleration by electrostatic waves. Two distributions which are apparently the result of the ion cyclotron acceleration are the pancake and conical distributions, illustrated in Fig. 5 and discussed respectively by HORWITZ et al. (1981a) and HORWITZ et al. (1982a). Conics are pitch angle distributions whose peak fluxes lie between pitch angles of $90°$ and either $0°$ or $180°$, and are frequently observed at low altitudes (SHARP et al., 1977; KLUMPAR, 1979; GORNEY et al., 1981) and also at synchronous orbit (HORWITZ, 1980). They are

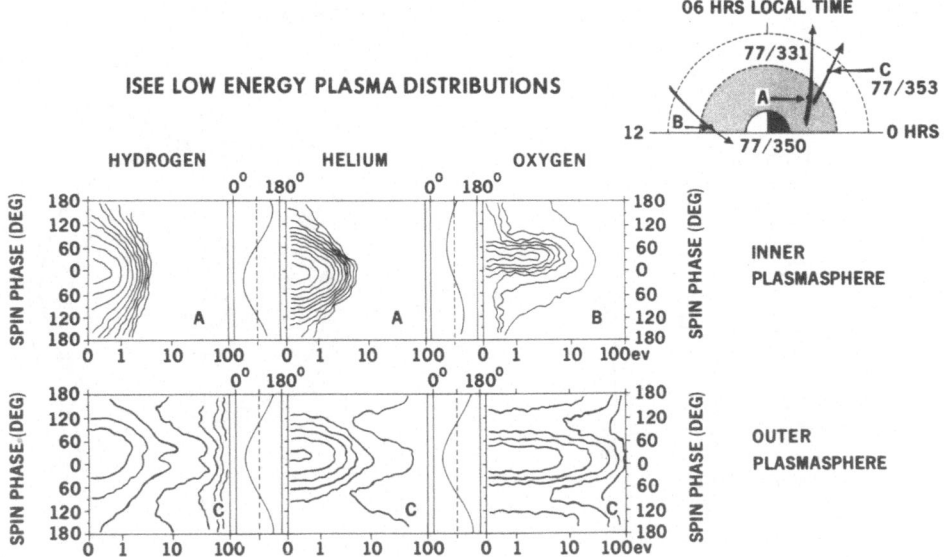

Fig. 3.   Examples of count rate contours for H^+, He^+, and O^+, plotted versus energy and spin angle, observed by ISEE-1 inside and outside the plasmasphere.

thought to result when upflowing ions are subjected to perpendicular acceleration by electrostatic waves at the local ion cyclotron frequency creating a pancake distribution of energized ions. As this group of ions continues to move up the magnetic field lines, conserving their magnetic moments in the process, their pitch angles fold toward the magnetic field line, creating the conical distribution at the observing location. Two basic types of conical distributions have been observed in low energy ions with ISEE-1: unidirectional conics (depicted in the upper left panel of Fig. 5) in which the ions are observed to come from one hemisphere only, with little or no flux from the opposite hemisphere, and the "symmetrical" conics (depicted in the lower left panel of Fig. 5) in which significant fluxes are observed to come from both hemispheres and the conical feature is seen in the fluxes from at least one hemisphere. HORWITZ *et al.* (1982a) have observed unidirectional conics in low-energy ions out to 18 R_E having relatively large pitch angles of 60°–80°, which implies that the accelerating electrostatic ion cyclotron turbulence must extend to approximately this altitude. Relative to the symmetric conics, unidirectional conics are found at larger geocentric distances and/or geomagnetic latitudes, that is, on field lines that are more likely to be open or at least highly distended.

The pancake distributions of warm ions have been observed both by ISEE-1 (right upper panel of Fig. 5) and SCATHA (right lower panel of Fig. 5). (Note that in contrast to ISEE-1, the SCATHA plots have upper energy cutoffs of 25 eV for He^+ and 6 eV for O^+.) As the ISEE-1 example of Fig. 5 shows, it is common to see such distributions in H^+ and/or He^+ while O^+ is field-aligned or below the instrument sensitivity level. They also tend to be observed in the vicinity of the plasmasphere and close to the magnetic equator. This confinement close to the magnetic equator is especially clear from the data of OLSEN (1981), who observed pancake distributions in both ions and electrons with an electrostatic analyzer aboard SCATHA.

LIGHT ION MASS SPECTROMETER (SCATHA)

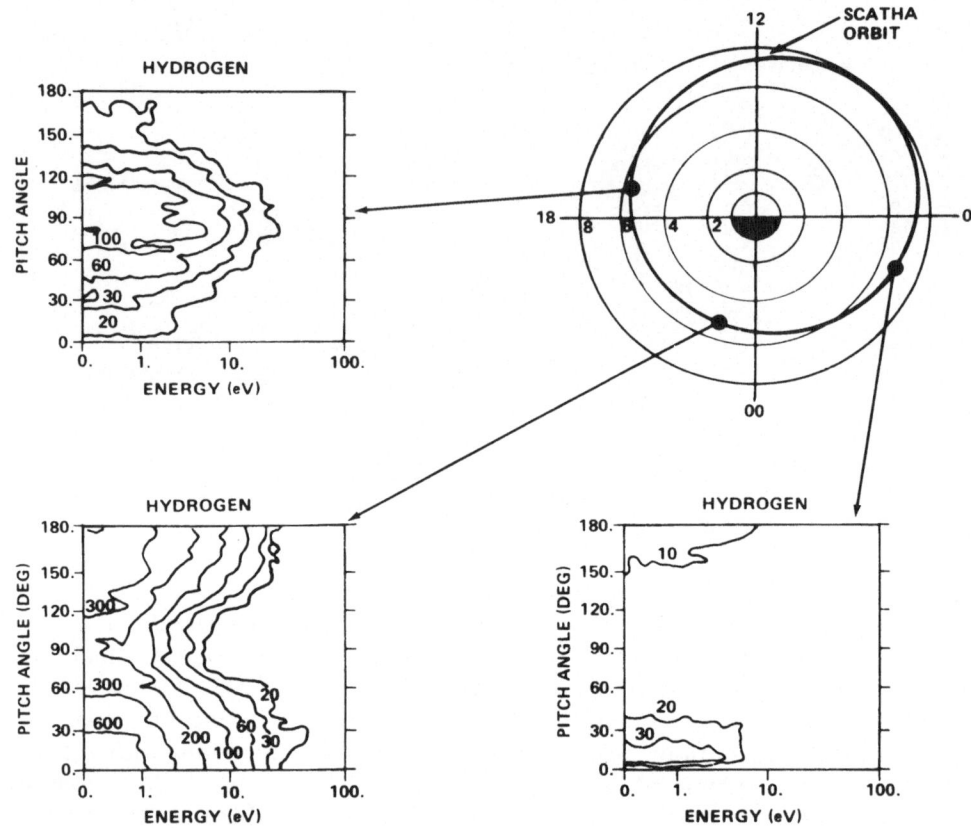

Fig. 4.   Similar to Fig. 3 but for observations by SCATHA.

In addition to energization processes apparently associated with cyclotron acceleration of cold ions to 10's of electron volts in energy and into the conical and pancake pitch angle distributions, there is some evidence for direct heating of the cold, isotropic component of the ion population in the outer plasmasphere region. An example of this is seen in Fig. 6, where the total plasma densities and cold ion temperatures are shown for an inbound ISEE-1 pass. The principal point here is the 'kink' observed in the temperatures about 0.7 R_E from the plasmapause, with no corresponding feature in the smooth density profile. It is possible that this elevation in the ion temperature was due to a local heat source (the hot ring current ion population may be such a source) acting on the outer part of the plasmasphere.

4.   Transport of Ionospheric Ions

One of the major unresolved topics in magnetospheric physics is the question of

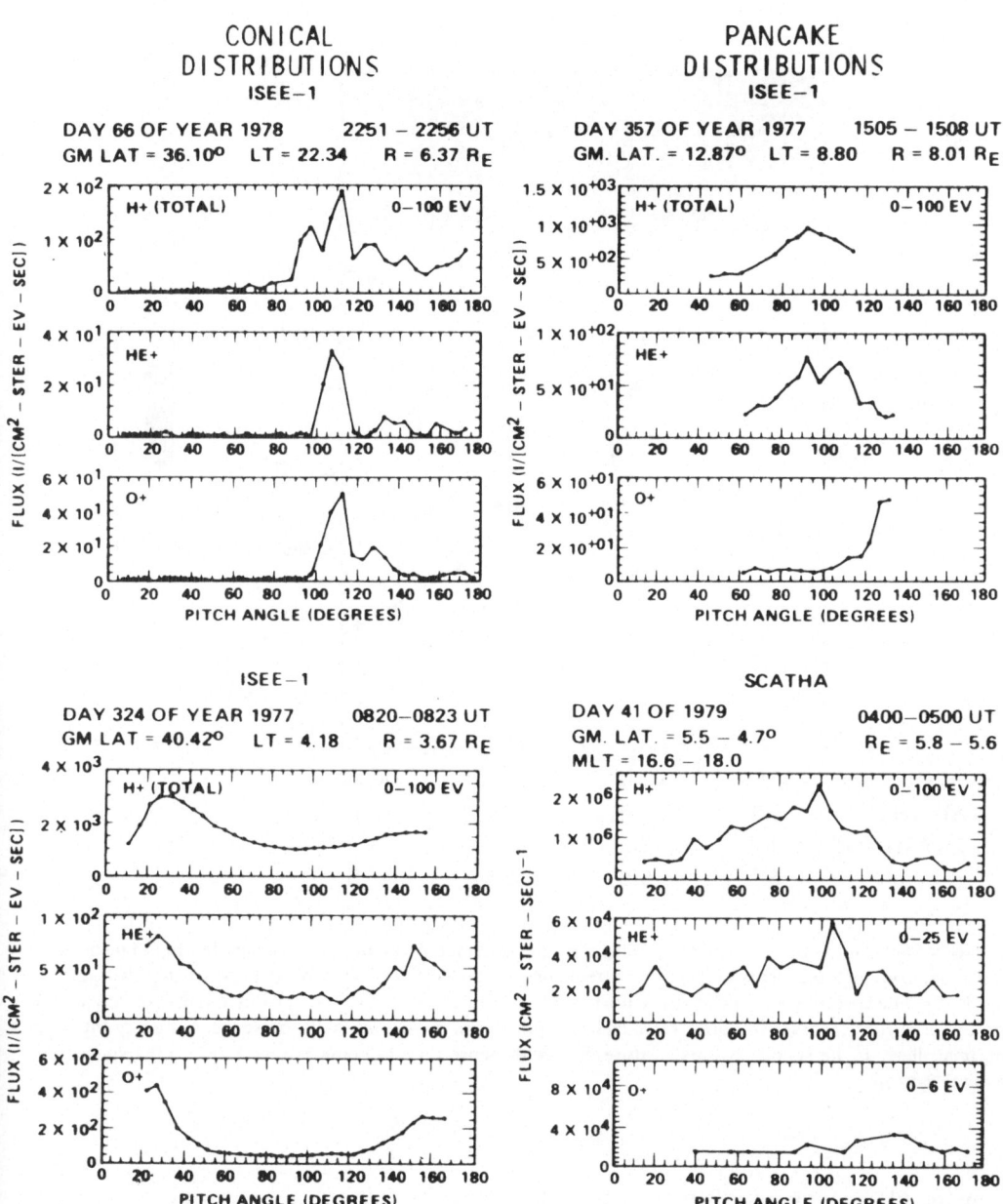

Fig. 5.   Conical and pancake-shaped pitch angle distributions of low-energy H$^+$, He$^+$, and O$^+$ 0–100 eV ion fluxes observed by ISEE-1 and SCATHA.

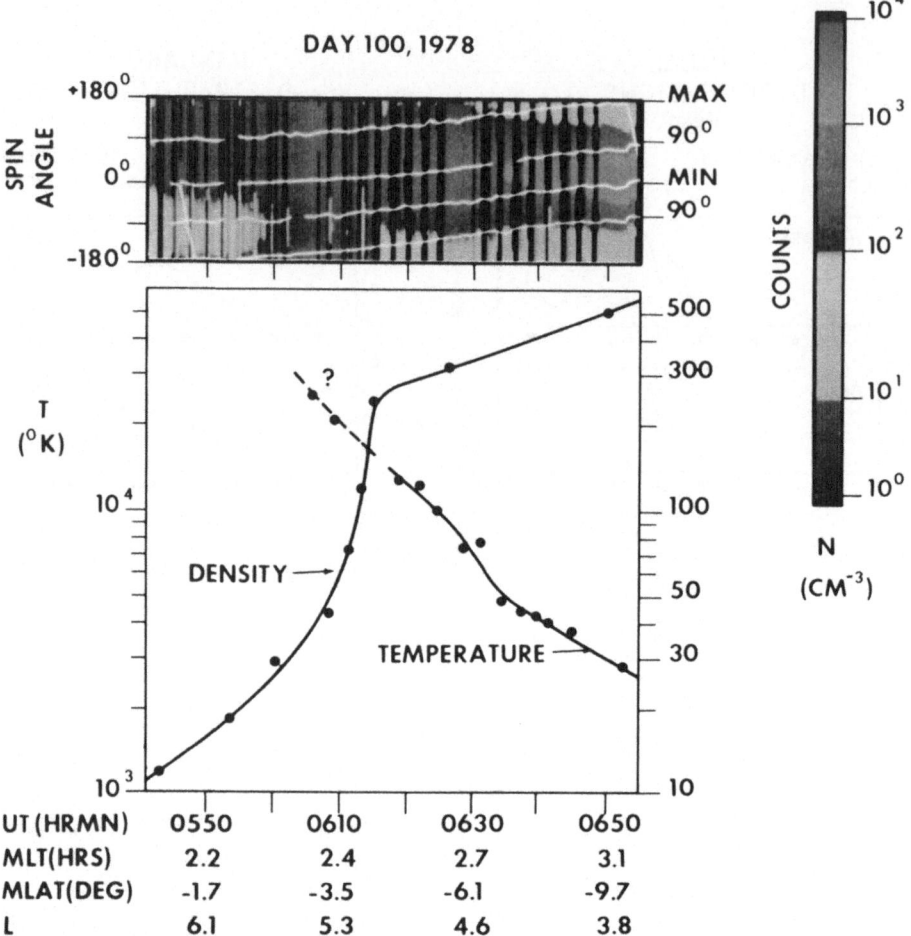

Fig. 6.   Thermal ion density and temperature profile observed on an inbound pass by ISEE-1. The density is derived from measurements of the plasma frequency by the ISEE-1 Plasma Wave Experiment (Gurnett *et al.*, 1978). The spectrogram in the upper portion of the figure depicts H$^+$ count rates versus spin angle and time; the abrupt increase in count rates near 0° spin angle, coinciding with the spacecraft entry into the plasmasphere, is the spectrogram signature of a cold, dense plasma being rammed due to the spacecraft orbital motion.

how the plasmasphere fills with plasma from the ionosphere. For several years there have existed theoretical models (e.g., Banks *et al.*, 1971) that have made specific predictions on the temporal and spatial evolution of the plasma distributions along a flux tube during the filling process, but it has not been possible to test most of the predictions with appropriate experimental data. Indeed, until recently, the principal direct evidence for plasmasphere filling came from sequential observations of radial plasma density profiles by ground-based whistler receivers and satellite-borne mass spectrometers from which it has been shown that the plasmasphere contracts during periods of high magnetic activity and expands during intervals of magnetic quiet (e.g., CARPENTER, 1967; CHAPPELL, 1972). An expansion of the plasmasphere occurs through the filling of those flux tubes that were depleted during a previous magnetically active period when their convection paths intersected the magnetopause, but which, during the quiet period, reside on convection paths which encircle the earth.

The context for measurements of spatial and temporal evolution of the plasma distributions can be seen by considering perhaps the most prominent filling model, proposed by BANKS et al. (1971), in which the filling process is initiated as an interaction near the magnetic equator between supersonic field-aligned plasma flows emerging from the two conjugate ionospheres at the feet of a flux tube depleted of plasma. The proposed filling model involves a shocklike interaction between the two flows, creating a high-density slab at the equator. The slab's edges are shock fronts that propagate down the magnetic field lines toward the topside ionosphere, and when these shock fronts arrive at the topside ionosphere, the plasma flow character changes from supersonic to subsonic and proceeds until diffusive equilibrium is attained. BANKS et al. (1971) estimated that for the $L = 5$ flux tube, the transition from supersonic to subsonic flow would take place about 22 hours after the filling process begins, and that diffusive equilibrium is attained roughly 10 days after the initiation of filling.

In order to make progress in the understanding of the plasmasphere filling process, HORWITZ et al. (1981b) examined thermal ion data measured on successive ISEE-1 passes during periods in which the magnetic activity decreased. Figure 7 shows density profiles versus L for the outbound ISEE-1 passes on December 4 (day 338) and 7 (341), 1977. These densities were derived from measurements of the electron plasma frequency by the plasma wave experiment (cf. GURNETT et al., 1978; CARPENTER et al., 1981). As the Kp history in the upper panel of Fig. 7 shows, the pass on December 4 was taken when Kp had the moderately high value of $\sim 3+$, while at the time of the December 7 pass, Kp had been low ($\lesssim 1$) for about the previous 24 hours.

To illustrate the characteristic features of the thermal ion distributions within the "filled" portion of the plasmasphere, Figs. 8a and 8b show, respectively, the 0–100 eV fluxes of H^+, He^+, and O^+ and a contoured He^+ distribution function, taken just inside the plasmapause at $L = 3.99$ and 4.07 on December 4, 1977. The umbrella-like spin-modulated flux curves of Fig. 8a are signatures of cold, essentially isotropic distributions moving relative to the spacecraft, owing primarily to the spacecraft orbital motion, with an offset from $0°$ in the flux peak due to the corotational plasma motion. The contours of the He^+ distribution function in Figure 8b show more clearly the nearly circular contours of an isotropic, cold ($kT \sim 0.5$ eV) ion distribution function being offset in the spacecraft reference frame by the orbital motion and corotation. As noted earlier, the $H^+ : He^+ : O^+$ order of dominance is characteristic of the plasmasphere.

The evolution of the ion distribution beyond $L \sim 4.2$, the approximate location of the December 4 plasmapause, is indicated in Figs. 9a and 9b, which contrast the distributions at roughly the same L shells on December 4 and 7. The H^+, He^+, and O^+ fluxes at $L = 4.6$ on December 4 and $L = 5.05$ on December 7 are plotted versus pitch

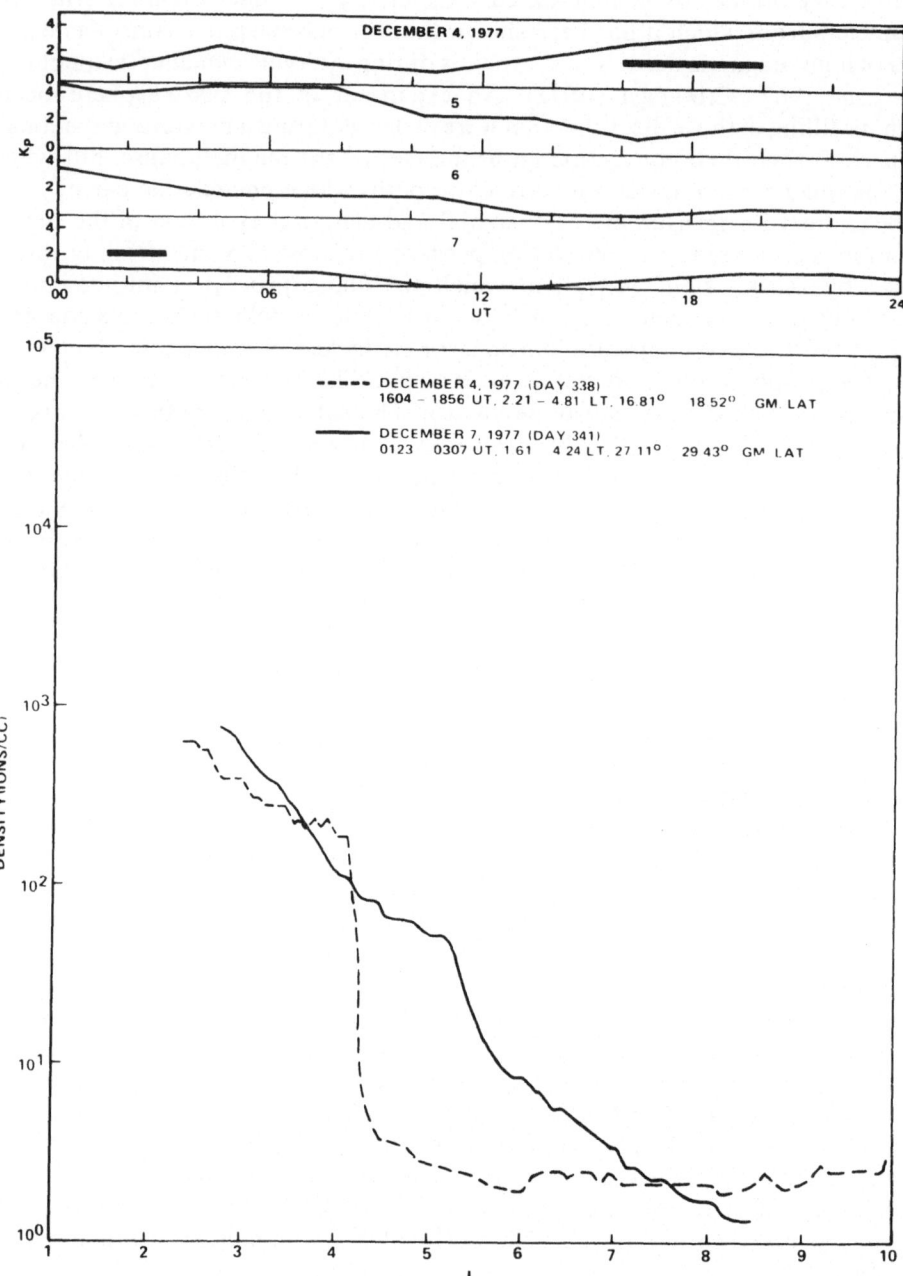

Fig. 7.   Plasma density profiles for two successive ISEE-1 outbound passes in the post-midnight sector, observed during a period when the magnetic activity level decreased, as seen in the *Kp* history shown in the upper panel.

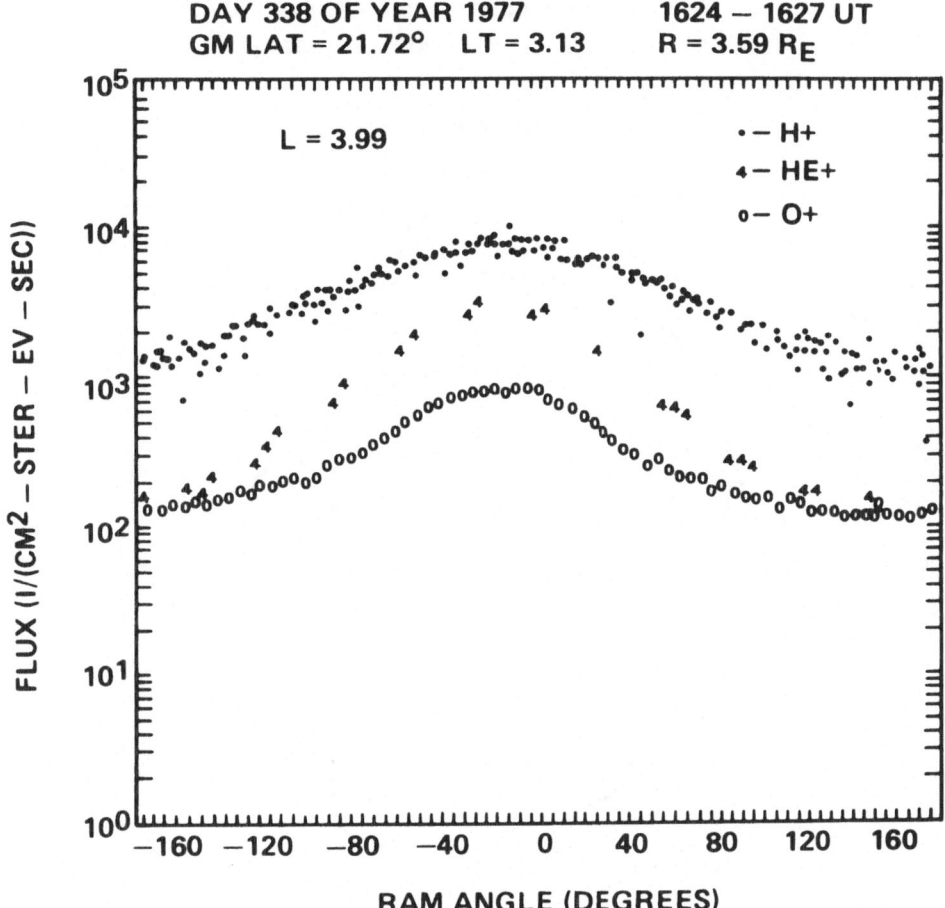

Fig. 8a.   Plot of 0–100 eV H$^+$, He$^+$, and O$^+$ fluxes versus spin angle, for ISEE-1 data observed just inside the plasmapause on December 4, 1977.

angle in 9a, since no ram angle anisotropies were evident (a reversal of the situation in Fig. 8a). The fluxes for both days were field-aligned but with an important difference: the December 4 observation showed a compositional order of dominance H$^+$:O$^+$:He$^+$, while the December 7 pass showed the H$^+$:He$^+$:O$^+$ ordering characteristic of the plasmasphere.

A further important difference was seen in the energy characteristics for the ions in these two cases, illustrated in Fig. 9b. Here the count rates for the incoming H$^+$ ions with pitch angles near 30° are plotted versus potential on the retarding potential analyzer. For the December 4 case, little fall-off of count rate was seen until the RPA potential reached 20 V, the count rate decreasing slowly with potential above this level. This situation is contrasted with the data of December 7 in which a sharp fall-off in count rate was observed at ~1–2 V. Thus the observed ion flux observed in the

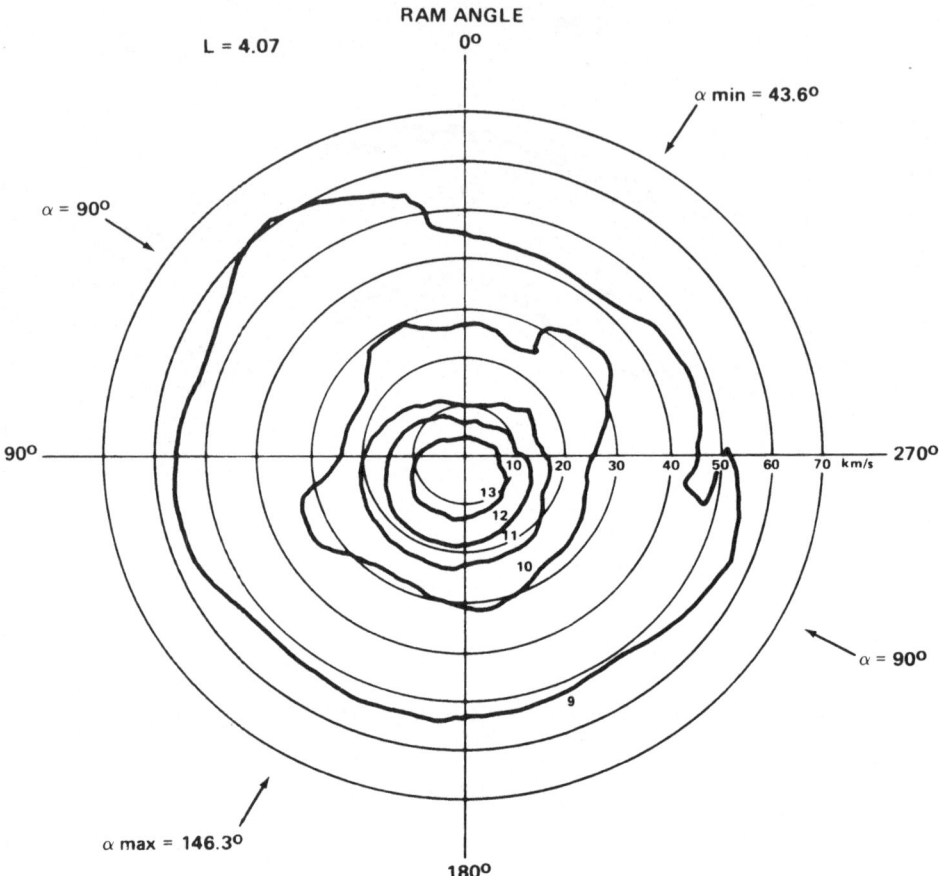

Fig. 8b.   A contoured He⁺ distribution function observed within the plasmasphere on December 4, 1977.

December 4 case was carried by ions with characteristic energies of the order 20 eV, whereas the ion flux for the December 7 case was mainly carried by ions of energies $\sim 1-2$ eV. This $1-2$ eV flow of H^+ and He^+ ions was evident on the December 7 pass until about $L = 7.6$, at which point an $H^+:O^+:He^+$ warm ($\sim 10-20$ eV) ion flow was observed to be dominant, similar to the December 4 observations presented in Fig. 9a.

A partial summary of the results from HORWITZ et al. (1981b) is shown in Fig. 10, which depicts the apparent characteristics of thermal plasma in the vicinity of the plasmasphere during magnetically active and quiet periods. It appears from the work of HORWITZ et al. (1982b) that much of the time during periods of higher magnetic

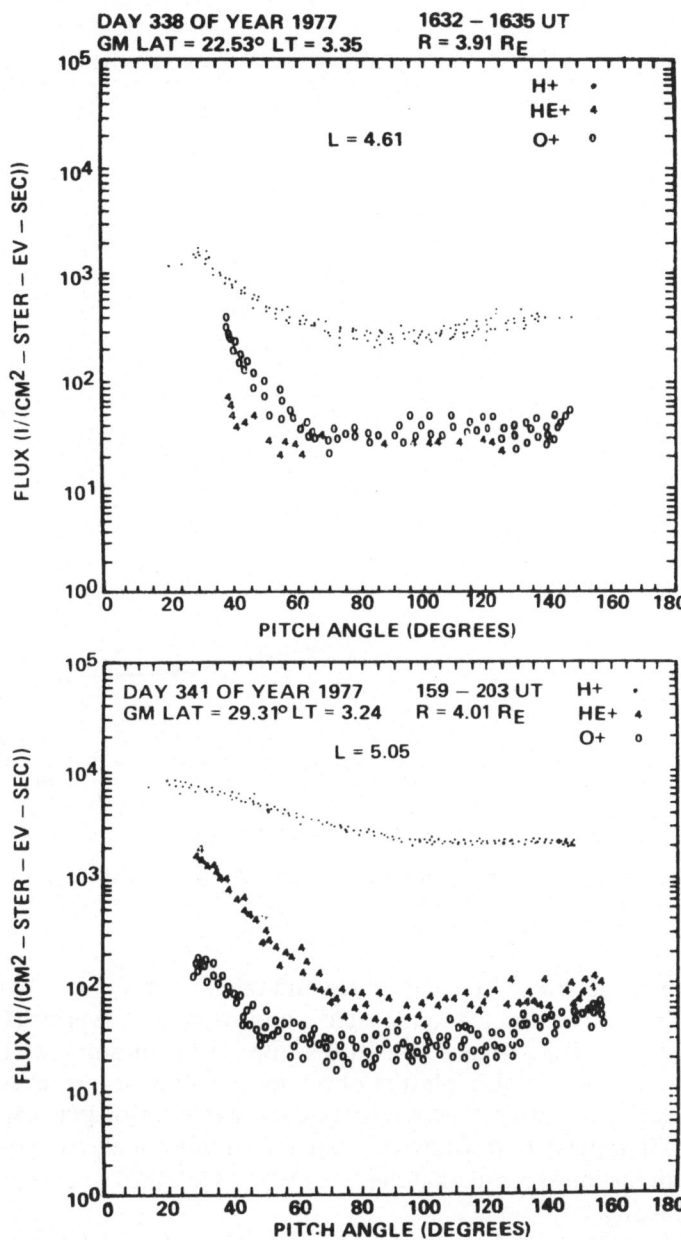

Fig. 9a.   Pitch angle distributions of $0-100$ eV H^+, He^+, and O^+ ion fluxes observed at roughly the same spatial location on the successive ISEE-1 outbound passes on December 4 and 7, 1977.

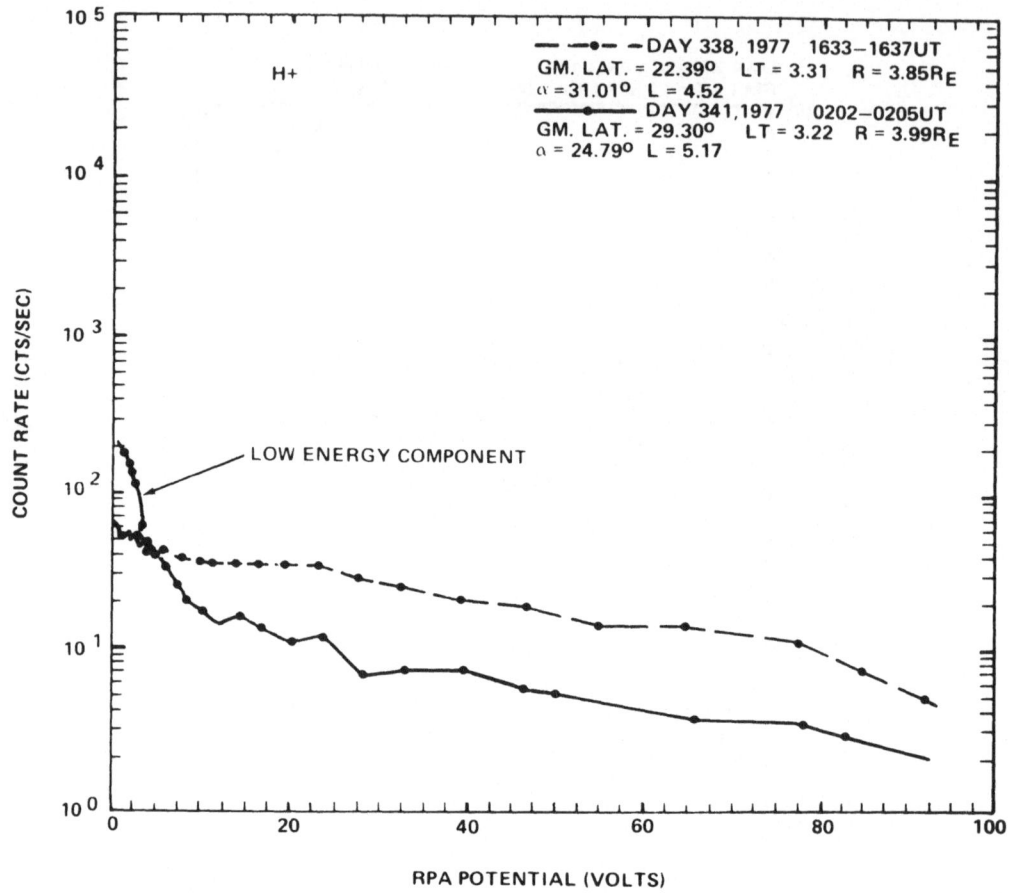

Fig. 9b. H^+ count rates versus potential on the ISEE-1 retarding potential analyzer (RPA) near 30° pitch angle for the cases shown on December 4 and 7.

activity, the edge of the "filled" plasmasphere and the inner boundary of the electron plasma sheet are approximately colocated, and that the warm (10's of eV) $H^+:O^+:He^+$ ordered field-aligned ion fluxes appear on plasma sheet field lines. During periods of quieting this plasma sheet inner boundary reaches to greater geocentric distances, leaving a gap between it and the active period plasmapause (lower panel of Fig. 10). It appears that within this gap, linked most likely to the mid-latitude light ion trough in the ionosphere, that the 1–2 eV H^+/He^+ field-aligned flows appear to fill the plasmasphere.

In addition to field-aligned plasma transport, of course, convection plays a major role in transporting low-energy plasma through the magnetosphere. In Fig. 11, convection flow vectors, measured by the LIMS instrument on SCATHA, are shown for the afternoon-dusk local time sector during a period of moderately high magnetic activity. Sunward convection flow velocities of up to 50 km/s were observed in

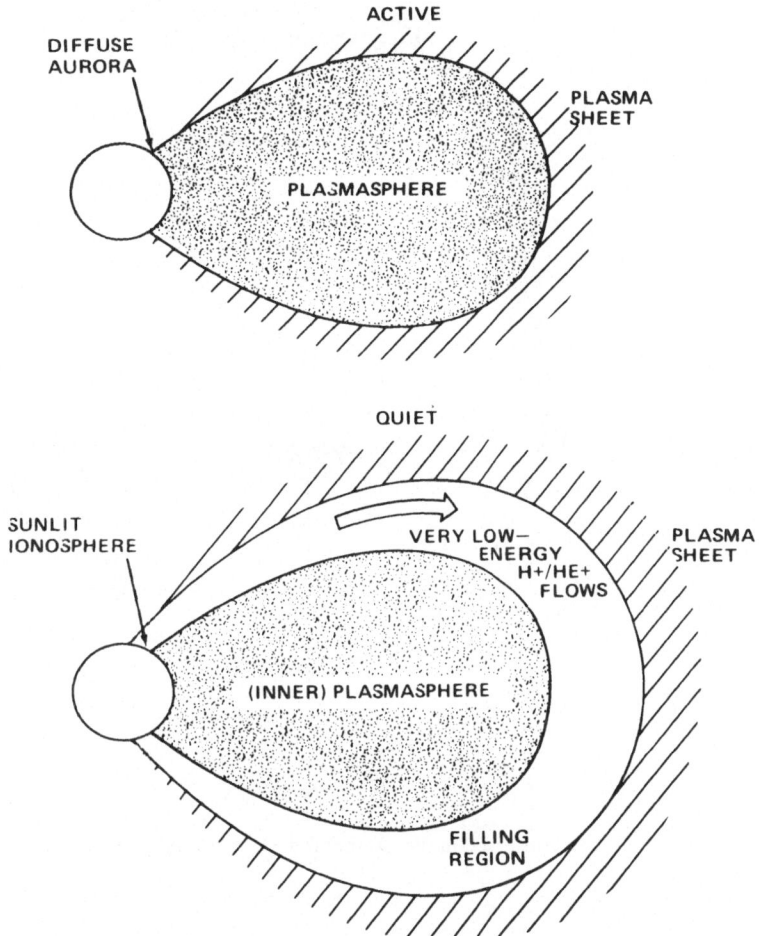

Fig. 10.   Schematic presentation of the relative locations of the plasmapause and the inner boundary of the plasma sheet-diffuse auroral flux tube during active periods and when activity has quieted (from HORWITZ *et al.*, 1981b).

association with a $Kp \sim 4^-$. Such strong sunward flows are believed to detach portions of the plasmaspheric bulge region and transport them toward the dayside magnetopause (cf. CHAPPELL, 1974).

Associated with the topic of transport of thermal ions along the field lines are some unusual compositional features of the plasmasphere seen by both ISEE-1 and SCATHA in terms of surprisingly high concentrations of He^+. In Fig. 12, data from SCATHA show count rate ratios of He^+ to H^+ for two passes during magnetically active days through the noon-to-midnight local time sector, including the dusk bulge region. If these count ratios approximate the density ratios (the ratios may be too high if the spacecraft is positively charged) then there would appear to be some instances of

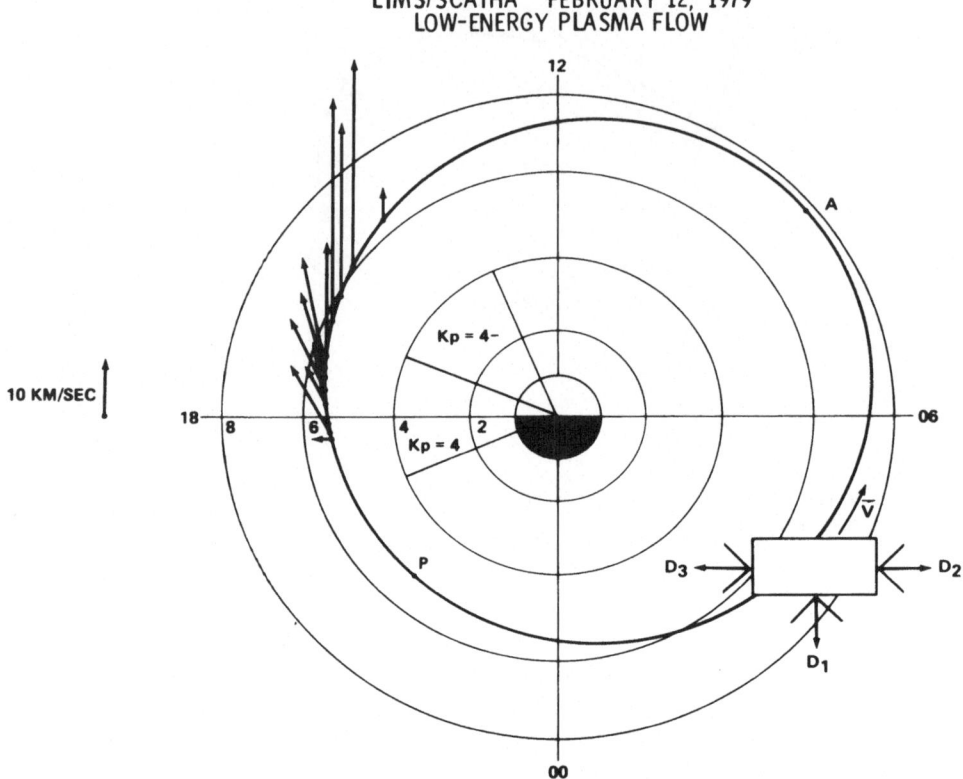

Fig. 11. Convection flow vectors observed in the afternoon-dusk sector by SCATHA during moderately high geomagnetic activity.

virtually pure He$^+$ regions (as in Fig. 12 lower panel between 0210 and 0230 UT where He$^+$/H$^+$ $\sim 10^2$). Waite and Horwitz (1981) observed, with ISEE-1, He$^+$ fractional concentrations of 10–50% and have examined the question of whether thermal diffusion-driven transport of He$^+$ from the topside ionosphere may act to enhance He$^+$ as apparently is the case with O^{++}. Preliminary findings indicate that thermal diffusion may in fact enhance the He$^+$ fractional concentrations, but it appears very difficult to achieve levels of 50%. The work of Waite and Horwitz (1981) modeled the steady state equilibrium distribution; it may be necessary to consider dynamical, non-equilibrium effects as possibilities for the high He$^+$ concentrations observed.

5. Loss of Low-Energy Plasma from the Magnetosphere

There are at least two general means and locations through which low-energy plasma might be lost from the magnetosphere: (1) through field-line reconnection and leakage across the dayside magnetopause and (2) streaming of the plasma out through

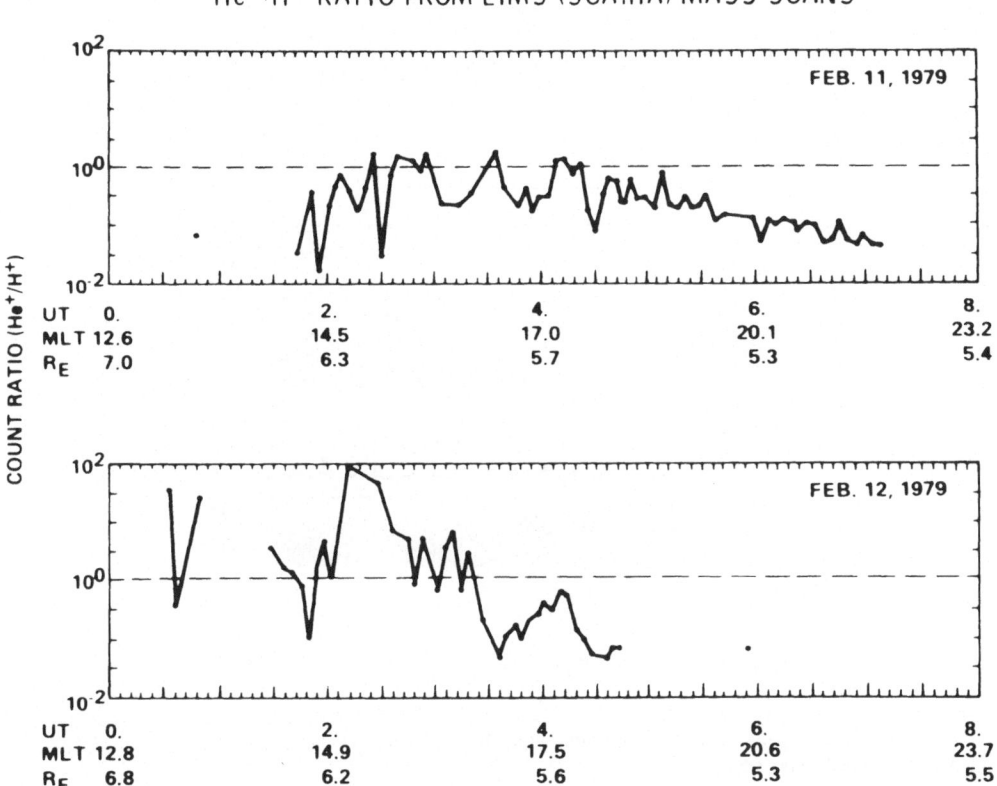

Fig. 12.   SCATHA-observed count rate ratios for thermal He^+ and H^+ in the noon-to-midnight local time sector.

the magnetotail lobes on field lines connected to the interplanetary magnetic field. In Fig. 13 are shown count rates for 0–100 eV fluxes of He^{++}, O^+, He^+, and H^+ observed during an inbound ISEE-1 crossing of the magnetopause near 0900 LT. It is evident that there is a sizable ionospheric origin (evidenced by the He^+ and O^+), low-energy ion component located immediately inside the magnetopause, while the He^{++} signifies the presence of the solar wind origin plasma of the magnetosheath. The presence of this low-energy plasma component so close to the magnetopause makes it seem likely that during such impulsive reconnection events as observed by RUSSELL and ELPHIC (1979) there should be leakage of this component out into the magnetosheath. Although this has not actually been observed in 0–100 eV ions as yet, PETERSON et al. (1981) have found evidence for energetic ($\sim$keV) He^+ and O^+ located in the magnetosheath. One strong possibility is that such cold ions are energized during a reconnection process in their transit from the magnetosphere to the magnetosheath. It may be noted in this context that FREEMAN et al. (1977) have proposed a mechanism for creating energetic ions of plasmaspheric origin in the magnetosphere, in which detached plasma regions

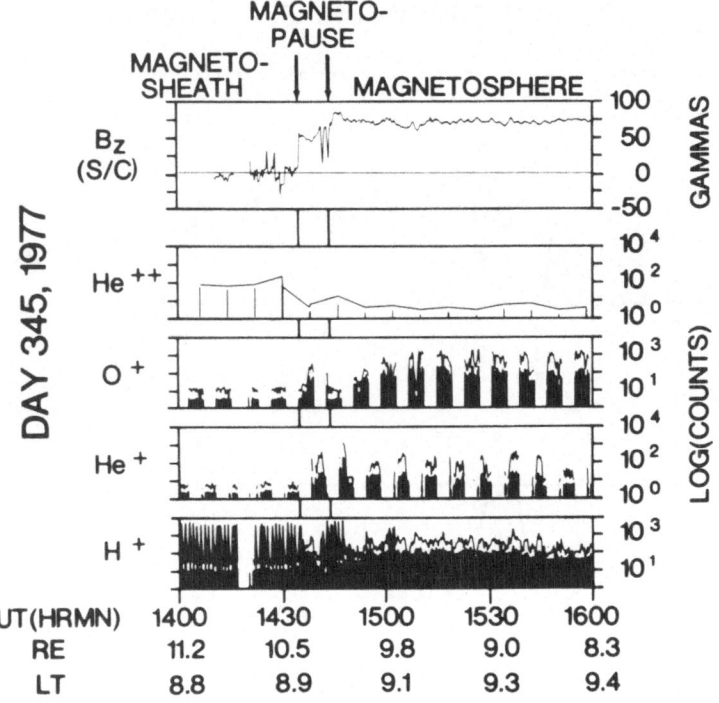

Fig. 13. Count rates for 0–100 eV He^{++}, O$^+$, He$^+$, and H$^+$ fluxes observed by ISEE-1 near the dayside magnetopause. The lines indicate peak count rates while the upper edges of the dark regions correspond to the count rates averaged over spin.

from the plasmasphere bulge region drift through the dayside reconnection region where they are accelerated and then convected over the magnetosphere through the magnetosheath along the flanks of the magnetotail. These ions are then convected back into the tail through the tail neutral line reconnection region (where they may again be energized) to circulate through the magnetosphere. In this model, leakage of ions during dayside merging might not result in permananent loss of these ions from the magnetosphere, but rather be an intermediate step in a circulation and energization pattern.

Figure 14 includes evidence for possible loss of magnetospheric low-energy ions via streaming out the magnetotail lobes. The four angle-time color spectrogram panels, from top to bottom, show count rates for 0–100 eV H$^+$ (or more accurately, total ion fluxes, with H$^+$ usually the dominant ion) plotted versus pitch angle and spin angle, and then He$^+$ and O$^+$ plotted versus spin angle. On each of the bottom three panels lines are drawn at the minimum and maximum pitch angles as well as 90° pitch angle crossings to indicate the general pitch angle character of the ion distributions. On the He$^+$ and O$^+$ panels the large white spaces are intervals where these ions were not sampled; indeed, the He$^+$/O$^+$ sampling was mutually exclusive.

At the top of Fig. 14 are indicated estimates from independent measurements as to

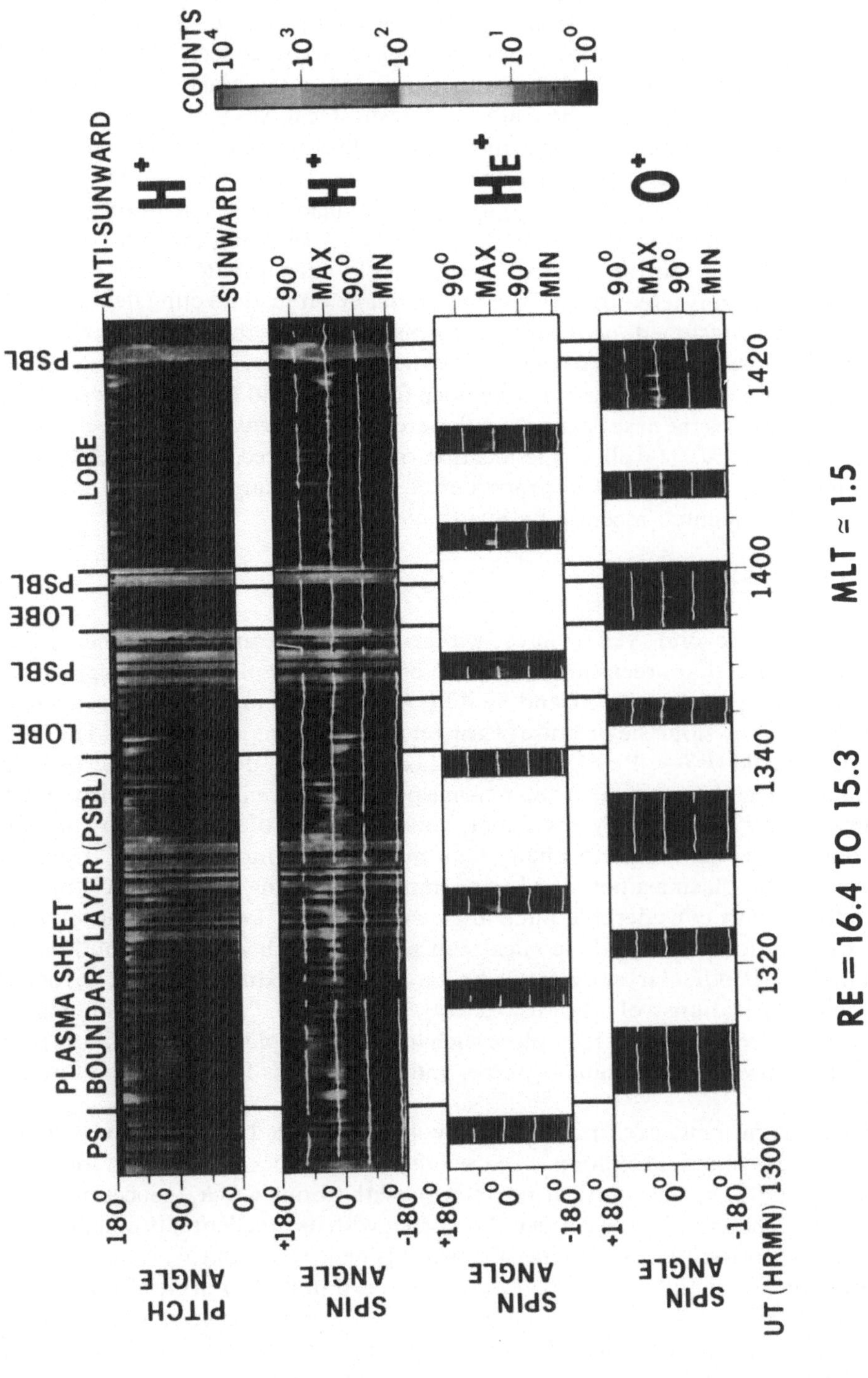

Fig. 14.   Spin angle-time color spectrograms for H$^+$, He$^+$, and O$^+$ 0–10 eV count rates observed by ISEE-1 deep in the magnetotail.

the spacecraft's location relative to the magnetotail structure, and it is evident that the spacecraft was variously located in the plasma sheet, the plasma sheet boundary layer, and the lobes. Drawing attention to the lobe regions, it can be seen that, although there is a decrease in the intensity of the fluxes compared to the plasma sheet and plasma sheet boundary layer regions, there are still measurable fluxes in the lobe regions. With MAX pitch angle corresponding to anti-sunward flow in the bottom three panels, it is evident that there are anti-sunward flows of low-energy H^+, He^+, and O^+ streaming out along the presumably open magnetotail lobe field lines. The observations were made moderately deep in the magnetotail at about $\sim 15-16 \ R_E$, within 1 1/2 hours of local midnight. There are at least two possibilities for the ultimate destiny of these ions: (1) they may be convected toward the tail midplane on reconnecting field lines before they pass the presumed neutral line downstream and thus become trapped on temporarily closed field lines in the plasma sheet, or (2) they may remain on, and exit the magnetosphere out along, the lobe open field lines, and thereby be lost. Whether scenario (1) or (2) is the case depends on the streaming velocity of the ions relative to the convection velocity, as well as the location of the occupied field line relative to last closed field line in the tail. The properties of such streaming ions in the magnetotail have been documented recently by SHARP et al. (1981).

6. Conclusions

The past several years have withnessed significant developments in the understanding and appreciation of the role of low-energy plasmas in magnetospheric dynamics. The present ISEE-1 and SCATHA results are major components of these developments and hold the promise of greater contributions in the future. The previous concept of a cold ($k T < 1$ eV), isotropic, Maxwellian, H^+-dominated thermal plasma population largely confined to the plasmasphere has given way to a view including significant, and occasionally dominant, concentrations of He^+ and O^+ in various locations within the magnetosphere, with measurable amounts of O^{++} and He^{++} observed in the plasmasphere, and important populations of warm (10's of eV) ion distributions with considerable pitch angle anisotropies. Two of the important pitch angle distributions observed—conical and pancake pitch angle distributions—may result from perpendicular energization by ion cyclotron frequency waves. Beyond these new characterizations of the low-energy plasma, the ISEE-1 and SCATHA observations are being used to explore such topics as the plasmasphere filling process, convection flows in the magnetosphere, and mechanisms for loss of the low-energy plasma.

In addition, new, exciting data is now being returned by the recently launched high-altitude Dynamics Explorer-1 spacecraft, which carries a low-energy ion detector. The spacecraft is in an elliptical polar orbit with about 4.6 R_E geocentric distance apogee and ~ 800 m altitude perigee. This orbit, with its excellent latitudinal coverage, especially when the precessing line of apsides is near the equator, is an almost ideal complement to the highly elliptical, near equatorial (initially) orbit of ISEE-1, which emphasizes radial coverage, and the near-geosynchronous orbit of SCATHA, which emphasizes local time coverage. The low-energy instrument, called the Retarding Mass Spectrometer (RIMS), is a progressively more sophisticated version of the detectors on

SCATHA and ISEE-1. With this instrument and the complement of experiments on the high and low-altitude DE spacecraft, it will be possible to make significant progress in a number of important topical areas such as the relationship of the plasmapause to the ionospheric light ion trough, the plasmaspheric filling process, characteristics of polar cusp plasma, and low-energy plasma in the auroral acceleration region, along with many other areas of investigation.

This research was supported at the University of Alabama in Huntsville by NASA/MSFC contracts NAS8-30563 and NAS8-33982 and at Marshall Space Flight Center by NASA RTOP 385-36-01. Dr. R. D. Sharp of the Lockheed Palo Research Laboratory is the principal investigator for the Plasma Composition Experiment on ISEE-1, and we thank him and the other co-investigators, H. Balsiger, E. Eberhardt, J. Geiss, G. Haerendel, R. G. Johnson, H. Rosenbauer, D. T. Young and E. G. Shelley, for their many valuable contributions to this experiment.

REFERENCES

ASHOUR-ABDULLAH and R. M. THORNE, Toward a unified view of diffuse auroral precipitation, *J. Geophys. Res.*, **33**, 4755, 1978.

BALSIGER, H., P. EBERHARDT, J. GEISS, and D. T. YOUNG, Magnetic storm injection of 0.9- to 16-keV/e solar and terrestrial ions into the high-altitude magnetosphere, *J. Geophys. Res.*, **85**, 1645, 1980.

BALSIGER, H., J. GEISS, and D. T. YOUNG, The composition of thermal and hot ions observed by the GEOS-1 and -2 spacecraft, this volume, pp. 195–230, 1982.

BANKS, P. M. and T. E. HOLZER, High-latitude plasma transport: the polar wind, *J. Geophys. Res.*, **74**, 6317, 1969.

BANKS, P. M., A. F. NAGY, and W. I. AXFORD, Dynamical behavior of thermal protons in the mid-latitude ionosphere and magnetosphere, *Planet. Space Sci.*, **19**, 1053, 1971.

CARPENTER, D. L., Relations between the dawn minimum in the equatorial radius of the plasmapause and *Dst*, *Kp*, and local *k* at Byrd Station, *J. Geophys. Res.*, **72**, 2969, 1967.

CARPENTER, D. L., Whistler evidence of dynamic behavior of the duskside bulge in the plasmasphere, *J. Geophys. Res.*, **75**, 3837, 1970.

CARPENTER, D. L., R. R. ANDERSON, T. F. BELL, and J. MILLER, Plasma density observations by whistlers and by a satellite-borne radio technique, *Geophys. Res. Lett.*, **8**, 1107, 1981.

CHAPPELL, C. R., Recent satellite measurements of the morphology and dynamics of the plasmasphere, *Rev. Geophys. Space Phys.*, **10**, 971, 1972.

CHAPPELL, C. R., Detached plasma regions in the magnetosphere, *J. Geophys. Res.*, **79**, 1861, 1974.

CORNWALL, J. M., Density sensitive instabilities in the magnetosphere, *J. Atmos. Terr. Phys.*, **38**, 1111, 1976.

FRANK, L. A., K. L. ACKERSON, R. J. DECOSTER, and B. G. BUREK, Three dimensional plasma measurements within the earth's magnetosphere, *Spac. Sci. Rev.*, **22**, 739, 1978.

FREEMAN, J. W., H. K. HILLS, T. W. HILL, and P. H. REIFF, Heavy ion circulation in the earth's magnetosphere, *Geophys. Res. Lett.*, **4**, 195, 1977.

GEISS, J. and D. T. YOUNG, Production and transport of O^{++} in the ionosphere and plasmasphere, *J. Geophys. Res.*, **86**, 4739, 1981.

GEISS, J., H. BALSIGER, P. EBERHARDT, H.-P. WALKER, L. WEBER, and D. T. YOUNG, Dynamics of magnetospheric ion composition as observed by the GEOS mass spectrometer, *Space Sci. Rev.*, **22**, 537, 1978.

GORNEY, D., A. CLARKE, D. R. CROLEY, J. F. FENNELL, J. M. LUHMANN, and P. F. MIZERA, The distribution of ion beams and conics below 8000 km, *J. Geophys. Res.*, **86**, 83, 1981.

GURNETT, D. A., F. L. SCARF, R. W. FREDERICKS, and E. J. SMITH, The ISEE-A and ISEE-B plasma wave investigation, *Geosci. Electron.*, GE-**16**, 225, 1978.

HOFFMAN, J. L. and W. H. DODSON, Light ion concentrations and fluxes in the polar regions during magnetically quiet times, *J. Geophys. Res.*, **85**, 626, 1980.

HOFFMAN, J. L., W. H. DODSON, C. R. LIPPINCOTT, and H. D. HAMMACK, Initial ion composition results from the ISIS satellite, *J. Geophys. Res.*, **79**, 4246, 1974.

Horwitz, J. L., Conical distributions of low-energy ion fluxes at synchronous robit, *J. Geophys. Res.*, **85**, 2087, 1980.

Horwitz, J. L., ISEE-1 observations of O^{++} in the magnetosphere, *J. Geophys. Res.*, **86**, 9225, 1981.

Horwitz, J. L., C. R. Baugher, C. R. Chappell, E. G. Shelley, and D. T. Young, Pancake distributions of warm ions observed by ISEE-1 *J. Geophys. Res.*, **86**, 3311, 1981a.

Horwitz, J. L., C. R. Baugher, C. R. Chappell, E. G. Shelley, D. T. Young, and R. R. Anderson, ISEE-1 observations of thermal plasma in the vicinity of the plasmasphere during periods of quieting magnetic activity, *J. Geophys. Res.*, **86**, 9989, 1981b.

Horwitz, J. L., W. K. Cobb, C. R. Baugher, C. R. Chappell, L. A. Frank, T. E. Eastman, R. R. Anderson, E. G. Shelley, and D. T. Young, On the relationship of the plasmapause to the equatorward boundary of the auroral oval and to the inner edge of the electron plasma sheet, *J. Geophys. Res.*, **87**, 9059, 1982b.

Horwitz, J. L., C. R. Baugher, C. R. Chappell, E. G. Shelley, and D. T. Young, Conical pitch angle distributions of very low-energy ion fluxes observed by ISEE-1, *J. Geophys. Res.*, **87**, 2311, 1982.

Johnson, R. G., Energetic ion composition in the earth's magnetosphere, *Rev. Geophys. Space Phys.*, **7**, 696, 1979.

Johnson, R. G., R. J. Strangeway, R. D. Sharp, and E. G. Shelley, Hot plasma composition results from the SCATHA spacecraft, this volume, pp. 287–306, 1983.

Klumpar, D. M., Transversely accelerated ions: An ionospheric source of hot magnetospheric ions, *J. Geophys. Res.*, **84**, 4229, 1979.

Lennartsson, W., E. G. Shelley, R. D. Sharp, and R. G. Johnson, Some initial ISEE-1 results on the ring current composition and dynamics during the magnetic storm of December 11, 1977, *Geophys. Res. Lett.*, **6**, 483, 1979.

Lundin, R., B. Hultqvist, N. Pissarenho, and A. Zakarov, The plasma mantle: Composition and other characteristics observed by means of the Prognoz-7 satellite, *Kiruna Geophysical Institute*, preprint 81:1, 1981.

Olsen, R. C., Equatorially trapped plasma populations, *J. Geophys. Res.*, **13**, 11235, 1981.

Peterson, W. K., E. G. Shelley, G. Haerendel, and G. Paschmann, Energetic ion composition in the subsolar magnetopause and boundary layer, **87**, 2147, 1982.

Peterson, W. K., R. D. Sharp, E. G. Shelley, R. G. Johnson, and H. Balsiger, Energetic ion composition of the plasma sheet, *J. Geophys. Res.*, **86**, 761, 1981.

Russell, C. T. and R. G. Elphic, ISEE observations of flux transfer events at the dayside magnetopause, *Geophys. Res. Lett.*, **6**, 33, 1979.

Sharp, R. D., Positive ion acceleration in the R_E altitude range, in *Proceedings of the Chapman Conference on the Formation of Auroral Arcs*, 1980.

Sharp, R. D., R. G. Johnson, and E. G. Shelley, Observations of an ionospheric acceleration mechanism producing energetic (keV) ions primarily normal to the geomagnetic field, *J. Geophys. Res.*, **82**, 3324, 1977.

Sharp, R. D., D. L. Carr, W. K. Peterson, and E. G. Shelley, Ion streams in the magnetotail, *J. Geophys. Res.*, **86**, 4639, 1981.

Sharp, R. D., A. G. Ghielmetti, R. G. Johnson, and E. G. Shelley, Hot plasma composition results from the S3–3 spacecraft, this volume, pp. 167–193, 1982a.

Sharp, R. D., W. L. Lennartsson, W. K. Peterson, E. G. Shelley, and R. G. Johnson, Hot plasma composition results from the ISEE-1 spacecraft, this volume, pp. 231–261, 1982b.

Shawhan, S. D., C.-G. Fälthammer, and L. P. Block, On the nature of large auroral zone electric fields at 1-R_E altitude, *J. Geophys. Res.*, **83**, 1049, 1978.

Shelley, E. G., Heavy ions in the magnetosphere, *Spac. Sci. Rev.*, **6**, 465, 1979.

Shelley, E. G., R. D. Sharp, R. G. Johnson, J. Geiss, P. Eberhardt, H. Balsiger, G. Haerendel, and H. Rosenbauer, Plasma composition experiment on IEEE-A, *ISEE Trans. on Geosci. Elec.*, GE-**16**, 266, 1978.

Waite, J. H., Jr. and J. L. Horwitz, He^+ in the terrestrial plasmasphere, *Trans. Am. Geophys. Union*, **62**, 990, 1981.

Young, D. T., Ion composition measurements in magnetospheric modeling, *AGU Geophysical Monograph Series*, **21**, 346, 1979.

Energetic Ion Composition in the Earth's Magnetosphere, edited by R. G. Johnson, 287–306.

Hot Plasma Composition Results from the SCATHA Spacecraft

R. G. Johnson,* R. J. Strangeway,* E. G. Shelley,*

J. M. Quinn,* and S. M. Kaye**

Lockheed Palo Alto Research Laboratory, Palo Alto, California 94304, U.S.A.
**Plasma Physics Laboratory, Princeton University, Princeton, New Jersey 08544, U.S.A.*

(Received July 19, 1982)

The SCATHA spacecraft provides hot plasma (0.1–32 keV) composition measurements near the equatorial plane in the L-shell range from about 5.3 to 8.3. The SCATHA mass spectrometer has provided the first routine pitch angle distribution measurements as a function of the ion mass near the equatorial plane and has doubled the upper energy range of previous plasma composition measurements. Ion pitch angle distributions are often found to be highly anisotropic, temporally/spatially structured, and mass dependent. During geomagnetic storms, hot plasma ions of ionospheric origin are found to be a major and frequently dominant component of the ion number and energy densities in the outer ring current region of the magnetosphere. Following magnetic substorm-injection events, energy-dispersed drifting ion clouds observed at the SCATHA orbit often contain large fluxes of O^+ ions and provide insight into the spatial/temporal history of the ions. The composition of the intense warm (about 10–500 eV) ion fluxes trapped within a few degrees of the magnetic equator has been found to be dominated by H^+ ions at the measured energies above 100 eV.

1. Introduction

The SCATHA spacecraft was launched in January 1979 into a high altitude elliptical and nearly equatorial orbit to investigate spacecraft charging at the high altitudes (hence SCATHA) and to investigate the plasma and wave environments which lead to spacecraft charging. A wide range of particle and field instrumentation was included in the payload, as well as ion and electron guns and a set of engineering experiments for spacecraft charging studies. Details on the instrumentation and on the program objectives are contained in a report by STEVENS and VAMPOLA (1978).

The SCATHA spacecraft is in a nearly geosynchronous orbit with apogee at 7.8 R_e, perigee at 5.3 R_e. and an inclination of 7.9°. The orbital period is 23.7 hours, resulting in an eastward drift of the groundtrack by about 5° per day. The spacecraft is spin stabilized with a spin period of one minute. The spin axis lies in the orbital plane and is perpendicular to the sun-earth line typically within 5°.

The SCATHA payload includes the Lockheed ion mass spectrometer for hot plasma composition measurements in the energy range 0.1 to 32 keV/q. The mass spectrometer consists of three separate analyzer and detector units (CXA 1, CXA 2, and CXA 3) with the view directions all at 11° to the normal to the spacecraft spin axis. The energy steps for these units are given in Table 1, along with the energy ranges of four broadband electron spectrometers which form a part of the instrument. The mass spectrometer is normally operated either in a "sweep" mode which scans in 32 steps through the mass range from 1 to about 32 AMU at each selected energy step or in a "lock" mode in which the instrument locks onto a selected mass value and scans rapidly in energy. Additional details on the instrument and its operating modes are contained in the report by STEVENS and VAMPOLA (1978).

Extensive hot plasma composition measurements have been and are still being obtained with the SCATHA spacecraft, including coordinated data acquisition with several other spacecraft (BALSIGER, 1981; JOHNSON, 1981). The SCATHA composition measurements complement and extend other composition measurements in this ($L = 5$–8) region of the magnetosphere in three principal areas: 1) pitch angle distribution measurements over nearly the full range of pitch angles are routinely obtained, 2) the pitch angle resolution is about one-third (5° FWHM) of previous measurements, and 3) the upper energy range (32keV) is twice that of previous measurements, thus providing information on the upper end of the hot plasma distributions and more complete information on the energy density of the plasma components. The only other extensive hot plasma composition measurements in the region of geostationary satellites are from the GEOS-2 spacecraft which was launched 7 months prior to SCATHA and is also still providing composition data (BALSIGER et al., 1982)

This paper is both a review and a progress report on the hot plasma composition results from the SCATHA spacecraft. It covers the results principally from studies on pitch angle distributions, sub-storms, large magnetic storms, and warm equatorially trapped ions.

2. Pitch Angle Studies

KAYE et al. (1981a) have conducted a study of transient O^+ and H^+ ion fluxes ("bursts") observed at the SCATHA orbit. These bursts are observed over relatively short time scales (minutes) and generally have distinctive pitch angle distributions. Twenty-two days of data between 21 March and 26 April 1979 were examined.

Figure 1 from KAYE et al. (1981a) shows a typical O^+ burst for three energy

Table 1.   SCATHA (P78–2): Lockheed ion mass spectrometer (SC-8) detector characteristics.

Detector	Particle	Energy per unit charge, keV/q
CXA 1	ions	0.10, 0.13, 0.17, 0.21, 0.27, 0.35, 0.45, 0.58
CXA 2	ions	0.75, 0.96, 1.23, 1.58, 2.03, 2.61, 3.35, 4.30
CXA 3	ions	5.5, 7.1, 9.1, 11.7, 15.1, 19.4, 24.9, 32.0
CMEA	electrons	0.09–0.27
CMEB	electrons	0.39–1.23
CMEC	electrons	1.8–5.6
CMED	electrons	8.3–26.7

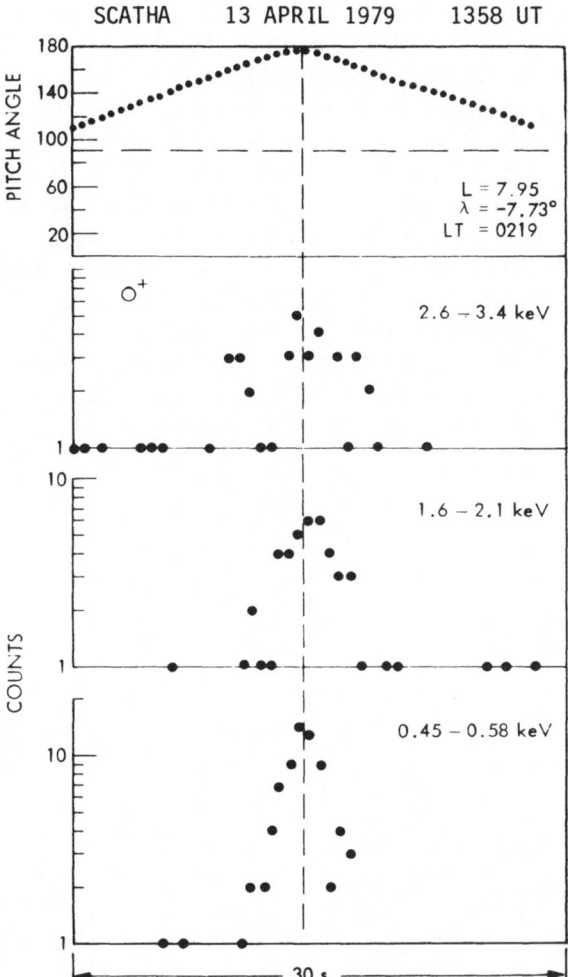

Fig. 1. O⁺ burst on April 13, 1979, at 1358 UT: 30 s of count rate data for 0.45–0.58, 1.6–2.1, and 2.6–3.4 keV are plotted in the bottom three panels, and the pitch angle being samples as a function of time is given in the top panel. The vertical dashed line indicates the time when the maximum field-aligned pitch angle (in this case, ~ 175°) was being sampled. The horizontal dashed line in the top panel represents 90° pitch angle.

intervals. The pitch angle of the observed particles is shown in the upper panel. The strong field alignment of the ions is evident although the counting statistics are a limiting factor, particularly at the higher energies. Although most of the bursts had a field-aligned structure, conic pitch angle distributions were also observed. Some bursts showed different pitch angle structure as a function of energy.

During the 22 days of data, 88 O⁺ bursts and 6 H⁺ bursts were directly observed. The H⁺ ions are seldom seen in conjunction with the O⁺ bursts. However, due to limitations of the spectrometer in the "lock" mode at the lower energies and to the data selection criteria, Kaye *et al.* (1981a) conclude that the data are not inconsistent with the existence of bursts dominated by H^+ ions.

The temporal structure of the ion bursts can only be determined to within one spin period of the spacecraft. Most of the bursts were only a few minutes in duration, and particularly noteworthy, they were not observed from the conjugate hemisphere. This implies that strong velocity space diffusion had occurred or that the bursts were spatially such narrow structures that ions mirrored in the conjugate hemisphere were not seen again at the spacecraft due to ion drift normal to the magnetic field or due to relative motion of the spacecraft and the magnetic field. Evidence for rapid pitch angle scattering may be inferred from the results on the burst pitch angle widths shown in Fig. 2. It is seen that there is a wide spread in the widths of the bursts and that the mean width of the bursts is large compared to the loss cone width (about 5° at SCATHA altitudes). Thus, if the bursts are of ionospheric origin, rapid pitch angle diffusion has

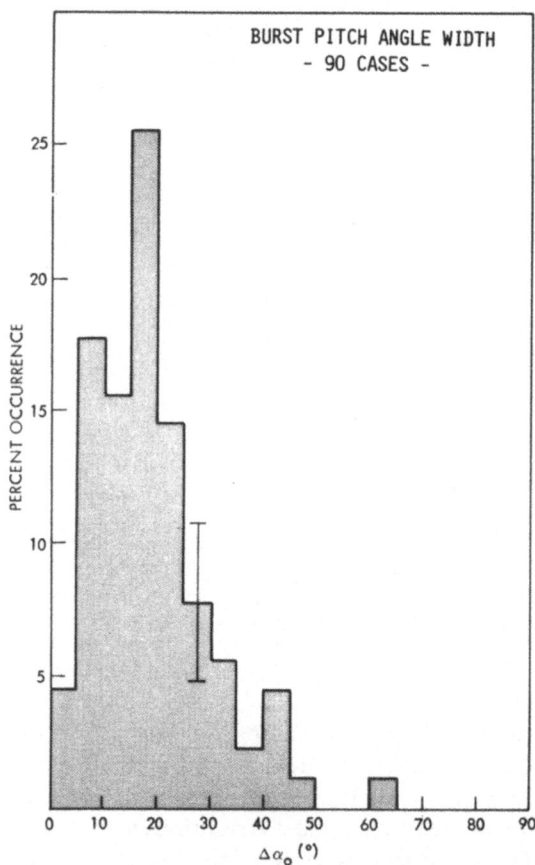

Fig. 2.   Percent occurrence of the burst pitch angle width. The width is defined to be the difference in equatorial pitch angles at the peak and at the edge of the burst.

occurred in the single transit from the ionosphere to the near equatorial location of the SCATHA. Such rapid pitch angle diffusion could largely remove the burst-like signature of the ions mirrored in the conjugate hemisphere if it were also occurring there.

At times, field-aligned electrons were observed simultaneously with field-aligned ion bursts. Figure 3 shows an example of an O^+ burst accompanied by field-aligned electrons observed in the lowest energy channel (0.09−0.27 keV). The field-aligned electrons observed with an ion burst are typically in the lowest energy channel, but occasionally electrons are also observed in the next highest channel ($E = 0.39−1.23$ keV). The simultaneous observation of field-aligned ions and electrons from the same hemisphere makes quasi-static parallel electric fields on a time scale of a few minutes an unlikely source of the ion bursts. However, rapidly varying field-aligned electric fields as discussed by Sharp *et al.* (1980) are a possible source of these ions and electrons. In

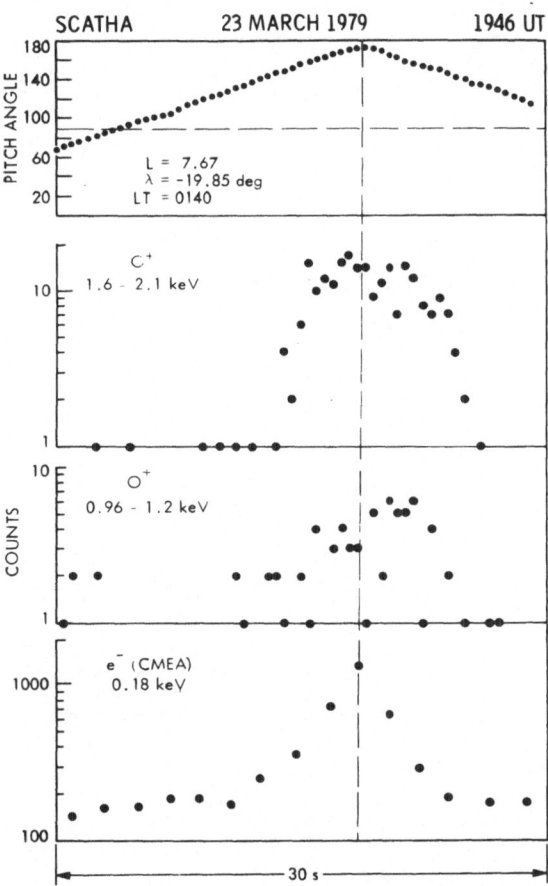

Fig. 3. Same as Fig. 1 for the event on March 23, 1979, at 1946 UT. During this event, field-aligned O^+ in the energy ranges 0.96−1.2 and 1.6−2.1 keV were seen to come from the same hemisphere as low-energy, field-aligned electrons (bottom panel). The most field-aligned pitch angle sampled in this case was 172°.

addition, it should be noted that ion and electron acceleration transverse to the magnetic field at low altitudes would also be consistent with the observations at SCATHA altitudes.

The spatial distribution of the bursts in local time and L-shell sectors is shown in Fig. 4. The shaded region shows the satellite coverage during the time of the burst study. It is seen that no bursts are seen in the 0800 to 1600 local time sector although bursts are observed in the same L-shell regions in the 1600 to 2200 local time sector. The peak of the occurrence frequency of the bursts is seen at high L-shells in the 2400 to 0600 local time sector. It should be noted that the location of the peak occurrence frequency of the bursts does not coincide with the dusk to midnight locations of the peak in the occurrence frequency of upward flowing ion events as observed with the S3-3 spacecraft at altitudes below 8,000 km (GHIELMETTI *et al.*, 1979). Obviously it is of interest to extend the burst study to cover the periods when the SCATHA apogee is in the dusk to midnight sector.

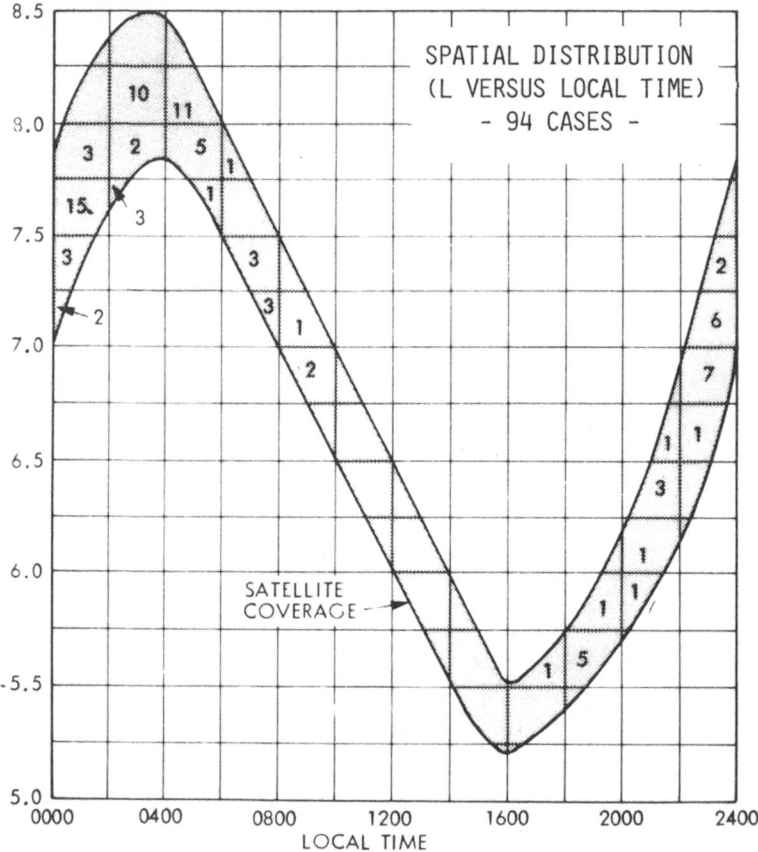

Fig. 4. The spatial distribution (L versus LT) of the ion bursts. The gray shaded region represents the total spatial region accessible to the satellite for the time period from March 21, 1979, to April 16, 1979. The numbers represent the total numbers of bursts occurring in that particular spatial bin.

In another study utilizing the pitch angle coverage of the SCATHA mass spectrometer, KAYE *et al.* (1981b) have investigated the ion composition of "zipper" events (FENNELL *et al.*, 1981). In these events the ion fluxes below a certain energy are peaked along the magnetic direction while the fluxes above this energy are peaked perpendicular to the field direction. The alternation between these two flux signatures when viewed on energy-time gray scale spectrograms gives the characteristic appearance of a zipper. Figure 5 from KAYE *et al.* (1981b) shows the ion mass spectrometer results for one zipper event. The left hand panel shows the H^+ energy distributions within $\pm 30°$ of the field-aligned directions and within $\pm 30°$ of perpendicular to the field. The right hand panel shows the same information for O^+ ions. In this zipper event, the transition energy between mainly field-aligned and mainly perpendicular fluxes as identified from the FENNELL *et al.* (1981) spectrograms is about 3 keV. It is obvious from Fig. 5 that below this energy the field-aligned fluxes are dominant for both O^+ an H^+ ions. Above this transition energy, the 90° H^+ flux is dominant over the field-aligned H^+ flux; whereas they are of comparable values for the O^+ ions, although of much lower intensities than for the H^+ ions. The signature of the combined O^+ and H^+ fluxes is thus consistent with results of FENNELL *et al.* (1981) which had no mass discrimination for the fluxes. Although roughly isotropic oxygen fluxes above the transition energy characterized six of the eight zipper events studied, the other two cases showed the same transition in flux for both protons and oxygen.

The number densities of both ion species were used by KAYE *et al.* (1981b) to give a statistical picture of the ion composition of zipper events. The number densities were

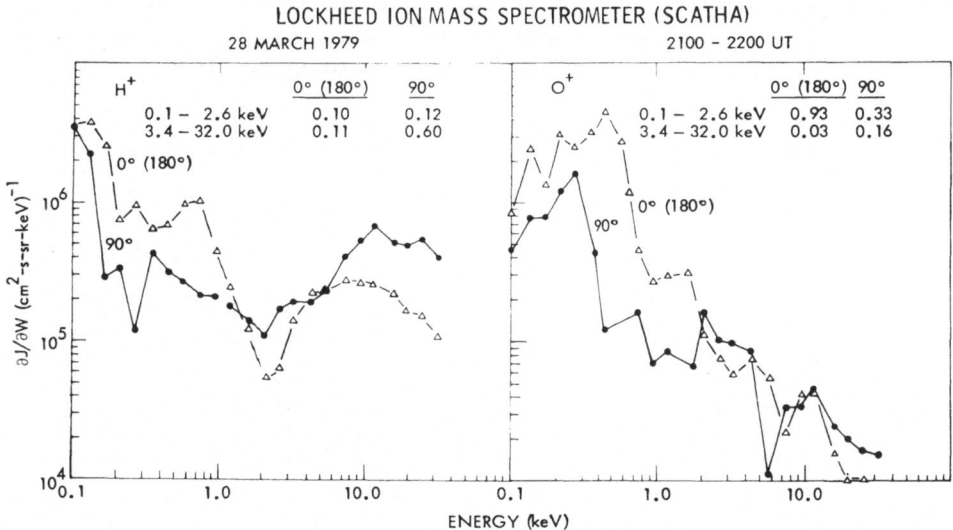

Fig. 5.   Energy spectrums for H^+ (left panel) and O^+ (right panel) for March 28, from 2100 to 2200 UT. Plotted in each panel is the differential number flux versus energy for 0° (180°) $\pm 30°$ pitch angle particles (dashed lines) and for 90° $\pm 30°$ pitch angle particles (solid lines). Shown in the top right hand portion of each panel are the number densities for the different pitch angle ranges below and above the transition energy (in this case, ~ 3 keV).

calculated for perpendicular and parallel fluxes above and below the transition energy. The results are shown in Fig. 6. The density ratio of O^+ to H^+ for perpendicular fluxes is plotted as a function of this ratio for parallel fluxes. Below the transition energy, it is seen that in most cases oxygen is the more abundant species in both the parallel and perpendicular components. Above the transition energy, it is seen that, with one exception, H^+ densities are comparable to or larger than the O^+ densities in both the parallel and perpendicular directions.

To summarize the results of Kaye *et al.* for zipper events, the ion populations above and below the zipper transition energy show marked differences in composition. The primarily parallel fluxes below the transition energy show a large proportion of ions of ionospheric origin, with O^+ being the dominant ion. Above the transition energy some O^+ is present, but most of the ions are protons and the proton fluxes are mainly perpendicular to the field. This suggests different sources for the two ion populations—a direct ionospheric source for the parallel lower energy fluxes and probably the plasma sheet for the higher energy fluxes since they have trapped distributions.

Using the combination of good pitch angle coverage with the routine SCATHA spacecraft crossings of the geomagnetic equator in the vicinity of geosynchronous orbit, Quinn and Johnson (1982) have investigated the mass composition of the highly trapped, warm ion population within a few degrees of the equator. Olsen (1981), using ion data without mass discrimination from the UCSD Charged Particle Experiment onboard SCATHA, reported intense fluxes of warm (0.01–0.5 keV) ions trapped within a few degrees of the magnetic equator. Quinn and Johnson (1982) analyzed the composition of eight of these events at energies greater than $100 \, eV/q$ and found protons to be the dominant constituent. Figure 7 presents pitch angle distributions from two of the events studied. In both cases the H^+ fluxes in the $90° \pm 15°$ pitch angle

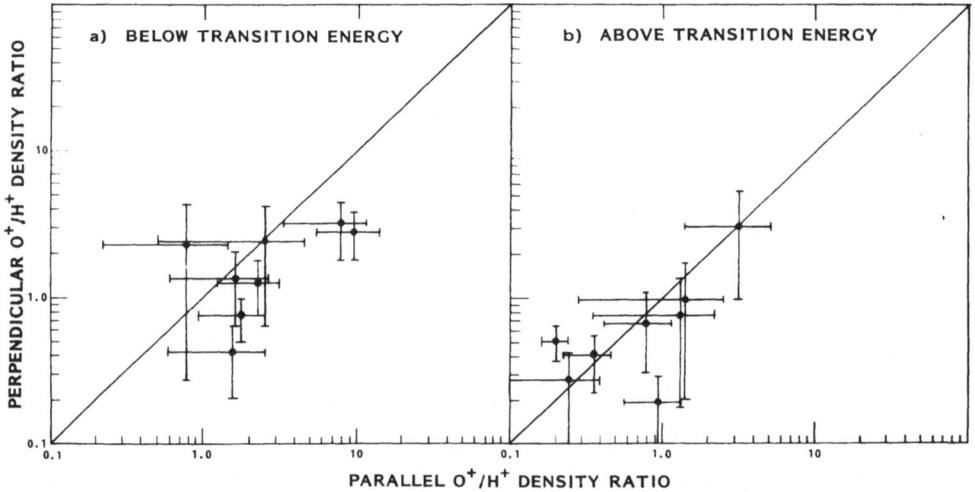

Fig. 6. O^+/H^+ density ratios in the perpendicular direction are plotted as a function of those in the parallel direction below the transition energy (left panel) and above the transition energy (right panel).

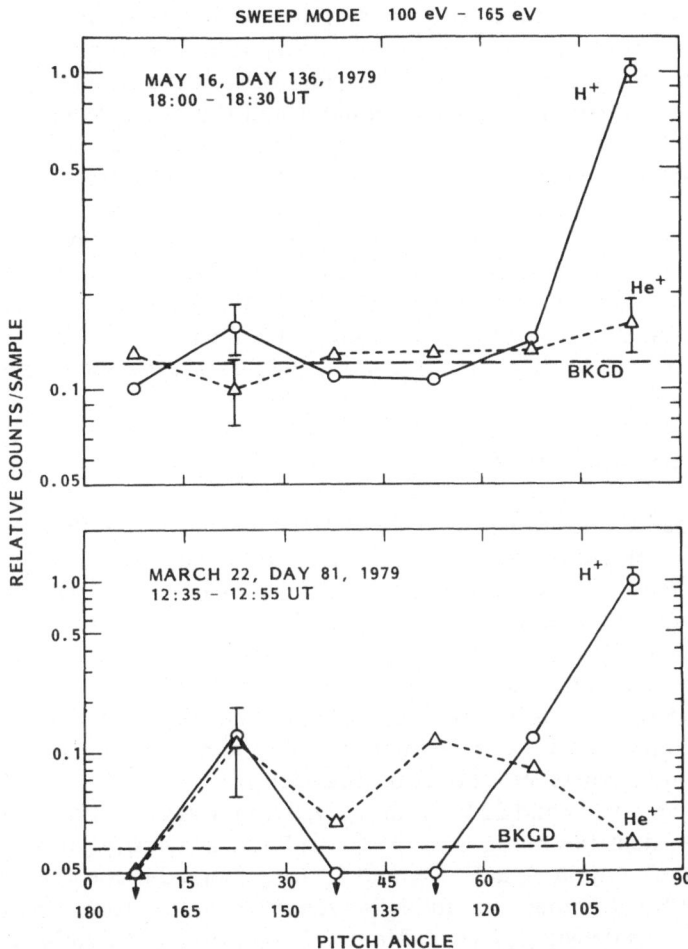

Fig. 7. Pitch angle distributions for equatorially trapped ions on 16 May 1979 and 22 March 1979.

bin are more than a factor of five higher than at other pitch angles.

The physical processes responsible for trapping these protons are not well understood. Recent observations and theoretical analyses involving He$^+$ ion cyclotron resonance suggest that there may at times be a substantial population of He$^+$ ions trapped near the equator (YOUNG et al., 1981; MAUK et al. 1981; MAUK, 1982). The data presented in Fig. 7, and the other six cases analyzed by QUINN and JOHNSON show no evidence that He$^+$ comprises a significant fraction of the equatorially trapped ions above 100 eV. In addition, neither O$^+$ nor He$^+$ had peak fluxes at 90° pitch angle in the events studied. It should be emphasized that these composition measurements were made for selected cases in which an intense equatorially trapped ion population extended above $E/q = 100$ eV. HORWITZ et al. (1981) have found an He$^+$ component in "pancake distributions"; however, their observations were made at magnetic latitudes as great as 30° (where fluxes at 90° pitch angle correspond to 34° equatorial pitch angle in a dipole field). Young et al. (1981) have observed warm He^+ ions in association with intense ion cyclotron waves, but with one exception their observations were more than 10° from the magnetic equator and thus were generally observing a different population than the "equatorially trapped ions" discussed by Olsen and by Quinn and Johnson which are confined to within a few degrees of the magnetic equator.

It is concluded that the most likely source of the H^+ ions is the nearby ionsophere/plasmasphere, as opposed to the solar wind or ionospheric ions transported inward from the magnetotail plasma sheet. Also, it was observed that these equatorially trapped ions were depleted and replenished on time scales shorter than one day. However, the energization, transport, and loss processes for these ions remain to be investigated.

3. Magnetic Storm Studies

Hot plasma composition studies have been conducted for the first five major ($D_{ST} < -90\,nT$) magnetic storms of 1979 for which data were available from the SCATHA spacecraft (Johnson and Kaye, 1980; Johnson, 1981). Hourly D_{ST} values for the five storms studied are shown in Fig. 8. It can be seen that the five storms denoted by the circled numbers vary considerably in their magnitudes, sudden commencements (or lack thereof), onset times to main phase peaks, and decay characteristics. However, as will be discussed later, they are all similar in that the O^+ (ionspheric) number density and energy density in the hot plasma are typically large and often dominant for several hours during each storm.

For the storms on 21 and 22 February, Fig. 9 from Johnson (1981) shows that O^+ and H^+ number densities along with the O^+/H^+ energy density ratios in the energy range 0.1 to 32 keV and at 32 keV. D_{ST} and K_p are presented in the bottom panel. It is seen that O^+ is the dominant ion following the onset of the storms, becoming more than 80% of the ion density for several hours. In addition, O^+ dominates the energy density in the instrument energy range during these periods. The O^+ energy density at 32 keV is also frequently comparable to and sometimes larger than that for H^+ near the peaks of the storm. The hatched segments labeled MS are periods when the spacecraft was in the magnetosheath plasma.

The SCATHA investigations of the magnetic storms on 21 and 22 February are also a part of a coordinated study of these storms utilizing ion composition data from seven spacecraft (Balsiger, 1981; Johnson, 1981). Further details on this coordinated study are reported in this volume by Balsiger et al.

The magnetic storm on 3–4 April 1979 was the largest of the storms investigated (Johnson, 1981). The hot plasma characteristics and D_{ST} for this storm are shown in Fig. 10. The O^+ density and energy density in the 0.1 to 32 keV range are seen to be dominant for about 12 hours near the peak of the storm, and during this time the O^+ energy density at 32 keV is also comparable to or larger than that for H^+. The average energies for O^+ and H^+, as seen in the second panel from the top in Fig. 10, are quite variable during the storm and this feature is typical of the five storms studied.

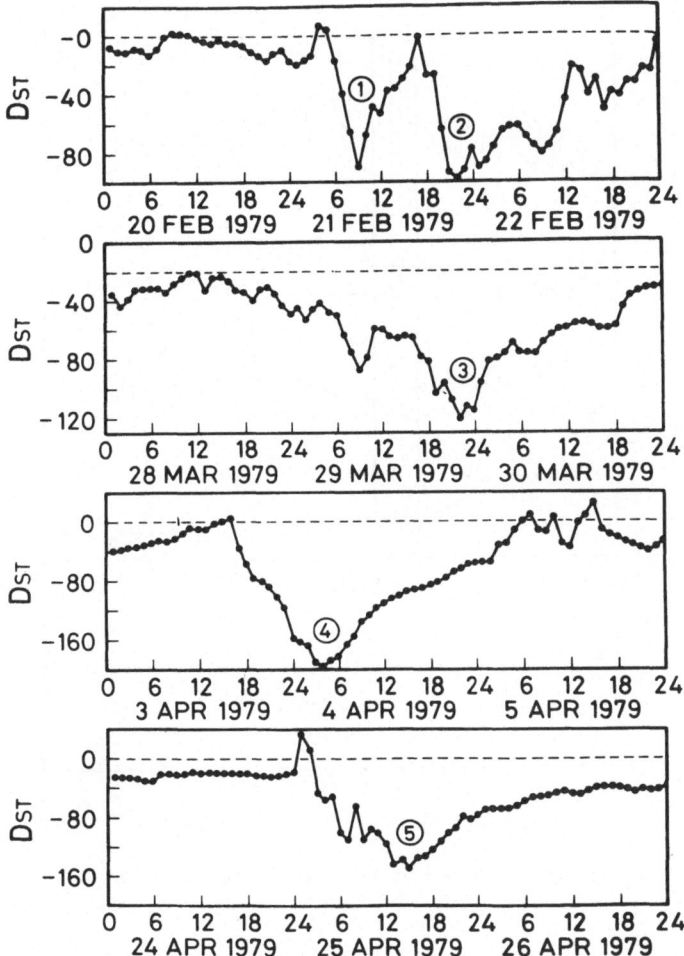

Fig. 8. D_{ST} values for the first five major magnetic storms of 1979 during which hot plasma composition measurements were obtained with the SCATHA spacecraft.

Figure 11 shows the hot plasma composition data for the magnetic storm on 25 April 1979. The ion fluxes and D_{ST} were relatively stable and low for many hours prior to the onset of the storm near 2400 hours on 24 April. The O^+ and H^+ number densities are shown for two energy groups to indicate that the number densities (and thus the energy densities) in the 1–32 keV range are dominant during the storms. However, the densities in the 0.1–0.8 keV range are not negligible. Although quite variable, the O^+/H^+ energy density ratio shows the O^+ energy density to be generally larger than or comparable to that of H^+ in the 0.1–32 keV range as well as at 32 keV.

Although the SCATHA data for the 28–30 March 1979 time period are incomplete due to noisy data tapes, there are several interesting features evident in the available data. Energy density information for O^+ and H^+ is shown in two energy

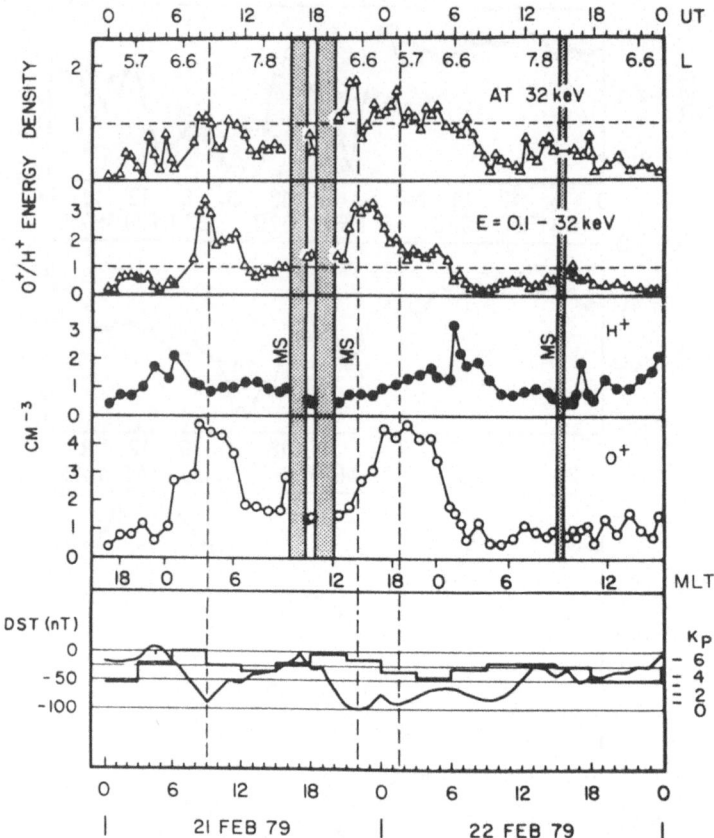

Fig. 9.   SCATHA hot plasma composition data during the magnetic storms on 21–22 February 1979.

intervals, along with D_{ST}, in Fig. 12. The ring current buildup, as evidenced by D_{ST}, is seen to be sporadic and occurs over a long time period—about 1.5 days. Prior to the ring current enhancements near 1200 hours on 28 March and near 0500 and 1500 hours on 19 March, the energy density of the O^+ ions in the 0.1 to 32 keV range is comparable to that in the H^+ ions in the L-shell range from 5.5 to about 6.6. If the populations near $L = 5.5$ contributed significantly to the subsequent ring current enhancements as a result of this population being convected to lower L-shells, e.g., near $L = 4$ as suggested by Lyons and Williams (1980), then clearly the ionospheric ion component (O^+) would be a significant contributor to the ring current energy in this storm. The number densities observed (not shown in Fig. 12) at $L = 5.5$ near 1100 hours on 28 and 29 March were, respectively, 1.9 and 2.3 cm^{-3} for O^+ and 0.9 and 1.0 for H^+.

Another feature evident in Fig. 12 is that the energy density in the range 0.1–15 keV (the approximate range of earlier mass spectrometers on spacecraft) for L = 5.5 to 6.6 is typically less than half of that in the 0.1 to 32 keV range. However, as

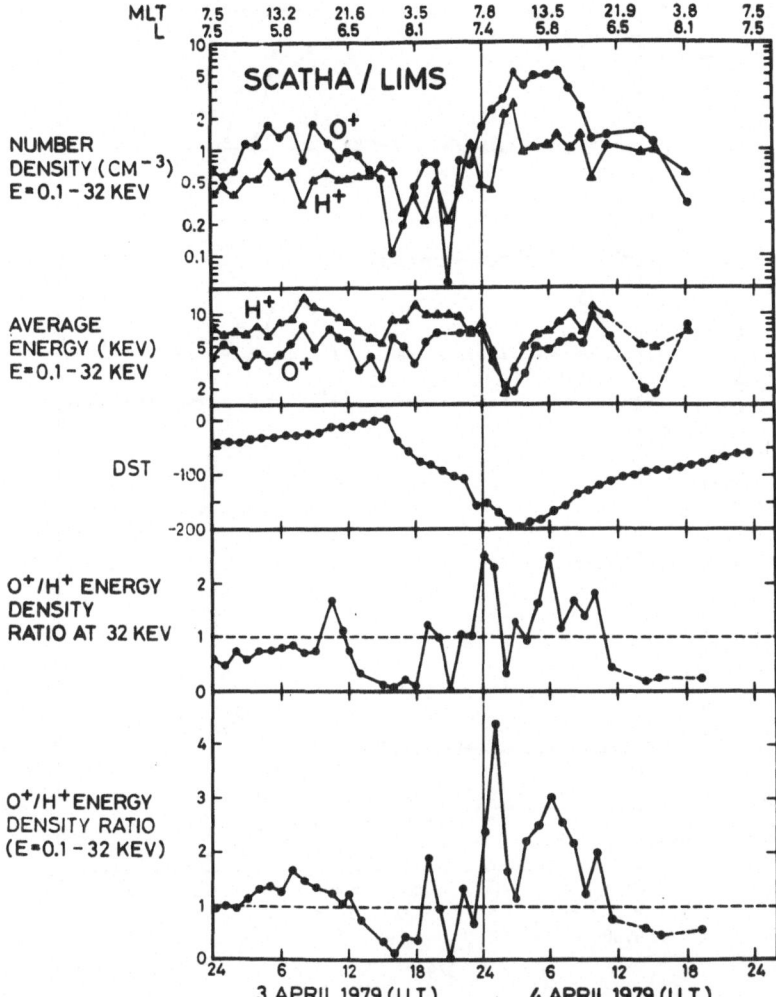

Fig. 10. Characteristics of the hot plasma composition observed with the SCATHA spacecraft during the 3–4 April 1979 magnetic storm.

seen by the top panel in this figure, the relative contributions to the energy densities in these ranges by O^+ and H^+ are very similar.

Based on these magnetic storm studies, it is concluded that during major geomagnetic storms near the peak of the solar activity cycle, ions of ionospheric origin in the energy range $0.1-32$ keV are a major and frequently dominant component of the ion number density and energy density in the outer ring current region of the magnetosphere. In addition, these ionospheric ions also serve as a source of ions for further convection into the inner ring current regions, particularly during storms which require several hours for the ring current to reach its maximum strength.

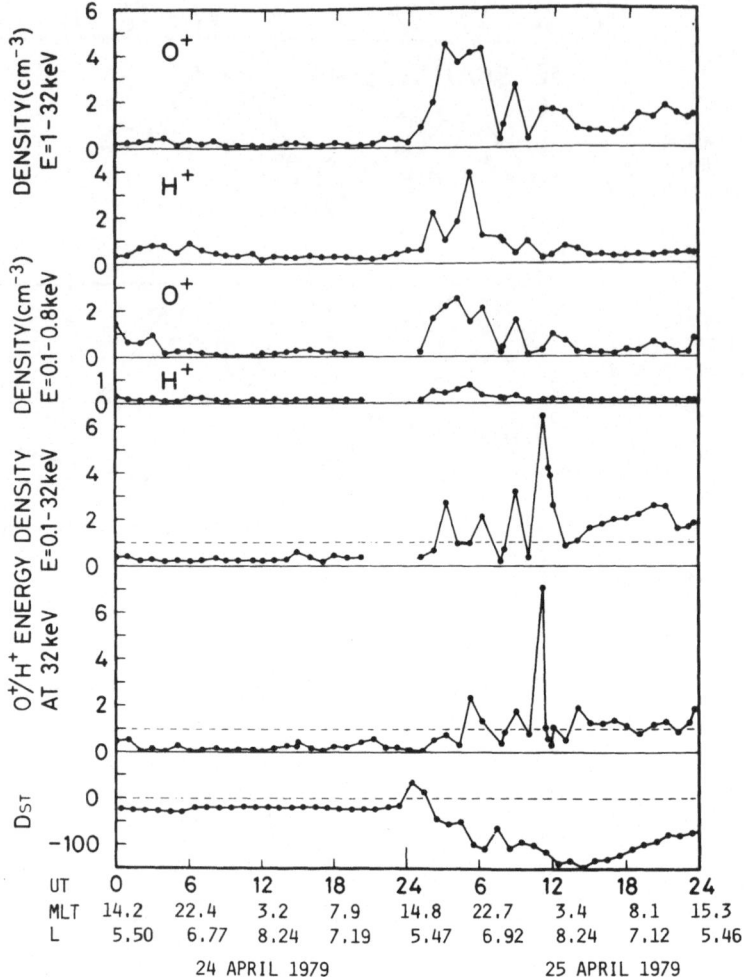

Fig. 11.   Characteristics of the hot plasma composition observed with the SCATHA spacecraft during the magnetic storms on 24–25 April 1979 magnetic storm.

4.   Substorm-Related Studies

The injection and dispersion of hot "plasma clouds" in the earth's magnetosphere have been investigated rather extensively without ion mass discrimination at the geostationary satellite orbit (DeForest and McIlwain, 1971; Mauk and McIlwain, 1974; Hultqvist *et al.*, 1981). These and other observations indicate that hot plasmas are injected into the magnetosphere during each substorm and that a low energy injection boundary of the substorm-injected plasma is generally located in the dusk to midnight region in the L-shell range from about 5 to 8, depending on local time and K_p (Mauk and McIlwain, 1974). Since the ion mass spectrometer on the SCATHA

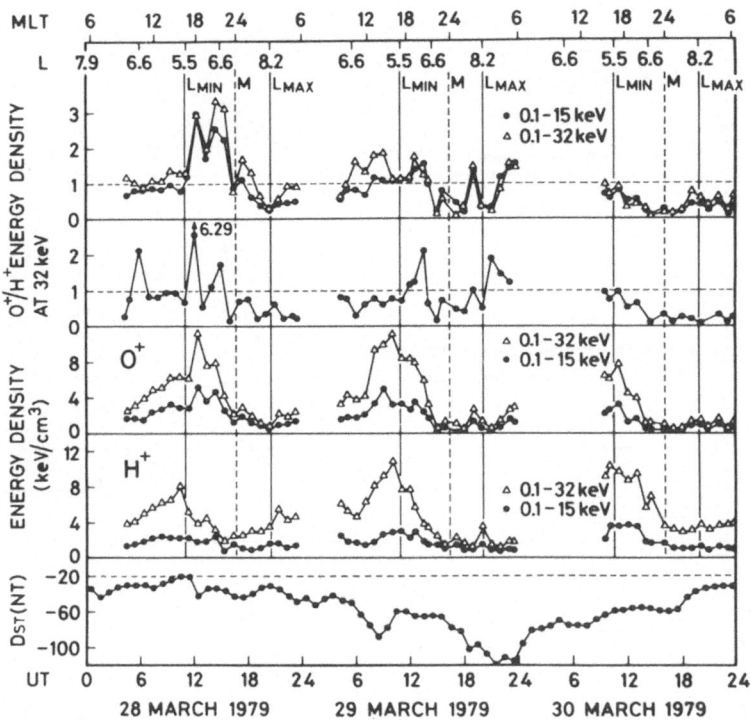

Fig. 12. Characteristics of the hot plasma composition observed with the SCATHA spacecraft during the magnetic storms on 28, 29, and 30 March 1979.

spacecraft provides measurements up to 32 keV, it is adequate to investigate the energy dispersion and other characteristics of the substorm-injected plasmas as a function of the ion mass. Recently, JOHNSON et al. (1981) have investigated the ion composition of two ion dispersion events which are associated with substorm plasma injection.

A brief synopsis of the two days studied by JOHNSON et al. is given before describing the ion dispersion in detail. Figure 13 summarizes the geomagnetic conditions and the particle data for 22 March (day 81) 1979. Shown in Fig. 13 from bottom to top are hourly D_{ST} values, three-hour K_p values together with the H-component of magnetic field as measured at Fort Chipewyan (G. Rostoker, private communication), and lastly hourly averages of the number densities as measured at the SCATHA orbit for both O^+ and H^+ over the energy range $0.1-32$ keV. At 0826 UT there was a sudden commencement (SC) as indicated by the small sudden decrease and recovery in the ground magnetogram. At this time, D_{ST} was increasing and continued to do so until about 1055 UT when a substorm occurred. Prior to 1100 UT the particle fluxes at SCATHA were quite low. Protons were the dominant ions with a number density of roughly 0.5 cm^{-3}. Later the densities decrease again, settling down to slightly higher values than before the substorm.

Figure 14 shows a summary of data for the second dispersion event studied, which occurred on 7 June (day 158) 1979. The format for this figure is the same as Fig. 13,

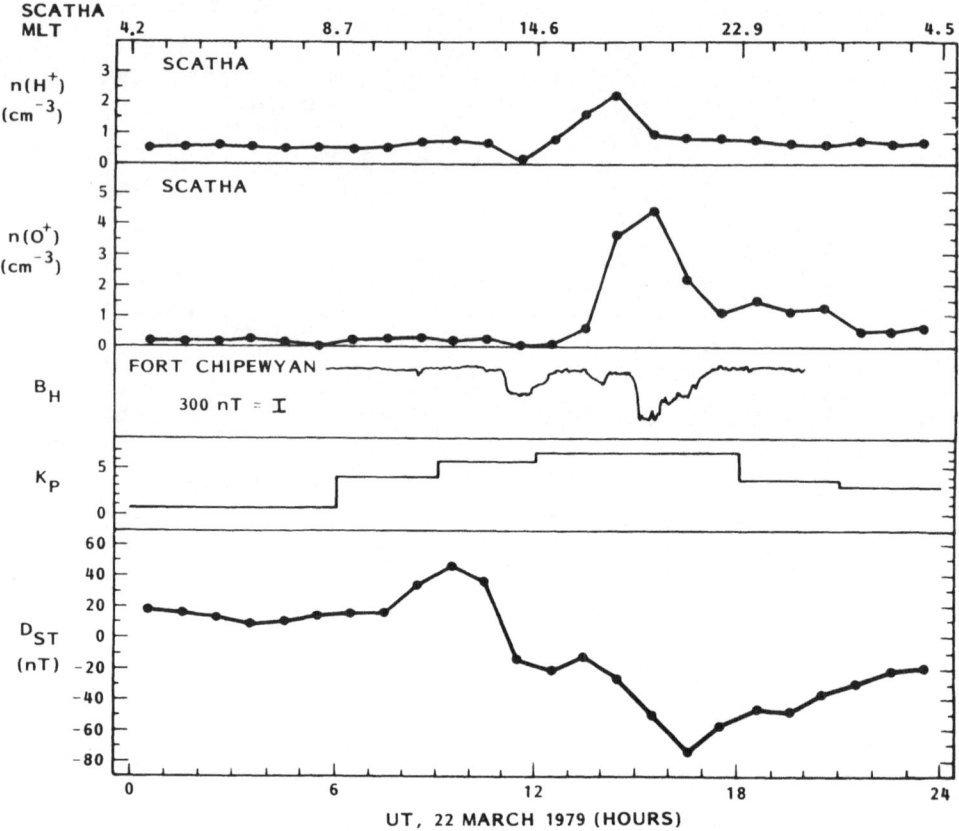

Fig. 13. O^+ and H^+ ion densities in the 0.1 to 32 keV range from the SCATHA spacecraft and magnetic field perturbation data during 22 March 1979.

except that the x-component of magnetic field as measured at Fort Churchill is shown. This day was much more disturbed than 22 March, and there was considerable activity on the previous day as indicated by the Fort Churchill magnetogram (not shown). Interestingly, D_{ST} shows a minimum near 0600 UT with a subsequent slow recovery. Also the ion fluxes as measured at the SCATHA orbit were quite high prior to this minimum in D_{ST}. This day is much more complicated than the 22 March, yet the ion fluxes on this day show a "clean" dispersion event near 0600 hours as will be discussed below.

Although the number densities are useful for summarizing the ion composition data, the actual signature of the ion injection as observed on board SCATHA is best shown using an energy-time spectrogram. Figure 15 shows color spectrograms to display the ion composition of the dispersion events on 22 March and 7 June 1979. The color code displays the differential number fluxes for O^+ and H^+ ions, and the intensity scales corresponding to the colors are shown at the right side of each panel. The ion energy from 0.1 to 32 keV is plotted downward as the ordinate. The absicca presents

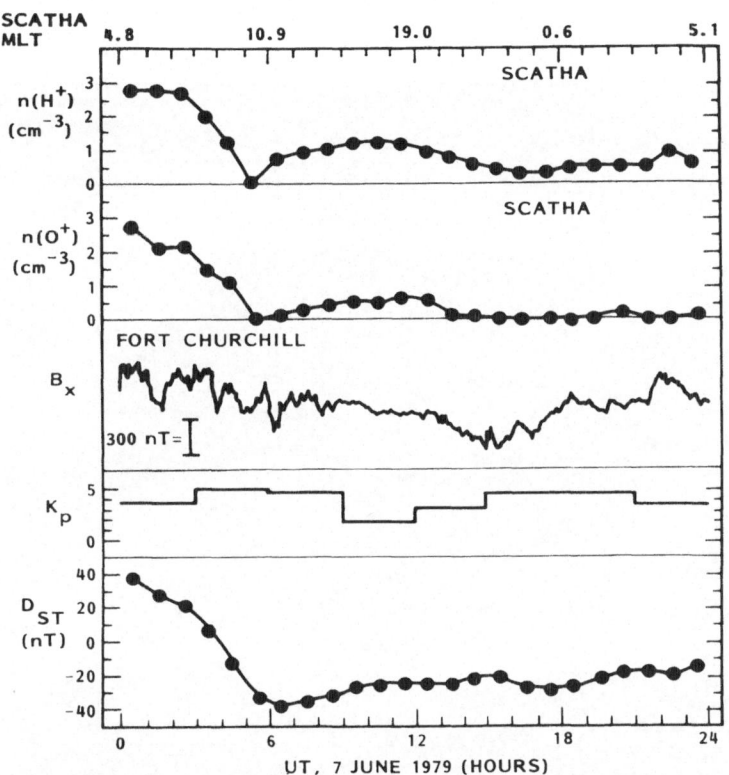

Fig. 14. O^+ and H^+ ion densities in the 0.1 to 32 keV range from the SCATHA spacecraft and magnetic field perturbation data during 7 June 1979.

the universal time (UT) along with the spacecraft ephemeris values for magnetic L-value (L), magnetic local time (MLT), magnetic latitude ($MLAT$), radial distance (R), geographic local time (LT), and geographic latitude (LAT).

Looking first at the fluxes on 22 March, there is a data dropout between 1037 and 1122 UT, which unfortunately includes the time of the first substorm onset (1055 UT). Nevertheless prior to this time there are reasonable fluxes of both protons and oxygen and after the dropout the fluxes have decreased. Subsequently, quite high ion fluxes are again observed, but the increase in flux is first detected at the higher energies, and both O^+ and H^+ ions show a characteristic dispersion in their energy spectrums. It is also interesting to note that both species show a dispersion ridge rather than a dispersion edge. In other words, at a particular time the flux level is first seen to increase as a function of increasing energy and then show a subsequent decrease. This is somewhat indicative of a limited temporal and spatial extent to the injection of the ions along the drift paths which map to the SCATHA orbit.

It is also seen that later in the dispersion event on 22 March the fluxes of O^+ become quite large, more than 0.25 times the H^+ fluxes, which, because of the velocity differences, indicates that oxygen is the dominant species. This was also apparent in the

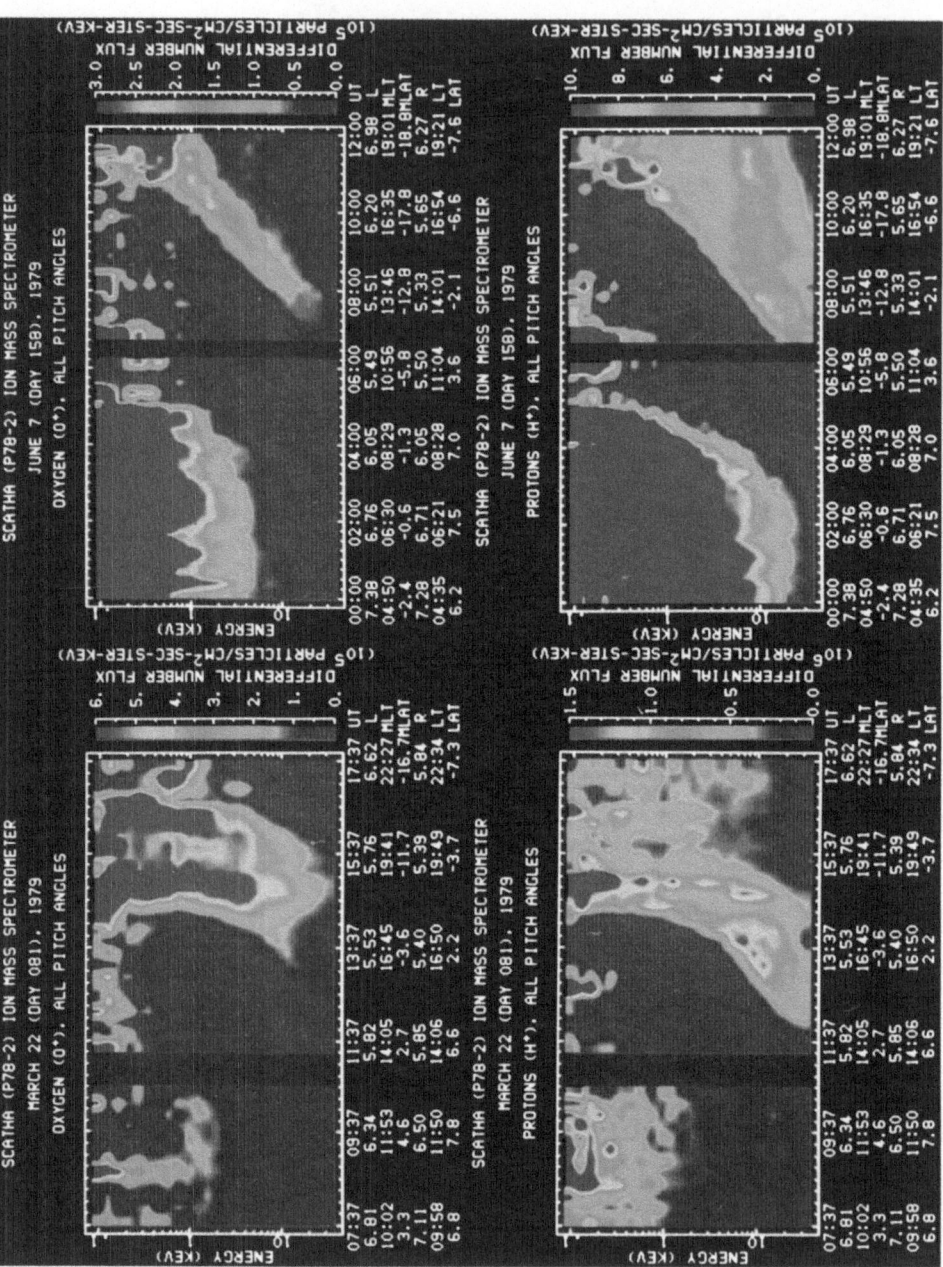

Fig. 15.   Energy-time spectrograms for O$^+$ and H$^+$ ions observed with the SCATHA spacecraft on 22 March 1979 and 7 June 1979. The panels are described in the text.

density plots shown in Fig. 13.

For 7 June 1979, the color spectrograms show O^+ and H^+ fluxes for the first 12 hours of the day. There is a data dropout between 0615 and 0645 UT. After 0300 UT, both proton fluxes and oxygen fluxes are seen to decrease, and just prior to the data gap, very little flux is present. During the first 6 hours, D_{ST} has been gradually decreasing (see Fig. 14). Following 0645 UT, a dispersion ridge is observed in both H^+ and O^+. This ridge is much more marked in the oxygen spectrums than in the proton spectrums. The prominent dispersion ridge in the O^+ flux may be a signature of a limited temporal and spatial extent to the injection of oxygen into the drift paths intersecting the SCATHA orbit. The proton spectrums show the same signature of a dispersion ridge, but, in addition, there is much more structure at higher energies. One interpretation is that the initial dispersion event is associated with direct injection of ionospheric plasma early in the event followed by convection of a different plasma population from greater radial distances. The probable source of this plasma is the plasma sheet. A similar conclusion was drawn by KAYE et al. (1981b) when discussing the composition of zipper events in which the different plasma sources result in different pitch angle signatures.

In each of these two cases, the dispersion event is not just a flux signature superimposed on an existing particle population. It appears that in addition to the injection of plasma in the nightside, there is a removal of plasma from the dayside, similar to that reported by HULTQVIST et al. (1981). At this stage of magnetospheric modeling, it is not possible to utilize this and the above detailed features in the data. However, much work has been carried out concerning the evolution of convection boundaries and particle drift paths within the magnetosphere (HAREL et al., 1981a, b; SPIRO et al., 1981), and in the future it may be possible to follow in greater detail the evolution of substorm-injected particle populations. Additionally, multiple spacecraft studies will also be valuable in the interpretation of such phenomena.

The authors thank R. D. SHARP and J. B. REAGAN for their many contributions to the SCATHA ion composition experiment and to E. HERTZBERG, L. A. HOOKER, and T. C. SANDERS for their important contributions and dedication to the design and testing of the ion composition spectrometer. One author (R.G.J.) thanks J. GEISS, H. BALSIGER and D. T. YOUNG for valuable discussions and support which significantly influenced the analysis of SCATHA data, while the author was a Visiting Professor at the University of Bern. Portions of this work were supported by the National Science Foundation under grant ATM-8119340 and the Office of Naval Research under Contract N00014-76-C-0444.

REFERENCES

BALSIGER, H., Composition of hot (0.1 to 16 keV/e) as observed by the GEOS and ISEE mass spectrometers and inferences for the origin and circulation of magnetospheric plasma, *Adv. Space Res.*, 1, 289, 1981.
BALSIGER, H., J. GEISS, and D. T. YOUNG, The composition of thermal and hot ions observed by the GEOS-1 and -2 spacecraft, this volume, pp. 195–230, 1983.
DEFOREST, S. E. and C. E. McILWAIN, Plasma clouds in the magnetosphere, *J. Geophys. Res.*, 76, 3587, 1971.
FENNELL, J. F., D. R. CROLEY, JR., and S. M. KAYE, Low-energy ion pitch angle distributions in the outer magnetosphere: Ion zipper distributions, *J. Geophys. Res.*, 86, 3375, 1981.
GHIELMETTI, A. G., R. D. SHARP, E. G. SHELLEY, and R. G. JOHNSON, Downward flowing ions and evidence for injection of ionospheric ions into the plasma sheet, *J. Geophys. Res.*, 84, 5781, 1979.

HAREL, M., R. A. WOLF, P. H. REIFF, R. W. SPIRO, W. J. BURKE, F. J. RICH, and M. SMIDDY, Quantitative simulation of a magnetospheric substorm, 1, Model logic and overview, *J. Geophys. Res.*, **86**, 2217, 1981a.

HAREL, M., R. A. WOLF, R. W. SPIRO, P. H. REIFF, C.-K. CHEN, W. J. BURKE, F. J. RICH, and M. SMIDDY, Quantitative simulation of a magnetospheric substorm, 2, Comparison with observations, *J. Geophys. Res.*, **86**, 2242, 1981b.

HORWITZ, J. L., C. R. BAUGHER, C. R. CHAPPELL, E. G. SHELLEY, and D. T. YOUNG, Pancake pitch angle distributions in warm ions observed with ISEE 1, *J. Geophys. Res.*, **86**, 3311, 1981.

HULTQVIST, B., B. APARICIO, H. BORG, R. ARNOLDY, and T. E. MOORE, Decrease of keV electron and ion fluxes in the dayside magnetosphere during the early phase of magnetospheric disturbances, *Planet. Space Sci.*, **29**, 107, 1981.

JOHNSON, R. G., Review of the hot plasma composition near geosynchronous altitude, *Proceedings of Spacecraft Charging Technology 1980 Conference*, N. J. Stevens and C. P. Pike, eds., NASA CP-2182, 412, 1981.

JOHNSON, R. G. and S. M. KAYE, Magnetospheric plasma composition observations up to 32 keV during geomagnetic storms, *EOS*, **61**, 1080, 1980.

JOHNSON, R. G., R. J. STRANGEWAY, and J. M. QUINN, Observations of mass dependent features in energy-time spectrograms at geosynchronous altitudes, *EOS*, **62**, 995, 1981.

KAYE, S. M., R. G. JOHNSON, R. D. SHARP, and E. G. SHELLEY, Observations of transient H^+ and O^+ bursts in the equatorial magnetosphere, *J. Geophys. Res.*, **86**, 1335, 1981a.

KAYE, S. M., E. G. SHELLEY, R. D. SHARP, and R. G. JOHNSON, Ion composition of zipper events, *J. Geophys. Res.*, **86**, 3383, 1981b.

LYONS, L. R. and D. J. WILLIAMS, A source for the geomagnetic storm main phase ring current, *J. Geophys. Res.*, **25**, 523, 1980.

MAUK, B. H., Helium resonance and dispersion effects on geostationary Alfven/ion cyclotron waves submitted to *J. Geophys. Res.*, 1982.

MAUK, B. H. and C. E. MCILWAIN, Correlation of K_p with substorm-injected plasma boundary, *J. Geophys. Res.*, **81**, 3193, 1974.

MAUK, B. H., C. E. MCILWAIN, and R. L. MCPHERRON, Helium cyclotron resonance within the earth's magnetosphere, *Geophys. Res. Lett.*, **8**, 103, 1981.

OLSEN, R. C., Equatorially trapped plasma populations, *J. Geophys. Res.*, **86**, 11235, 1981.

QUINN, J. M. and R. G. JOHNSON, Composition measurements of equatorially trapped ions near geosynchronous orbit, *Geophys. Res. Lett.*, **9**, 777, 1982.

SHARP, R. D., E. G. SHELLEY, R. G. JOHNSON, and A. G. GHIELMETTI, Counter-streaming electron beams at altitudes of ~ 1 R_E over the auroral zone, *J. Geophys. Res.*, **85**, 92, 1980.

SPIRO, R. W., M. HAREL, R. A. WOLF, and P. H. REIFF, Quantitative simulation of a magnetospheric substorm, 3, Plasmaspheric electric fields and evolution of the plasmapause, *J. Geophys. Res.*, **86**, 2261, 1981.

STEVENS, J. R. and A. L. VAMPOLA, Description of the space test program P78-2 spacecraft and payloads, *Air Force Space Division Report* TR-78-24, 1978.

YOUNG, D. T., S. PERRAUT, A. ROUX, C. DEVILLEDARY, R. GENDRIN, A. KORTH, G. KREMSER, and D. JONES, Wave-particle interactions near Ω_{He^+} observed on GEOS 1 and 2, 1. Propagation of ion cyclotron waves in He^+-rich plasma, *J. Geophys. Res.*, **86**, 6755, 1981.

Energetic Ion Composition in the Earth's Magnetosphere, edited by R. G. Johnson, 307–351.
Copyright © 1983 by Terra Scientific Publishing Company (TERRAPUB), Tokyo.

Composition of the Hot Magnetospheric Plasma as Observed with the PROGNOZ-7 Satellite

R. Lundin,* B. Hultqvist,* N. Pissarenko,** and A. Zacharov**

*Kiruna Geophysical Institute, Kiruna, Sweden
**Space Research Institute, Soviet Academy of Sciences, Moscow, U.S.S.R.

(Received March 8, 1982)

Measurements of the ion composition from PROGNOZ-7 in the 0.2 to 17 keV/q energy range have revealed new and exciting data from regions of the magnetosphere previously unexplored by spacecraft containing mass-analyzing particle spectrometers. During the ~7.5 months mission time of the PROGNOZ-7 satellite the measurements have been particularly suited for studies of the hot plasma in the high latitude boundary layer of the magnetosphere. However, the satellite also made excursions into other interesting regions of the magnetosphere and ion composition studies have been performed in the boundary layer near the subsolar point, in the ring current region and on auroral field lines in the altitude range of 2–5 earth radii.
    The purpose of this report is to review the PROGNOZ-7 ion composition results, to describe the existing volume of information and the way it is represented, and finally to discuss some physical implications of the results.

1. Introduction

PROGNOZ-7 was launched on 30 October 1978 into a highly eccentric orbit with an apogee of 203,000 km, perigee ~ 500 km and orbital inclination 65° relative to the earth's equatorial plane and with the line of apsides forming an angle of 88° with the ecliptic plane. The orbital period was ~ 4 days. The spin period was ~ 2 minutes and the spin axis was always pointing within ~ 10° of the sunward direction during its mission period of ~ 7.5 months. At launch the inbound part of the orbit was at a local time of about 1900 and the inbound orbit moved into the dayside after launch. The operation of the satellite was terminated on 12 June 1979.

In Fig. 1, orbit 27 is shown projected on the Geocentric Solar Ecliptic (GSE) XZ-coordinate plane. During this orbit the PROGNOZ-7 orbital plane was close to the noon meridional plane. In this figure the "average" magnetopause location, with variations in the dipole tilt not taken into account, is shown (Fairfield magnetic field model used). The actual magnetopause crossings during this orbit are marked by MP_{in} (inbound) and MP_{out} (outbound). This figure gives a rough idea of the various regions of the magnetosphere that PRONGOZ-7 encountered. Those regions were: 1) the "nightside" high latitude boundary layer (the plasma mantle); 2) the "dayside" boundary layers (the low-latitude boundary layer, the entry layer and the cusp region);

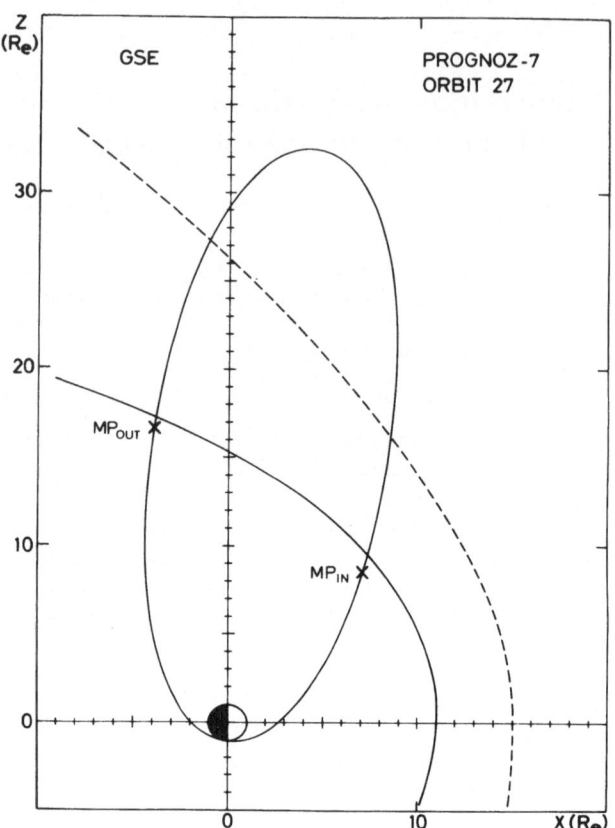

Fig. 1.   An example of a PROGNOZ-7 orbit projected down on the Geocentric Solar Ecliptic (GSE) *XZ*-plane. The "average" magnetopause and bow-shock are marked by the solid curve and dashed curve respectively. The actual magnetopause crossings during this orbit are marked by MP$_{in}$ (inbound) and MP$_{out}$ (outbound).

3) the nightside auroral region at altitudes of a few earth radii; and 4) the ring current region.

The very high eccentricity of the PROGNOZ-7 orbit, however, limits the observation time in the magnetosphere to about 12 to 20 hours of the 98 hours orbit period. This is of course a strong limitation for studies in the magnetosphere (e.g. the ring current). Although other research satellites, such as HEOS-2 and HAWKEYE have had similar orbits, PROGNOZ-7 was the first one carrying mass-analyzing energetic ion spectrometers to the high latitude regions of the magnetosphere. It was the second spacecraft that was equipped for making energetic ion composition studies of the magnetospheric boundary layers. That knowledge about the energetic ion composition is of vital importance for understanding magnetospheric processes has become more and more evident during the past decade. After the first discoveries by SHELLEY *et al.* (1972) of large amounts of precipitating O$^+$ ions and by JOHNSON *et al.*

(1977) of large amounts of trapped O^+ ions near $2R_e$ during magnetic storms, there has been an increasing number of observations indicating that the traditional view on the earth's ionosphere as "passive" in a magnetospheric environment dominated by plasma of solar wind origin, is indeed misleading. In situ ion composition measurements in the magnetospheric plasma at high altitudes, carried out by means of the GEOS, ISEE, PROGNOZ, and SCATHA satellites, have clearly demonstrated the occurrence of substantial fluxes of ionospheric ions (e.g. O^+) not only in the plasma sheet and ring current (GEISS et al., 1978; LENNARTSSON et al., 1979, 1981; BALSIGER et al., 1980; LUNDIN et al., 1980; PETERSON et al., 1981a) but also in the boundary layers (LUNDIN et al., 1979, 1981a; HULTQVIST et al., 1979; PETERSON et al., 1981b) and in the high-latitude lobes (SHARP et al., 1981), i.e. everywhere in the magnetosphere. The strong dominance of O^+ ions during some magnetic storms (e.g., BALSIGER et al., 1980; LENNARTSSON et al., 1981; JOHNSON, 1981; and LUNDIN, 1981a) suggests that magnetospheric processes, at least sometimes, are very strongly coupled to the ionosphere, even to the extent that ionospheric conditions may control the transfer of energy into the magnetosphere.

That ionospheric O^+ ions are ejected locally with high fluxes into the magnetosphere along auroral latitude magnetic field lines has been shown by the S3-3 measurements (SHELLEY et al., 1976; GHIELMETTI et at., 1978 and others). Whether these aurora-associated field-aligned ion beams are sufficient as source for the ionospheric component of the hot magnetospheric plasma appeared uncertain for some time. For source/loss reasons very wide latitude intervals had to be inferred in order to supply the magnetosphere with O^+ ions at a rate corresponding to the observed ring current densities. This supply problem seems to have been removed by recent PROGNOZ-7 observations, which show that almost the entire auroral oval ionosphere (from $L \sim 4-15$) may act as a strong source for O^+ ions during magnetic storms (LUNDIN et al., 1981b).

In this report we will give a brief summary of the PROGNOZ-7 ion composition results from the magnetospheric regions previously discussed and its relevance to present theoretical models of the magnetosphere.

2. Instrumentation

The PROMICS-1 experiment on PROGNOZ-7 (PROGNOZ Magnetospheric Ion Composition Spectrometers) was a joint Swedish-Soviet experiment designed and built at Kiruna Geophysical Institute but with most of the check-out and operational work performed by the Space Research Institute of Moscow.

PROMICS-1 consisted of three energetic ion composition spectrometers (ICSs) for the energy range 0.2–17 keV/q and mass (M/q) range $\sim 1-20$ AMU. Two ICS:s were oriented perpendicular to the spin axis, thereby scanning the ecliptic YZ-plane, and a third ICS which covered the energy range 0.65–5.1 keV/q was oriented 25° with respect to the spin axis. The experiment also comprised three energy-per-charge (E/q) spectrometers using cylindrical electrostatic analyzers and channel multipliers, measuring positive ions and electrons from ~ 10 eV to 50 keV. The ICSs consisted of crossed-field analyzers (Wien filters) placed in front of hemispherical electrostatic energy analyzers. The high voltage control system made it possible to use the ICSs for mass-scanning at fixed energy settings as well as for energy-scanning at fixed mass settings. This latter way of operation turned out to be very useful when the instrument was operating in the low-bitrate mode.

The PROMICS-1 experiment had two basic modes of operation. One mode was a low speed mode which used a low bitrate (~ 8 bps). In this mode, data were taken during ~ 10 seconds and were read out over periods of either ~ 82 s or ~ 164 s. To increase the time resolution for H^+ and O^+ ions in this mode, an integral flux mode was used which provided flux measurements (in particles per $cm^2 s$ sr over the energy range $\sim 0.2-17$ keV/q) for both ion species every ~ 20 seconds. The other mode (high speed) was only used about 1.5 hours per orbit (~ 98 hours orbit period). The data acquisition rate in high speed mode was about 500 times higher than that in low speed. The main characteristics of the PROMICS-1 experiment are summarized in Table 1. Figure 2 shows the periods for which data are available. The solid oblique line marks the time of perigee passage and the two broken lines on each side of it indicate when the satellite passed the altitude 100,000 km, which is of the order of the distance to the magnetopause. As can be seen in the figure, data are available for almost the entire period from a few days after launch to the end of operation in June 1979.

3.　Observations in the Nightside High Latitude Boundary Layer—the Plasma Mantle

As a result of the interaction between the solar wind and the magnetosphere a boundary layer inside the magnetosphere is formed. From the first observations by e.g. Hones *et al.* (1972), Rosenbauer *et al.* (1975), and Haerendel *et al.* (1975), the boundary layer appeared to be quite similar to the magnetosheath, and it was therefore expected to be dominated by magnetosheath plasma. Although there has been some indications that ionosperic O^+ ions are present in the distant tail lobe (Frank *et al.*,

Table 1

Spectrometer and orientation	Energy range (keV)	Energy levels high speed	Energy levels low speed	Energy bandwidth (FWHM)	Field of view (FWHM)	Conversion factor (cm²sr keV cts/part)
ICS-D2 (90°)	0.20–1.57	8	2	0.12	6° × 8°	2.13 10^{-4} E (keV)
ICS-D1 (90°)	2.14–16.9	8	2	0.044	5° × 6°	5.31 10^{-5} E (keV)
ICS-D6 (25°)	0.65–5.08	8	2	0.056	5° × 7°	1.09 10^{-4} E (keV)
IS D3 (90°)	0.02–30.	64	16	0.082	5° × 13°	1.09 10^{-4} E (keV)
ES D4 (90°)	0.03–48.	128	16	0.055	4° × 14°	2.6 10^{-5} η (E) E (keV)
D5 (25°)-(Same as D4)						
ES (e-mode)	0.02–45.	128	8	0.057	4° × 14°	2.8 10^{-5} η (E) E (keV)
IS (p-mode)	0.15–45	128	8	0.057	4° × 14°	2.8 10^{-5} η (E) E (keV)

Abbreviations: ICS, positive ion composition spectrometer; IS, positive ion spectrometer (E/q); ES, electron spectrometer; FWHM, full width half maximum; η, detector efficiency (η for electrons varies between ≈ 1 and ≈ 0.5 and η is about 0.6 for positive ions).

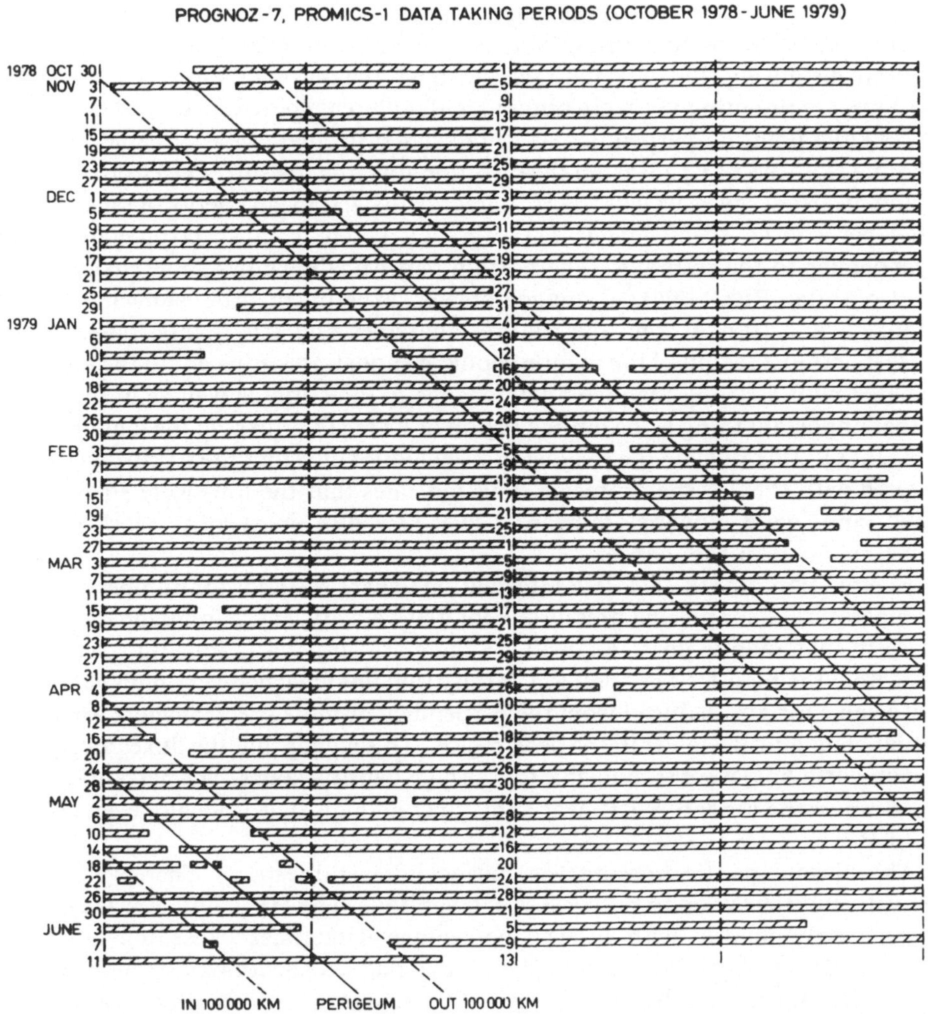

PROGNOZ-7, PROMICS-1 DATA TAKING PERIODS (OCTOBER 1978-JUNE 1979)

IN 100 000 KM   PERIGEUM   OUT 100 000 KM

Fig. 2.   Shows the periods for which PROGNOZ-7 data from the PROMICS-1 experiment are available. Empty areas mark data gaps. The solid oblique line gives the time of perigee passage and the two broken lines on each side of it mark the time when the satellite passed the altitude 100,000 km which is of the order of the distance to the magnetopause.

1977), it was essentially not until the PROGNOZ-7 and ISEE observations that direct evidence for the presence of significant amounts of ionospheric ions in the boundary layers was obtained (see e.g. LUNDIN *et al.*, 1979; HULTQVIST *et al.*, 1979; and PETERSON *et al.*, 1981b). In a recent paper by LUNDIN *et al.* (1981a), the mantle observations from PROGNOZ-7 have been extensively discussed. We will in this section show a few of the mantle examples.

The first example is from one of the largest magnetic storms in terms of peak Dst-value (-197 UT) in the lifetime of PROGNOZ-7. The satellite entered the mantle at

2025 UT on 3 April and passed the magnetopause at ~ 0010 UT on 4 April. K_p varied between 7 and 8 during this time period. Figure 3 shows the integral fluxes of H^+, O^+, and electrons during this outbound pass through the plasma mantle. The data in the two frames representing O^+ were obtain with different spectrometers: the upper one (0.2–17 keV) pointing perpendicular to the sun-earth line (see e.g. Table 1) and the lower one (1.1–3.8 keV) 25° from the direction towards the sun. The electron integral flux and mean energy as well as the H^+ integral flux and mean energy of ions (from the energy per charge spectrometer D3) were all taken from spectrometers looking perpendicular to the spin axis. The bottom panel finally shows some magnetic field data from the PROGNOZ-7 magnetometer experiment (the sunward, i.e. B_x, component and $|B|$). Among the features characterizing this mantle pass, the very intense innermost region of the mantle should be noted. It is adjacent to the lobe and is associated with a strong magnetic field change in direction as well as in magnitude. Another important thing is the limited region of strong O^+ fluxes in the mantle with no significant O^+ flux outside the magnetopause. Notice also that the 0^+ flux was mainly measured with the 25° spectrometer. This indicates that the ions were strongly field aligned and were flowing tailward. The O^+ flux frequently reached above 10^6 $\text{cm}^{-2}\text{s}^{-1}\text{sr}^{-1}$ in the mantle.

Figure 4 contains the NTPVB information (where N stands for number density, T for temperature, P for plasma pressure, V for flow velocity, and B for magnetic field) for this storm time mantle pass. As in Fig. 3 there are large variations (in space and/or time) within the mantle. The total ion density (full line) reached very high values. The density value for H^+ deduced from the perpendicular mass spectrometers were about an order of magnitude lower in the peak region. This means that the fluxes along the x-axis (sun-earth line) were responsible for the high density values. The percentage of O^+ ions, given in the second panel, was sometimes so high that it locally even exceeded that of H^+.

This plasma mantle crossing was unusual in several respects. There were number density fluctuations of up to two orders of magnitude. The He^+ content was unusually high (the only mantle with a He^+ percentage sometimes exceeding 10%). The antisunward flow in the mantle as well as in the magnetosheath was strongly beamed.

The very intense, high density, region of magnetosheath-like plasma at the inner edge of the mantle had characteristics resembling more those in the magnetosheath than those in other parts of the mantle. However, the fact that this intense region with associated magnetic field change (diamagnetic decrease and strong reorientation) was found at the inner edge of the mantle makes it unlikely that the satellite came into the magnetosheath for the limited time interval between ~ 2030 and ~ 2110 UT. This therefore suggests that high density magnetosheath plasma elements may penetrate the magnetopause further upstream from the high latitude boundary layer and form high density regions in the plasma mantle. Such a penetration of solar wind plasma elements into the magnetosphere was first proposed by Lemaire (1977) and its consequences were further discussed by Lemaire and Roth (1978) and Lemaire *et al.* (1979). Although some boundary layer observations have been interpreted as possible "inclusion" events (e.g. Paschmann *et al.*, 1978), it was essentially not studied in detail before the PROGNOZ-7 observations in the plasma mantle (Lundin and Aparicio, 1981).

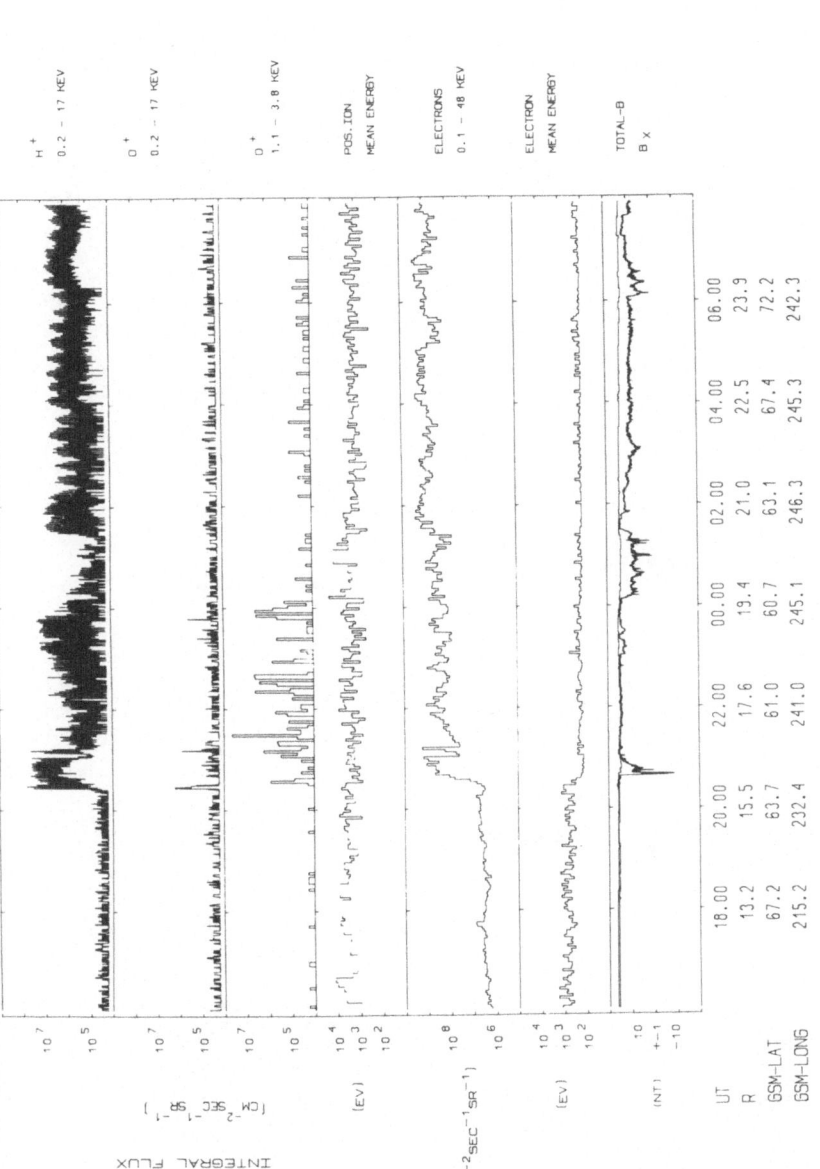

Fig. 3. Integral flux data from the plasma mantle crossing of PROGNOZ-7 on 3–4 April, 1979, during a magnetic storm. The two top panels show integral fluxes for H$^+$ and O$^+$ over the energy range 0.2–17 keV taken from the perpendicular ICS:s (scanning the ecliptic YZ-plane). The third panel shows the O$^+$ flux in the energy range 1.1–3.8 keV as taken from the ICS looking in the sunward direction (25° with respect to the satellite spin axis). The fourth panel from the top gives the average energy for ions, as deduced from the perpendicular E/q ion spectrometer. The fifth and sixth panels give the integral flux and average energy of electrons in the energy range 0.1–48 keV taken from the perpendicular electron spectrometer. The bottom panels show the magnetic field component in the sunward direction and the radial magnitude (logarithmic scale used) as taken from the on-board magnetometer. The time and space coordinates (geocentric radial distance in earth radii, latitude and longitude in GSM) are given along the horizontal axis.

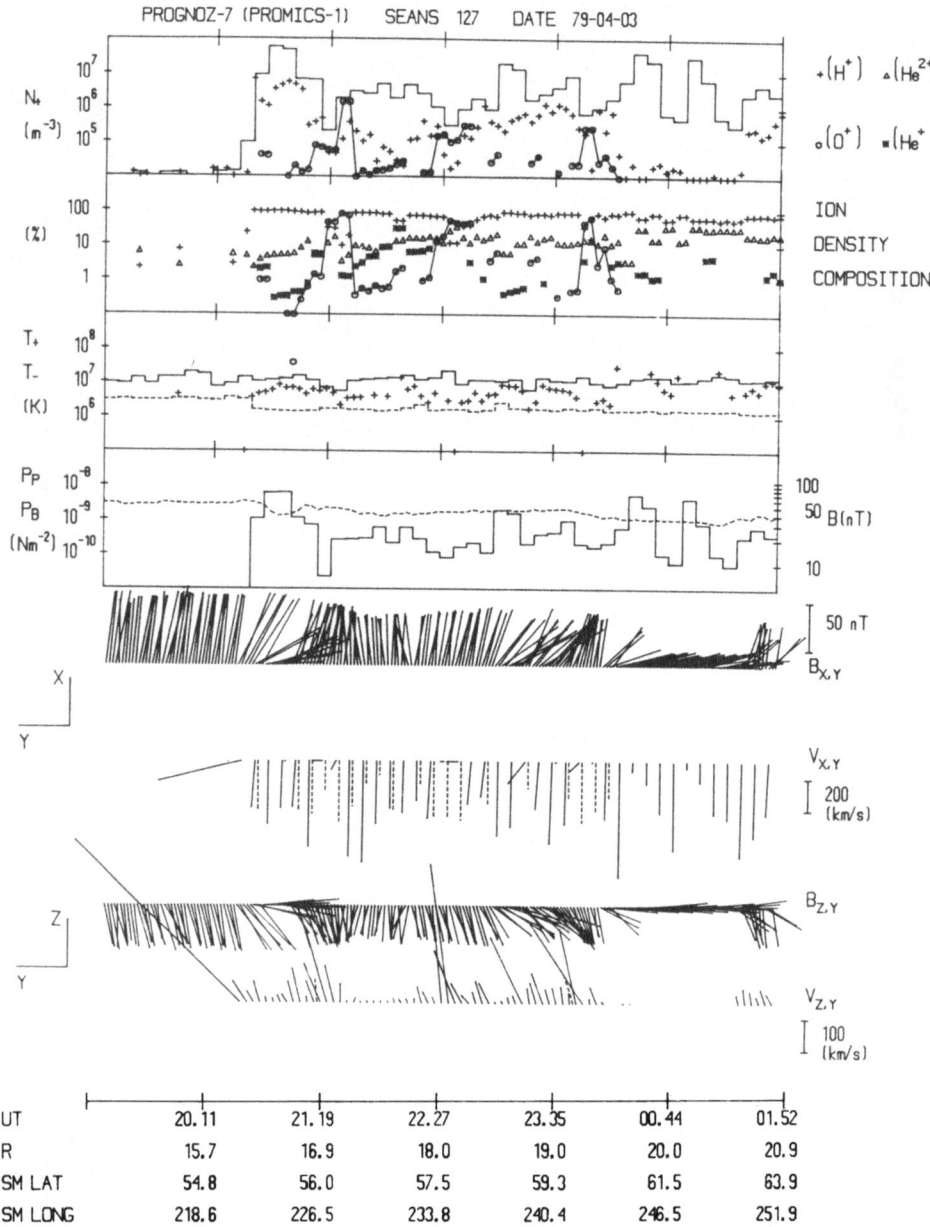

Fig. 4.   Plasma parameters and magnetic field data (in SI units) for the plasma mantle crossing on 3–4 April 1979. The upper panel shows the ion number density (N_+) as deduced from the E/q spectrometers and assuming the ions were all protons (solid line). Plus signs (+) represent the density of H^+ as deduced from the perpendicular ICSs (assuming isotropy) and circles ($\bigcirc$) represent the number density of O^+ using all ICSs. The second panel from the top represents the percentages of the four major ion constituents with respect to the total number density (logarithmic scale used). The third panel shows the temperatures of ions (solid line) and electrons (broken line) as deduced from the E/q electron and ion spectrometer data fitted onto

The "inclusion" structures showed more or less "activity". Figure 4 is an example of an "active" structure. The notation "active" was chosen simply because the penetration structure still showed high density, high flow velocity and no O^+ content, as in the magnetosheath, apparently not having lost too much of its excess momentum, but was at the same time associated with very high fluxes of ionospheric O^+ in other parts of the mantle. The flow vectors of H^+ (full lines) and O^+ (broken lines), given in Fig. 4, also show that the mantle plasma was essentially flowing field aligned and tailwards with about the same velocity for both ion constituents.

Our next example of a plasma mantle encounter, shown in Fig. 5, was also obtained during a magnetic storm which reached its peak negative Dst of -98 nT around 2100 UT on 21 February 1979. It also represents an example with strong O^+ fluxes in the plasma mantle. The O^+ percentage locally reached up to $\sim 10\%$. Notice that here also a very sharp decrease of the O^+ density was observed at the magnetopause (around 0200 UT) with no measurable O^+ flux outside it. The interval between ~ 23 and 24 UT was characterized by a very high plasma density (even higher than after 0200 UT) and magnetosheath-like composition and flow, again more similar to the magnetosheath than to the mantle. There were, however, several features in this region which differed considerably from that found further out in the magnetosheath, such as magnetic field direction and magnitude, which suggest that it was part of the mantle and not of the magnetosheath.

The energy spectra of the individual ion species as well as the energy spectra taken from the E/q positive ion and electron spectrometers are shown in Fig. 6 for this second mantle example. Notice the gradual increase of the ion flux and the gradual hardening of the ion spectra when going from the lobe region into the mantle. The presence of He^{2+} throughout the entire mantle region indicates, as expected, a strong contribution of solar wind ions to the mantle plasma. As can be seen in Fig. 5, He^{2+} was in general more abundant than O^+ in this mantle. The gradual softening of the ion spectra towards the lobe, first described by ROSENBUER et al. (1975), is most likely a result of convection, caused by the dawn-dusk electric field which separates the ions according to their initial velocities along the magnetic field lines. The O^+ ions were found in the mantle only in the highest energy channels. There is also present in Fig. 6 a high energy component of the H^+ population. We will return later to a discussion of some of the spectral characteristics.

Figure 7 shows one of the most unusual mantle passages of PROGNOZ-7 in terms of the time it took to pass the mantle and in terms of the apparent mantle thickness ($\sim 11.5 \, R_e$). The passage occurred during a series of small storms (Dst varied between -35 and -70 nT and K_p varied between $3+$ and 5). Except for the outermost region, the mantle was characterized by fairly high O^+ fluxes. Notice here that the most intense

---

Fig. 4.   (cont.)

Maxwellians. In the same panel the "perpendicular" H^+ ($+$) and O^+ ($\bigcirc$) temperatures have been plotted. The fourth panel shows the ion plasma pressure (solid line) and magnetic field pressure (dotted line).

The low part of the NTPVB-plot gives the magnetic field and flow velocity components in the XY and YZ Solar Ecliptic coordinate planes. Solid line of the flow velocity represents the H^+ flow vector and broken line gives the O^+ flow vector. The time and space coordinates (in Solar Magnetic, SM, coordinates) are given along the horizontal axis.

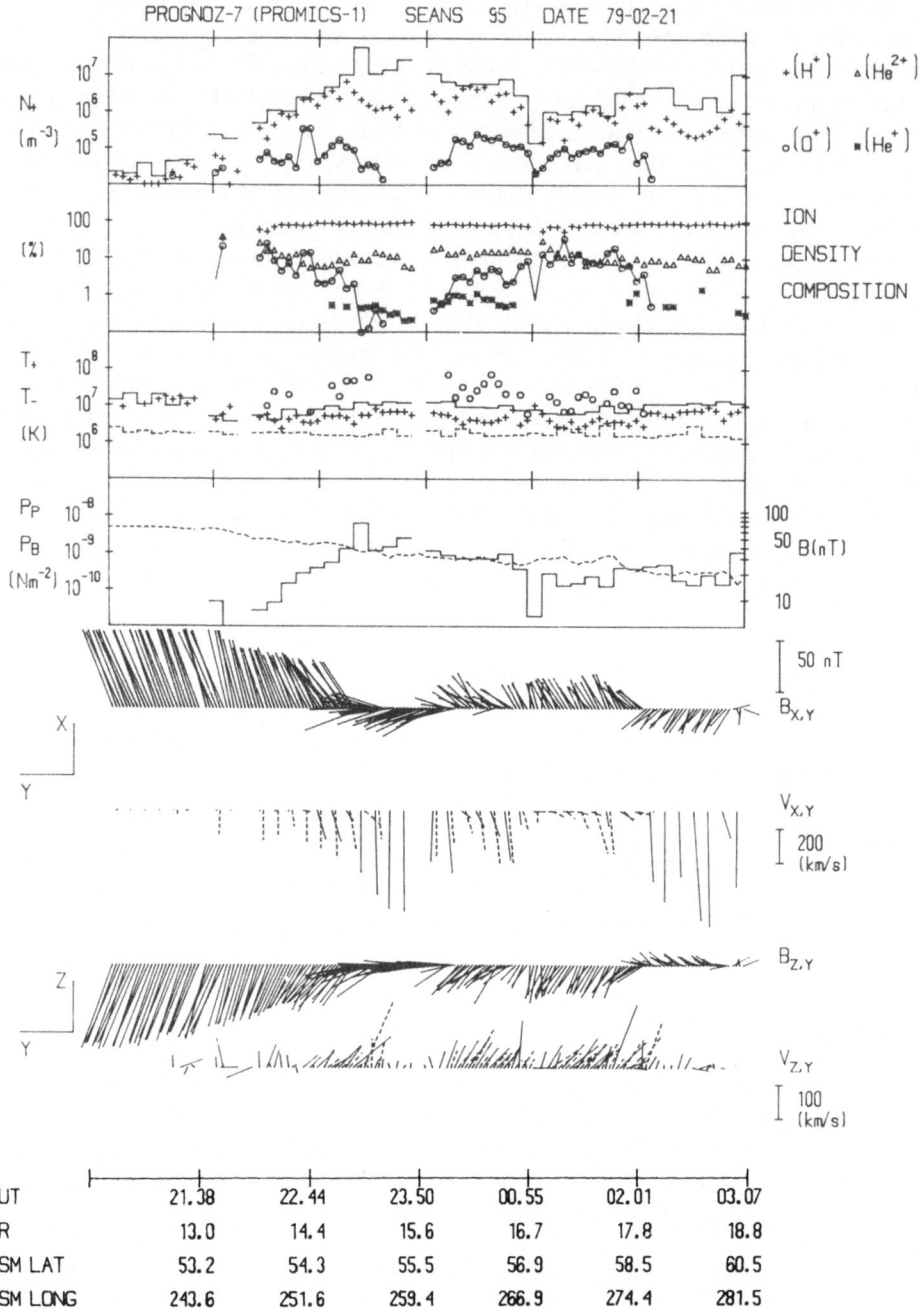

Fig. 5. Plasma parameters and magnetic field data from the PROGNOZ-7 mantle crossing on 21–22 February, 1979. The format is the same as in Fig. 4.

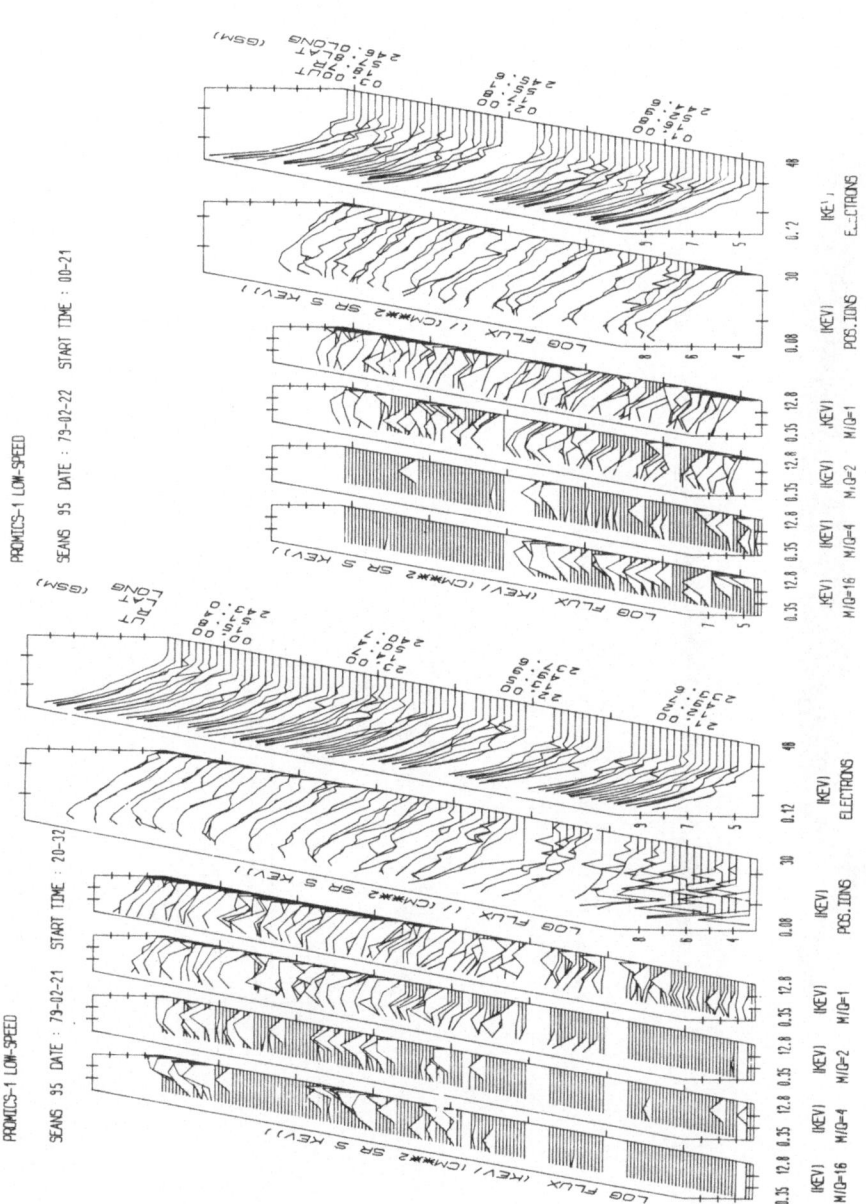

Fig. 6.   Spectrograms (flux vs energy vs time) taken from the perpendicular spectrometers for the mantle crossing on 21–22 February, 1979. To the left, individual four point energy spectra for the four major ion constituents are depicted (using differential energy flux units). To the right, 16 point energy spectra for positive ions and electrons (E/q spectrometers), using differential flux units, are plotted. Time and space coordinates (in GSM-coordinates) are given along the inclined vertical axis.

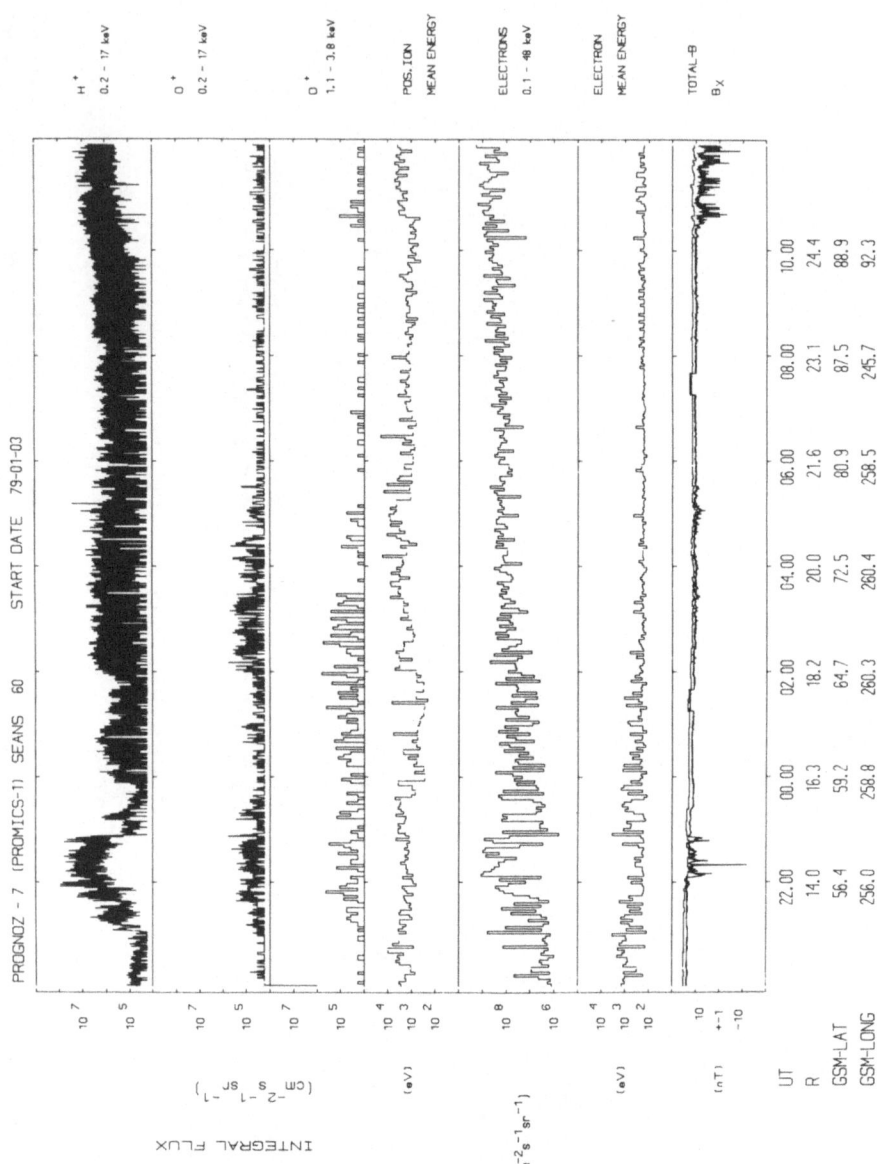

Fig. 7. Integral flux data from the PROGNOZ-7 plasma mantle crossing on 3–4 January, 1979. The format is the same as in Fig. 3.

mantle fluxes, as measured by the "perpendicular" spectrometers, were found around $\sim 22-23$UT when a significant magnetic field disturbance occurred as well. In Fig. 8, showing the NTPVB-plot from this innermost part of the mantle, we can see that this region is again characterized by a high density, high β (~ 1) plasma with magnetosheath-characteristics. A difference between this region and the ones previously discussed is, however, that the flow velocity was much reduced as compared to that in the magnetosheath and that considerable fluxes of O^+ were present within the structure as well. This type of structure, denoted as stagnant penetration structures in the LUNDIN and APARICIO (1981) paper, has therefore characteristics similar to what one would expect from a plasma element that has lost most of its excess momentum after the penetration of the magnetopause.

The energy spectra of the various ions from the mass spectrometers, the total ion spectrum and the electron spectrum are shown in Fig. 9. Some characteristics to be noted there are the following.

The total ion spectrum was considerably harder in the "penetration" structure (~ 2200 UT) than in the magnetosheath further out. Also some of the electron spectra were hotter than the magnetosheath ones.

The total ion spectrum was sometimes double peaked in large regions of the mantle due to low energy protons forming the peak at lower energy and mainly O^+ ions the higher energy one (see e.g. ~ 0220 UT on 4 January). Notice also that during the time period ~ 0200 to 0530 UT on 4 January the more energetic ion peak contained mainly O^+ in the inner part and mainly H^+ in the outer part. The intermediate interval was characterized by a mixing of H^+ and O^+ in the peak. No He^{2+} ions could be observed in the high energy peak. An obvious conclusion is, therefore, that this energetic ion component (peaking between 10 and 20 keV) originated in the ionosphere. The temporal/spatial shift of the ion peak versus mass can be understood in terms of convection due to the dawn dusk electric field which drives the slowest ions (O^+) more deeply into the mantle. The fact that both O^+ and H^+ (for the energetic component) peak at about the same energy suggests that the acceleration mechanism is not mass dependent (e.g. a parallel potential drop).

In the majority of the PROGNOZ-7 passages through the northern hemisphere mantle, a flow approximately along the magnetic field lines in the direction out of the northern hemisphere has been observed, except possibly in the magnetosheath-like regions previously mentioned. But there are also mantle crossings in which virtually no ordered flow has been found. One such mantle crossing is shown in Fig. 10. This outbound mantle passage on 7–8 January 1979 occurred during a small magnetic storm which started in the middle of 7 January and reached a peak negative Dst value of -94 nT between 20 and 21 UT. K_p was, however, only 4 and 3+ during the mantle encounter. In Fig. 10 we can see that there was practically no regular flow in the plasma mantle (except in a narrow region near 01 UT). The computer evaluation of flow vectors produced low and very variable values in agreement with what is expected for a stagnant plasma.

The ion densities in the mantle of Fig. 10 are quite low as compared to the other mantle examples shown. The O^+ percentage was on average higher than in the previous cases and was also, in contrast to the other cases shown above, mostly higher

R. Lundin, B. Hultqvist, N. Pissarenko, and A. Zacharov

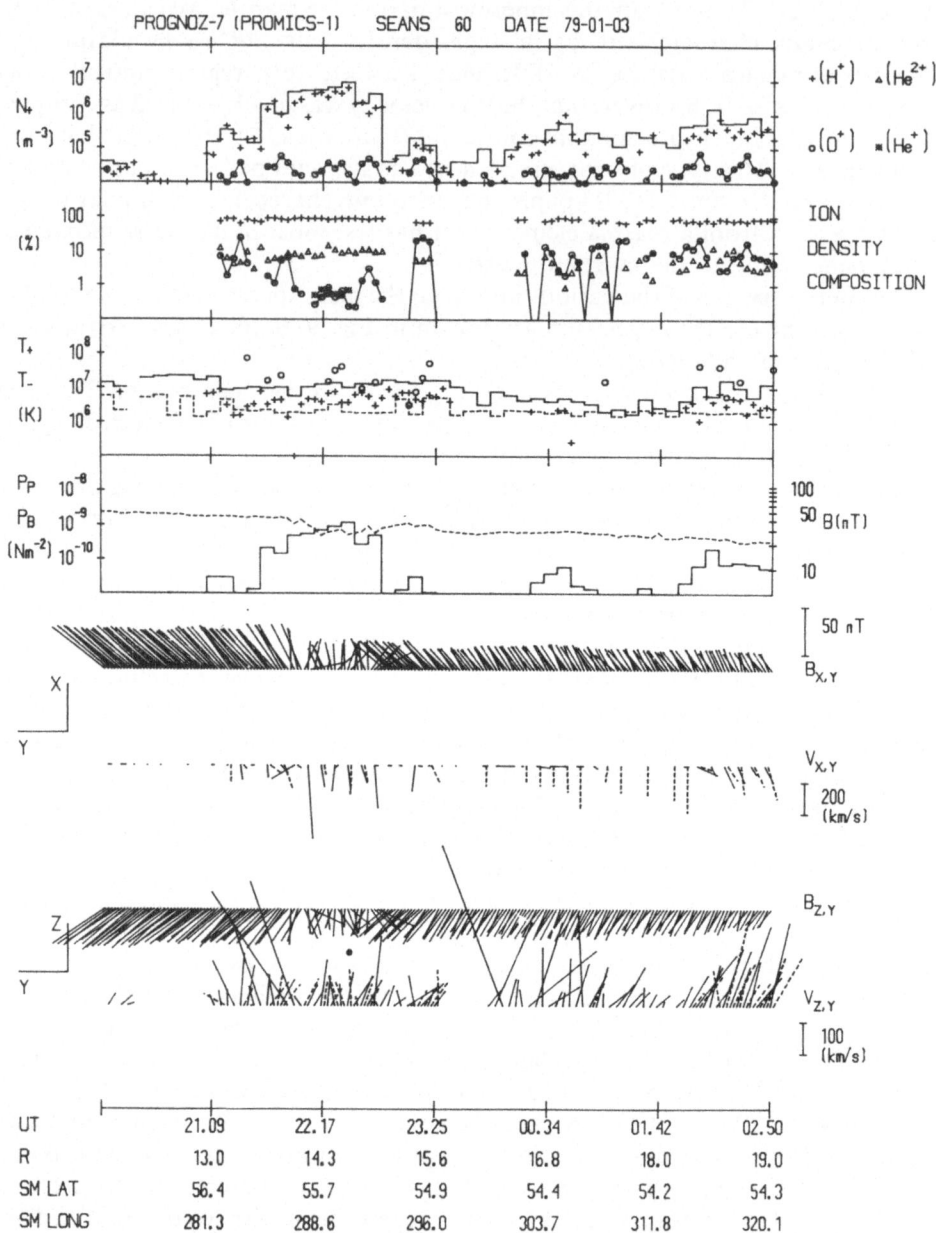

Fig. 8. Plasma parameters and magnetic field data from the plasma mantle crossing on 3–4 January, 1979. The format is the same as in Fig. 4.

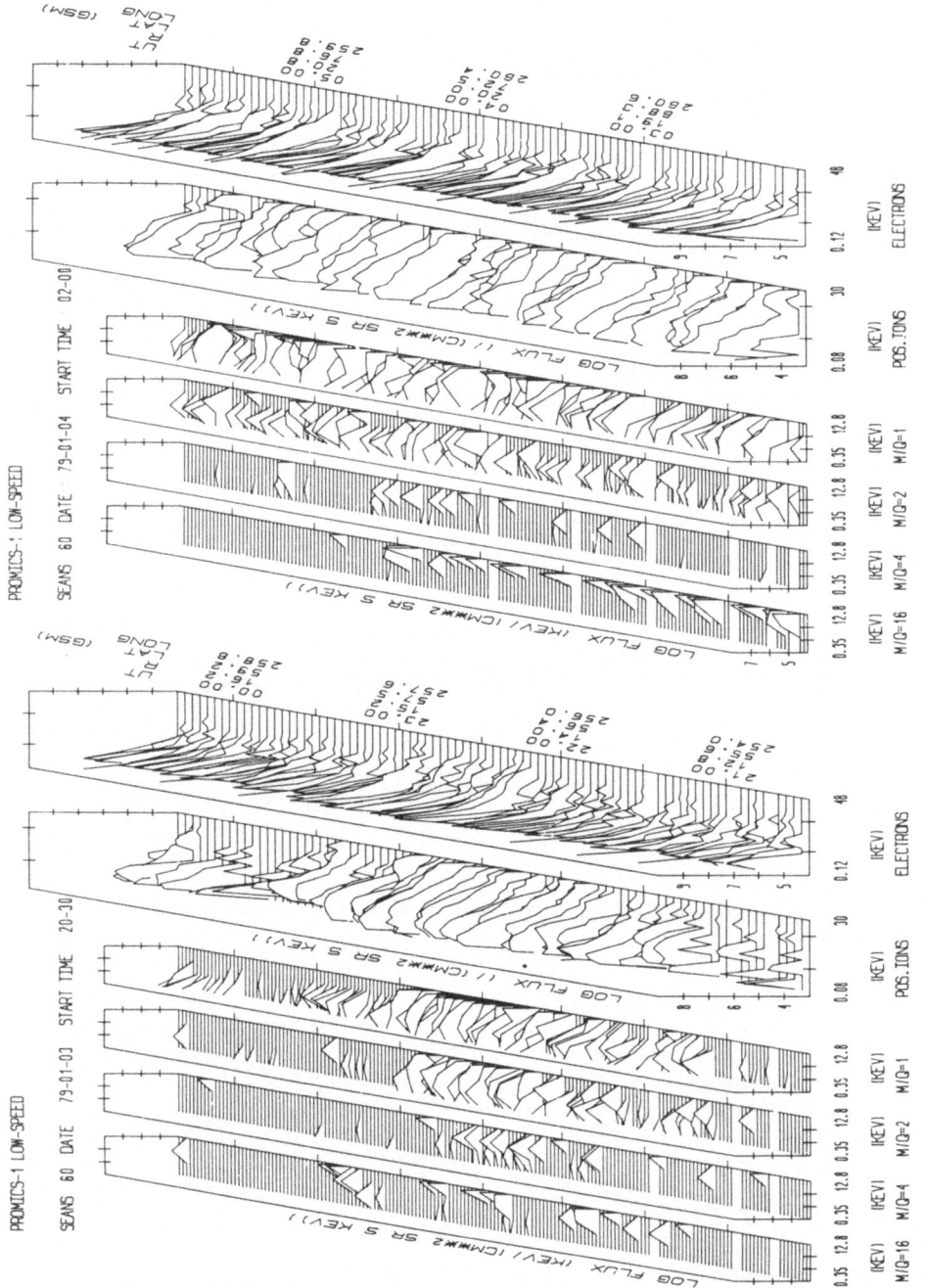

Fig. 9.   Ion and electron spectra from the plasma mantle crossing on 3–4 January, 1979. The format is the same as in Fig. 6.

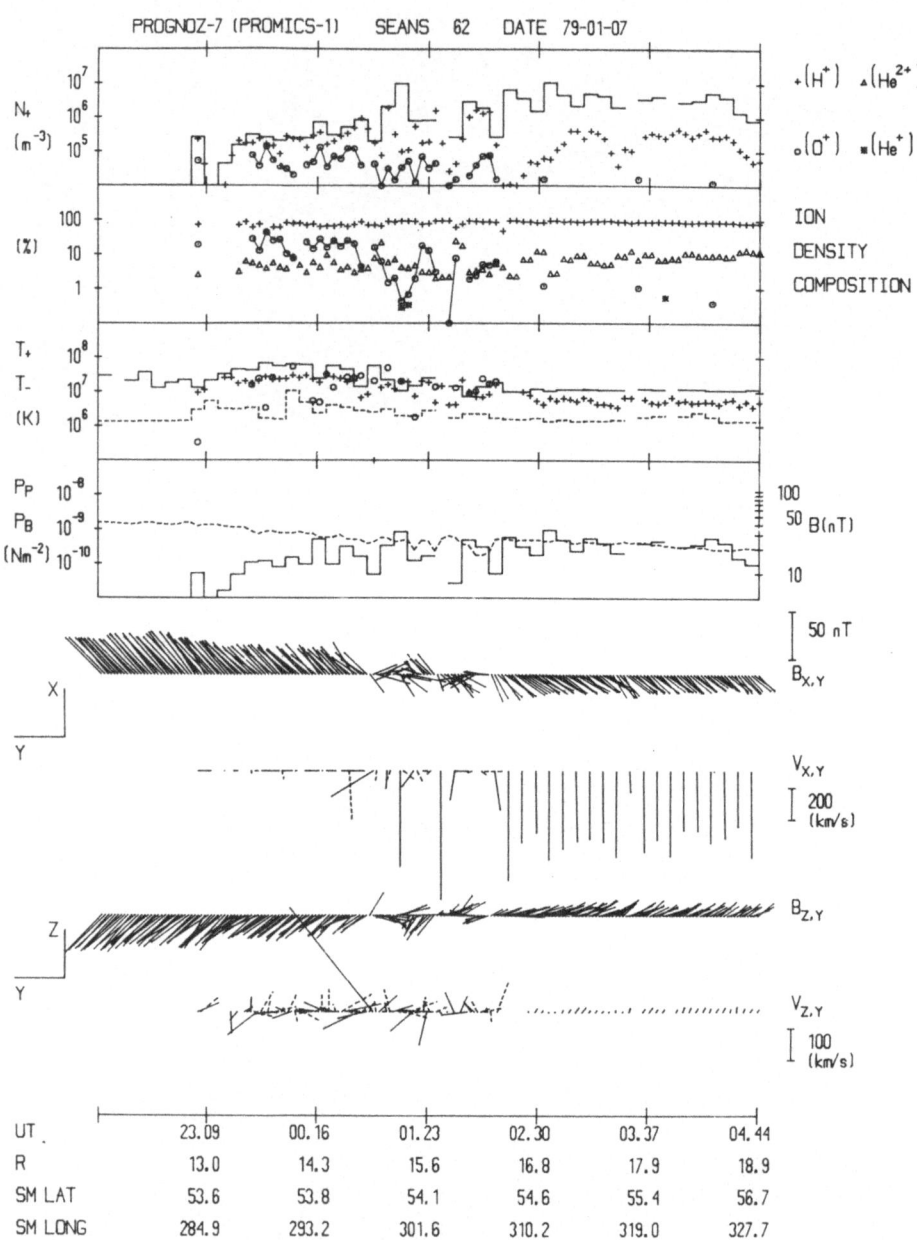

Fig. 10.   Plasma parameters and magnetic field data from the PRONGOZ-7 plasma mantle crossing on 7–8 January, 1979. The format is the same as in Fig. 4.

than that for He^{2+}. More important is the fact that both the ion and electron temperatures were higher in this mantle than in any of the previous ones. This is shown by the temperature frame in Fig. 10, but is even more evident when studying the energy spectra of Fig. 11. The ion differential flux spectrum peaks in the several keV range and is very different from the "normal" ion mantle spectrum. Also the electron energy spectrum was much harder than in the earlier cases and showed significant fluxes even above 10 keV in parts of the mantle. This type of mantle observation is very different from what one expects to find in present models of the magnetosphere. It seems hard to understand how a hot and stagnant plasma can exist on mantle field lines, generally believed to be open. The fact that these stagnant plasma mantle observations occurred mainly along the magnetospheric flanks might indicate that the magnetic field lines were actually closed in this region. However, a few of these cases were found so close to the nightside polar region that a closure of these magnetic field lines is inconceivable without a strong revision of present magnetospheric magnetic field models. A second possibility could be that a "pseudo trapping" of plasma occurs along the mantle field lines, either as a result of bent magnetic field geometries or as a consequence of reflecting electrostatic potentials. Finally, the hot and stagnant plasma may be the result of a strong turbulence process which heats and randomizes the plasma in the mantle. At present it seems difficult to interpret these observations in terms of existing models. A detailed study of stagnant plasma mantle cases, using both HEOS-2 and PROGNOZ-7 data, has been initiated.

We have seen that in all the four examples of mantle passages presented above, O^+ ions have constituted a significant, although not dominant, fraction of the number density. All these cases represent more or less disturbed magnetospheric conditions. In quiet conditions much weaker mantles with little or no O^+ content are observed. By weak we mean here small thickness and/or low density. The relation is, however, not a very clear one with regard to the total number density and thickness of the mantle. This can be seen from Fig. 12 in which the product of the thickness of the mantle and the peak proton flux recorded during the passage is plotted for the 32 mantle passages of PROGNOZ-7 that have been analysed. The peak proton flux was then the integral flux (0.2–17 keV) taken from the "perpendicular" mass spectrometers. As a magnetic activity parameter, the sum of K_p over a 12 hour period centered on the mantle passage has been plotted. The dots are for the nightside mantles and triangles represent flank mantles. As can be seen, the exclusion of the flank mantles does not change the relation between the two variables. A positive correlation can be seen. It is, however, not a very strong one. The scatter is large especially at low K_p values and fairly good mantles are found also at low activity level.

If a similar scatter diagram is prepared for the O^+ ion flux, the diagram in Fig. 13 is obtained. Although the degree of scatter is large also in this case, especially for low K_p values, the correlation is much better. The correlation coefficient for an exponential dependence of the sum of K_p versus the product of peak flux and thickness is as high as 0.82. Again there is no obvious difference between the flank mantles and the nightside ones.

The difference between the point distributions in Fig. 12 and 13 is obviously consistent with the existence of a H^+ source fairly independent of magnetic activity in

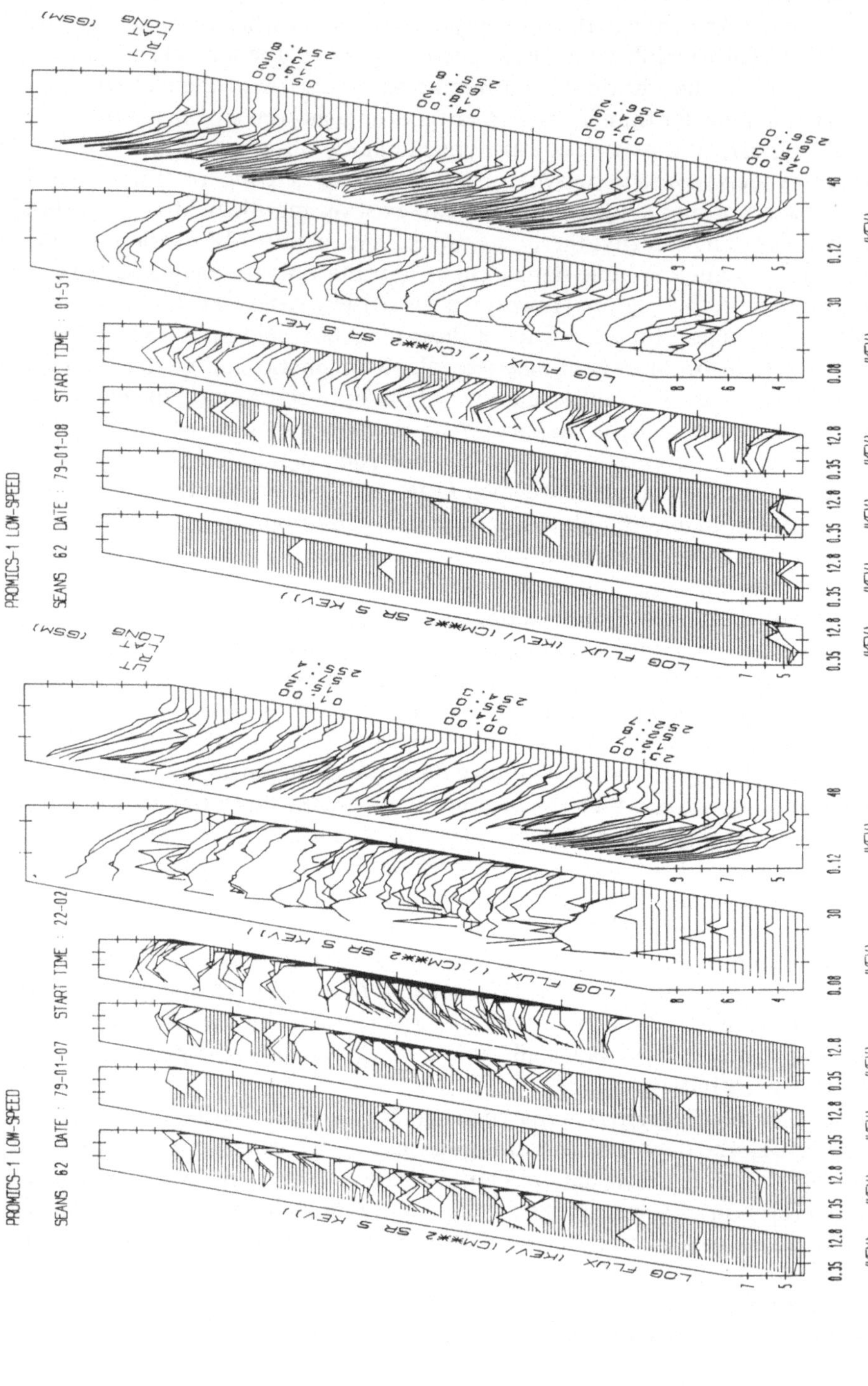

Fig. 11.   Ion and electron spectra from the plasma mantle crossing on 7–8 January, 1979. The format is the same as in Fig. 6.

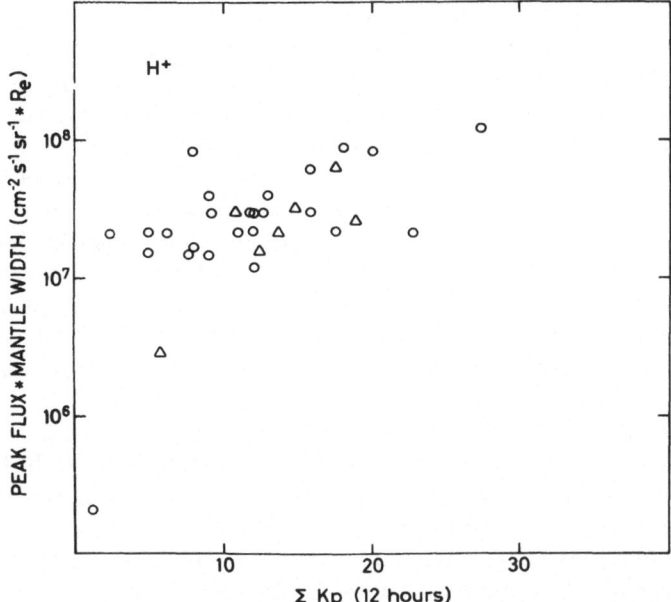

Fig. 12.   Diagram showing the peak flux of protons (H^+) times the estimated width of the plasma mantle versus magnetic activity for 32 mantle passages. The mantle width was taken as the difference in radial distance of the inner and outer (magnetopause) boundaries of the mantle along the satellite trqzuctory. As magnetic activity parameter the sum of K_p over a 12 hour period has been taken.

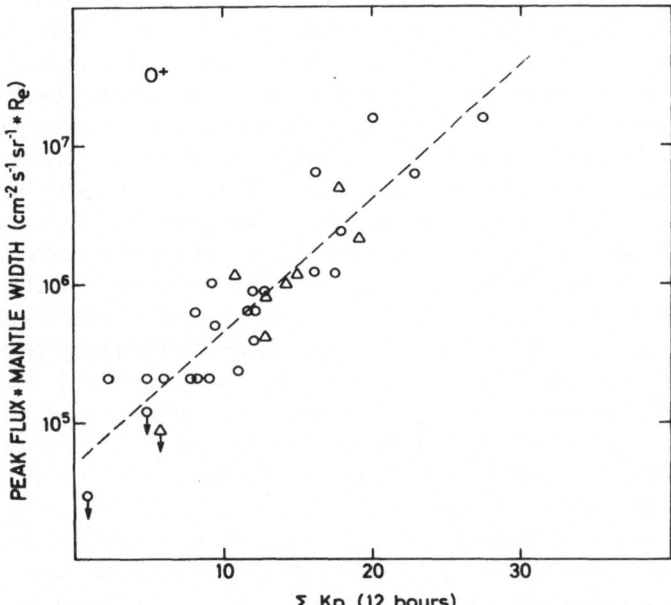

Fig. 13.   Same as Fig. 12 but for O^+ ions. The broken line represents an exponential least squares fit of the data points with a correlation coefficient of 0.82.

addition to the activity dependent one(s).

The existence of magnetosheath-like regions within the mantle, even as far away from the magnetopause as at the inner edge of the mantle, seems to have been reported first by Lundin and Aparicio (1981). In some of these cases one may dispute whether the satellite was in the mantle or had made an excursion into the magnetosheath. However, because of a number of arguments, including the existence of O^+ in some of these regions, sometimes very special flow and magnetic field characteristics, the frequent occurrence of a magnetosheath-like region at the transition between mantle and lobe, and the occurrence of a variety of degrees of "magnetosheath-likeness" in both particles and fields among the observed $\beta \gtrsim 1$ events, we find it difficult to believe that the satellite was in the magnetosheath during these events. We believe instead that they were part of the mantle, but obviously special parts, frequently associated with strong currents.

At the edges of these dense regions the density of ionospheric ions (O^+) generally peaks, indicating that particularly intense Birkeland currents flow at the borders of the regions (see e.g., Fig. 4). Some high-β regions in the mantle are associated with fairly slow flow and contain a significant percentage of O^+ ions in the otherwise magnetosheath-like plasma. In these regions a diamagnetic decrease of the magnetic field intensity is generally found but only small directional changes. They tend to occur at greater depths in the mantle than the more "active" extreme, which, together with some of the other characteristics, indictates that these less "active" regions are simply older evanescent structures of the first-mentioned kind.

A consequence of the penetration of magnetosheath plasma elements into the magnetosphere may be that ionospheric ions become accelerated in the current circuit generated by the penetrating elements. The observations indicate such a direct interaction with the ionosphere. The only theory which seems to account for these observations is the impulsive penetration theory first proposed by Lemaire (1977). We will present more evidence for the impulsive penetration in the next section.

Finally, we shall in this section discuss briefly the source process for the energetic O^+ ions in the mantle and the flow characteristics there. In Fig. 14 we have plotted an example of the energy spectrum and pitch angle distribution of H^+, He^{2+}, and O^+ from a deep nightside plasma mantle when data was taken in high speed mode. Notice that each ion species has formed its own conical distribution, with stronger folding around the field line the heavier the ions. All ions also move with roughly the same parallel velocity ($\sim 130-150$ km/s). This is most likely a velocity dispersion effect caused by the dawn-dusk electric field which separates ions with respect to their parallel velocities due to the $\varepsilon \times B$ drift. The same effect causes the earthward softening (decrease in velocity) of the mantle ion spectra as was first observed from HEOS-2 by Rosenbauer et al. (1975). This type of feature appears to be fairly typical of the deep nightside mantle where, most likely, the dawn-dusk electric field has "organized" the mantle plasma according to the parallel velocity of individual ions. This evidently indicates that very little thermalization of ions occurs in the deep nightside mantle and that the transport of plasma there can frequently be described by single particle particle motion.

The presence of O^+ in the mantle is usually highly variable and, as previously

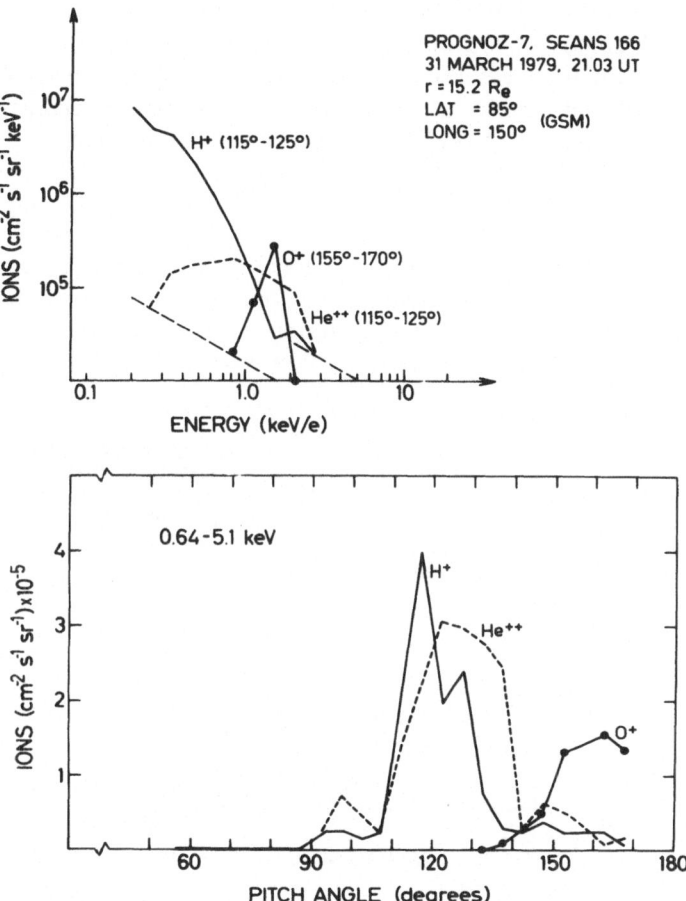

Fig. 14. An example of ion energy spectra and pitch angle distributions in the deep nightside plasma mantle.

discussed, dependent on the disturbance level. In the course of a mantle passage, usually of the order of a few hours long, the mantle characteristics may change considerably. A study of the systematic radial dependence of the mantle characteristics may obviously be affected by this variability. The mantle shown in Fig. 7, 8, and 9 appears to be one with little or no variation. The double peaked ion spectrum in Fig. 9, discussed previously, is such an example where stable conditions seems to have prevailed for at least three hours. The energetic ion peak, containing mainly O^+ in the inner part and H^+ in the outer part, is clearly of ionospheric origin. Under the assumption that the ionospheric ions were accelerated upwards from the dayside auroral ionosphere, the spatial separation versus mass of the ion beam in the mantle could have been due to a dawn-dusk electric field of $\sim 0.3 - 0.5$ mV/m. This is in quite good agreement with what one would expect at this altitude over the polar cap.

Figure 15 shows another high-speed data example of H^+, He^{2+}, and O^+ energy spectra and pitch angle distributions, but now in a part of the mantle very close to the cusp. The magnetopause was crossed inbound some 50 minutes prior to the time period contained in Fig. 15. Notice the very broad beam for all ion constituents and the very energetic appearance of the O^+ spectrum with high fluxes present up to the instrument energy limit of 17 keV. To explain such a broad and energetic upward flowing ionospheric ion beam at this altitude, nonadiabatic scattering or transverse acceleration has to be inferred. Alternatively, these O^+ ions do not originate directly from the ionosphere but from some other region with hot magnetospheric O^+, e.g. the dayside ring current near the magnetopause. However, there are several reasons for these O^+ beams to be due to direct extraction and acceleration of ions out of the ionosphere. One reason may be that the O^+ flow has a different speed as compared to

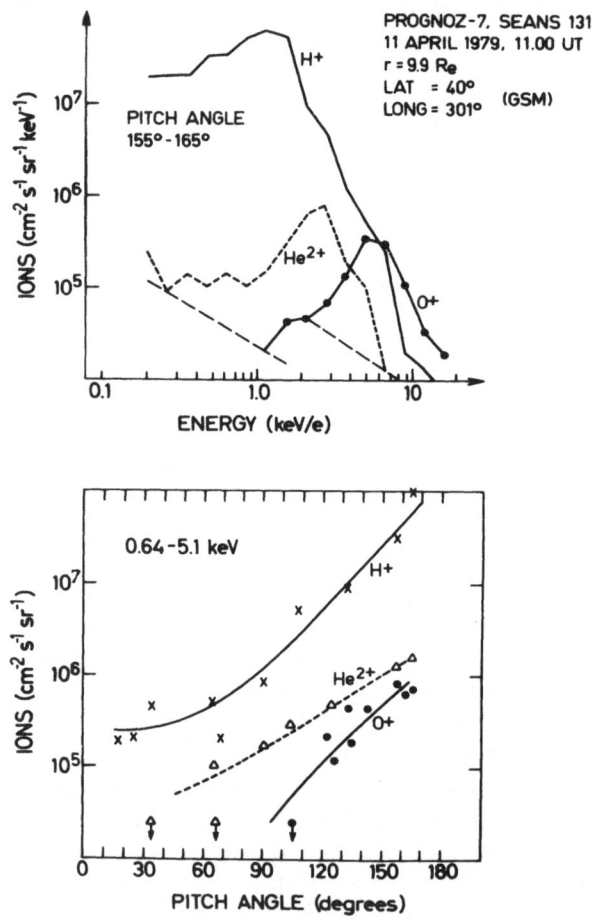

Fig. 15.   Energy spectra and pitch angle distributions of a broad ion beam at an altitude of $\sim 57,000$ km in a part of the plasma mantle near the cusp.

the solar wind ions (H^+ and He^{2+} having about twice the velocity of O^+) and also a different flow direction, an example of which can be seen in Fig. 16. Such "hot" and broad upward flowing O^+ beams have been found on auroral field lines at much lower altitudes as well (LUNDIN *et al.*, 1981b; SHELLEY, 1979).

In Fig. 16 we have plotted the flow vectors for H^+ (open vector) and O^+ (filled vector) and the magnetic field orientation (solid line) for an inbound passage through the dayside high latitude boundary layer near the cusp. The flow vectors for various times along the satellite orbit have been projected onto the XY and YZ-solar ecliptic planes. The broken curve shows the satellite trajectory in Solar Magnetic coordinates. Notice in this figure that while the O^+ ions appear to flow mainly along the magnetic field lines, the H^+ cross field component is comparable to the parallel component especially in the outer part of the boundary layer. A consequence of this is therefore not only that the O^+ ions and the bulk of H^+ ions originate in different source regions, but also that magnetosheath plasma must have had direct access to the outer boundary layer.

To summarize the results of the PROGNOZ-7 ion composition measurements in the plasma mantle:

1)   The plasma mantle frequently contains an appreciable amount of terrestrial

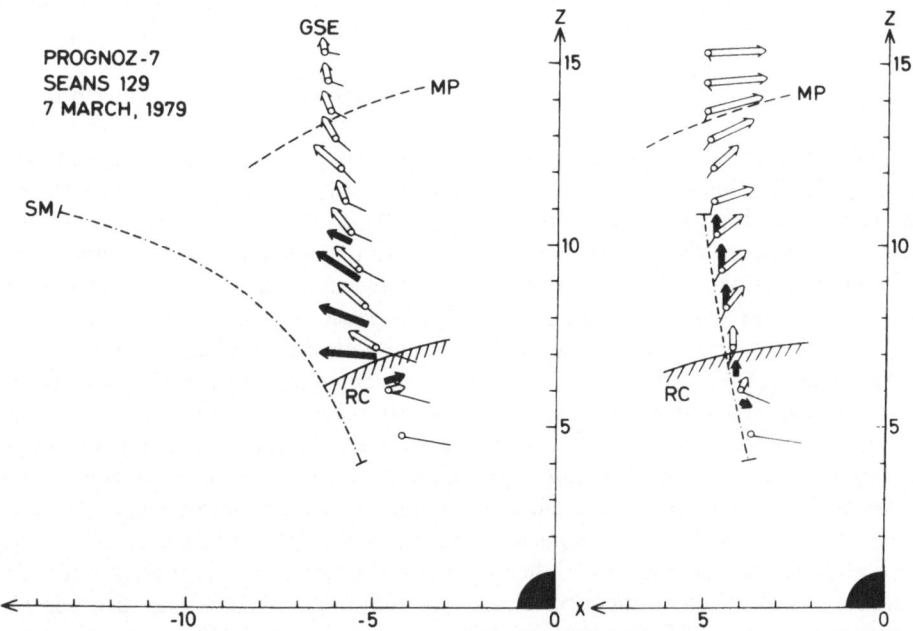

Fig. 16.   Inbound passage through the dayside high latitude boundary layer showing projections of the flow vectors for H^+ (open symbols), O^+ (filled symbols) and the magnetic field (solid line) onto the YZ- and XZ-solar ecliptic planes. The flow velocity scale is linear and the biggest flow vector in the YZ-projection corresponds to about ~ 200 km/s whereas the biggest on the XZ-projection corresponds to ~ 300 km/s. The satellite trajectory in solar magnetic (SM) coordinates is marked by the dashed line. The location of the magnetopause is indicated by MP.

ions. O^+ number densities up to 10% of the total ion density are not uncommon, but generally the O^+ abundance is a few percent or less.

2)   The O^+ abundance is usually much higher than the He^+ abundance, indicating that the source for the terrestrial ions in the mantle is predominantly the low-altitude ionosphere.

3)   The ionospheric ions frequently have energies in the several keV-range in the mantle (up to ~20 keV have been identified).

4)   Far tailward of the cusp, the mantle is usually connected with strong velocity dispersion effects due to the dawn-dusk electric field.

5)   The ionospheric ion content in the mantle is strongly dependent on the magnetic activity level.

6)   The nightside magnetopause appears to be a more solid boundary for O^+ ions than is expected on the basis of the open magnetosphere model.

7)   Magnetosheath-like ($\beta \gtrsim 1$) plasma elements may exist deep inside the plasma mantle. These high β regions are usually associated with upward flowing beams of ionospheric ions at their edges.

8)   Fairly broad and hot ionospheric ion beams in the dayside (near cusp) part of the mantle, inconsistent with adiabatic motion from a low altitude source, suggest that scattering and/or transverse acceleration processes are present at higher altitudes.

9)   The mantle plasma sometimes has a very low or totally absent flow, especially near the flanks. It is then also generally much hotter than normal.

4.   Observations in the Dayside Boundary Layer near the Subsolar Point

The dayside boundary layer near the subsolar point is usually a very narrow region inside the magnetopause which was fairly poorly investigated before the recent high time resolution measurements with ISEE. This region has its particular interest because it is believed to be an important, if not the major, site of momentum exchange between the solar wind plasma and the magnetosphere. In the "classical" reconnection theory of DUNGEY (1961), the subsolar magnetopause is the region where the earth's magnetic field merges with the solar wind magnetic field, when the interplanetary magnetic field (IMF) is pointing southward, thereby enabling transfer of energy into the magnetosphere. Later CROOKER (1979, 1980), using a three-dimensional model of the magnetosphere, proposed that merging occurs along a line which, depending on the B_y component, may extend to quite high magnetic latitudes. One of the consequencies of reconnection taking place at the dayside magnetopause should be high speed jets of plasma flowing away from the separatrix region of the merging line. The discovery of these high speed jets during the "right" IMF conditions was therefore believed to be an important proof for reconnection being the major energy transfer process at the magnetopause. The ISEE observations by PASCHMANN et al. (1979) therefore seemed to have put the reconnection hypothesis on fairly firm grounds. The only problem was that, despite diligent search, only a few examples were found from a large amount of data.

A competing process for energy transfer into the magnetosphere—the viscous type of interaction—was first proposed by AXFORD and HINES (1961). The process was in the beginning perhaps mainly a name, but it has recently met renewed interest after the proposal by LEMAIRE (1977) of impulsive penetration of plasma elements through the magnetopause as an important process for transferring solar wind energy into the magnetosphere. An interesting feature of the impulsive penetration theory is that the IMF-conditions for penetration agree with those for reconnection (LEMAIRE *et al.*, 1979). The most important difference is that the solar wind has to be irregular for penetration to occur. As already mentioned, the dayside boundary layer is usually very thin. In the low bitrate mode of PROMICS-1 these boundary layer crossings usually gave rise to only a few data formats. Fortunately we have obtained data from five dayside boundary layer crossings at SM latitudes below $\sim 40°$ when the experiment was working in the high bitrate mode. Three of these also contained the magnetopause crossing.

Figure 17 is an example of an inbound magnetopause crossing at a local time of ~ 1100 UT and a latitude of $\sim 50°$ GSM ($\sim 24°$ SM). The top panel shows ion integral fluxes from the "perpendicular" ICSs for H^+, He^{2+}, He^+, and O^+ with pitch angles given by the panel right underneath. Notice that the "perpendicular" ICSs approximately scan the ecliptic YZ-plane. The third and fourth panels give the ion fluxes and pitch angle from the ICS looking in the forward direction towards the sun (ICS-D6). The bottom panel finally depicts the electron flux measurements from the "perpendicular" electron spectrometer for energy intervals of ~ 10 eV to 300 eV (upper histogram curve), 0.9 to 1.9 keV (middle curve) and 21 to 43 keV (lower histogram curve). Time, MT (local Moscow time; MT = UT + 3 hours) and orbital parameters (radial distance and GSM latitude and longitude) are also given at 10 minutes intervals.

The magnetopause (at ~ 0619 MT i.e. 0319 UT) can easily be distinguished as a marked change in the flux pattern of solar wind ions (H^+ and He^{2+}) and a drastic hardening of the electron flux. Notice also that the magnetopause represents a sharp composition boundary with very little O^+ on the magnetosheath side. Some He^+ ions appear to be present on the magnetosheath side, as the $25°$ ICS data show. This is, however, most likely a consequence of the limited mass resolution of the instrument, where the He^{2+} background in the He^+ channel is up to 5%. However, we cannot exclude that ions with M/q around 4 were present as well. Recent ion composition measurements in the solar wind with an ICS with higher mass resolution on PROGNOZ-8 indicate that ions with $M/q \sim 4$ at times are present in the solar wind. The plasma characteristics in this boundary layer are quite different from those based on the assumption of a mixture of magnetosheath and magnetosphere plasma. A more appropriate description of this boundary layer, particularly evident in the He^{2+} and electron data, is that of an incomplete mixing of plasma of different origins, where regions of magnetosheath-like plasma are separated by regions of more magnetosphere-like plasma. The magnetosheath-like structures in the boundary layer, characterized by a high He^{2+} content and a high antisunward flow, were frequently associated with strong beams of He^+ ions. The flux intensity in these beams was sometimes about 10 times higher for He^+ than for O^+ which is usually the more abundant ionospheric ion species at low polar altitudes. This indicates that the source for these ions is most likely the plasmasphere, where up to 15% of the thermal ions may

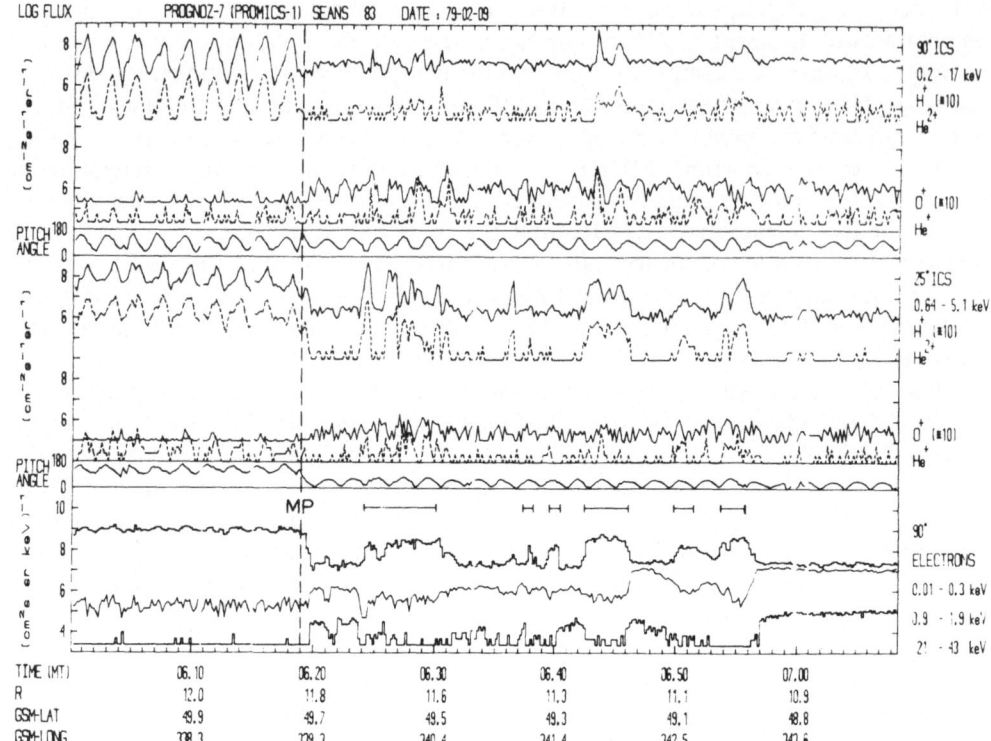

Fig. 17.   An example of an inbound passage through the magnetopause near local noon ($\sim$1100 LT) at
$\sim$24° latitude (SM). The top panel shows ion integral fluxes from the "perpendicular" ICSs for H$^+$, He^{2+},
He$^+$, and O$^+$ with pitch angles given by the panel right underneath. The third and fourth panels show
integral fluxes and pitch angle for the ICS pointing towards the sun (25° with respect to the spin axis). The
bottom panel contains electron fluxes from the "perpendicular" electron spectrometer (pitch angles given by
the upper pitch angle panel) for energies 0.01–0.3 keV (upper histogram curve), 0.9–1.9 keV (middle curve),
and 21–43 keV lower histogram curve). The magnetopause location is indicated by MP. "Inclusion"
structures are marked by horizontal bars in the electron flux panel.

be He$^+$ (YOUNG *et al.*, 1977; GEISS *et al.*, 1978). Notice also that no such beams were
observed outside the magnetopause. The acceleration of the He$^+$ ions therefore
appears to be a magnetospheric phenomenon. The pitch angle panels also demonstrate
that the He$^+$ beams flow essentially *perpendicular* to the local magnetic field. Such an
assymmetry in the pitch angle distribution, suggestive of a convection motion, is also
evident in the O$^+$ flux throughout the entire boundary layer.

A comparison of the low energy electron ($\sim$0.01–0.3 keV) and ion data indicates
that the "inclusion" structures are magnetosheath-like with regard to the electrons as
well. The more energetic electrons (21–43 keV), however, demonstrate that the
boundary layer is not just "inclusion" structures embedded in a background of
magnetospheric plasma as e.g. the O$^+$ data suggests.

Figure 18 shows energy spectra obtained with the three ICSs approximately every 10 seconds for H^+, He^{2+}, He^+, and O^+ in a time interval around the magnetopause crossing (~ 0619 MT). As can be seen, the magnetopause stands out clearly in this data form too. The ion spectra that appear in the figure immediately on the magnetosphere side of the magnetopause are similar to those found well beyond the boundary layer in the magnetosphere. What is of particular interest in this figure is the magnetosheath-like region after 0625 MT. It contains not only a high energy "magnetospheric" component but also a low energy ionospheric one (O^+ and He^+), which cannot be found on the sheath side of the magnetopause. An interesting energy dispersion effect in the H^+ ions, and to some extent also in the He^{2+} ions, is that the plasma of magnetosheath origin besides streaming along the magnetic field lines also has a perpendicular flow component. The flow can be seen as higher fluxes into the detector pointing close to the sun direction (D6) than into those pointing perpendicularly to the sun-earth line (D1D2).

Figure 19 contains energy spectra for the various ion species from a time period succeeding the one in Fig. 18. Notice the time interval with very strong He^+ flux around 0643 MT. There the H^+ spectrum peaks at ~ 0.2 keV, the He^+ one at ~ 1 keV and the O^+ one at ~ 4 keV. This suggests that ions of plasmaspheric origin have been accelerated to the same velocity. Notice also that the accelerated "plasmaspheric" ions are observed in a region with enhanced flux of magnetosheath ions (e.g. He^{2+}).

We mentioned previously that both the He^+ and O^+ ions appeared to have a flow component perpendicular to the magnetic field direction. In Fig. 20, velocity contour plots for He^+ and O^+ ions are displayed together with corresponding data from the E/q spectrometers (assuming $M/q = 1$). We can see that indeed a strong perpendicular, dawnward flow is present. Notice that a field aligned ion component (mainly H^+ but also some He^{2+}) is added to the total ion distribution, whereas He^+ and O^+ ions flow mainly perpendicular to the magnetic field direction. If we interpret the perpendicular flow of ~ 200 km/s as a convection motion, it corresponds to an electric field of ~ 8 mV/m pointing radially outward. The direction of this electric field is of particular interest since it is the direcion required to drive the boundary layer plasma in the direction of the magnetosheath flow just outside the magnetopause. The motion of the plasma across the magnetic field lines will build up polarization charges along the boundary layer like in an MHD-dynamo. The electric field thus developed will point radially outward on the dawn-side and radially inward at dusk. As a result of this radial electric field, cold, "detached", plasma of plasmaspheric origin, which may reach the dayside boundary layer, will drift with the convection velocity towards dawn or dusk.

Since the "temperature" of the thermal plasmaspheric ion population is expected to be of the order of a few eV, the drifting population should have a very "spiky" appearance. That this sometimes is the case can be seen in Fig. 18 (at ~ 0624 MT) where the H^+ energy spectrum peaked at ~ 0.25 keV, that of He^+ ions at ~ 0.9 keV and the O^+ one at ~ 3.0 keV. The temperature of the H^+ beam in this example must have been very low since a rough estimate from the peak gives a temperature of ~ 20 eV which is about the measurement threshold for the ICS at this energy.

In most cases, however, the temperature of these beams is somewhat higher, indicating that some local heating of the plasmaspheric ions occurs in the boundary

334   R. Lundin, B. Hultqvist, N. Pissarenko, and A. Zacharov

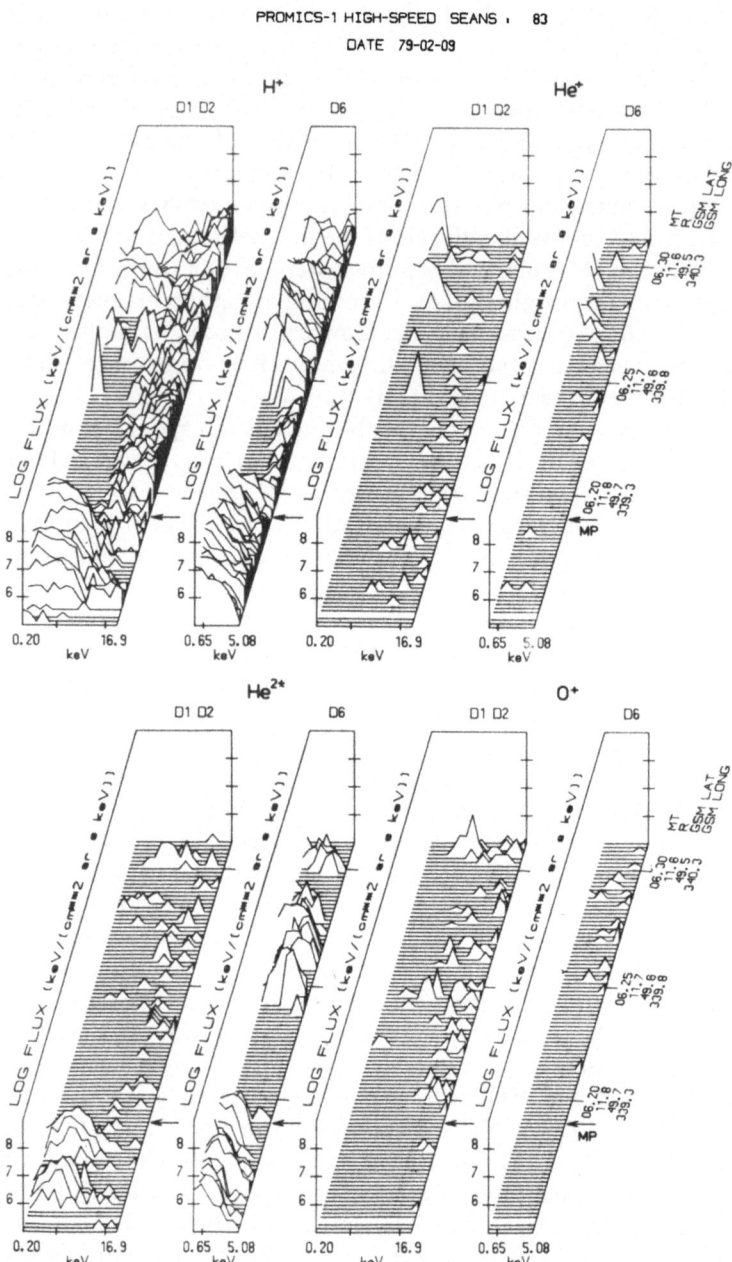

Fig. 18.   Energy spectra for H⁺, He²⁺, He⁺, and O⁺ plotted versus time and orbital parameters for each
∼ 10 s from the "perpendicular" (D1, D2) and sunward oriented (D6) ICSs. The data correspond to the time
period around the magnetopause crossing contained in Fig. 17. The magnetopause is marked by MP.

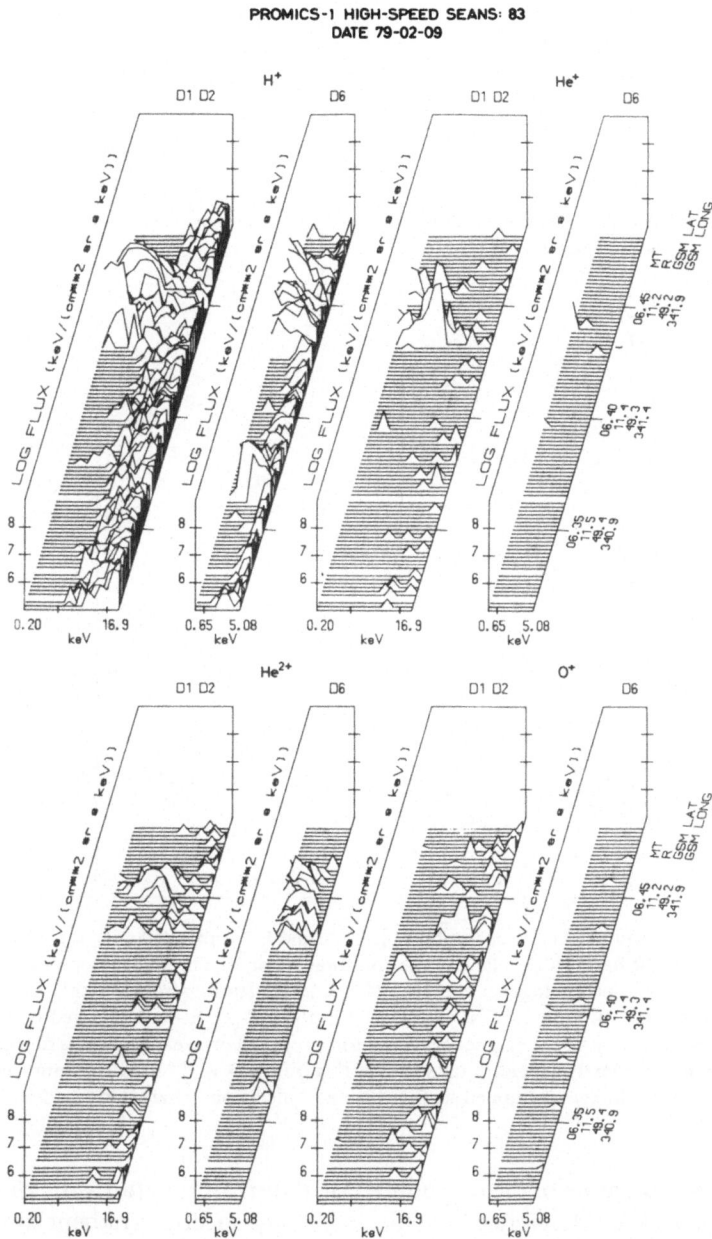

Fig. 19.   Energy spectra for H^+, He^{2+}, He^+, and O^+ plotted versus time and orbital parameters for each ~ 10 s from a time period succeeding the one in Fig. 18, i.e., fairly deep in the boundary layer contained in Fig. 17.

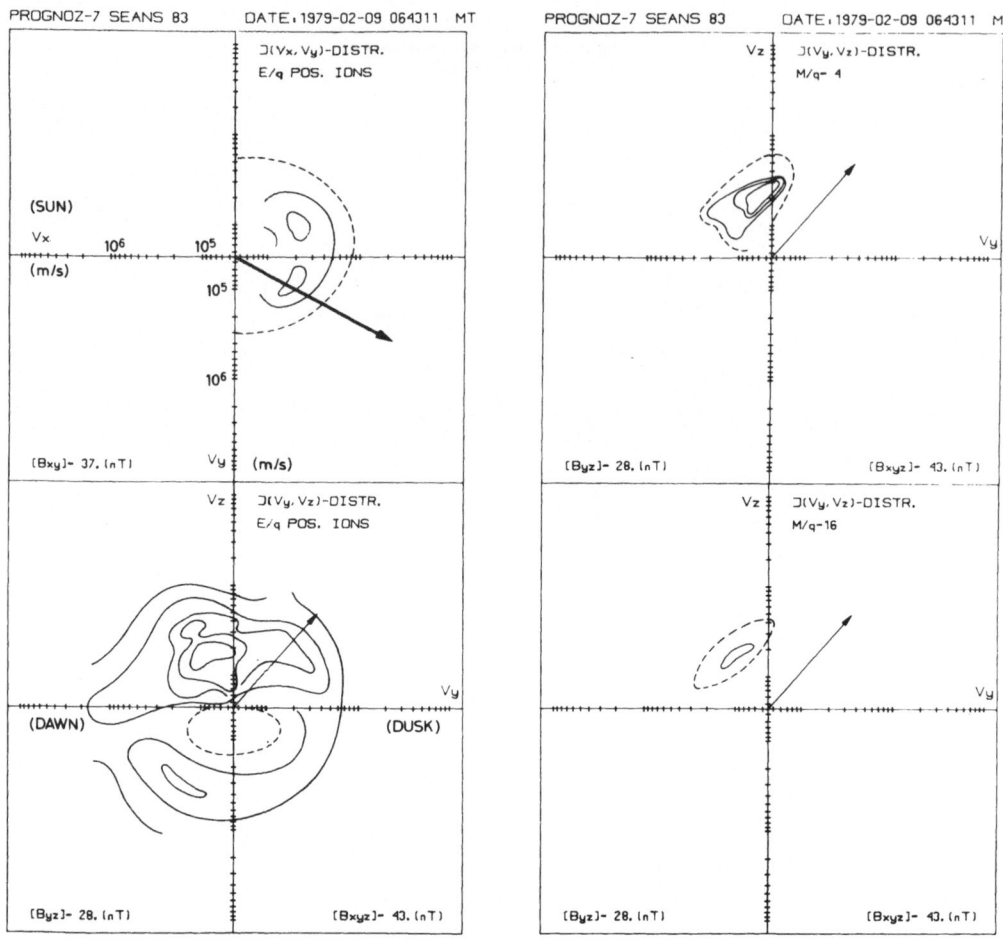

Fig. 20.   Velocity contour plots for ions (total E/q, He$^+$, and O$^+$) projected onto the XY-(upper left) and YZ-(lower left and right hand side) solar ecliptic coordinate planes. The velocity contours are plotted for equal differential energy fluxes in logarithmic levels of 1, 3, 10, etc. starting from 10^6 keV cm^{-2}s^{-1}sr^{-1}keV^{-1} in the XY-projection and 10^5 keV cm^{-2}s^{-1}sr^{-1}keV^{-1} in the YZ-projections (marked by broken curves). The magnitude and direction of the magnetic field vector, projected onto each plane, is marked by an arrow symbol. The velocity is given by a logarithmic scale starting from $\sim 4 \times 10^4$ m/s. The data represent the time period with strong He$^+$ fluxes (contained in Fig. 19) and "inclusion" characteristics (see Fig. 17).

layer as well. An example of the one dimensional distribution function for the beam in Fig. 20 is given in Fig. 21. Notice here the very sharp spectral gradient towards higher velocities for H$^+$, indicating quite a low temperature of this ion beam too. The total number density was about 6.3 cm^{-3} for H$^+$, ~ 0.7 cm^{-3} for He$^+$, ~ 0.1 for O$^+$, and ~ 0.01 for He^{2+}. The number density composition ($\sim 88\%$ H$^+$, $\sim 10\%$ He$^+$, $\sim 1.5\%$ O$^+$) demonstrates a plasmaspheric origin for these ions (see e.g. YOUNG et al., 1977, and GEISS et al., 1978). Even the very small He^{2+} content ($\sim 0.2\%$) may be of

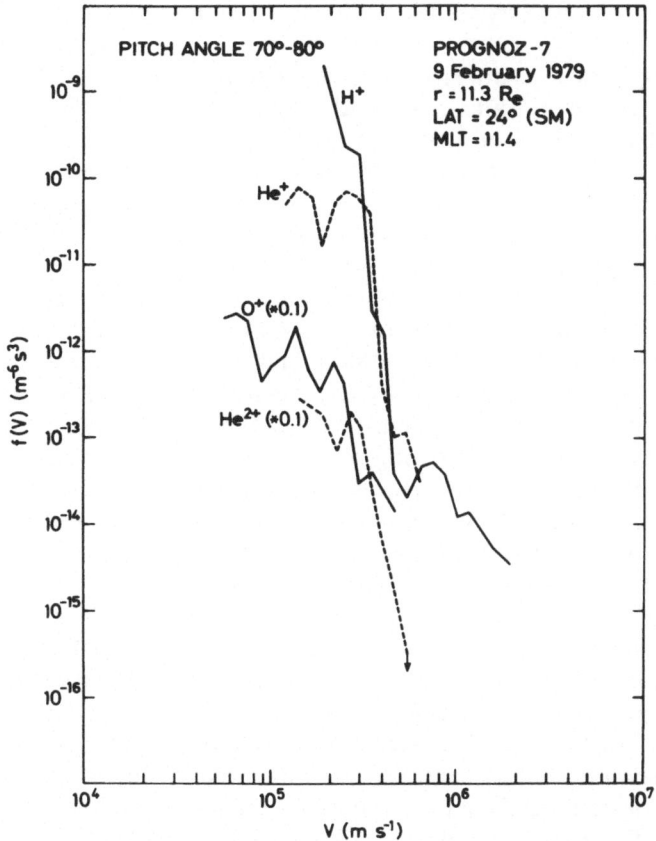

Fig. 21.   One dimensional distribution function for the transverse flowing ion beam shown in Fig. 20. The phase space densities for O^+ and He^{2+} have been multiplied by 0.1.

plasmaspheric origin (YOUNG et al., 1977). In fact the temperature of the He^{2+} ions in this beam was ~ 230 eV, which is much too low to be of solar wind origin.

The examples of PROGNOZ-7 dayside boundary layer observations presented above therefore provide further evidence in favor of impulsive penetration of magnetosheath plasma into the magnetosphere, as proposed by LEMAIRE (1977) and discussed also by HEIKKILA (1979). Not only does the prediction that "clouds" of magnetosheath plasma shall be observed within the magnetosphere agree with our observations but, perhaps more important, the predicted polarization electric field within the boundary layer (e.g. HEIKKILA, 1978, 1979) agrees astonishingly well with what we observe. A more detailed analysis, including additional cases, will be presented in the near future. However, so far our findings are consistent with a polarization electric field in the boundary layer pointing radially outward at dawn and radially inward at dusk, whenever "inclusion" events are observed.

5.   Observations on Auroral Field Lines at Altitudes of a Few Earth Radii

That ionospheric ions are ejected locally with high fluxes and quite high energies into the magnetosphere along auroral latitude magnetic field lines was first demonstrated by the S3-3 measurements (Shelley et al., 1976). Further statistical studies by e.g. Ghielmetti et al. (1978), Gorney et al. (1981), and Collin et al. (1981) have revealed that the occurrence frequency of upward flowing ion (UFI) events maximizes at the statistical auroral oval, exhibits a strong maximum at dusk and increases with magnetic disturbance level.

Most of what has been learned about upward beamed ionospheric ions originates from observations below an altitude of $\sim 8{,}000$ km. A few studies have been made in the magnetotail by Sharp et al. (1981) and Formisano et al. (1981) and at geostationary or synchronous orbits by Geiss et al. (1978) and Kaye et al. (1981a, b). To some extent the "low" altitude ($< 8{,}000$ km) and high altitude ($> 5 \; R_e$) observations have revealed conflicting results with regard to the source regions and acceleration mechanisms for the upstreaming terrestrial ions. For instance, the S3-3 results have suggested an H^+ dominance in the UFI-beams (Ghielmtti et al., 1978) whereas at synchronous orbits O^+ has been found to be the dominant ion species in the low energy part of "zipper" events (Kaye et al., 1981b).

PROGNOZ-7 provided scans through magnetic field lines connected to the auroral oval at fairly low to intermediate altitudes about every 4 days. These auroral oval crossings were located at radial distances of roughly ~ 3 to 6 earth radii, that is in an altitude range previously unexplored by satellites carrying mass analyzing instruments for hot plasma. The satellite moved almost straight up towards ecliptic north during the outbound passage (see e.q., Fig. 1) and traversed the auroral oval field lines in about an hour. Only data obtained in the high speed mode are suitable for auroral studies. Of the 11 auroral zone crossings that provided high speed data only 5 contained well defined cases of UFI-beams. The reason for this was that UFI-beams in the 3 to 6 R_e radial distance range were found to be very narrow (within $\sim 10°$) except in one passage during a magnetic storm. The beams could therefore in general be observed only when one of the ICSs came very close to 180° in pitch angle. The existing PROMICS-1 data base for studies of UFI-beams is thus quite limited.

Figure 22 shows an example of such an auroral zone passage when the sunward oriented ICS-D6 (25° with respect to the spin axis) reached close to 180° in pitch angle. This figure contains the integral fluxes of H^+, He^{2+}, He^+, and O^+ ions and of electrons for three different energy ranges versus time and orbit parameters for an outbound pre-midnight auroral zone passage. The format of this figure is the same as that of Fig. 17, except that the electron fluxes for the two lowest energy ranges are from the 25° electron spectrometer while those for the highest energies are taken from the 90° electron spectrometer. Notice in Fig. 22 the change from a depleted loss cone to a "source" cone for the H^+ ions around 2150 MT. The satellite at that time entered an extended region in which ionospheric ions were energized and ejected out into the magnetosphere. As can be seen, H^+, He^+, and O^+ ions were present in the UFI-events with highly variable abundance ratios. The O^+ content in the beams was mostly higher than the H^+ content. Occasionally the H^+ ions even had a depleted loss cone while the O^+ ions

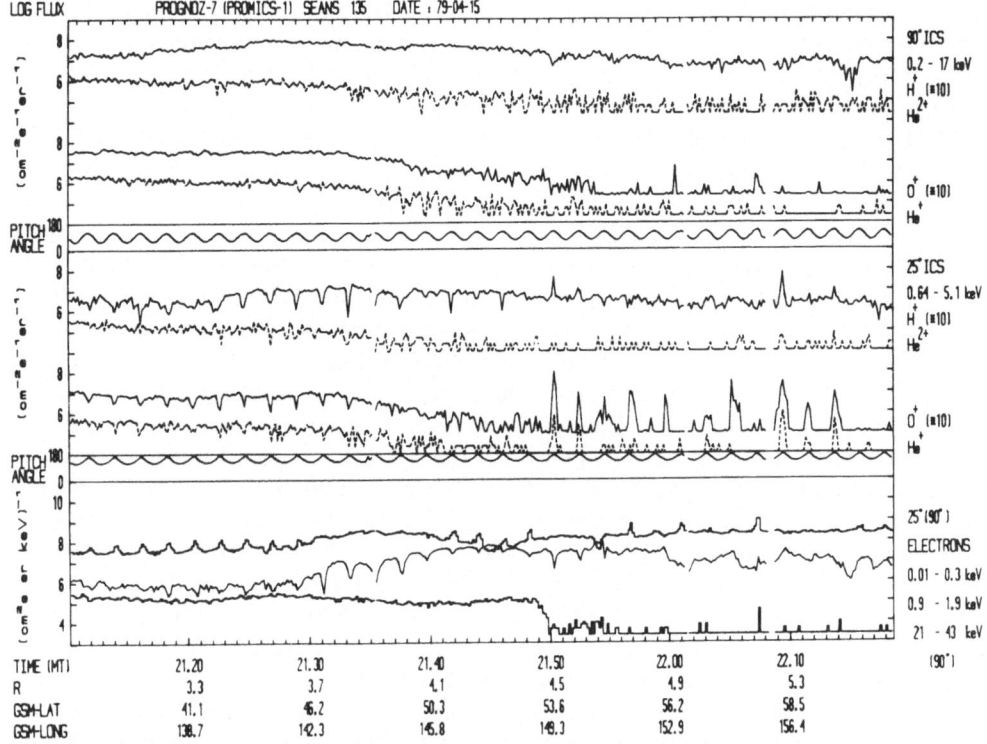

Fig. 22.   PROGNOZ-7 auroral zone passage showing integral fluxes for the various ion constitutents (H^+, He^{2+}, He^+, and O^+) and electron fluxes at different energy intervals plotted versus time and orbital parameters. The top panel shows ion data from the perpendicular ICSs with pitch angles given in the panel below. The third panel contains the ion flxues from the 25° ICS with the corresponding pitch angle shown in the fourth panel. The bottom panel depicts the electron fluxes from ∼10 eV–300 eV (upper curve), 0.9–1.9 keV (middle curve) and 21–43 keV (lower histogram curve). The lowest electron energies are taken from the 25° spectrometer whereas the highest energies are from the perpendicular spectrometer.

showed a "source" cone (at ∼2155 and ∼2157 MT). Notice also that when the UFI-beam lacked H^+ ions it also lacked He^+ ions, that is a "pure" O^+ beam was observed. On the other hand, whenever He^+ ions were observed they were associated with ionospheric H^+ ions.

Although the UFI-beams were highly variable with respect to the ion composition, some systematic features were observed during this auroral zone passage. For instance, a high H^+ and He^+ abundance was usually associated with a fairly high acceleration energy. Conversely, a "pure" O^+ beam was mainly observed when the acceleration was less than about one keV. Notice finally in Fig. 22 that equatorward of the UFI-events low energy upward flowing electrons frequently occurred. Although a full pitch angle distribution is required to determine if these electrons were related to the downward part of the Birkeland current system or were simply a part of a counterstreaming electron population (see e.g. SHARP *et al.*, 1980), the high flux and high mean energy ($\sim 100 - 300$ eV) in these electron beams seem to preclude that they were simply atmospherically back-scattered electrons. Further analysis of these electron beam observations will be made in the near future.

In Fig. 23 we have plotted the pitch angle distribution and energy spectrum for one of the ion beams containing H^+, He^+, and O^+ ions. As can be seen, the beams are very narrow, which is consistent with adiabatic motion from a source located far down in the ionosphere. Because of the field of view ($\sim 10°$) of the ion composition spectrometer and lacking any further knowledge about the characteristics of the acceleration process, we cannot say how far from the source the observation was made. If the acceleration process for the ionospheric ions was isotropic or transverse, the beam width indicates a source altitude of only a few thousand kilometers (see e.g. the arrows in Fig. 23 marking the "mirroring altitude"). If, on the other hand, the acceleration process acted mainly along the magnetic field lines the source region could have been almost anywhere between the ionosphere and the satellite, provided sufficient densities of ionospheric plasma were available along that particular field line. The very high amount of O^+ in the beam limits, however, the possible source location considerably.

As has been pointed out by MOORE (1980), charge exchange processes prohibit O^+ ions from escaping upwards unless they are preaccelerated before reaching the neutral hydrogen/oxygen cross-over altitude (located at 500–1,500 km). This does not mean that all the acceleration has to take place at very low altitudes, but rather that a considerable outflow of ions must have started there.

A significant feature in Fig. 23 is the difference in peak energy for H^+ and O^+. Mass dispersion effects like the one in Fig. 23 have been observed in a number of high altitude UFI-beams when both H^+ and O^+ have been present, and always with the O^+ peak energy being higher than the H^+ peak energy. In this case there is no doubt that the origin of the O^+ and H^+ beams was the ionosphere; the difficulty is rather to explain why the O^+ ions have gained more energy than the H^+ ions.

One explanation is that the difference in energy gain is due to an acceleration mechanism working in an altitude range where ionospheric O^+ ions dominate in the main part and the H^+/O^+ ratio increases upwards. An alternative way of interpreting this mass-dispersion is in terms of a mass-dependent acceleration mechanism which strongly favours the acceleration of heavy ions. In Fig. 23 we also see that the He^+ ion flux peaks at a higher energy than the H^+ flux, in apparent agreement with a mass-dependent efficiency for the acceleration.

While Fig. 23 was an example of a narrow UFI-beam, the ions seemingly moving adiabatically upward, Fig. 24 shows a much broader ion beam which cannot in a simple way be related to an adiabatic transport from a low altitude source. In fact, both the much broader energy distribution and the dispersed beam characteristics suggest that the beam has undergone a substantial diffusion in velocity space before reaching the observation point. Notice also that this UFI-beam, completely dominated by O^+ ions, is very energetic with a significant flux present above ~ 10 keV. Figure 24 also demonstrates that these energetic ions clearly originate in the UFI-event and are not reflected ions. Most of the UFI-events observed during this early morning (~ 03 local time) auroral zone passage had fairly broad beams, although many of them were much

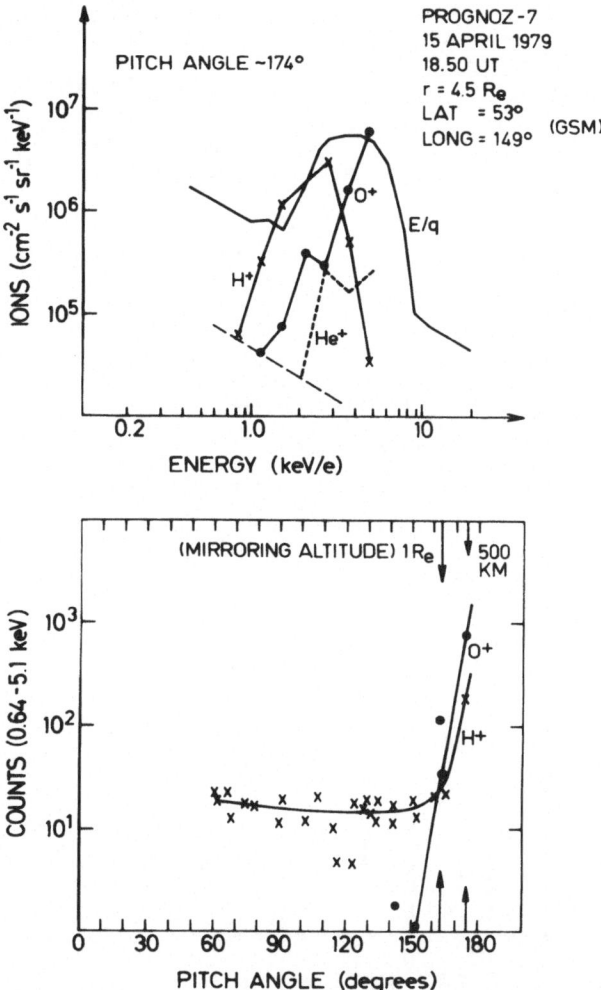

Fig. 23. Energy spectra and pitch angle distributions for a narrow upward flowing ion beam at ~23,000 km altitude. The energy spectra for H⁺, He⁺, and O⁺ are plotted together with ion E/q spectra from the electrostatic analyzer. The width of the loss cone (at 500 km) and the mirroring altitude at 1 earth radius are given in the pitch angle distribution diagram.

less energetic than that of Fig. 24. An obvious conclusion from this observation is therefore that ionspheric ions may be accelerated to very high energies (~ 10 keV) during what seems to be strong turbulence at altitudes below some 20,000 km. Another important aspect is the very high flux of some of the ion beams observed during this passage. If projected down to the low altitude ionosphere, escape fluxes of up to 10^9 cm^{-2}s^{-1} were measured. Such high escape fluxes can only be obtained from an upper ionosphere which is very hot and dense (BANKS and HOLZER, 1969, and LYONS, 1980), unless it is due to a non-stationary depletion of the upper ionosphere.

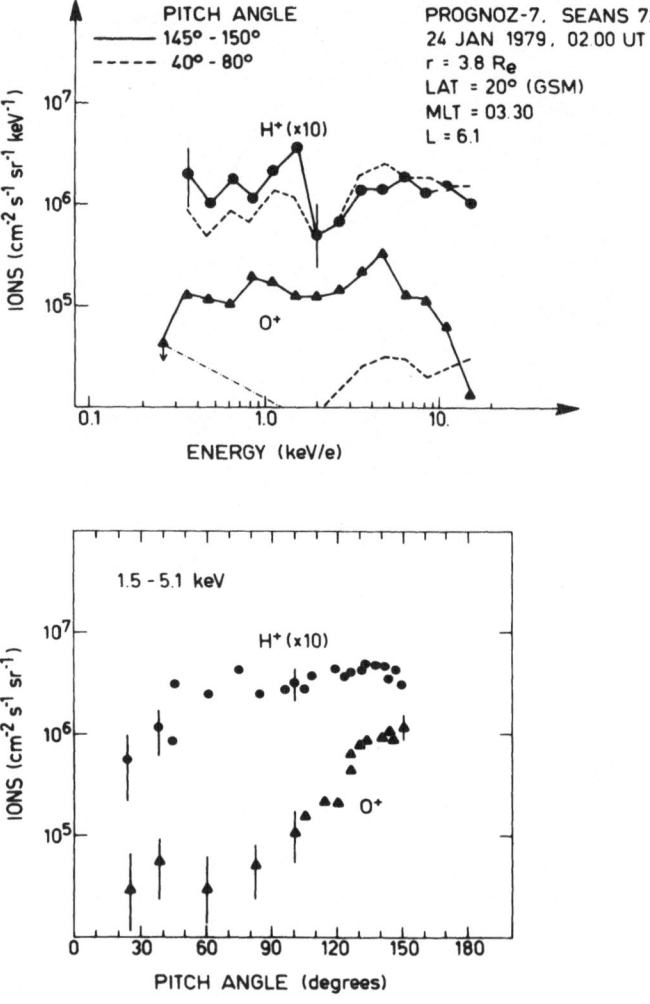

Fig. 24.   Energy spectra and pitch angle distributions for a broad upward flowing ion beam at ~ 18,000 km altitude. The energy spectra for two pitch angle intervals are indicated. Notice that the H⁺ values in both the pitch angle and energy spectrum diagrams have been multiplied by 10.

The above presented examples of both "adiabatic" and "nonadiabatic" UFI-beams measured from PROGNOZ-7 demonstrate the rather variable situation, with the acceleration process sometimes involving heating or transverse acceleration as important mechanisms and sometimes not. At higher altitudes, in magnetospheric regions connected to the dayside and flank (morningside) auroral oval, ion heating seems to be more common. This may then be related to the presence of energetic ions (both H^+ and O^+) and fairly broad ion beams sometimes observed in the flank mantle (Lundin et al., 1981a, Example 9) and possibly also to the observation that ion "conics" are more frequently found on the dayside (Gorney et al., 1981). We may also conclude from these examples that O^+ and H^+ UFI-beams occurred together, the O^+ ions had acquired more energy in the acceleration process than the H^+ ions. Differences in energy gain of more than two kilovolts have been observed. A similar relation of the mean energy of O^+ being higher than that for H^+ have also been reported by COLLIN et al. (1981).

6.   Observation in the Ring Current Region

The first evidence that the ionosphere might be a major contributor to the hot magnetospheric plasma was, as previously mentioned, the unexpected discovery by the Lockheed group of large fluxes of O^+ ions in the energy range 0.7–12 keV precipitating into the auroral zone during magnetic storms (SHELLEY et al., 1972). The Bern group with its mass spectrometer on GEOS-1 was the first to make direct composition measurements in the hot magnetospheric plasma near the equator (GEISS et al., 1978; BALSIGER et al., 1980). They already concluded from their first measurements that the ionosphere as source was of comparable importance to that of the solar wind for the energetic magnetospheric ions. Later, measurements from the GEOS-2, SCATHA, and ISEE satellites (e.g. YOUNG, 1980; LENNARTSSON et al., 1981; and JOHNSON, 1981) have demonstrated that terrestrial ions are practically always present at energies below ~ 20 keV and that terrestrial ions may dominate even at fairly modest activity levels. During major storms they may completely dominate.

A rather extreme case of O^+ dominating over almost the entire dayside magnetosphere was observed during the magnetic storm which started on 21 February 1979. The data shown in Fig. 25 was obtained in the early part of the storm that culminated in the hour 21–22 UT ($Dst \sim -98$ UT). A somewhat smaller peak ($Dst \sim -90$ UT) occurred between 08 and 09 UT, that is shortly before the time period contained in Fig. 25. As can be seen in Fig. 25, O^+ was the dominating magnetospheric ion species almost all the way out to the magnetopause. At smaller distances from the earth the O^+ dominance increased and in the innermost part of the ring current the O^+ number density for energies up to 17 keV was two orders of magnitudes higher than the H^+ number density. Notice also the increasing He^+ abundance towards the innermost part of the ring current (up to $\sim 10\%$). In the period 1540–1630 UT the background count rates due to MeV electrons in the radiation belts were very high and are not eliminated from the data shown in Fig. 25. Outside this region the background count rates are not important for the present discussion.

Also on the nightside, O^+ had the highest number density out to $\sim 6 R_e$. The satellite passed out of the plasmasheet/ring current very quickly through its upper boundary. However, it can be seen that H^+ dominated the uppermost part of the plasmasheet at $\sim 40°$ SM-latitude and ~ 02 MLT.

We may thus conclude that during the February storm in 1979 the entire dayside magnetosphere was filled with keV ions of ionospheric origin already in the early phase of the storm. LYONS and MOORE (1981) have demonstrated that charge exchange may increase significantly the O^+/H^+ density ratio near the equatorial plane in a few hours for equatorial pitch angles below $\sim 10°$. The data shown in Fig. 25 is an average over broad pitch angle sectors which only rarely include very small pitch angles. Charge

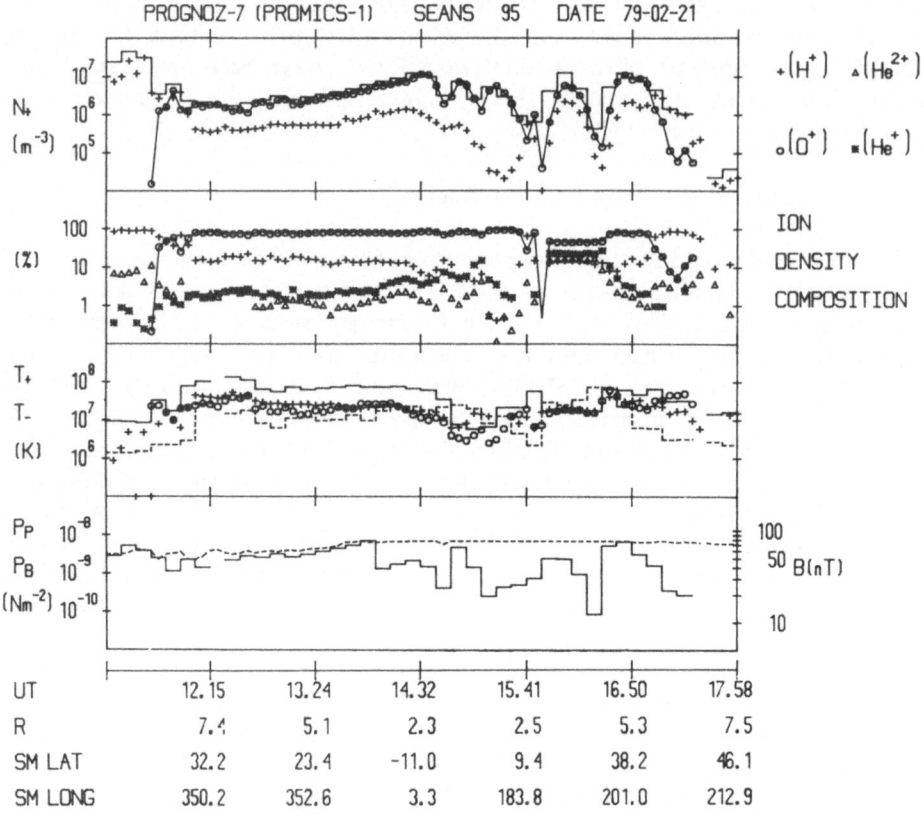

Fig. 25.   An example of a storm time ring current when O$^+$ dominated completely over the entire dayside magnetosphere. Between ~1545 and 1635 UT, radiation belt MeV electrons produced a background that dominated the detectors. This background has not been eliminated in the figure. The format of the plasma parameters is the same as in Fig. 4 except for the magnetic field and flow vectors which are not included here.

exchange is, therefore, not expected to be of importance for the high O$^+$/H$^+$ density ratios in Fig. 25, which consequently are determined by the source strength in the first hand.

An example of ring current profiles as determined from **PROGNOZ**-7 during prestorm and late recovery phase is shown in Fig. 26. The figure shows energy densities for H$^+$, He$^+$, and O$^+$ in the 0.2–17 keV energy range, and for the 0.1–45 keV range under the assumption that only protons were measured, plotted versus L-value, geomagnetic latitude and magnetic local time. An interesting characteristic of these radial profiles is the sharp composition change occurring around $L \sim 4$, where the H$^+$ density drops towards background values inside $L \sim 4$. This innermost part of the ring current, usually the peak density region during magnetic storms, is only refilled by "fresh" magnetospheric plasma during storms. For charge exchange reasons, but also from observed and inferred decay rates in the ring current, Lyons and Evans (1976)

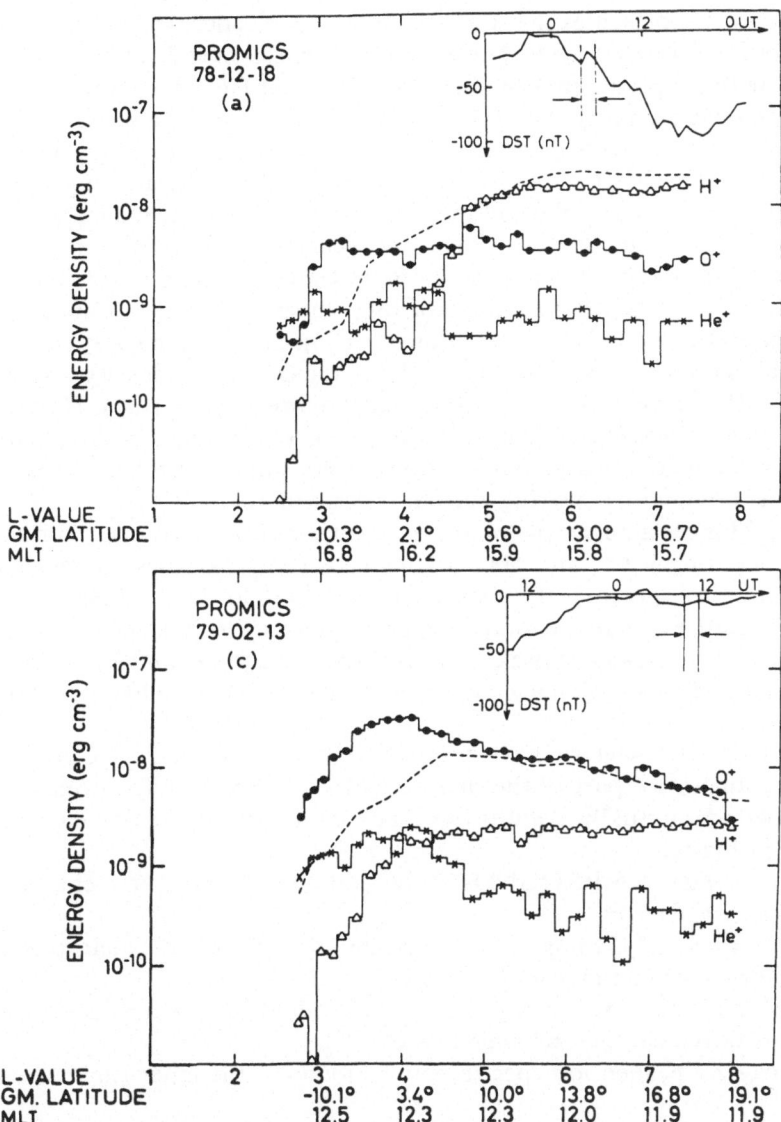

Fig. 26.   Energy densities of H$^+$, He$^+$, and O$^+$ in the energy interval 0.2–17 keV vs orbital parameters (L-value, magnetic latitude, and local time) for two different PROGNOZ-7 passes through the ring current. The broken curves indicate energy densities calculated from the E/q spectrometers (0.1–45 keV) under the assumption that only protons were measured. Included in each panel is a graph showing the Dst values during the pass, marked by the arrows.

proposed that the innermost part of the ring current, for ions below ~ 50 keV, is not dominated by protons but by some other ion species. Figure 26, after Lundin et al. (1980), clearly demonstrates that the proton density has decayed as compared to O^+ and He^+ ions. Notice that the He^+ density is fairly constant vs L-value which is in good agreement with the predictions by Tinsley (1976) that He^+ ions should be much less affected by charge exchange than H^+ ions. However, except for a very small interval in the innermost part in the upper diagram, O^+ is much more abundant than He^+ also inside $L \sim 4$. Notice also in the lower figure that again there was a strong dominance of O^+ out to high L-values on the dayside, and now even when the magnetic disturbance level was low. Evidently O^+ ions are generally "favored" in the dayside magnetosphere as has been found also from the GEOS measurements (Young, 1980).

Figure 27 gives another example of ring current profiles, the upper one taken during a magnetic storm. Notice that this storm injection was the only case where H^+ was more abundant than He^+ inside $L \sim 4$, but the O^+ energy density was still higher than that for H^+ below $L \sim 5$ (the region of maximum energy density). The fact that the O^+ energy density was even higher than the 0.1–45 keV energy density obtained from the electrostatic analyzer assuming protons as constituent (broken curve) indicates that O^+ was dominating the ring current at least up to 45 keV.

Figure 28 gives an example of ion spectra taken from a ring current passage. It again shows a strong O^+ and He^+ dominance at low L-values. A very significant feature in these ion spectra is the peaked characteristics for O^+ and He^+ found at low L-values. As can be seen in the total ion spectra, going up to 30 keV in this panel, there is a sharp peak at ~ 20 keV, a peak which is entirely due to O^+ and He^+. Traces of this peak can be found also at higher L-values but then the peak most likely contains H^+ as well.

Peaked O^+ ion spectra have been observed in several PROGNOZ-7 passes through the innermost part of the ring current, and then usually during moderately disturbed periods. Actually Dst was less than -15 nT during the ring current passage illustrated in Fig. 28.

To summarize, the PROGNOZ-7 ion composition measurements in the ring current have revealed that:

1) O^+ ions may completely dominate the dayside magnetosphere during magnetic storms for energies up to ~ 17 keV.

2) O^+ and He^+ are usually the most abundant ion species inside $L \sim 4$ (as a consequence of the charge exchange loss of H^+).

3) Strongly peaked ion spectra may be found in the inner ring current.

7.   Summary

The observational results presented above are summarized below in very brief form.

7.1   Plasma mantle

1)   Ionospheric ions represent generally a few percent to about ten percent of the number density.

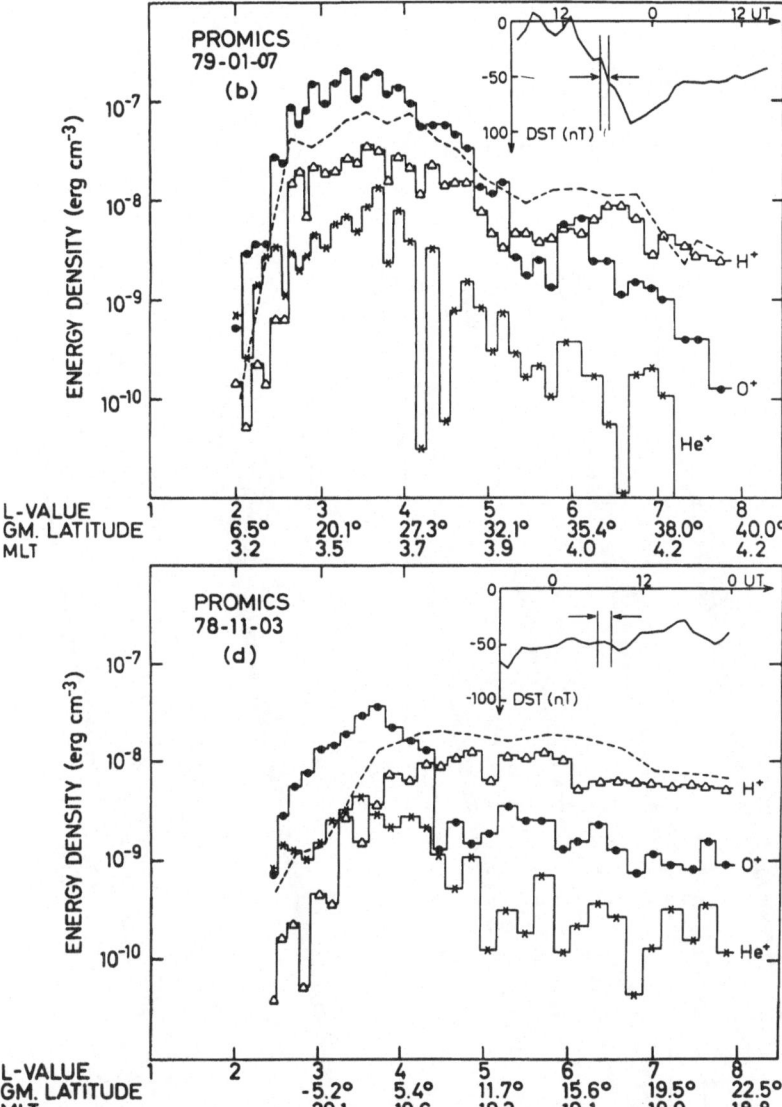

Fig. 27. Energy densities of H⁺, He⁺, and O⁺ versus orbital parameters for two different PROGNOZ-7 passes through the ring current. The format is the same as in Fig. 26.

2) The ionospheric ion content is strongly activity dependent.
3) Particle motions in the deep nightside mantle are adiabatic.
4) Hot stagnant mantles are sometimes observed at the flanks.
5) Penetrated solar wind plasma elements are frequently observed.
6) The magnetopause is a fairly solid border for ionospheric ions.

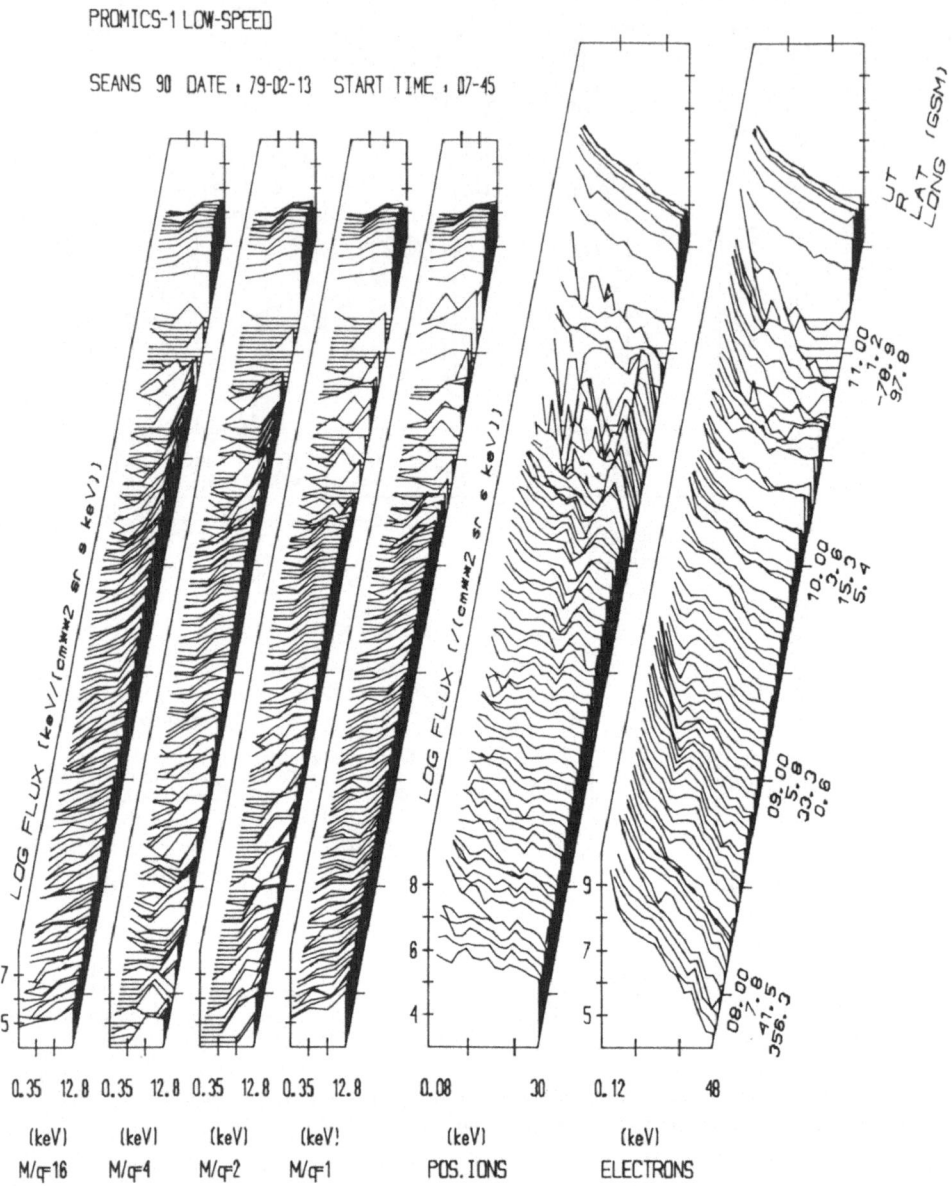

PROMICS-1 LOW-SPEED

SEANS 90 DATE : 79-02-13 START TIME : 07-45

Fig. 28. Spectrogram plot for an inbound passage through the ring current (same as the one shown in Fig. 26), showing peaked O^+ and He^+ spectra in the innermost part. The format is the same as in Fig. 6.

7.2   Dayside boundary layer

1)   The boundary layer consists of poorly mixed magnetosheath and magnetosphere plasma.

2)   Observations of plasma characteristics support impulsive penetration of magnetosheath plasma into the magnetosphere.

3)   The electric field in the boundary layer has been found to have a strong radial component.

7.3   Auroral particles at 2–5 R_e altitude

1) O^+ ions generally dominate UFI beams below ~ 1 keV.

2)   Heavier ions (e.g. O^+ and He^+) may gain several keV more energy than H^+ in UFI-beams.

3)   Narrow beams consistent with adiabatic motion from a few thousand km altitude to 5 R_e have been seen.

4)   Broad UFI-beams observed at fairly high altitudes ($\sim 3 R_e$) indicate a strong scattering or perpendicular acceleration at great altitudes.

7.4   Ring current

1) O^+ ions may completely dominate the dayside magnetospheric plasma during magnetic storms for energies up to 17 keV.

2) O^+ and He^+ are generally more abundant than H^+ in the quiet time ring current inside $L \sim 4$.

3)   Strongly peaked ion spectra are frequently found in the inner ring current.

The authors are very much obliged to the magnetic field experimenters on PROGNOZ-7, Principal Investigator Dr E. Eroshenko, IZMIRAN, for providing the magnetic field data.

The contributions by the technical staffs at Kiruna Geophysical Institute and Swedish Space Corporation to the success of the PROMICS experiment are gratefully acknowledged. The Swedish part has been financed partly be means of grants from the Swedish Board for Space Activities.

REFERENCES

AXFORD W. I. and C. O. HINES, A unifying theory of high latitude geophysical phenomena and geomagnetic storms, Can. J. Phys., 39, 1433, 1961.

BALSIGER, H., P. EBERHARDT, J. GEISS, and D. T. YOUNG, Magnetic storm injection of 0.9 to 16 keV/e solar and terrestrial ions into the high altitude magnetosphere, J. Geophys. Res., 85, 1645, 1980.

BANKS, P. M. and T. E. HOLZER, High latitude plasma transport: The polar wind, J. Geophys. Res., 74, 6317, 1969.

COLLIN, H. L., R. D. SHARP, E. G. SHELLEY, and R. G. JOHNSON, Some general characteristics of upflowing ion beams over the auroral zone and their relationship to auroral electrons, J. Geophys. Res., 86, 6820, 1981.

CROOKER, N. U., Dayside merging and cusp geometry, J. Geophys. Res., 84, 951, 1979.

CROOKER, N. U., The half-wave rectifier response of the magnetosphere and antiparallel merging, J. Geophys. Res., 85, 575, 1980.

DUNGEY, J. W., Interplanetary magnetic field and auroral zones, Phys. Rev. Lett., 6, 47, 1961.

FORMISANO, V., S. ORSINI, and M. CANDIDI, Observations of ionospheric oxygen in the geomagnetic tail by ISEE-2, Adv. Space Res., 1, 313, 1981.

FRANK, L. A., K. L. ACKERSON, and D. M. YEAGER, Observations of atomic oxygen (0⁺) in the earth's magnetotail, *J. Geophys. Res.*, **82**, 129, 1977.

GEISS, J., H. BALSIGER, P. EBERHARDT, H. P. WALKER, L. WEBER, D. T. YOUNG, and H. ROSENBAUER, Dynamics of magnetospheric ion composition as observed by the GEOS mass spectrometer, *Space Sci. Rev.*, **22**, 537, 1978.

GHIELMETTI, A. G., R. G. JOHNSON., R. D. SHARP, and E. G. SHELLEY, The latitudinal diurnal and altitudinal distributions of upflowing energetic ions of ionospheric origin, *Geophys. Res. Lett.*, **5**, 59, 1978.

GORNEY, D. J., A. CLARKE, D. R. CROLEY, J. F. FENNEL, J. M. LUHMANN, and P. F, MIZERA, The distribution of ion beams and conics below 8,000 km, *J. Geophys. Res.*, **86**, 83, 1981.

HAERENDEL, G. and G. PASCHMANN, Entry of solar wind plasma into the magnetosphere, in *Physics of the hot Plasma in the Magnetosphere*, edited by B. Hultqvist and L. Stenflo, p. 23, Plenum Press, New York, 1975.

HEIKKILA, W. J., Electric field topology near the dayside magnetopause, *J. Geophys. Res.*, **83**, 1071, 1978.

HEIKKILA, W. J., Impulsive penetration and viscous interaction, in *Magnetospheric Boundary Layers*, ESA-SP-148, European Space Agency, Paris, France, 1979.

HONES, E. W., Jr., Plasma flow in the plasma sheet and its relation to substorms, *Radio Sci.*, **8**, 979, 1973.

HULTQVIST, B., R. LUNDIN, I. SANDAHL, N. PISSARENKO, and A. ZACHAROV, First observations of a heated plasma component in the plasma mantle, in *Magnetospheric Boundary Layers*, ESA SP-148, European Space Agency, Paris, France, 1979.

JOHNSON, R. G., A review of the hot plasma composition near geosynchronous altitude, in *Spacecraft Charging Technology*-1980, edited by N. J. Stevens and C. P. Pike, Conference Proceedings, NASA CP-2182, p. 412, 1981.

JOHNSON, R. G., R. D. SHARP, and E. G. SHELLEY, Observations of ions of ionospheric origin in the storm-time current, *Geophys. Res. Lett.*, **4**, 403, 1977.

KAYE, S. M., R. G. JOHNSON, R. D. SHARP, and E. G. SHELLEY, Observations of transient H⁺ and O⁺ bursts in the equatorial magnetosphere, *J. Geophys. Res.*, **86**, 1335, 1981a.

KAYE, S. M., E. G. SHELLEY, R. D. SHARP, and R. G. JOHNSON, Ion composition of zipper events, *J. Geophys. Res.*, **86**, 3383, 1981b.

LEMAIRE, J., Impulsive penetration of filamentary plasma elements into the magnetospheres of the Earth and Jupiter, *Planet. Space Sci.*, **25**, 887, 1977.

LEMAIRE, J. and M. ROTH, Penetration of solar wind plasma elements into the magnetosphere, *J. Atmos. Terr. Phys.*, **40**, 331, 1978.

LEMAIRE, J., M. J. RYCROFT, and M. ROTH, Control of impulsive penetration of solar wind irregularities into the magnetosphere by the interplanetary magnetic field direction, *Planet. Space Sci.*, **27**, 47, 1979.

LENNARTSSON, W., E. G. SHELLEY, R. D. SHARP, R. G. JOHNSON, and H. BALSIGER, Some initial ISEE-1 results on the ring current composition and dynamics during the magnetic storm of December 11, 1977, *Geophys. Res. Lett.*, **6**, 483, 1979.

LENNARTSSON, W., R. D. SHARP, E. G. SHELLEY, R. G. JOHNSON, and H. BALSIGER, Ion composition and energy distribution during 10 magnetic storms, *J. Geophys. Res.*, **86**, 4628, 1981.

LUNDIN, R. and B. APARICIO, Observations of penetrated solar wind plasma elements in the plasma mantle, *Planet. Space Sci.*, 1981 (in press).

LUNDIN, R., L. R. LYONS, and N. PISSARENKO, Observations of the ring current composition at $L < 4$, *Geophys. Res. Lett.*, **7**, 425, 1980.

LUNDIN, R., B. HULTQVIST, N. PISSARENKO, and A. ZACHAROV, The plasma mantle: Composition and other characteristics observed by means of the PROGNOZ-7 satellite, *Space Sci. Rev.*, 1981a (in press).

LUNDIN, R., B. HULTQVIST, E. DUBININ, A. ZACHAROV, and N. PISSARENKO, Observation of outflowing ion beams on auroral field lines at altitudes of many earth radii, submitted for publication to *Planet. Space Sci.*, 1981b.

LUNDIN, R., I. SANDAHL, B. HULTQVIST, A. GALEEV, O. LIKHIN, A. OMELCHENKO, N. PISSARENKO, O. VAISBERG, and A. ZACHAROV. First observations of the hot ion composition in the high latitude magnetospheric boundary layer by means of PROGNOZ-7, in *Magnetospheric Boundary Layers*, ESA-SP-148, European Space Agency, Paris, France, 1979.

LYONS, L. R., Generation of large-scale regions of auroral currents, electric potentials, and precipitation by the divergence of the convection electric field, *J. Geophys. Res.*, **85**, 17, 1980.

LYONS, L. R. and D. S. EVANS, The inconsistency between proton charge exchange and the observed ring current decay, *J. Geophys. Res.*, **81**, 6197, 1976.

LYONS, L. R. and D. J. WILLIAMS, A source for the geomagnetic storm main phase ring current, *J. Geophys. Res.*, **25**, 523, 1980.

LYONS, L. R. and T. E. MOORE, Effects of charge exchange on the distribution of ionospheric ions trapped in the radiation belts near synchronous orbit, *J. Geophys. Res.*, **86**, 5885, 1981.

MOORE, T. E., Modulation of terrestrial ion escape flux composition (by low-altitude acceleration and charge exchange chemistry), *J. Geophys. Res.*, **85**, 2011, 1980.

PASCHMANN, G., W. SCKOPKE, G. HAERENDEL, J. PAPAMASTORAKIS, S. J. BAME, J. R. ASBRIDGE, J. T. GOSLING, E. W. HONES, Jr., and E. R. TECH, ISEE plasma observations near the subsolar magnetopause, *Space Sci. Rev.*, **22**, 717, 1978.

PETERSON, W. K., R. D. SHARP, E. G. SHELLEY, R. G. JOHNSON, and H. BALSIGER, Energetic ion composition of the plasma sheet, *J. Geophys. Res.*, **86**, 761, 1981a.

PETERSON, W. R., E. G. SHELLEY, G. HAAERENDEL, and G. PASCHMANN, Energetic ion composition in the subsolar magnetopause and boundary layer, Lockheed Palo Alto Res. Lab., Preprint, 1981b.

ROSENBAUER, H., H. GRÜNWALDT, M. D. MONTGOMERY, G. PASCHMANN, and N. SCKOPKE, HEOS 2 plasma observations in the distant polar magnetosphere, *J. Geophys. Res.*, **80**, 2723, 1975.

SHELLEY, E. G., R. G. JOHNSON, and R. D. SHARP, Satellite observations of energetic heavy ions during a geomagnetic storm, *J. Geophys. Res.*, **77**, 6104, 1972.

SHELLEY, E. G., R. D. SHARP, and R. G. JOHNSON, Satellite observations of an ionospheric acceleration mechanism, *Geophys. Res. Lett.*, **3**, 654, 1976.

SHELLEY, E. G., Ion composition in the dayside cusp: injection of ionospheric ions into the high latitude boundry layer, in *Magnetospheric Boundary Layers*, ESA-SP-148, European Space Agency, Paris, France, 1979.

SHARP, R. D., E. G. SHELLEY, R. G. JOHNSON, and A. G. GHIELMETTI, Counter streaming electron beams at altitudes of ~ 1 R_e over the auroral zone, *J. Geophys. Res.*, **85**, 92, 1980.

SHARP., R. D., D. L. CARR, W. K. PETERSON, and E. G. SHELLEY, Ion streams in the magnetotail, *J. Geophys. Res.*, **86**, 4639, 1981.

TINSLEY, B. A., Evidence that the recovery phase ring current consists of helium ions, *J. Geophys. Res.*, **81**, 6193, 1976.

YOUNG, D. T., Synoptic studies of magnetospheric composition, Paper submitted to the "philosophisch-naturwissenschaftlichen Fakultät" of the University of Bern in partial fulfillment of the require ments for obtaining the "venia docendi", Bern, Feb. 1980.

YOUNG, D. T., J. GEISS, H. BALSIGER, P. EBERHARDT, and A. GHILMETTI, Discovery of He^{2+} and O^{2+} ions of terrestrial origin in the outer magnetosphere, *Geophys. Res. Lett.*, **4**, 561, 1977.

Energetic Ion Composition in the Earth's Magnetosphere, edited by R. G. Johnson, 353–367.

Initial Hot Plasma Composition Results from the Dynamics Explorer

E. G. Shelley,* H. Balsiger,** P. Eberhardt,** J. Geiss,** A. Ghielmetti,** R. G. Johnson,* W. K. Peterson,* R. D. Sharp,* B. A. Whalen,*** and D. T. Young****

Lockheed Palo Alto Research Laboratory, Palo Alto, California 94304, U.S.A.
**Physikalisches Institut, University of Bern, Bern, Switzerland*
***Herzberg Institute of Astrophysics, National Research Council of Canada, Ottawa, Canada*
****Los Alamos National Laboratory, Los Alamos, New Mexico 87545, U.S.A.*

(Received July 19, 1982)

Initial observations from the energetic ion composition spectrometer on the DE-1 spacecraft have found mass or charge dependent processes to be operating in the vicinity of an inverted-V event in the auroral acceleration region, above the polar cap ionosphere, and in the mid-altitude dayside cusp. Separate color spectrograms for the different ion species provide a synoptic picture of these processes that can contribute significantly to our understanding of both the solar wind and the ionosphere as sources of energetic magnetospheric ions.

1. Introduction

The energetic ion composition spectrometer (EICS) on the Dynamics Explorer (DE-1) spacecraft is very similar to a family of spacecraft borne spectrometers using the same basic ion optics (GEOS-1, GEOS-2, ISEE-1, and the *AMPTE*/-CCE). The earlier instruments are described in detail by Balsiger *et al.* (1976) and Shelley *et al.* (1978) and the EICS is described in detail by Shelley *et al.* (1981a). The spectrometer covers the energy per charge range from zero (limited only by the relative effects of spacecraft ram velocity and spacecraft charging) to approximately 17 keV/e and the mass per charge range from 1 to approximately 150 amu/e. It has a high sensitivity (~ 1 cm^2-sr-eV) and a mass resolution $M/\Delta M$ of ~ 10 at focus with reduced resolution at higher energies and masses.

To accommodate the more spatially and temporally structured plasma regions encountered by the DE-1 spacecraft in the auroral zones, the instrument's measurement cycle period was significantly shortened relative to those of the previous spectrometers of this type. In addition, a high current channel electron multiplier was used to accommodate a wider dynamic range of count rates. The instrument control electronics were also modified to more efficiently utilize the faster instrument measurement cycle capabilities in performing mass, energy, and pitch angle distribution measurements. As with the ISEE-1 spectrometer, specialized modes have been developed to support various geophysical studies. The specific regions of interest include the low latitude plasmaspheric region, the auroral zone, the dayside cusp and the polar cap. These modes differ in their relative concentration on the mass-energy ranges covered and their temporal resolution. Some of the modes will be described in more detail in association with the observations discussed below.

The Dynamics Explorers − 1 and − 2 were launched into coplanar polar orbits on 3 August 1981. The DE-1 orbit is eccentric with an apogee of approximately 4.6 R_E geocentric distance and a low altitude perigee. The initial apogee was over the northern polar region and changes in latitude at a rate of about one degree every three days.

One of the principal objectives of the DE mission is to study the interaction between the cold ionospheric plasmas and the hot magnetospheric plasmas. One important manifestation of this interaction is the acceleration of ionospheric ions to energies of several keV in the suroral zone and their injection into the equatorial magnetosphere. This process was first observed by the ion mass spectrometer on the S3-3 spacecraft (SHELLEY et al., 1976b). (See review paper by SHARP et al. in this volume for more details on the S3-3 results). While the S3-3 measurements have provided us with many valuable clues to this process, they left many unanswered questions on which the DE-1 measurements are expected to shed a great deal of light. With the two spacecraft in coplanar orbits, we can obtain measurements at two points on the same field line and thus have an additional handle on the auroral ion acceleration process. The higher apogee and faster spin rate of the DE-1 spacecraft also extends the measurements to much higher altitudes ($\simeq$ 22,000km versus $\simeq$ 8,000km) and provides significantly higher spatial and temporal resolution. Furthermore the instrumentation on the DE spacecraft provides global auroral imaging and much more extensive coverage of both particles and waves. In particular the ion composition spectrometer on DE has much greater sensitivity than did the S3-3 spectrometer and extends the energy coverage into the important range below 500eV.

The Dynamics Explorer Mission approach has been to coordinate the moding of the instruments on both spacecraft so that all investigators are concentrating on the same set of geophysical problems at the same time rather than the more traditional approach of performing more or less independent survey experiments. While the operations have been coordinated, the initial results presented here will involve primarily the data from the EICS instrument alone. Many coordinated data analysis projects are underway; however due to their more complex nature they do not lead to publishable results as rapidly as those studies involving a single data set. For a more detailed discussion of the Dynamics Explorer mission and the associated scientific instuments, see the set of papers contained in the special issue of Space Science Instrumentation (HOFFMAN, 1981). For initial scientific results from the Dynamics Explorer mission see the special September, 1982 issue of Geophysical Research Letters.

2.   Observations and Initial Results

2.1   Auroral acceleration

Some of the earliest EICS results on the auroral acceleration process are shown in Fig. 1 (SHELLEY et al., 1981b). The data were acquired when the spacecraft was entering the high latitude auroral zone from the polar cap at about 1900 MLT with an altitude of approximately 11,000 km. The data are presented in the form of

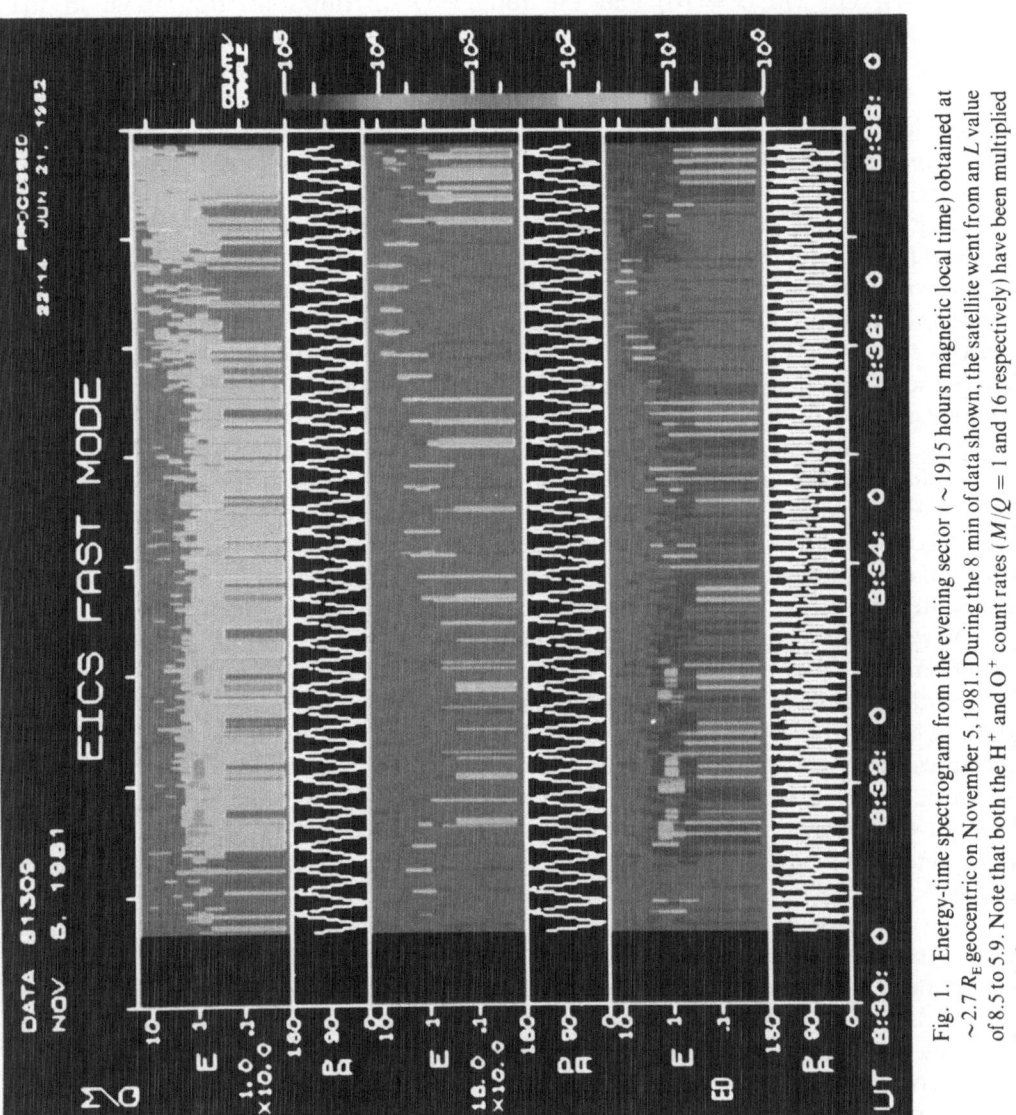

Fig. 1.   Energy-time spectrogram from the evening sector ($\sim$1915 hours magnetic local time) obtained at $\sim$2.7 R_E geocentric on November 5, 1981. During the 8 min of data shown, the satellite went from an L value of 8.5 to 5.9. Note that both the H^+ and O^+ count rates ($M/Q = 1$ and 16 respectively) have been multiplied by ten before plotting. See text for a description of the format.

energy-time spectrograms similar to those typically used for fast plasma instruments; however, one key difference is that the count rate, coded as indicated by the color bar at the right of the figure, is approximately proportional to differential number flux rather than the more conventional differential energy flux. The count rates for H^+ ($M/Q = 1$), O^+ ($M/Q = 16$) and the total E/Q distributions are shown in the top, middle, and bottom spectrograms respectively. The E/Q count rate is derived from an independent detector following the electrostatic analysis section of the spectrometer but preceeding the mass analysis section. The logarithmic energy scale extending from $10eV/e$ to $17keV/e$ is indicated in the left margin of each spectrogram. The panel below each spectrogram indicates the pitch angles for the respective measurements.

For these auroral studies the EICS was programmed to lock onto the H^+ mass channel for 6sec (approximately one spin) while rapidly cycling through a set of eight interleaving energy steps spanning the full range from $10eV/e$ to $17keV/e$ every $15°$ of spacecraft rotation. On the succeeding spin the same sequence of measurements were performed for O^+. The next two spins again sequentially sampled H^+ and O^+ but for a second set of interleaved energy steps. Thus, during a single 3sec (1/2spin) period one obtains an 8×12point energy-pitch angle distribution measurement for either O^+ or H^+ plus a similar measurement of the total ion (E/Q) distribution. Full 8×24 and 15×24point, two dimensional distributions are obtained for both species in 12sec and 24sec, respectively. For half of the spins only seven energy steps are covered and the eighth is devoted to a background measurement. Since each species is sampled on only one half of the spins the time scale from each spin is expanded in the spectrograms to eliminate data gaps, thus the time scale has an uncertainty of ± 6sec.

Several features are clear in these data. First, both H^+ and O^+ show a strong enhancement of flux moving upward along the magnetic field ($PA = 180°$). Second, the differential fluxes are frequently peaked in energy. Third, both the O^+ and H^+ fluxes show at least three examples of "inverted-V" structures, i.e., the energy of the peak flux increases and then decreases as the spacecraft moves in latitude. These inverted-V structures are a common feature in both upflowing ions (CLADIS and SHARP, 1979) and downflowing auroral electrons (FRANK and ACKERSON, 1971) and are generally interpreted as having resulted from an upward directed electrostatic field parallel to the magnetic field.

To demonstrate the relative characteristics of the upflowing O^+ and H^+ ion distributions we show in Fig. 2 the pitch angle distributions for the ions at the peak energy near the midpoint of the inverted V observed near 0636:30 in Fig. 1. We note that both the O^+ and H^+ are strongly peaked near the field line but that the H^+ also contains a lower intensity, nearly isotropic component.

Another way of looking at the distributions is shown in Fig.3a and b where the energy spectra are plotted for ions near $180°$ (Fig. 3a) and ions trapped near $90°$ (Fig. 3b). Two features stand out. First, there is essentially no trapped O^+ flux at any energy while the trapped H^+ spectrum is much softer than the field aligned component. Second, the field aligned components of both H^+ and O^+ are strongly peaked in energy with very similar energy spectra except for a lower intensity component in the H^+ similar to the trapped H^+ component. These data suggest a source of ionospheric H^+

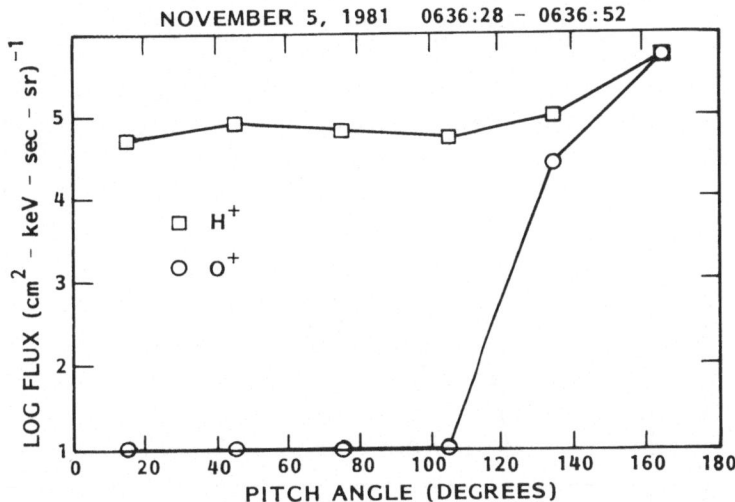

Fig. 2. Pitch angle distributions for ions with energies of $\sim 7\,\text{keV}$ near the peaks of the "inverted-V" structures at 0636:30 UT shown in Fig. 1.

and O^+ which has been accelerated by a potential drop of 5 to 6keV below the spacecraft plus a trapped component of H^+ which has not experienced such an acceleration.

Plots similar to those in Fig. 2 and 3 were examined for other times throughout the period from about 0635 to 0638 UT and all followed the same general pattern. The results are summarized in Fig. 4 where both the integral flux and the mean energy of the upflowing component (pitch angles $= 160°-180°$) of the O^+ and H^+ ions are plotted. Two interesting features are apparent. The O^+ and H^+ flux intensities (lower panel) appear to vary independently. This is consistent with the statistical results from the S3-3 data (COLLIN et al. 1981) where they found no systematic correlation between the O^+ and H^+ peak fluxes. Neither the relative flux variations between H^+ and O^+ nor the absolute variations of either with energy (see upper panel) are consistent with simple electrostatic acceleration of a constant low-energy source feeding the acceleration region from below. One possibility is a source which is resonantly driven, for example by electrostatic ion cyclotron heating below the acceleration region, and a spatially or temporally varying wave power spectral density. Another possibility is that downward accelerated electrons are heating the ionosphere locally and altering the relative scale heights of H^+ and O^+. This latter process could be accompanied by a variation in the low altitude extent of the accelerating electrostatic field. It may be possible to resolve this question by utilizing the full DE-1 and DE-2 data sets including waves, fast plasma, electron, and cold plasma data both above and below the acceleration region. Such studies are presently only in their formative stages but show great promise.

In contrast to the relative flux intensities, the mean energies of the two species (upper panel) track very closely over the entire event. The upturn in the H^+ mean energy near 0638 resulted from the much harder isotropic component of H^+ which was

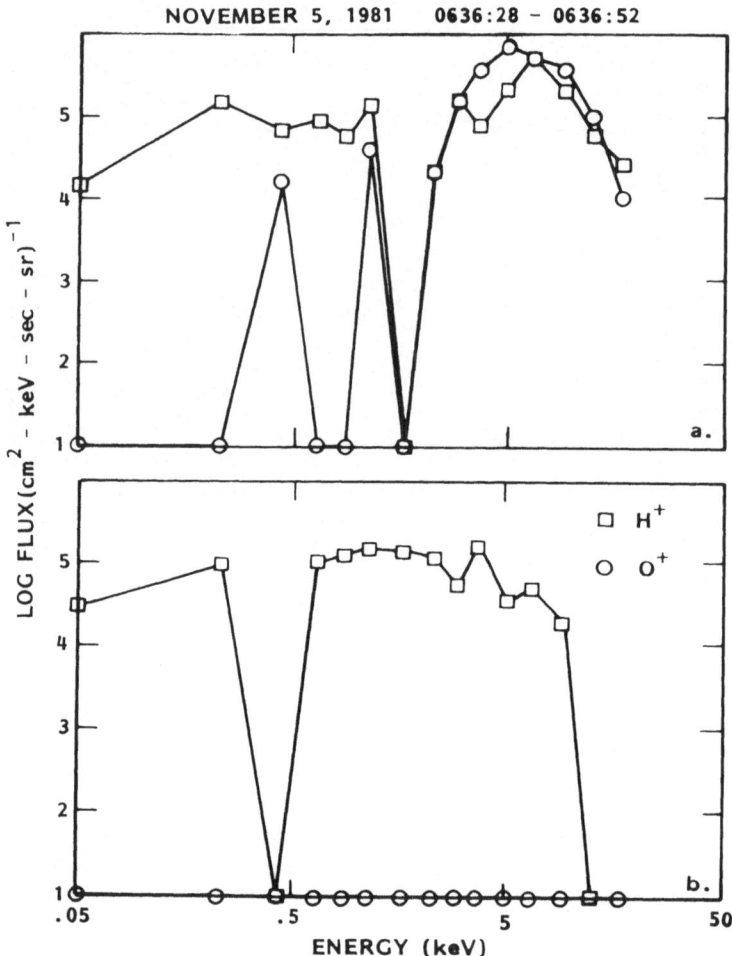

Fig. 3.   Energy spectra for ions with pitch angles near 180° (a) and 90° (b) during the same time period as Fig. 2.

encountered at this time (see Fig. 1). This close relationship between the mean energies of the two species is not consistent with the results of COLLIN et al. (1981) who found that statistically the mean energy of the upflowing O^+ exceeds that of the H^+ by a factor of about 1.7. This may simply be an unusual event, however, it may be that comparable energies are not unusual when both species occur with comparable intensities and that events which are primarily O^+ result in greater accelerations than do events dominated by H^+. The answer to this question could have important implications on the acceleration process involved. The answer, must await a much more extensive statistical analysis of the EICS data.

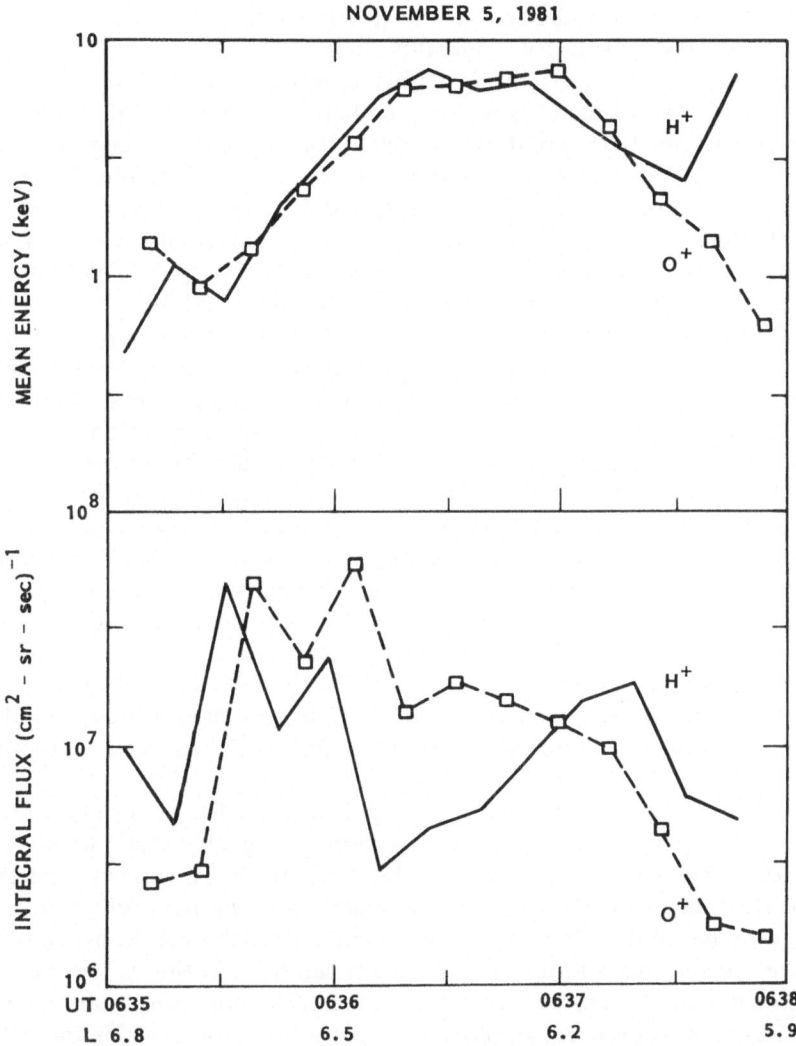

Fig. 4.   Mean energy (upper panel) and integral flux (lower panel) of the upflowing ions observed on 5 November, 1981. Universal Time (UT) and L shell (L) of the data are indicated below.

2.2   Polar cap ion acceleration

It had long been assumed that the only significant flow of ions out of the high latitude polar cap ionosphere was the polar wind which consists primarily of low energy ($\simeq 1$ eV) H^+ and He^+ (AXFORD, 1968; HOFFMAN and DODSON, 1980; BRINTON et al., 1971). By contrast, the initial results from ion composition measurements on the DE-1 spacecraft suggest that heavier ions (O^+, O^{++}) as well as the lighter ions (He^+ and H^+) are frequently observed streaming upward over much of the polar cap with energies in the range from about 10 eV to a few hundred eV (Shelley *et al.*, 1982, Chappell *et al.*, 1982, Gurgiolo and Burch, 1982).

Figure 5 is a set of color spin-phase vs. time spectrograms for data obtained by the EICS between 1230 and 1306 UT on 20 September, 1981 while the DE-1 spacecraft was located over the northern polar cap ($\sim 90°$ latitude) at a geocentric distance of approximately 4.6 R_E. By contrast to the energy vs. time spectrograms shown in Fig. 1, here the ordinate is spin phase with $0°$ corresponding to the spacecraft velocity or ram direction. At this time the magnetic field direction is at approximately $90°$ spin phase. The five spectrograms starting from the top are for H^+ ($M/Q = 1$), O^+ ($M/Q = 16$), O^{++} ($M/Q = 8$), He^+ ($M/Q = 4$) and E/Q. The color coding is the same as was described for Fig. 1, but note that the O^{++} and He^+ count rates have been multiplied by ten to emphasize their lower flux levels. For these polar cap studies the EICS was programmed to rapidly cycle between the four mass components while remaining locked on a single energy step for 6 sec ($\simeq 1$ spin). The energy range from 0 eV to 2 keV was covered by 16 retarding potential analysis (RPA) steps below 100 eV plus seven electrostatic analysis steps ranging up to 2 keV. The energy step sequence is indicated by the bottom panel plus the RPA/ESA flag between the upper two spectrograms. When the flag is on (white) the step indicator range is from 0 to 100 eV; when the flag is off (black) the range is from 0 to 17 keV as indicated in the right margin. A total measurement sequence, including one spin of background, requires 192 sec ($\simeq 32$ spins). The background, which is negligible over the polar cap, is color coded between the second and third spectrograms. The flag above the E/Q (labeled ED) spectrogram indicates the programmed sensitivity of the E/Q measurement. When the flag is on (white) the sensitivity is reduced by about a factor of ten relative to the other measurements.

Figure 5 clearly shows the existence of highly directed fluxes of all four ion species sampled, streaming upward approximately along the geomagnetic field. (The very narrow feature observed at approximately $180°$ in both the H^+ and O^+ spectrograms during the RPA measurements is a contaminant resulting from photoionization of residual neutral gas in the RPA section of the analyzer each time the instrument looks directly into the sun and should be ignored.) It can be seen that O^+ is generally the dominant ion during this period and in fact was the dominant ion streaming out of the polar cap during most periods sampled. Most of the flux is observed in the RPA range ($\lesssim 100$ eV) but it clearly extends to several 100's of eV. To investigate the distribution functions in more detail, the period centered at about 1241 UT, when the fluxes appear to be relatively stable, was selected. Figure 6 shows the count rate vs. retarding potential for the four ions. It should be possible to determine the flow speed, temperature and density for each species by independently fitting a Maxwellian distribution to each of these curves; however, due to the variability of the fluxes during the measurement cycle, a reasonable fit was obtained only for the O^+. The result was a bulk flow of $\simeq 24$ km/sec (46 eV), a temperature of $\simeq 3$ eV and a density of $\simeq 0.3$ cm^{-3}.

Table 1 shows the results of comparing the knee potential (100% transmission), the cut-off potential (10% transmission), and the plateau count rates (upward flux) for the four ion species. The values illustrated for these parameters are considered characteristic of the accelerated polar cap ion fluxes examined to date. That is, O^+ is

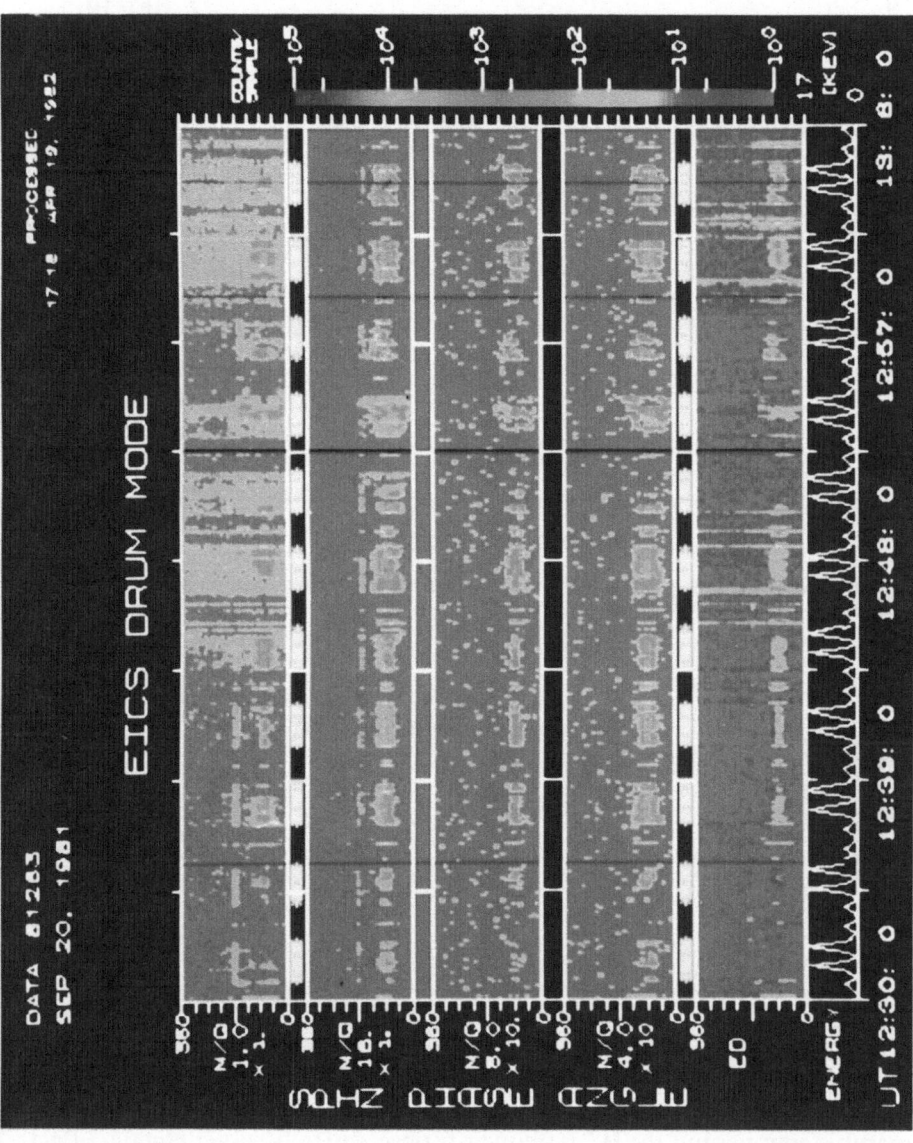

Fig. 5. Spin phase angle-time spectrograms acquired at $\sim 4.6\ R_E$ over the northern polar cap on September 20, 1981. Note that both the O^{++} and He^+ count rates ($M/Q = 8$ and 4 respectively) have been multiplied by ten before plotting.

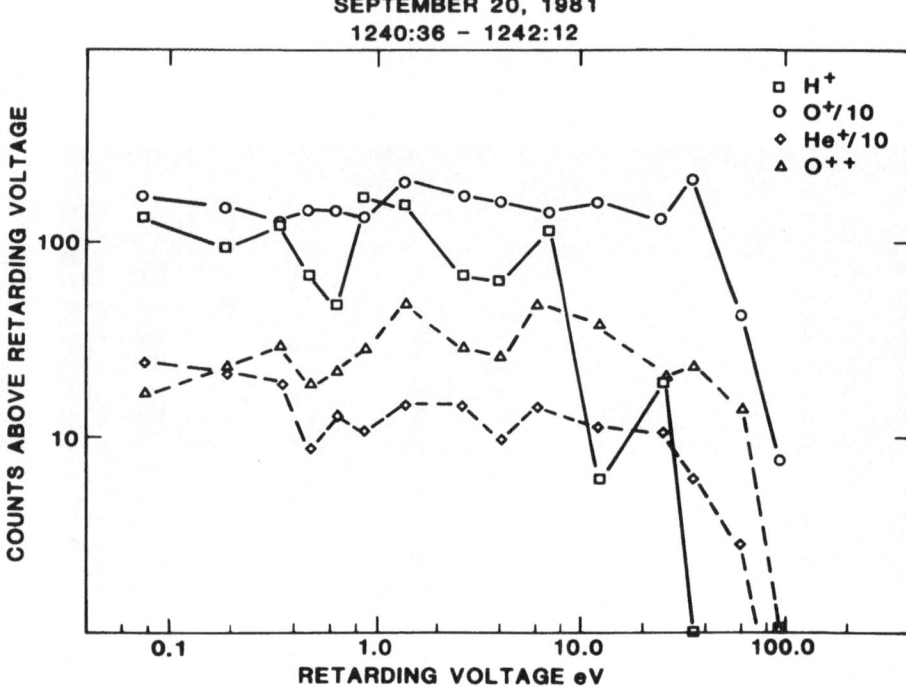

Fig. 6. Count rate vs. retarding potential analyzer voltage for ions streaming out of the polar cap ionosphere.

Table 1. Parameters derived from the data in Fig. 6.

Species	(RPA Voltage) Transmission		Upward Flux
	100%	10%	(cm²-sec)⁻¹
O^+	35 ± 5	75 ± 10	$7 \times 10^5 \pm 15\%$
He^+	25 ± 10	75 ± 10	$7 \times 10^4 \pm 20\%$
H^+	7 ± 3	20 ± 7	$5 \times 10^4 \pm 40\%$
O^{++}	35 ± 15	80 ± 10	$1 \times 10^4 \pm 35\%$

generally dominant followed in order by H^+ and He^+ with comparable intensities. O^{++} is frequently observed at a level of a few % of the O^+. The three heavier species have comparable energies but the H^+ is frequently observed to be less energetic. Effects of spacecraft charging, thus far, limit the ability to perform detailed quantitative assessments of the relative velocities and densities.

Based on approximately 9 hrs of polar cap observations, SHELLEY *et al.* (1982) concluded that fluxes comparable to those given in Table 1 are present approximately

50% of the time. They estimated that such a polar cap source ($\simeq 10^{25}$ ions/sec) would be significant in maintaining the plasma sheet population. They also concluded that these ions were the source of the tail lobe ion streams reported by SHARP *et al.* (1981) and BAME *et al.* (1968) but were not consistent with polar wind expectations. These initial polar cap observations raise many questions concerning the polar wind mechanism and magnetosphere ion sources but much more extensive and detailed studies will be required to answer those questions.

2.3   Dayside polar cusp observations

The DE-1 orbit and instrumentation are well suited to study the low to mid-altitude dayside cusp region. Although no detailed analysis of the EICS data have yet been performed for data acquired in the cusp region, a survey of these data have confirmed earlier observations and suggest that this data set will in fact contribute significantly to our understanding of both the solar wind and ionospheric source processes.

Figure 7 is a set of two spectrograms showing data acquired during a pass through a mid-altitude ($\simeq$ 21,000 km) dayside cusp on 24 October, 1981 (see Table 2 for orbit parameters for this interval). The presentation in Fig. 7 is similar to that for Fig. 5. except that the ordinates are pitch angle rather than spin phase. Here from top to bottom, the spectrograms are for H^+ ($M/Q = 1$), O^+ ($M/Q = 16$), He^{++} ($M/Q = 2$), O^{6+} ($M/Q = 2.7$) and E/Q respectively. At lower latitudes one observes moderate intensity isotropic H^+, plus a mix of soft isotropic and field aligned O^+. At approximately 1005 UT the spacecraft enters the cusp as indicated by the intensification of the H^+ and He^{++}. The O^{6+} flux levels are near background and would require integration for any quantitative analysis. One clearly sees the softening of the energy distributions of both the H^+ and He^{++} at higher latitudes. This has previously been observed and explained in terms of velocity dispersion resulting from poleward convection of the cusp field lines (SHELLEY *et al.*, 1976a; REIFF *et al.*, 1977). The upward streaming O^+ ions in the form of beams and conics are also clearly visible. They intensify in the heart of the cusp region but extend well into the polar cap. These measurements provide significantly improved temporal and spatial resolution compared to those previously reported for the S3-3 observations (SHELLEY *et al.*, 1979) and should lead to a better understanding of the acceleration process.

Figure 8 shows a distinct feature in the mid-altitude dayside cusp H^+ and He^{++} energy-pitch angle distributions. The instrument mode and data presentation are similar to those in Figure 1 except that H^+ and He^{++} rather than H^+ and O^+ are being sampled. Both species show a V like structure. Close examination reveals that as the pitch angle increases, the energy also increases. This feature has been explained by BURCH *et al.* (1982) as resulting from time of flight dispersion caused by poleward convection of ions originating from a relatively localized source region. A comparison of the relative energy-pitch angle distributions for H^+ and He^{++} confirm that indeed they have the same velocity dependence as would be expected from the model of Burch *et al.*

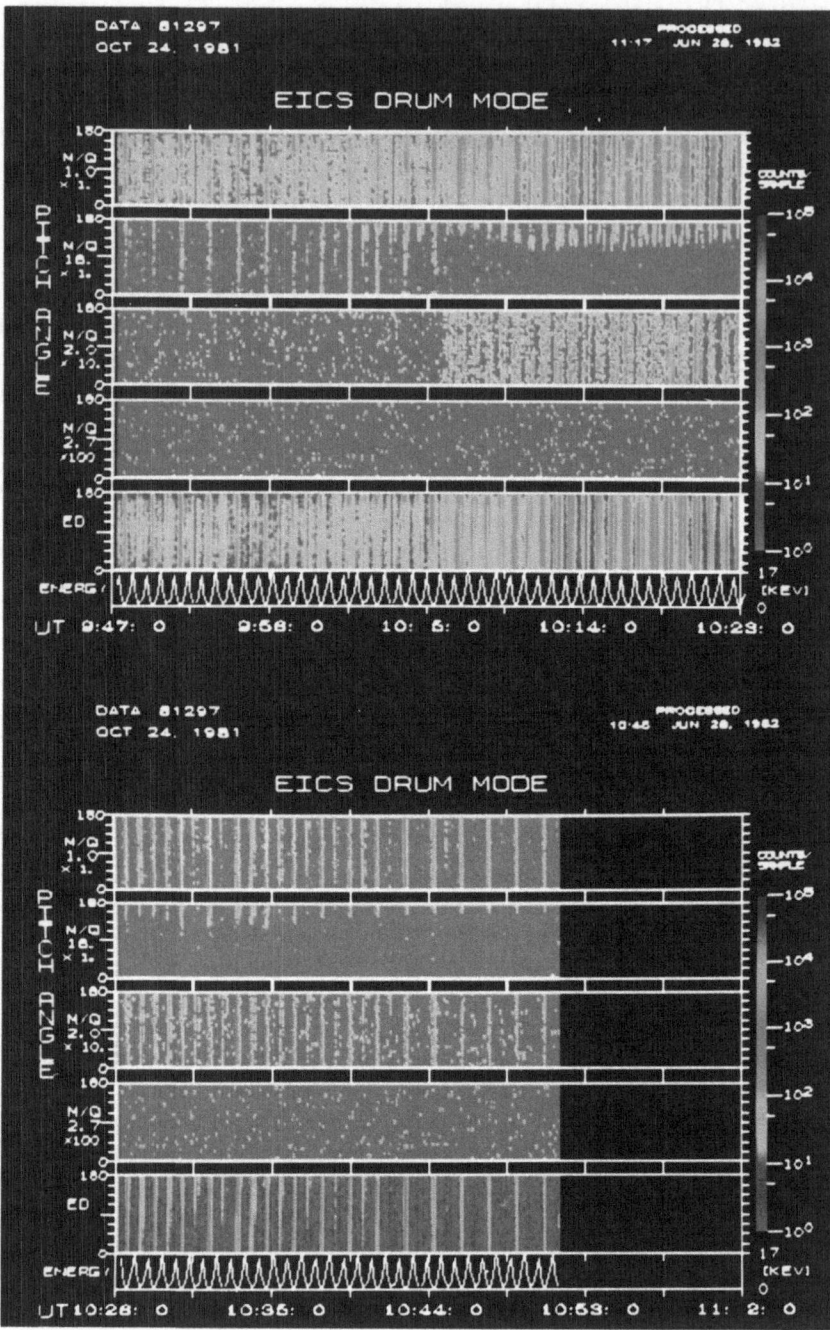

Fig. 7. Pitch angle-time spectrograms from the cusp region acquired on October 24, 1981. Geocentric distance, magnetic local time and L values for four times during this interval are given in Table 2. Note that the He^{++} data ($M/Q = 2$) have been multiplied by ten, and the O^{6+} data ($M/Q = 2.7$) have been multiplied by 100 before plotting.

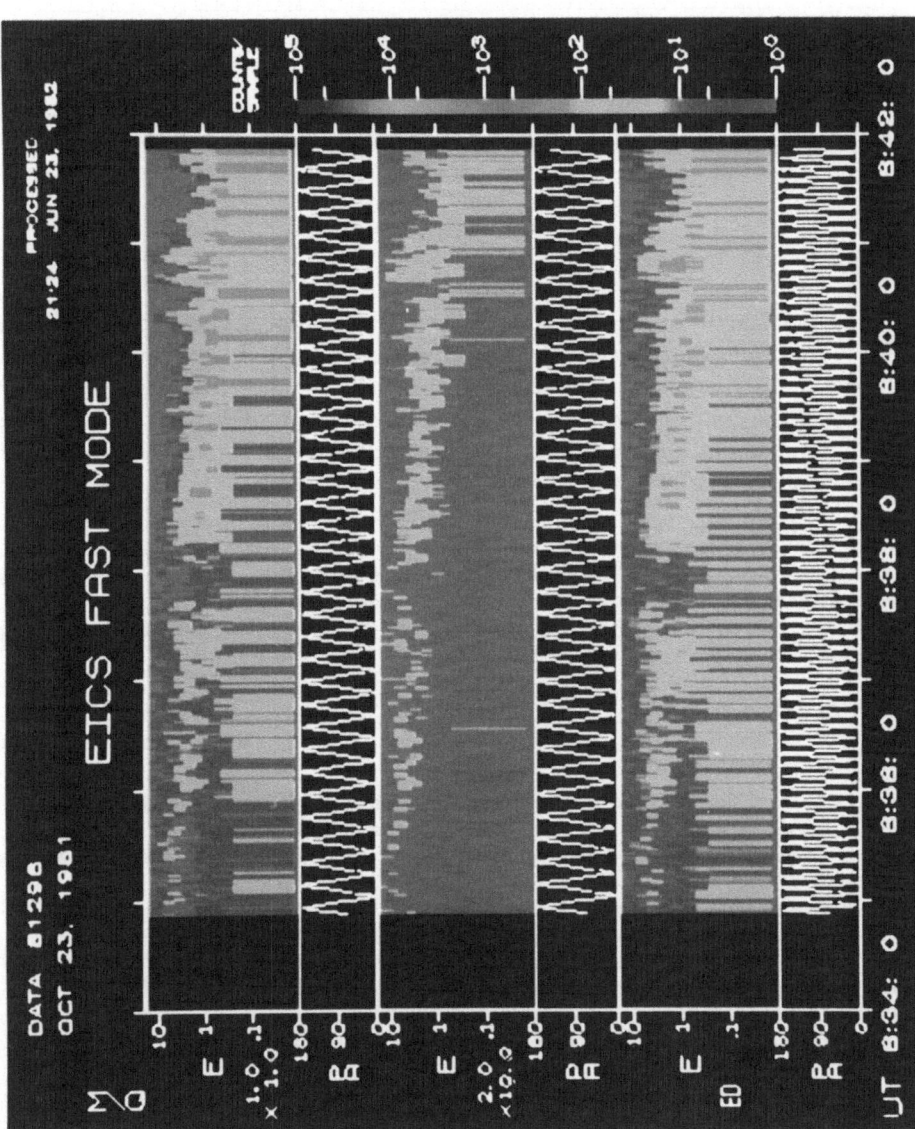

Fig. 8. Energy-time spectrogram from the dayside cusp (~ 0957 magnetic local time 4.2, R_E, L value 9.7) on October 23, 1981. Note that the He^{++} data ($M/Q = 2$) have been multiplied by ten before plotting.

Table 2.　Orbit parameters for the period displayed in Fig. 7.

Universal time	Geocentric distance (R_E)	Magnetic local time	L value
0947	4.0	0947	10.6
1005	4.2	0957	16
1028	4.4	1013	29
1047	4.6	1032	51

3.　Comments

At this early stage of the DE mission the level of data analysis is relatively limited but already some new discoveries have been made and much improved data on previously observed phenomena have been obtained. Significant progress in our understanding of the coupling between the hot magnetosphere plasmas and the dense cold ionospheric plasmas and neutral atmosphere will surely result as the analysis of these combined data sets from the two DE spacecraft proceeds.

We want to take this opportunity to express our appreciation to E. Hertzberg, T. Sanders, D. Simpson, J. Drake, D. Carr, E. Teague, E. Barnes, C. Gustafson, and C. Freeman for their energetic efforts in various phases of the construction and operation of the instrument and in data analysis.

This work was supported by NASA contract NAS5–25694, and by the Swiss Federal Science Foundation under grant 2.170.78 and 2.467.79.

REFERENCES

AXFORD, W. I., The polar wind and the terrestrial helium budget, *J. Geophys. Res.*, **73**, 6855, 1968.

BALSIGER, H., P. EBERHARDT, J. GEISS, A. GHIELMETTI, H. P. WALKER, D. T. YOUNG, H. LOIDL, and H. ROSENBAUER, A satellite-borne ion mass spectrometer for the energy range 0 to 16 keV, *Space Science Instrum.*, **2**, 499, 1976.

BAME, S. J., E. W. HONES, JR., S. I. AKASOFU, M. D. MONTGOMERY, and J. R. ASBRIDGE, Geomagnetic storm particles in the high latitude magnetotail, *J. Geophys. Res.*, **73**, 907, 1968.

BRINTON, H. C., J. M. GREBOWSKY, and H. G. MAYR, Altitude variation of ion composition in the mid latitude trough region: Evidence for upward plasma flow, *J. Geophys. Res.*, **76**, 3738, 1971.

BURCH, J. L., P. H. REIFF, R. A. HEELIS, J. D. WINNINGHAM, W. B. HANSON, C. GURGIOLO, J. D. MENIETTI, R. A. HOFFMAN, and J. N. BARFIELD, Plasma injection in the mid-altitude polar cusp, *Geophys. Res. Lett.*, **9**, 921, 1982.

CHAPPELL, C. R., J. L. GREEN, J. F. E. JOHNSON, and J. H. WAITE, JR., Pitch Angle variations in magneotspheric thermal plasma—initial observations from Dynamics Explorer-1, *Geophys. Res Lett.*, **9**, 937, 1982.

CLADIS, J. B. and R. D. SHARP, Scale of electric field along magnetic field in an inverted *V* event, *J. Geophys. Res.*, **84**, 6564, 1979.

COLLIN, H. L., R. D. SHARP, E. G. SHELLEY, and R. G. JOHNSON, Some general characteristics of upflowing ion beams over the auroral zone and their relationship to auroral electrons, *J. Geophys. Res.*, **86**, 6820, 1981.

FRANK, L. A. and K. L. ACKERSON, Observations of charged particle precipitation into the auroral zone, *J. Geophys. Res.*, **76**, 3612, 1971.

GURGIOLO, C. and J. L. BURCH, DE-1 observations of the polar wind—a heated and unheated component, *Geophys. Res. Lett.*, **9**, 945, 1982.

HOFFMAN, J. H. and W. H. DODSON, Light ion concentrations and fluxes in the polar regions during magnetically quiet times, *J. Geophys. Res.*, **85**, 626, 1980.

HOFFMAN, R. A., Editor, Dynamics Explorer, *Space Sci. Instrum.* **5**, 345–573, 1981.

REIFF, P. H., T. W. HILL, and J. L. BURCH, Solar wind plasma injection at the dayside magnetospheric cusp, *J. Geophys. Res.*, **82**, 479, 1977.

SHARP, R. D., D. L. CARR, W. K. PETERSON, and E. G. SHELLEY, Ion streams in the magnetotail, *J. Geophys. Res.*, **86**, 4639, 1981.

SHARP, R. D., A. GHIELMETTI, R. G. JOHNSON, and E. G. SHELLEY, Hot plasma composition results from the S3–3 spacecraft, this volume, pp. 167–193, 1983.

SHELLEY, E. G., R. D. SHARP, and R. G. JOHNSON, He^{++} and H^+ flux measurements in the dayside cusp: Estimates of convection electric field, *J. Geophys. Res.*, **81**, 2363, 1976a.

SHELLEY, E. G., R. D. SHARP, and R. G. JOHNSON, Satellite observations of an ionospheric acceleration mechanism, *Geophys. Res. Lett.*, **3**, 654, 1976b.

SHELLEY, E. G., R. D. SHARP, R. G. JOHNSON, J. GEISS, P. EBERHARDT, H. BALSIGER, G. HAERENDEL, and H. ROSENBAUER, Plasma composition experiment on ISEE-A, *IEE Trans. Geosci. Electron.*, **GE-16**, 266, 1978.

SHELLEY, E. G., Ion composition in the dayside cusp: Injection of ionospheric ions into the high latitude boundary layer, *Proceedings of Magnetospheric Boundary Layers Conference, Alphbach, ESA SP-148*, p. 187, August 1979.

SHELLEY, E. G., D. A. SIMPSON, T. C. SANDERS, E. HERTZBERG, H. BALSIGER, and A. GHIELMETTI, The energetic ion mass spectrometer (EICS) for the Dynamics Explorer-A, *Space Sci. Instrum.*, **5**, 443, 1981a.

SHELLEY, E. G., W. K. PETERSON, J. GEISS, H. BALSIGER, and B. A. WHALEN, Preliminary results from the energetic ion composition spectrometer on Dynamics Explorer-A Spacecraft, *EOS, Trans. Am. Geophys. Union*, **62**, 996, 1981b.

SHELLEY, E. G., W. K. PETERSON, A. G. GHIELMETTI, J. GEISS, The polar ionosphere as a source of energetic magnetospheric plasma, *Geophys. Res. Lett.*, **9**, 941, 1982.

Energetic Ion Composition in the Earth's Magnetosphere, edited by R. G. Johnson, 369–421.

Experimental Determination of Geomagnetically Trapped Energetic Heavy Ion Fluxes

Walther N. Spjeldvik* and Theodore A. Fritz**[1]

*Cooperative Institute for Research in Environmental Sciences, University of Colorado, Boulder, Colorado 80309, U.S.A.
**Space Environment Laboratory, NOAA/ERL, Boulder, Colorado 80303, U.S.A.

(Received March 8, 1982)

The observed spatial, directional and spectral characteristics of geomagnetically trapped heavy ions are reviewed. It is found that ions heavier than protons are significantly abundant in the earth's radiation belts, not only at low (ring-current) energies, but also at multi-MeV energies. Observations both at the geomagnetic equator and at low altitudes on the corresponding L-shell show the important roles played by the heavy ions. The heavy ion-to-proton flux ratio can exceed unity when compared at equal total ion energy. This is particularly the case at several MeV energies and in the outer radiation zone. In contrast, comparisons at equal energy per nucleon generally favor protons. The trapped fluxes of helium ions, of carbon, nitrogen and oxygen (CNO) ions, and of ions heavier than fluorine all exhibit high degrees of pitch angle anisotropy. Many of the observed radiation belt particle distribution characteristics are in agreement with first-order (quiet time) theoretical predictions, although some of the finer details await further research.

1. Introduction

The preceding decade has been one of significant strides in experimental determination of the magnetospheric energetic ion fluxes. Emphasis has been placed on the precise identification of the different geomagnetically trapped ion species. In parallel there has been important work done in consolidation and refinement of the fundamental theory for the radiation belt processes. The experimental observations have led to the realization that ions heavier than protons can dominate the trapped ion population at some energies and locations (e.g., Shelley *et al.*, 1972; Fritz and Wilken, 1976; Lundin *et al.*, 1980), and theoretical studies of ion charge state altering processes have demonstrated the effects of high energy chemical reactions in the radiation belts (e.g., Cornwall, 1972; Spjeldvik and Fritz, 1978a).

Early observations that detected energetic helium ions (e.g., Krimigis and Van Allen, (1967) and heavier (CNO) ions (Krimigis *et al.*, 1970; Van Allen *et al.*, 1970) in the MeV-range as well as at low (keV) energies (i.e., Shelley *et al.*, 1972) were

---

[1]Now at: Los Alamos National Laboratory, Los Alamos, New Mexico 87545, U.S.A

subsequently confirmed and extended (e.g., Sharp et al., 1974a, b, 1976a, b, 1977a, b; Shelley et al., 1974, 1976a, b, 1977; Johnson et al., 1974, 1975, 1977, 1978; Fritz and Williams, 1973; Fritz, 1976; Fritz et al., 1977; Fritz and Spjeldvik, 1978, 1979; Spjeldvik and Fritz, 1978a, b; Blake, 1973; Blake et al., 1973; Fennell et al., 1974; Hovestadt et al., 1978; MaSung et al., 1980; Scholer et al., 1979).

The qualitative picture that has emerged is one of quite variable ion composition at low energies (observations from a few tens of eV to several tens of keV), depending on geomagnetic activity, and of the persistent dominance of the heavy ions at typical radiation belt (MeV) energies. The ion composition in the intermediate energy range covering the transition from convection dominated particle energies ($\lesssim 30$ keV) to that fully controlled by diffusive processes ($\gtrsim 500$ keV) is not yet experimentally fully established, although for geomagnetically quiet (non-storm) conditions indirect evidence points to protons as the dominant ion species at ~ 800 keV per ion (Spjeldvik, 1977).

While all the relative comparisons alluded to above were made at equal total ion energy, it is sometimes physically meaningful to make such comparisons at equal ion energy per nucleon (E/A) or per ionic charge (E/Q). In such comparisons all heavy ion energies are shifted downward by factors of A and Q respectively, and whenever the ion spectra have negative slopes, the lighter ions such as protons are usually favored over the heavier ions.

Radiation belt ions are found to be highly anisotropically distributed, and for trapped ions the pitch angle distributions deep within the trapping region are found to be sharply peaked around an equatorial pitch angle $\alpha_0 = 90°$. When a sinusoidal fit, $j(\alpha_0) = j_\perp \sin^n \alpha_0$, is made to the observed distributions, n-values in the range 5 to 15 are deduced. This anisotropy varies with particle species, energy and radial location within the trapping region. For helium ions a systematic increase in the anisotropy with increasing ion energy (in the range 0.5 to 9 MeV/ion) has been determined (Fritz and Spjeldvik, 1982).

This article is organized into a number of sections: a section on steady-state observations on L-shells inside the quiet time average plasmapause location ($L \lesssim 5$), a section on radiation belt observations beyond $L \sim 5$ and a section on disturbed conditions during magnetic storms. Different sub-sections under these main divisions treat our observational knowledge of the different ion species from the point of view of equatorial observations and from low altitude observations. The paper does not treat the theoretical aspects of radiation belt physics (see reviews by Schulz, 1975; Spjeldvik, 1979) but the summary section demonstrates some of the implications of the experimental findings for the theory. Energetically, this review places the emphasis on the more energetic ions above a few hundred keV per ion.

2.   Steady-State Observations below $L \sim 5$

Except during major geomagnetic storms and their after-effects in the magnetosphere, the inner part of the trapping region on L-shells below $L \sim 5$ is found to exhibit a substantial degree of stability when viewed in energetic ($E \gtrsim 200$ keV) ions. When a steady-state for these particles is maintained, the situation is labeled "quiet", although substorms may occur and have effects on the lesser energy particles and/or in the outer parts of the magnetosphere. This observed steady-state aspect has led theoreticians to model the radiation belt "quiet time" structure using time-independent models in which, on the average, the radiation belt particle loss due to exospheric interactions is replenished by inward radial diffusive transport from a source region located well beyond $L = 5$. The radial diffusion process itself is a result of geomagnetic activity (including substorms), the effects of which are averaged over in the diffusion process (e.g., SCHULZ and LANZEROTTI, 1974). In the following we present observational evidence which to a large extent support this simplified modeling assumption, but where also some discrepancies are revealed.

2.1 Helium ions

Helium ions, either alpha particles (He^{++}) or singly ionized helium (He^+), were positively determined to be a radiation belt constituent during the first decade of in-situ space exploration (e.g, KRIMIGIS and VAN ALLEN, 1967). These measurements were made with the low altitude, polar orbiting Injun-4 spacecraft, and an example of the results obtained is shown in Fig. 1. For helium ions with $E = 2.09$ to 15 MeV/ion and $E = 3.89$ to 7 MeV/ion flux radial peaks are found just beyond $L \sim 3$. The right panel of Fig. 1 also shows the observed local pitch angle distribution which for L-shells corresponding to the radial flux maximum ($L = 2.8$ to 3.4) has a sharply peaked

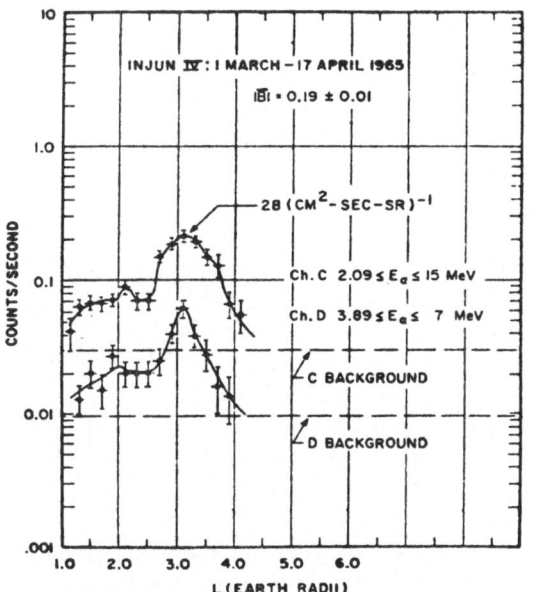
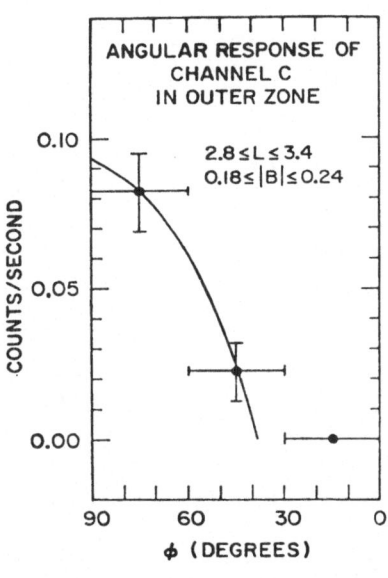

Fig. 1.   Left panel: Radiation belt observations of helium ions at 2.09–15 and 3.89–7 MeV/ion made with Injun-4 during March 1–April 17, 1965 at B = 0.19 ± 0.01 Gauss. Right panel: Angular response of 2.09–15 MeV/ion helium ions averaged over the L-shell range $L = 2.8$–3.4 and for $B = 0.21 \pm 0.03$ Gauss (from KRIMIGIS and VAN ALLEN, 1967).

angular distribution around local pitch angle $\alpha = 90$ degrees. Later investigation confirmed the continual presence of these ions (e.g., Fritz and Krimigis, 1969; Krimigis, 1970; Blake and Paulikas, 1970, 1972; Krimigis and Verzariu, 1973; Panasyuk et al., 1977; Rubin et al., 1977) at low altitudes; but the question remained as to the bulk of the ion population which, by virtue of the pitch angle anisotropy, had to be located in the vicinity of the equatorial plane.

The launch of Explorer 45 in November 1971 into a near equatorial orbit presented an opportunity to investigate directly the equatorial helium ion populations (e.g., Fritz and Williams, 1973; Spjeldvik and Fritz, 1978a; Fritz and Spjeldvik, 1978). These measurements were made for L-shells in the range $L \sim 2$ to $L \sim 5$. Measurements were also made with the spacecraft OV1-19 in the inner radiation zone at $L \sim 1.5$ to $L \sim 2.1$ close to the equatorial plane (Blake et al., 1973). Together these data sets provide a fairly systematic mapping of the equatorial radiation belt helium ion fluxes below $L \sim 5$. Some of these results are depicted in Fig. 2 which shows the Explorer 45 observations of equatorially mirroring ($\alpha_0 = \pi/2$) ions at 1.16–1.74 and 1.74–3.15 MeV/ion for $L \sim 2.25$ to $L \sim 5$ during the geomagnetically quiet period June 1–15, 1972. Notice the flux peak location at $L \sim 3.25$ and that the radial distribution falls off by well over an order of magnitude in an interval $\Delta L \sim 1$ away from this maximum.

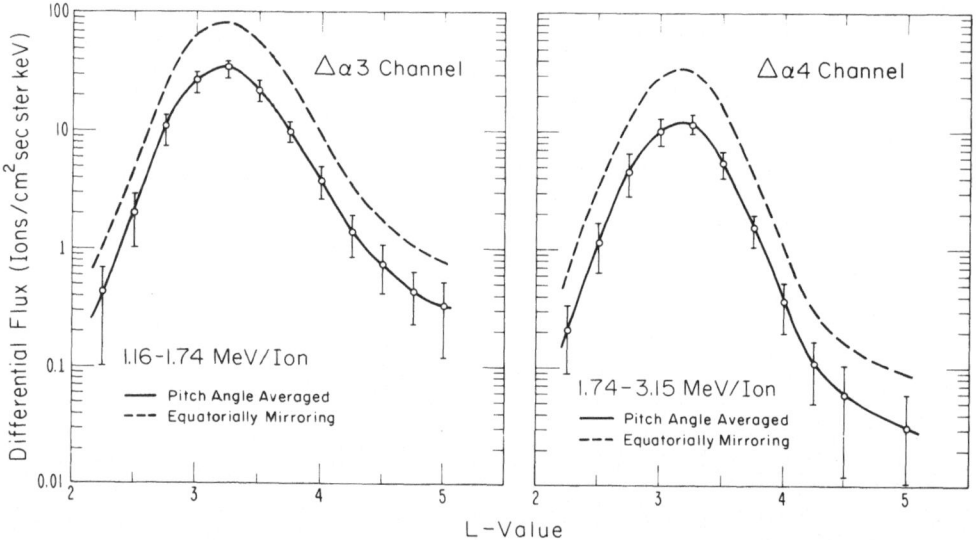

Fig. 2.   Radial profiles of energetic helium ions in the detector channels $\Delta\alpha 3$ and $\Delta\alpha 4$, which for an isotopic mass of 4 have energy passbands of 1.16–1.74 and 1.74–3.15 MeV per particle, respectively. The data have been averaged over the period June 1–15, 1972, and are mapped to the geomagnetic equator. The solid curves show pitch angle averaged ion fluxes as indicated by the Explorer 45 count rates, and the vertical error bars depict the standard deviation in the count statistics. The dashed curves show the corresponding calculated equatorially mirroring helium ion fluxes in these two detector channels (from Fritz and Spjeldvik, 1978).

Recently more data on energetic heavy ions in the radiation belts have become available. Figure 3 shows results from ISEE-1 for energetic helium ions averaged over a number of quiet periods from October 22, 1977 through January 25, 1978 (HOVESTADT et al., 1981). These data cover L-shells from $L \sim 1.5$ to $L \sim 6.7$ in six energy passbands from 0.39 to 2.18 MeV/ion for ^{4}He in either charge state. The radial profiles are found to vary smoothly (within the available count statistics) to well beyond $L = 5$ and demonstrate trapping at least to $L = 7$.

Spectrally, energetic helium ions exhibit a systematic variation from higher ($L \gtrsim 5$) towards lower ($L \lesssim 2$) L-shells. This is illustrated in Fig. 4 which depicts the spectra for 0.59 to 3.99 MeV/ion helium ions mirroring at the geomagnetic equator ($\alpha_0 = \pi/2$). Notice that these spectra are very steep at $L \gtrsim 4$, but that they become much harder with lower L-shells and are found to be nearly flat within this energy range around $L \sim 3$. This evolution with decreasing L-shells is expected from theoretical considerations; helium ions at the lower energies are lost through exospheric interactions at a faster rate than the more energetic ions at multi-MeV energies.

The equatorial data do not extend much below $L \sim 2$ because of data sampling problems associated with the Explorer 45 orbital speed and the steep L-gradients at the lower L-shells. Data are, however, available from the OV1-19 spacecraft, and Fig. 5 shows the observed helium ion spectra at a slightly off equatorial location for which $B/B_0 \sim 1.3$ (from BLAKE et al., 1973), where B and B_0 are the measured local and equatorial magnetic field induction on the same field line. Notice the spectral turnover from near flat spectra at $L \sim 2$ to spectra with positive spectral gradients ($\partial j/\partial E > 0$) at the lower L-shells, at least down to $L \sim 1.525$. Qualitatively these data consistently extend the spectral trend observed with Explorer 45.

The newer and more extensive helium ion observations also confirm the early results shown in the right panel of Fig. 1. Data from OV1-19 at $L = 1.8$ to 1.9 are depicted in Fig. 6, where the left panel depicts energetic helium and the right panel depicts energetic proton pitch angle distributions. For illustrative purposes the distributions are compared with $\sin^n \alpha$ distributions, and for helium ions $n \sim 11.3$ while for protons $n \sim 6.7$. Helium pitch angle distribution data are also available from Explorer 45 at the higher L-shells, and these data have been analyzed with several different techniques. FRITZ and WILLIAMS (1973) exploited the spin-asynchronous nature of the Explorer 45 data acquisition to deduce a value $n = 8$ at 0.91–2.0 Mev/ion. The sinusoidal functional form implies that $j(\alpha_0) = j_\perp(B/B_0)^{-n/2}$ (e.g., ROEDERER, 1970). Based on the orbital characteristics of this spacecraft, FRITZ and SPJELDVIK (1978) carried out a statistical study of the B/B_0-dependence of the helium ion fluxes. Some of their results are reproduced in Fig. 7 which covers L-shells from $L = 2.5$ to $L = 4$ and the two energy passbands of Fig. 2. The n-values deduced as a function of L are illustrated in Fig. 8 with error bars according to the available count statistics. From this indirect method n-values of 8 and 10 were determined for these two energy channels respectively. To lowest order, no pronounced L-dependence of n was deduced, but Fig. 8 does illustrate that some L-dependence may be permitted within the range of the observational error bars.

The question of radiation belt helium ion pitch angle anisotropy has been the subject of several recent papers. PANASYUK and VLASOVA (1981) have deduced a semi-empirical

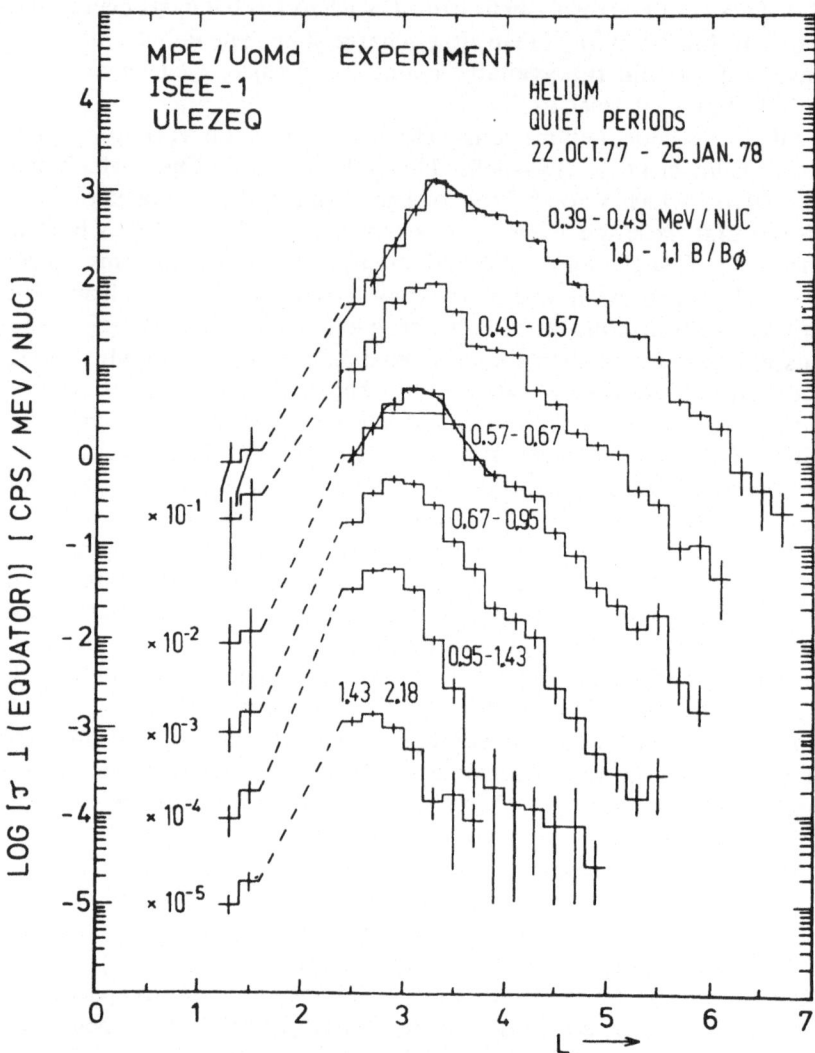

Fig. 3.   Radiation belt helium ion observations at 0.39 to 2.18 MeV/nucleon in six energy passbands versus
L-shell derived from the ULEZEQ-sensor on ISEE-1. The observations were made during October 22, 1977
through January 25, 1978 (from Hovestadt *et al.*, 1981).

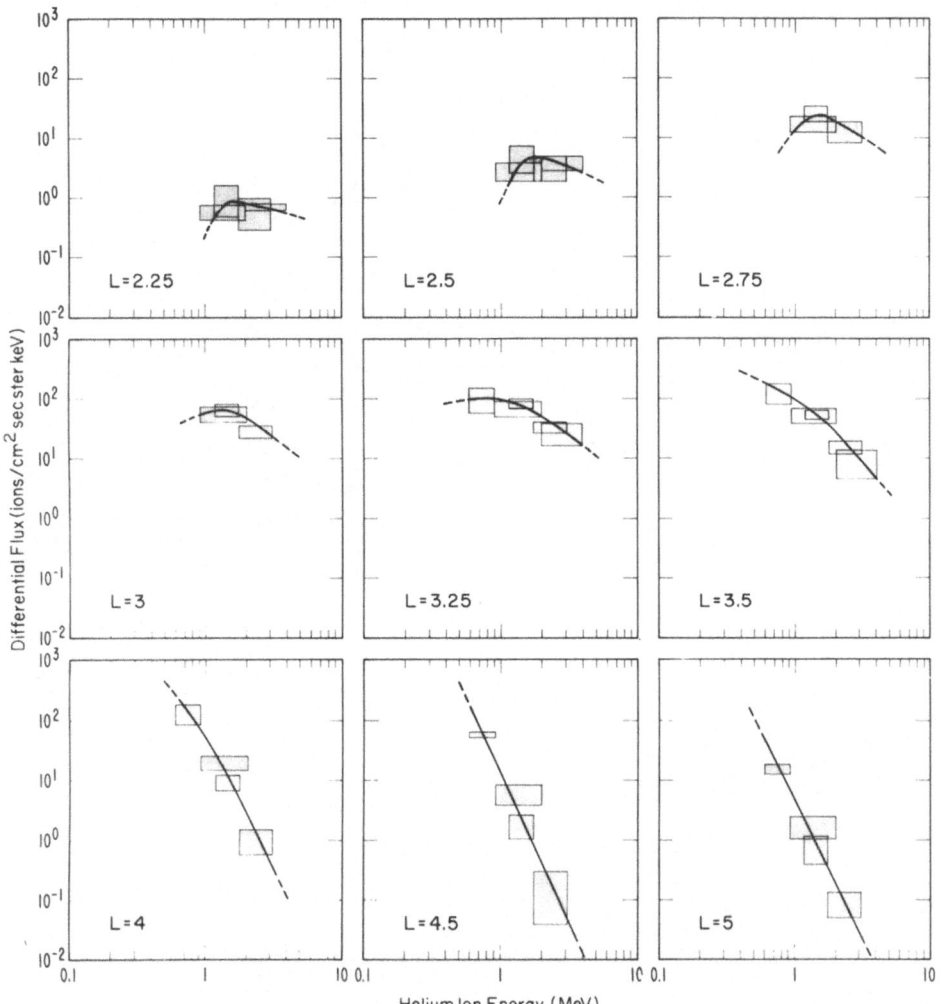

Fig. 4. Energy spectra of equatorial radiation belt helium ions deduced from mass selective ion observations on Explorer 45 during the geomagnetically quiet period June 1–15, 1972 (from FRITZ and SPJELDVIK, 1979).

relation $n = 41.7 - 10.6 \, L$ based on two orbital passes of the Soviet Molniya-2 spacecraft for a single detector passband 4–10 MeV/ion helium ions and heavier ions. The strict validity of this formula has been questioned (i.e., FRITZ and SPJELDVIK, 1982), and the latter authors demonstrated that the energetic helium ion data from a number of experiments are ordered better as function of ion energy than of the L-shell parameter. This finding is also supported by the work of BLAKE et al. (1980) who carried out a study of low altitude, outer zone ($L \sim 2$ to 5) heavy ions using data from the S3-2 satellite. A composite of available n-index values from equatorial helium ion

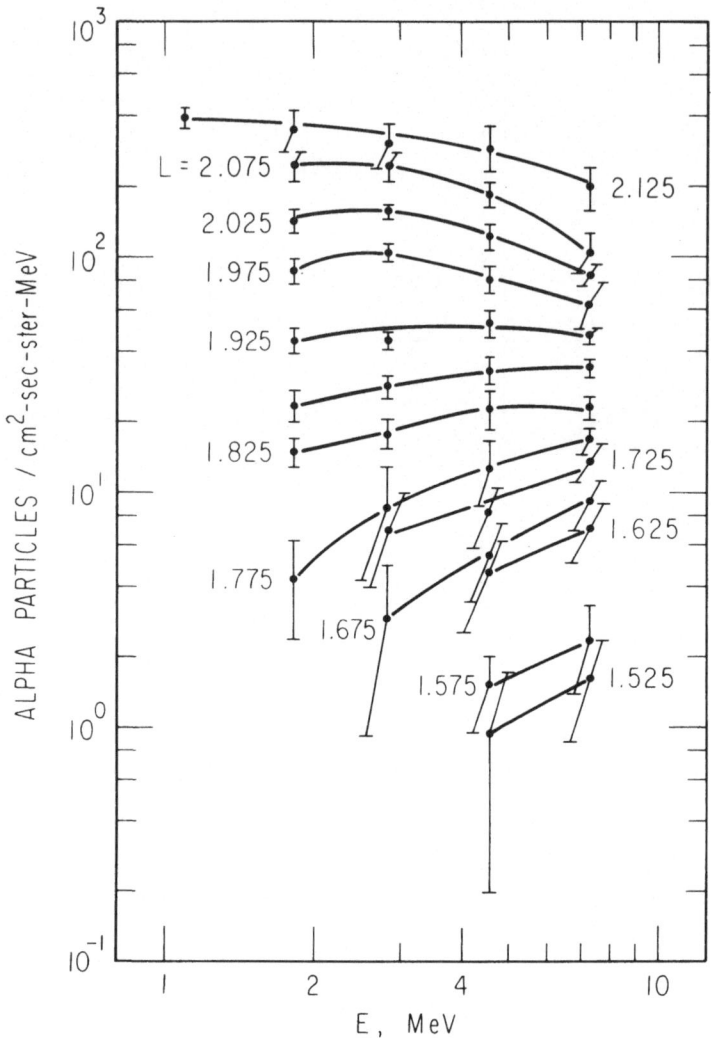

Fig. 5. Differential energy spectra of helium ions at L-shells from $L = 1.525$ to $L = 2.125$ for $B/B_0 = 1.3$ (from Blake *et al.*, 1973).

data is shown in Fig. 9, also including the results of Hovestadt *et al.* (1981), Fennell and Blake (1976), and Fritz and Spjeldvik (1979). When any possible L-shell variation in the range $L \sim 2$ to 5 are disregarded, one deduces an empirical energy dependence of n given by: $n(E) = 7 + 9.1 \log(E)$ with E in the range 0.5 to 9 MeV/ion. If there is an increase in helium ion pitch angle anisotropy with lower L-shells, then this formula represents an upper limit to the energy dependence of n. Panasyuk (1981) has presented an argument based on the break-down of trapping for these energetic helium ions to indicate that the ions measured must be alpha particles, e.g, in the second charge state.

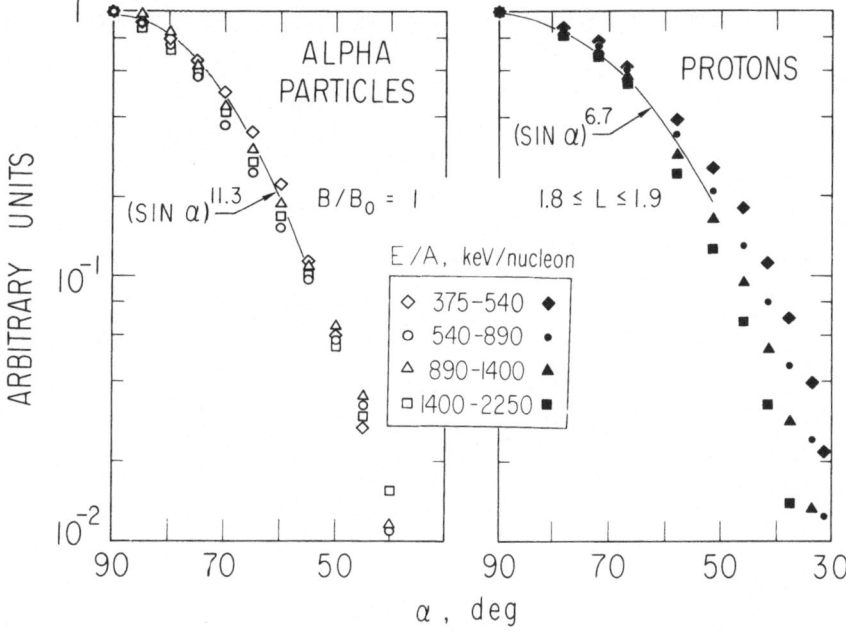

Fig. 6.   Equatorial pitch angle distribution of helium ions (left panel) and protons (right panel) at the same energies per nucleon. Curves of $\sin^n\alpha_0$ are shown for comparison (from BLAKE *et al.*, 1973).

2.2   CNO ions

Ions of carbon, nitrogen and oxygen (CNO ions) are found to be continually present at MeV energies in the radiation belts. These kinds of ions were first detected by instruments on the Injun 5 spaceraft in a near polar (inclination $\sim 81°$ to the equatorial plane), low altitude (apogee 2,525 km and perigee 644 km) orbit (KRIMIGIS *et al.*, 1970; VAN ALLEN *et al.*, 1970). The results for CNO ions at 0.31 to 14 MeV/nucleon (5 to 224 MeV/ion for oxygen ions) are depicted in Fig. 10 as a radial profile of the CNO ion count rate. The Injun 5 instrument was not able to distinguish between the different ions within the CNO group. The maximum count rate of these ions was found at $L \sim 3.2$ and corresponded to a passband integrated flux of 1.2 ions/cm²sec ster. Because of the great width of the energy passband these data are not easily converted to differential flux units; a flat spectrum over the entire passband would have yielded an upper limit of $j \sim 5.5 \times 10^{-6}$ ions/cm²sec ster keV for oxygen ions (say), while a steeply falling spectrum following a $E^{-\gamma}$ relation with $\gamma = 4$ (say) would have yielded $j = 440 \times (E/E_0)^{-4}$ ions/cm²sec ster keV with $E_0 = 1$ MeV. In either case, the differential flux at 0.31 MeV/nucleon (5 MeV/ion for oxygen ions) would be small. These authors also demonstrated the highly anisotropic nature of the CNO ion distributions. Relative comparison with simultaneously measured helium ions yields a flux ratio of $j(\text{CNO})/j(\text{He}) \sim 0.003$ above ~ 0.3 MeV/nucleon (~ 4.8 MeV/ion for oxygen ions) (e.g., VAN ALLEN *et al.*, 1970). These results pertain to the L-shell interval $L = 3$ to 3.5 and at small equatorial pitch angles near the atmospheric bounce loss cone where $B = 0.15$ to 0.2 Gauss.

Explorer 45 Helium Ion Observations June 1-15, 1972

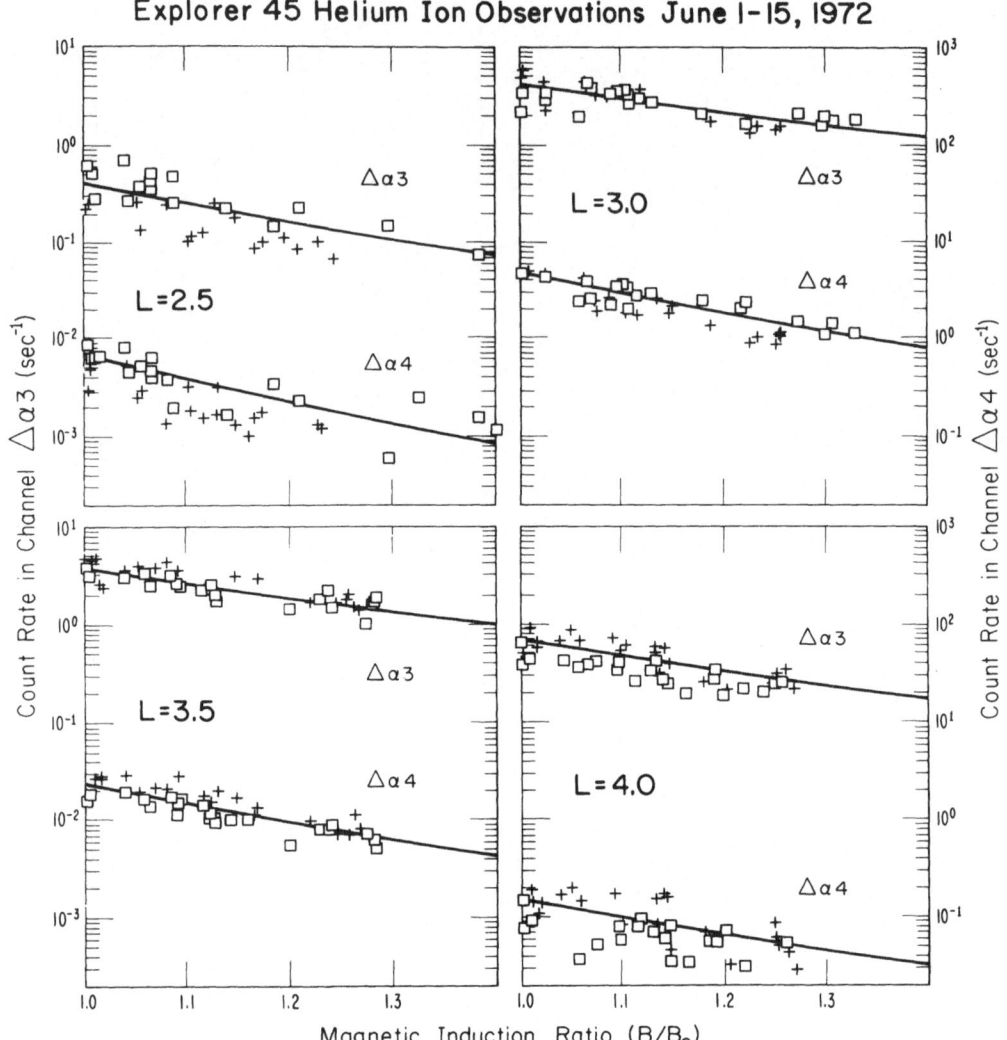

Fig. 7. Near equatorial helium ion observations at $E = 1.16-1.74$ and $1.74-3.15$ MeV/ion shown versus the magnetic induction ratio B/B_0 for the geomagnetically quiet period June 1–15, 1972. Crosses depict data for LT = 16.7 ± 1 hour and squares depict data for LT = 14.7 ± 1 hours (from Fritz and Spjeldvik, 1978).

Low altitude heavy ion measurements have also been made with the S3-2 satellite in a polar orbit with apogee $\sim 1,500$ km and perigee ~ 230 km (Blake et al., 1980), and their results are similar as to the shape of the CNO ion radial profile presented in Fig. 10, but the latter authors find the $j(\text{CNO})/j(\text{He})$ ratio to be $\sim 6.5 \times 10^{-5}$ at $L = 3$ to 3.5, $B = 0.25$ to 0.30 Gauss and at $E > 250$ keV/nucleon ($E > 4$ MeV/ion for oxygen ions), a factor of ~ 45 below the result of Van Allen et al. (1970). Thus it is reasonable

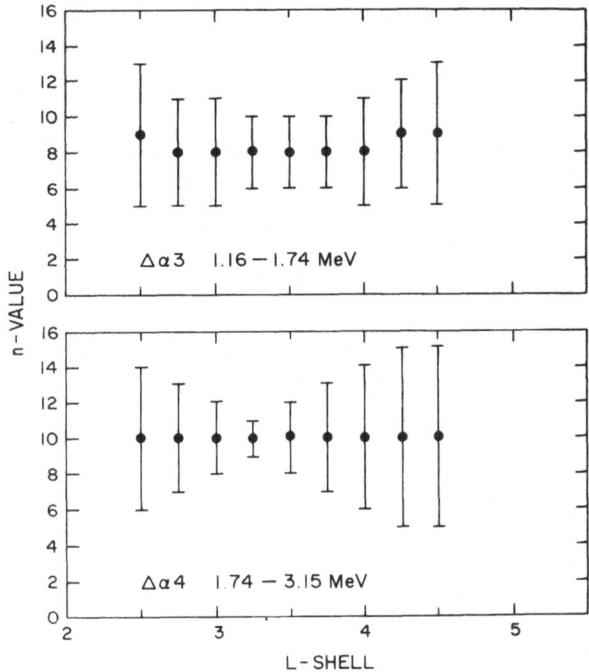

Fig. 8.   Pitch angle anisotropy index, n, versus L-shell for helium ions in the vicinity of the equatorial plane. Vertical errorbars depict the uncertainty in the deduced result (from data given by FRITZ and SPJELDVIK, 1978).

to surmise that there are significant temporal variations in the $j(CNO)/j(He)$ ratio between the Injun 5 measurements in the late 1968 and the S3-2 measurements in the early 1976. In an extensive study RANDALL (1973) has, in fact, demonstrated that the $j(CNO)/j(He)$ as well as the $j(He)/j(p)$ ratios can vary substantially, particularly in association with magnetic disturbances.

Equatorial observations of CNO ions were made with Explorer 45 between $L \sim 2$ and $L \sim 5$, and the steady-state radial flux profile for 1.82–4.8 MeV/ion CNO ions is shown in Fig. 11 representing an average over the geomagnetically quiet period June 1–15, 1972 (SPJELDVIK and FRITZ, 1978b). The instrumentation on this spacecraft also did not permit distinction between the different ion species with nuclear charge $Z \geqq 4$. It is worth noting that this radial profile of equatorially mirroring CNO ions shows substantially broader radial characteristics than that of simultaneously measured equatorial helium ion fluxes (Fig. 2).

An instrument that provided discrimination between the CNO ions was flown on the ISEE-1 spacecraft (i.e. HOVESTADT et al., 1978), and results from this experiment are depicted in Fig. 12. Ions of helium, carbon, nitrogen, oxygen and even heavier ions were detected. Of particular interest is the $j(C)/j(O)$ ion flux ratio which is considered an important indicator of the origin of the trapped radiation (e.g., BLAKE, 1973). The solar

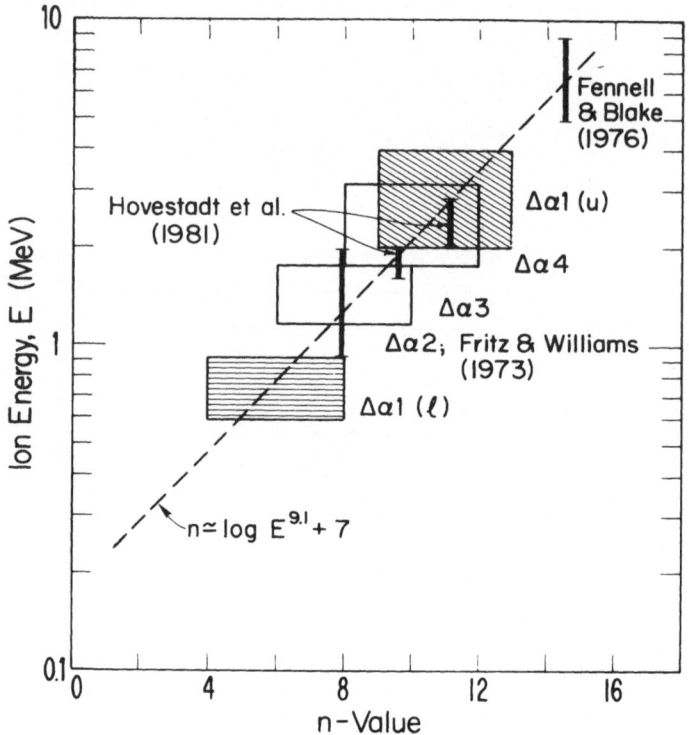

Fig. 9.   Variation of the helium ion pitch angle anisotropy with energy. The data are compiled from Fritz and Spjeldvik (1978, 1979) and other papers indicated. The empirical relation $n \simeq 7 + 9.1 \log E$ (with E in MeV) is deduced (from Fritz and Spjeldvik, 1982).

photospheric value of this abundance ratio is ~ 0.55 (e.g., Hauge and Engvold, 1970; Withbroe, 1971), and solar cosmic ray measurements have yielded an abundance ratio of ~ 0.5 for 9–35 MeV/nucleon particles (i.e., 144–560 MeV/ion for oxygen ions) indicating that the solar energetic particles have an ion abundance similar to the photospheric gas (Teegarden et al., 1973). Furthermore, Mogro-Campero (1972) has reported observations of C and O ions in the magnetosphere using a cosmic ray telescope aboard the OGO-5 spacecraft covering energies in the range 13 to 33 MeV/nucleon (208 to 528 MeV/ion for oxygen ions), and he reported $j(C)/j(O)$ ~ 0.5. Ions of these high energies have, however, very large gyroradii in the geomagnetic field and are possibly solar and galactic cosmic rays on transit through parts of the magnetosphere rather than stably trapped ions. On the other hand, the measurements show a radial maximum which could imply a trapping-like character. More research on hundreds of MeV ions would seem necessary. Thus every available indication points to a $j(C)/j(O)$ ratio of ~ 0.5 for an extraterrestrial ion source. In contrast, there is only a very low relative abundance of carbon ions in the terrestrial ionosphere: $j(C)/j(O) < 10^{-5}$ (Blake, 1973). Consequently, the elemental abundance

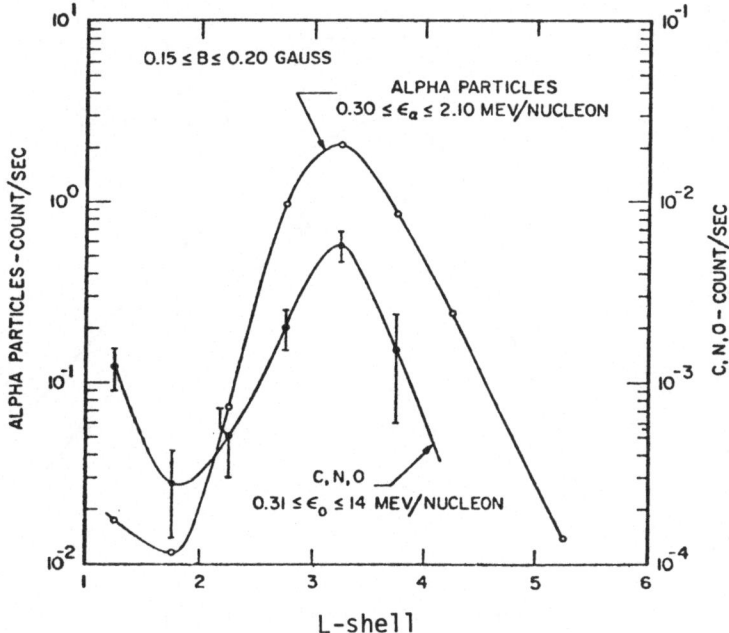

Fig. 10.   Radial profiles of CNO ions and helium ions obtained with the spacecraft Injun-5. The absolute directional intensities (in ions/cm^2 ster) at local pitch angles $\alpha = 90°$ may be obtained by multiplying the count rates by 200 (from VAN ALLEN et al., 1970).

measurements of HOVESTADT et al. (1978) are of great value in distinguishing between the ionospheric and the solar particle sources, and in the MeV range this ratio has now been established to be of the order of unity.

Data on carbon and oxygen ions are depicted in Fig. 13 (from HOVESTADT et al., 1981) in four energy passbands for each of the two ion species; for carbon the energy coverage extends from 0.4 to 1.51 MeV/nucleon (4.8 to 18.12 MeV/ion), and for oxygen ions the range is 0.39 to 1.41 MeV/nucleon (6.24 to 22.56 MeV/ion). Apart from the lower count statistics for these ions, it is clear that carbon and oxygen ions are present throughout the radiation belts in roughly comparable quantities. Figure 14 (from HOVESTADT et al., 1978) shows the deduced $j(C)/j(O)$ ratio at equal energy per nucleon in the range from ~ 0.4 to ~ 3 MeV/nucleon. The flux ratio is determined to vary from ~ 1.3 at ~ 0.45 MeV/nucleon to ~ 4.1 at ~ 1.2 MeV/nucleon with a mean value of ~ 2.5 over this energy range. These results pertain to the heart of the radiation belts at $L = 2.8$ to 3.8. When the fluxes of carbon and oxygen ions are compared at equal total ion energy one arrives at a ratio of typically 0.5. The measurements of HOVESTADT et al. (1978) thus clearly indicate an extraterrestrial source of the energetic trapped radiation belt heavy ions.

CNO ions are also highly anisotropically distributed in pitch angle at the equator. For geomagnetically quiet conditions the Explorer 45 data indicate that the CNO ion flux anisotropy index, n, has values of about 12; this is depicted in Fig. 15 where the

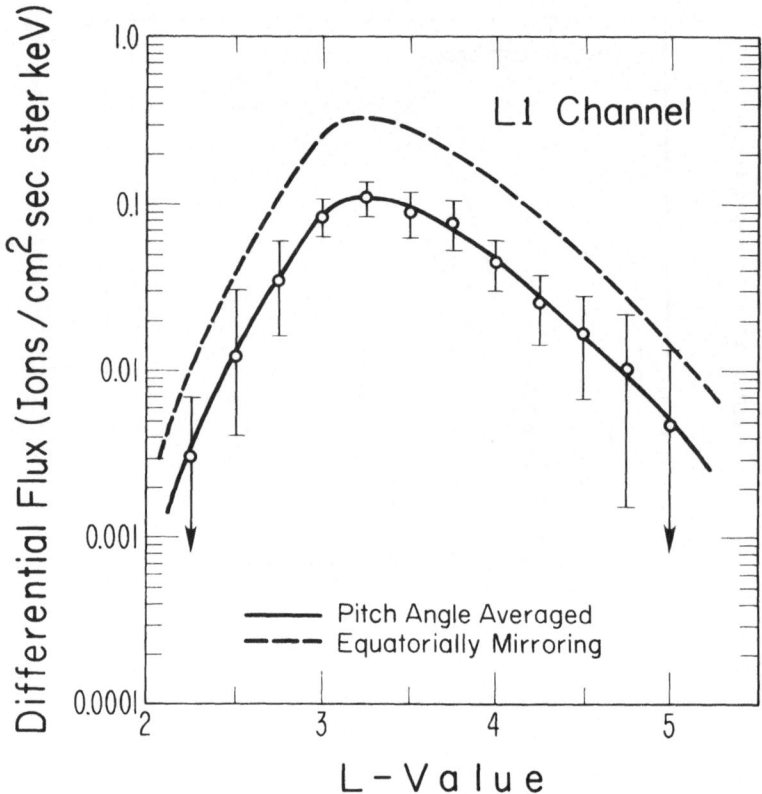

Fig. 11.   Radial profile of the Explorer 45 quiet time observations in channel L1 during the period June 1–15, 1972. The vertical errorbars indicate the standard deviation in the data set at each quarter integral L shell. The dashed line depicts the calculated equatorially mirroring flux at 1.82–4.8 MeV per ion (from Spjeldvik and Fritz, 1978c).

error bars result from the statistical uncertainty (e.g., Spjeldvik and Fritz, 1978c) which start to become substantial at $L \gtrsim 4$. The CNO ion pitch angle anisotropy is very substantial, and to lowest order no systematic L-shell variation of n is observed. Hovestadt *et al.* (1981) report that the pitch angle distributions fall off sharply down to an equatorial pitch angle of $\alpha \sim 39^{0}$ ($B/B_0 \sim 2.5$) with values of the anisotropy index, n, varying from typically $n \simeq 10$ for helium ions to $n \simeq 16$ for carbon and oxygen ions at 0.4 to 0.5 MeV per nucleon, and that the heavy ion fluxes at $B/B_0 > 2.5$ are subject to large temporal fluctuations. This is illustrated for $L = 2.8$–3.8 in Fig. 16. These data indicate that pitch angle distributions are organized following $(B/B_0)^{-\gamma}$ dependencies ($\sin^n \alpha_0 = (B/B_0)^{-\gamma}$; $n = 2\gamma$) where γ attains the values 4.8, 7.9, and 7.9 ($n = 9.6, 15.8,$ and 15.8) for helium, carbon, and oxygen ions respectively. At high B/B_0-values, deviations from this functional relationship are seen. These angular distributions pertain to $L = 2.8 - 3.8$.

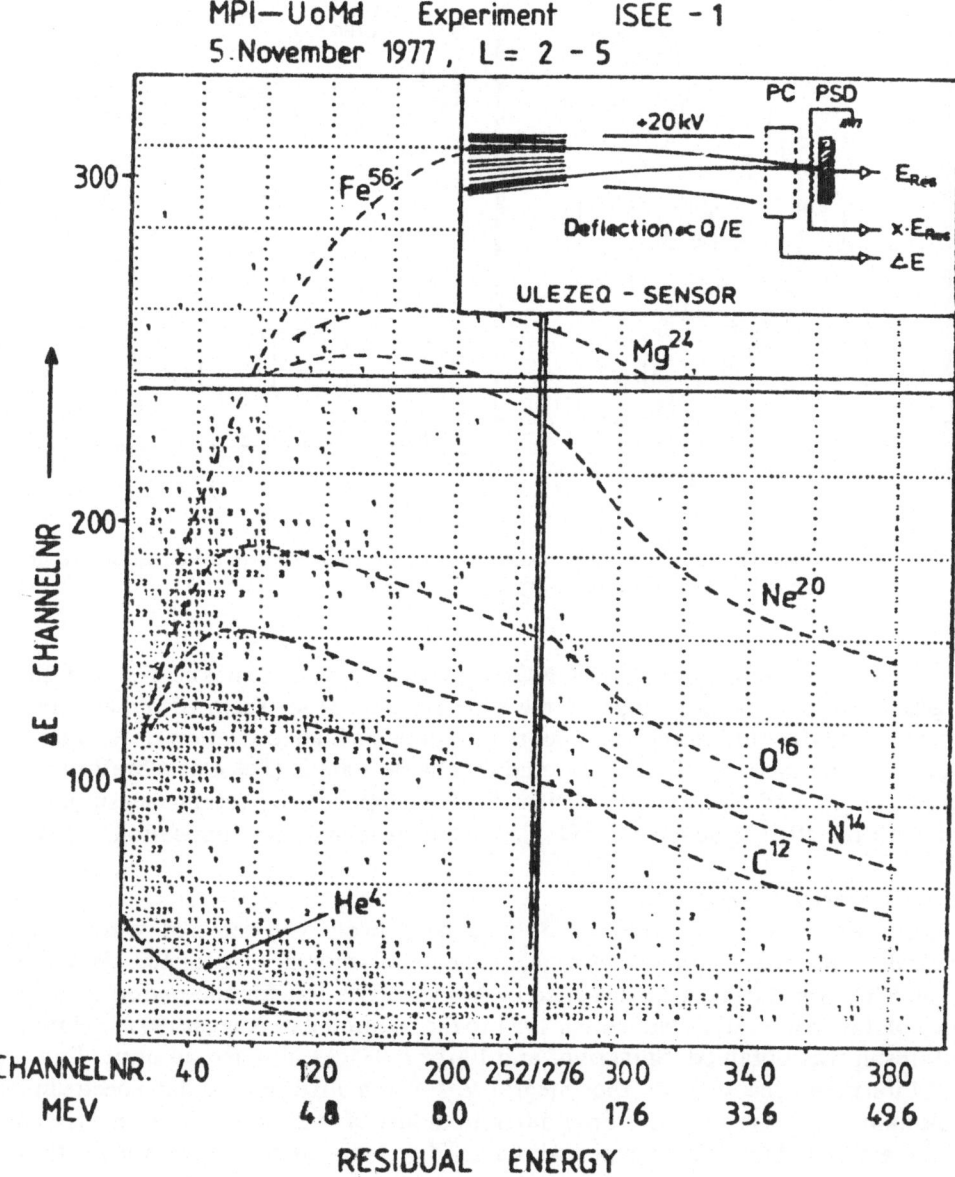

Fig. 12.   Heavy ion observations made with the ULEZEQ-sensor on ISEE-1 on November 5, 1977 at $L = 2$ to 5 (from HOVESTADT *et al.*, 1978).

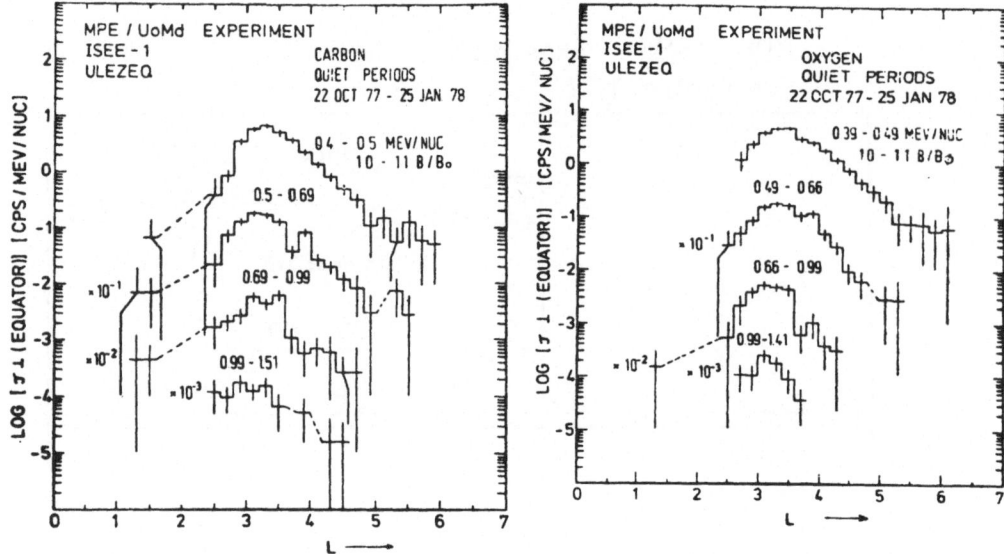

Fig. 13.   Observations of radial profiles of carbon and oxygen ions obtained with the ULEZEQ-sensor on
ISEE-1 during October 22, 1977 through January 25, 1978 (Hovestadt *et al.*, 1981).

As to the state of ionization of the CNO ions, Panasyuk (1981) has argued that these
energetic ions must be in a high charge state ($Q > +3$) for them to remain stably
trapped within the radiation belts. Theory also predicts high ionization states, and for
oxygen ions the theoretically most abundant charge state is $Q = 4$ at 4 MeV per ion
(i.e., Spjeldvik and Fritz, 1978b). This is also supported by recent measurements
made with the ISEE-1 spacecraft (D. Hovestadt, personal communication, 1980).

2.3   Ions heavier than oxygen

Ions with nuclear charge $Z > 8$ have been positively detected with several
spacecraft. The heavy ion detector telescope on Explorer 45 detected these ions
at $\gtrsim 10$ MeV/ion during a disturbance in August 1972, but for the period from the
launch of this spacecraft in November 1971 to the beginning of August 1972 essentially
a null-result was obtained. Spjeldvik and Fritz (1981c) estimated an upper limit of
$J = 0.1$ ions/cm^2 sec ster for the integral $Z \gtrsim 9$ ion flux. For a flat spectrum of
magnesium ions (say) an upper limit differential flux of $\sim 3 \times 10^{-6}$ ions/cm^2 sec ster
keV was estimated for this quiet period. Later, $Z \gtrsim 9$ ions were also observed with the
ATS-6 spacecraft.

The more advanced instrumentation flown on ISEE-1 permitted separate
measurements of the different $Z > 8$ ions. For L-shells in the range $L = 2.6$ to $L = 3.4$
Hovestadt *et al.* (1978) reported ratios of $j(\text{Ne-Si})/j(\text{O}) \sim 0.2$ and $j(\text{S-Fe})/j(\text{O}) < 0.02$
at energies of 0.4 to 1.5 MeV/nucleon (~ 9.6 to 36 MeV/ion for magnesium ions). The
fluxes of each of these ion species were, however, found to be very low during the
geomagnetically quiet period on November 5, 1977 in agreement with the Explorer 45
observations.

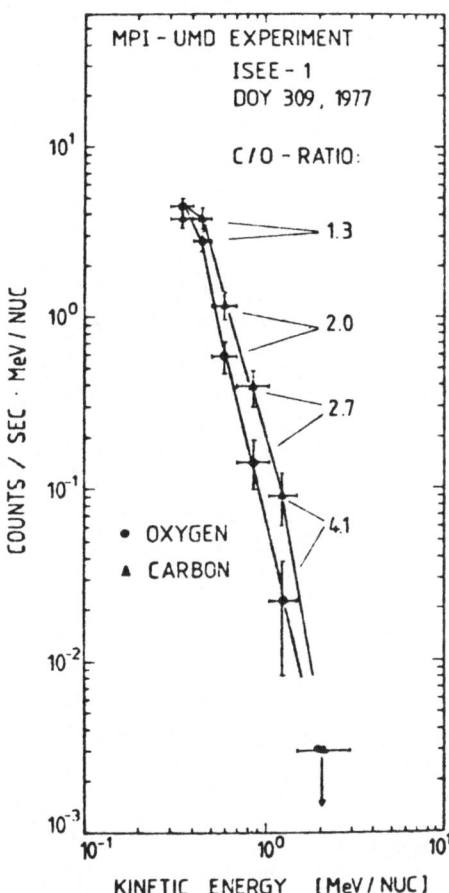

Fig. 14.   Radiation belt spectra of carbon ions and oxygen ions measured with the ULEZEQ-sensor on ISEE-1 at $L = 2.8$ to 3.8 (from HOVESTADT et al., 1978).

2.4   Comparison with protons

Protons are no longer considered the predominant radiation belt ion species at all energies and locations, and current research work is directed at establishing the relative roles of the various ion species. Figure 17 shows an early comparison between helium ions at 2.09 to 15 MeV/ion (0.52 to 3.75 MeV/nucleon) and protons at 0.52 to 4 MeV, from measurements with Injun 4 during March and April, 1965 (i.e., KRIMIGIS and VAN ALLEN, 1967). The peak integral fluxes in comparable passbands (when compared at equal energy per nucleon) were found to be 2.8×10^1 and 1.24×10^5 ions/cm²sec ster for helium ions and protons respectively at $L \sim 3$ and $B \sim 0.19$ Gauss. Thus $j(He)/j(p) \sim 2.3 \times 10^{-4}$ at equal energy per nucleon at this low altitude location.

FENNELL et al. (1974) report simultaneous measurements of helium ions and protons over a range of energies, 0.85–9 MeV/ion for helium ions and

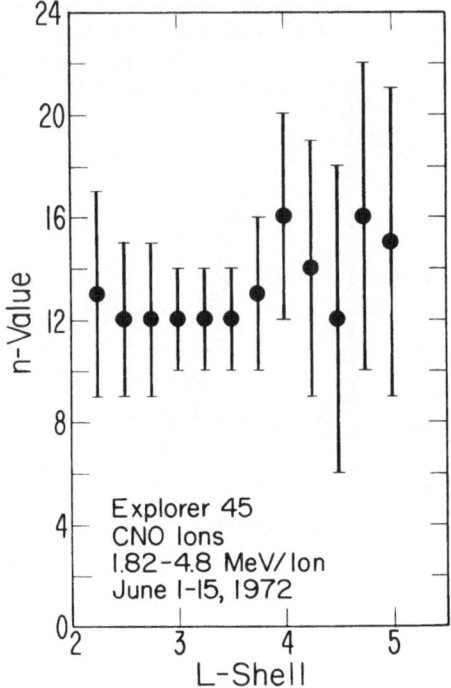

Fig. 15.   Pitch angle anisotropy index n, versus L-shell for Explorer 45 equatorial observations of CNO ions at 1.82–4.8 MeV/ion (from Spjeldvik and Fritz, 1978c).

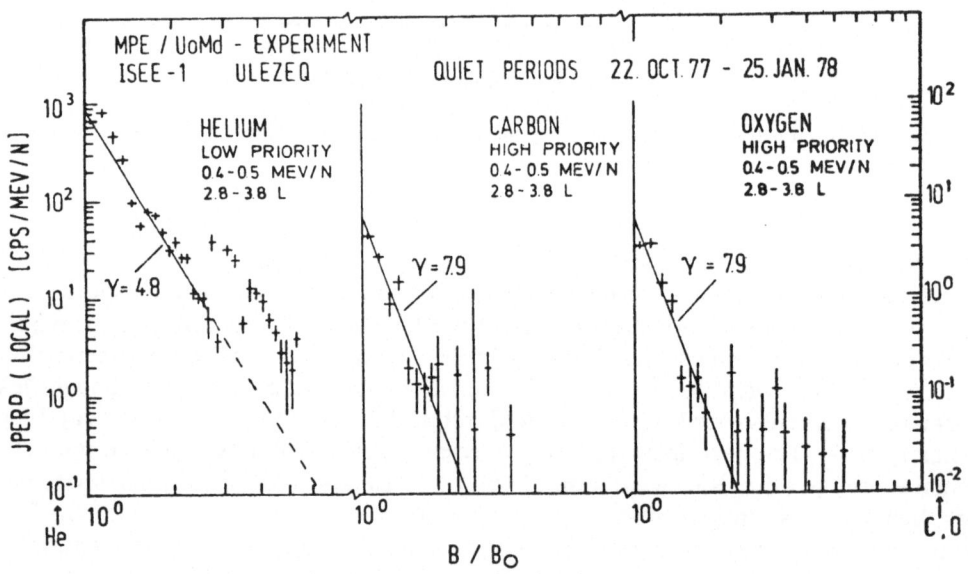

Fig. 16.   The observed B/B_0-dependence of 0.4–0.5 MeV/nucleon helium, carbon and oxygen ion fluxes on L-shells averaged over the range $L = 2.8$ to $L = 3.8$ (from Hovestadt *et al.*, 1981).

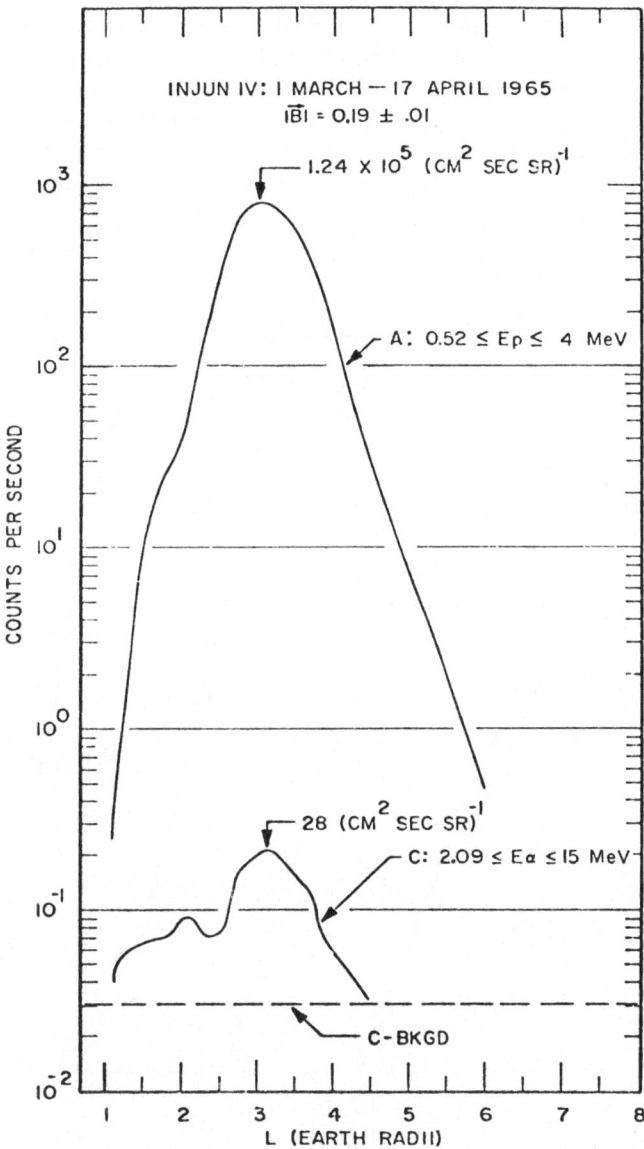

Fig. 17.   Proton (Channel A) and helium ion (Channel C) observations made with Injun-5 during March 1-April 17, 1965 (from KRIMIGIS and VAN ALLEN, 1967).

0.166–10.6 MeV for protons. Figure 18 depicts some of their results from the OV1-19 spacecraft which on March 18, 1969 was launched into a low altitude, polar orbit with apogee 5,796 km and perigee 471 km. To the extent of this low altitude data, a close correspondence between helium ion spectra and proton spectra was reported for the L-shell range from $L = 2.3$ to $L = 3.4$, and for B/B_0-values in the range 2.28 to 8.48. At equal total ion energy the $j(\mathrm{He})/j(\mathrm{p})$ ratio was found to vary from $\sim 10^{-1}$ at $L = 3.35 \pm 0.05$ to $\sim 3.5 \times 10^{-3}$ at $L = 2.35 \pm 0.05$. Since there may be variations in the helium ion and proton pitch angle distributions over this range of L-shells, it was not known whether the observed $j(\mathrm{He})/j(\mathrm{p})$ ion flux ratio (at equal total ion energy)

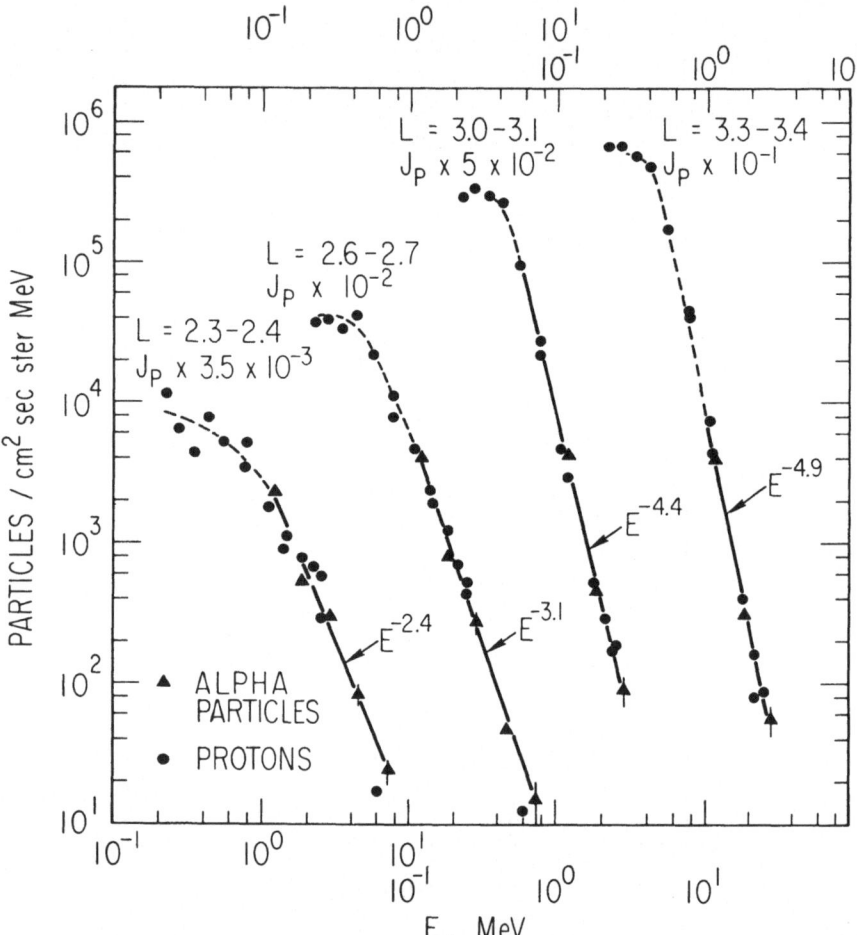

Fig. 18.   Proton and helium ion spectra plotted as differential flux versus total ion energy. The proton fluxes have been normalized to the helium ion flux level by multiplication of proton fluxes with factors of 0.0035, 0.01, 0.05, and 0.1 for the four spectra shown respectively, and the B/B_0-values are 2.28, 3.73, 6.11, and 8.48 at $L = 2.3, 2.6, 3.0$, and 3.3 respectively. The spectra at $L = 2.3 - 2.4$ and $3.0 - 3.1$ are to be referred to the lower energy scale. (from Fennell et al., 1974).

variation was due to the change in L-shell or the simultaneous change in B/B_0-value over the OV1-19 orbit or both.

Simultaneous helium ion and proton data were also obtained with Explorer 45. Quiet time radial profiles of equatorially mirroring protons ($B/B_0 = 1$, $\alpha_0 = \pi/2$) averaged over the period June 1–15, 1972 are shown in Fig. 19, and the energy

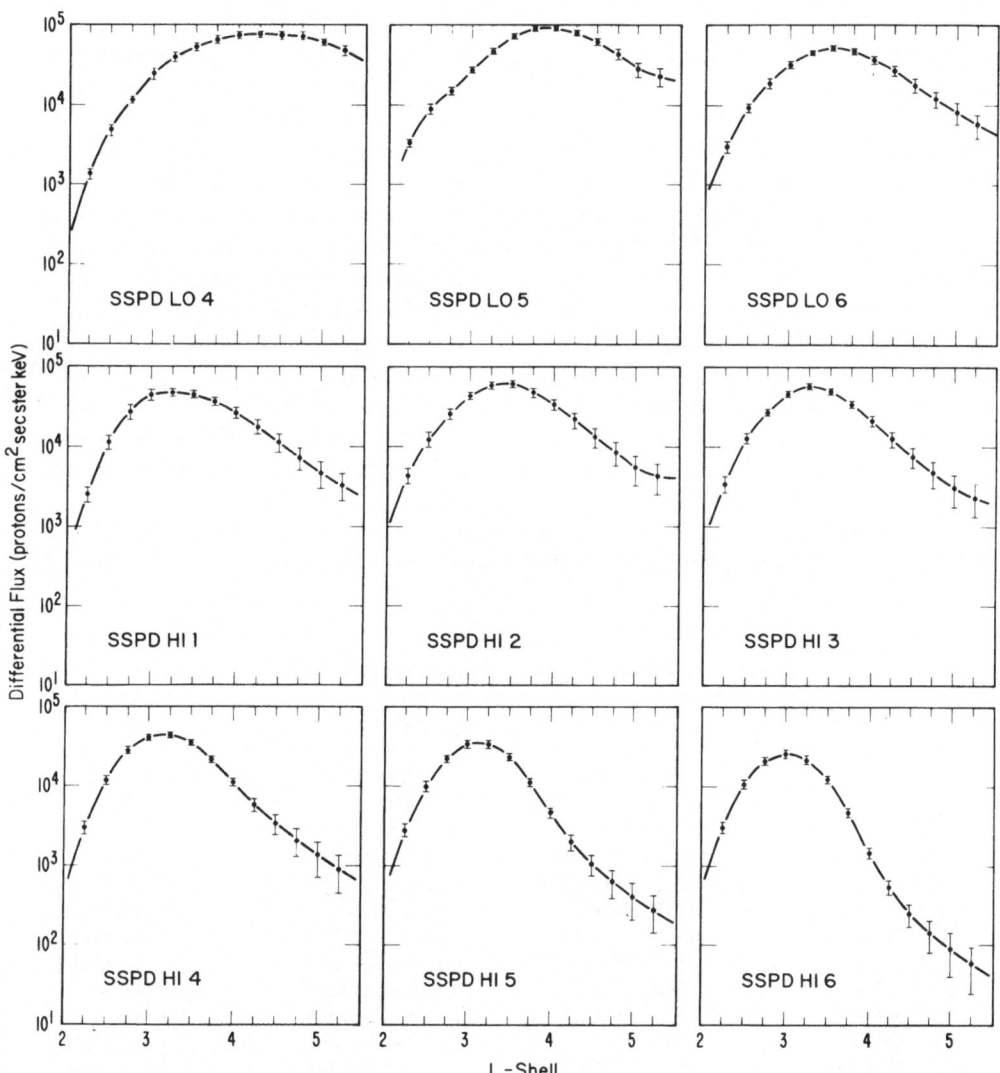

Fig. 19. Radial profiles of equatorially mirroring radiation belt protons. Standard deviations calculated at quarter-integral L shells are depicted as the vertical errorbars. The data represent averages over 60 Explorer 45 passes through the equatorial radiation belts June 5–15, 1972. Nine proton channels are shown, SSPDLO4 through SSPDHI6, see Table 1 (from FRITZ and SPJELDVIK, 1979).

passband identification is given in Table 1; the total energy range of the nine proton passbands in Fig. 19 covers from 78.6 to 872 keV. It can be seen that the peak in the radial proton flux profile is displaced towards lower L-shells with higher proton energy. The available count statstics is generally better than that of the (higher energy) helium ion and CNO ion data from Explorer 45.

The spectra of equatorially mirroring protons for the quiet period June 1–15, 1972 are shown in Fig. 20, covering the L-shell range from $L = 2.25$ to $L = 5$ and energies from 0.1 MeV to 20 MeV (for further details, see Fritz and Spjeldvik, 1979). As in the case of equatorial energetic helium ion spectra, the steep spectral shapes for protons are also found primarily at the higher L-shells. A considerable hardening of the proton spectra below $L \sim 3$ is evident, including a definite spectral turn-over (where $\partial j/\partial E > 0$) below ~ 600 keV at these L-shells. This figure also incorporates two additional proton energy channels (H7 and Int of Table 1) beyond that shown in Fig. 19.

When the proton pitch angle anisotropy is determined, one generally finds a lesser degree of anisotropy than for helium ions and CNO ions at comparable energies; one deduces n-values in the range $n = 2$ to $n = 8$. This would suggest that there is an ion mass dependence of n. Figure 21 shows a statistical result of plotting the proton flux versus B/B_0 for different energies and L-shells. It can be seen that there is a weak trend of finding higher anisotropy with increasing proton energy and decreasing L-shell. At the highest L-shell in this figure ($L = 5$) the count rates are quite low, and the corresponding statistical significance of n is less. Table 2 gives a listing of the n-values obtained from this statistical method together with estimates of the uncertainty in these numbers.

In comparing proton data with flux data on heavier ions, Fritz and Wilken (1976) pointed out that protons are not always the dominant ion above ~ 1 MeV/ion.

Table 1.　Explorer 45 proton detector passband responses.

Channel	Protons*	Helium ions[†] (He^4)	Oxygen ions[†] (O^{16})
		Passband	
SSPDLO-4	78.6–138.5 keV	94–156 keV	146–217 keV
SSPDLO-5	138.5–195.5 keV	156–227 keV	217–297 keV
SSPDLO-6	195.5–300 keV	227–323 keV	297–402 keV
SSPDHI-1	363.5–375 keV	1.16–1.18 MeV	4.5–4.6 MeV
SSPDHI-2	375–390 keV	1.18–1.21 MeV	4.6–4.7 MeV
SSPDHI-3	390–430 keV	1.21–1.25 MeV	4.7–4.8 MeV
SSPDHI-4	430–533 keV	1.25–1.37 MeV	4.8–5.3 MeV
SSPDHI-5	533–674 keV	1.37–1.52 MeV	5.3–5.7 MeV
SPDHI-6	674–872 keV	1.52–1.74 MeV	5.7–6.2 MeV
H7	872–1,600 keV	1.72–52 MeV	> 6.2 MeV
Int	3,300–22,000[†]keV	> 11.9 MeV	> 95 MeV

*From beam calibrations.
[†]Nominal values.

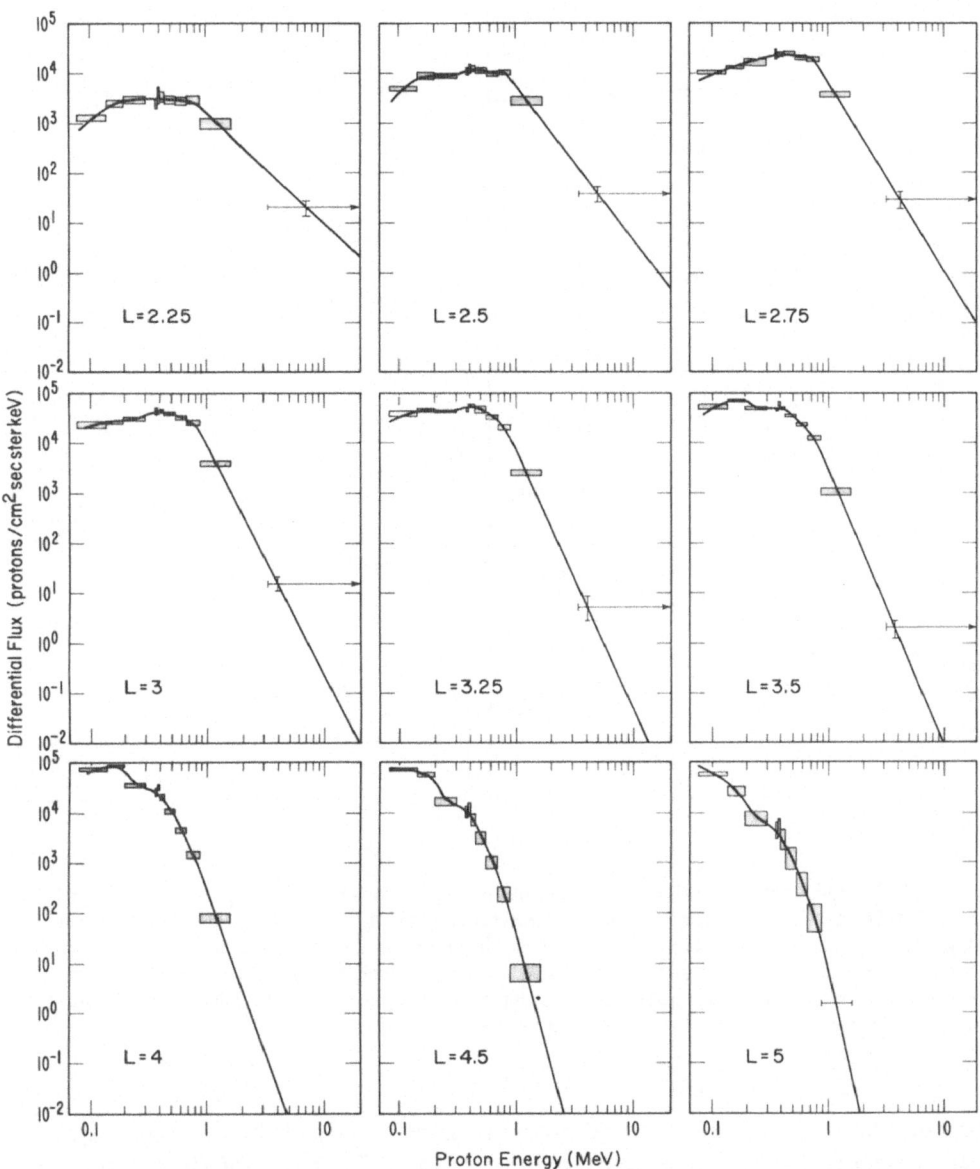

Fig. 20. Energy spectra of equatorial radiation belt protons in the range 0.1–20 MeV deduced from Explorer 45 observations during the geomagnetically quiet period June 5–15, 1972. The solid curve in each panel is drawn as the best fit to the data points for the differential energy channels and as an analytic approximation to the quasi-integral channel. The data are represented by the energy passband (horizontal extent) and the standard deviations (vertical extent); for details see FRITZ and SPJELDVIK (1979).

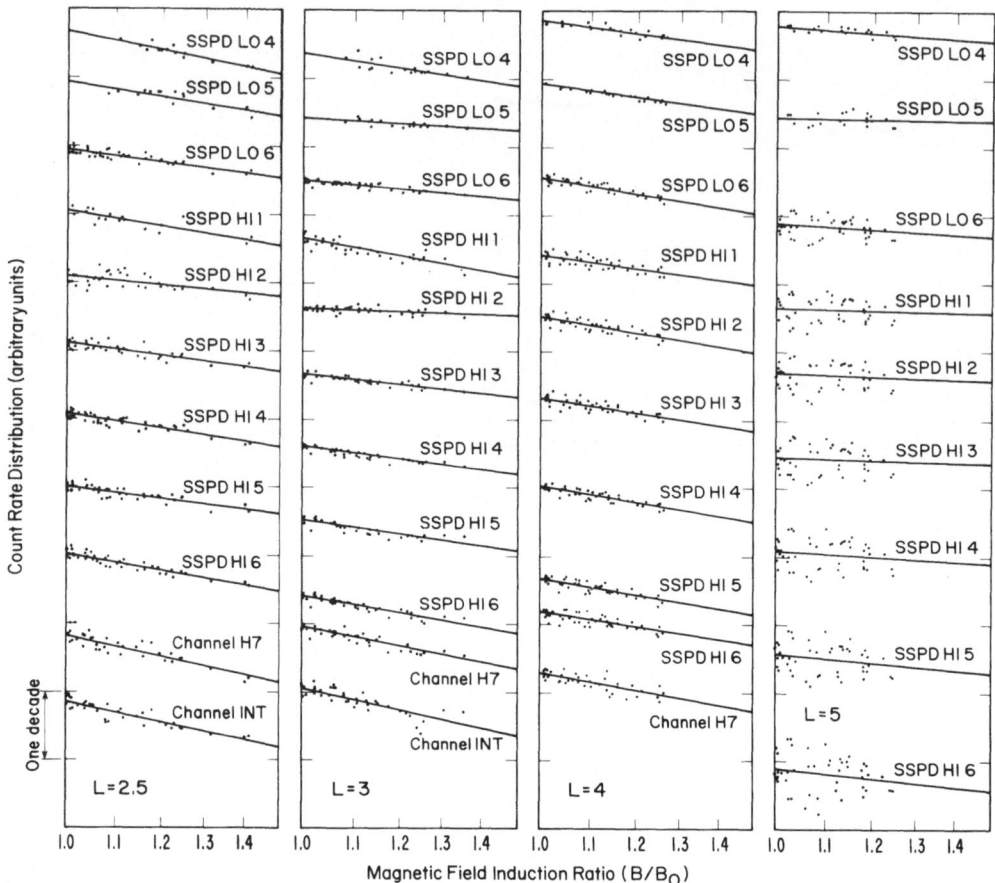

Fig. 21. B/B_0 dependence of locally mirroring radiation belt protons near the geomagnetic equator (June 5–15, 1972). The data are given for the 11 proton detector channels listed in Table 1. The fluxes measured in the highest energy channels drop below detectability at the outermost L shells. The solid lines drawn through the data points assume a pitch angle dependence $j = j_\perp \sin^n\alpha$ such that $j_\perp/j_{\perp 0} = (B/B_0)^{-n/2}$, where the subscript zero denotes equatorial quantities (e.g., FRITZ and SPJELDVIK, 1978). The exponent values are generally in the range $n = 2$–8 (from FRITZ and SPJELDVIK, 1979).

FRITZ and SPJELDVIK (1979) carried out a systematic study of the $j(\mathrm{He})/j(\mathrm{p})$ ratio for quiet geomagnetic conditions during June 1–15, 1972. Some of their results are depicted in Fig. 22 which shows the observed ratio at equal total ion energy in the range ~ 0.5 to ~ 5 MeV/ion. At these energies the helium ions are most likely to be in the second charge state (He^{2+}), and for this reason they are often labeled alpha-particles (e.g., CORNWALL, 1972; SPJELDVIK and FRITZ, 1978a). From the data in Fig. 22, which extends over the L-shell range $L \sim 2.25$ to $L \sim 5$, it is evident that there are rather large variations in the flux ratio both with ion energy and with L-shell. The value of $j(\mathrm{He})/j(\mathrm{p})$ at equal total ion energy well exceeds unity at the higher L-shells ($L \gtrsim 3.25$)

Table 2.   Statistical n-values for radiation belt protons during June 1–15, 1972.

Channel	$L = 2.5$	3	3.5	4	4.5	5
SSPDLO-4	7 ± 3	6 ± 2	5 ± 2	5 ± 2	4 ± 2	2 ± 3
SSPDLO-5	6 ± 4	2 ± 3	4 ± 2	5 ± 2	4 ± 2	1 ± 3
SSPDLO-6	5 ± 3	3 ± 3	4 ± 2	6 ± 3	5 ± 4	2 ± 4
SSPDHI-1	5 ± 3	6 ± 3	4 ± 2	5 ± 3	4 ± 4	1 ± 4
SSPDHI-2	5 ± 3	3 ± 2	4 ± 2	6 ± 3	4 ± 4	1 ± 4
SSPDHI-3	5 ± 3	4 ± 2	5 ± 2	5 ± 3	3 ± 4	1 ± 4
SSPDHI-4	5 ± 2	5 ± 2	5 ± 1	6 ± 2	5 ± 3	2 ± 3
SSPDHI-5	5 ± 2	5 ± 2	5 ± 1	6 ± 2	5 ± 3	3 ± 4
SSPDHI-6	6 ± 2	6 ± 2	6 ± 2	5 ± 2	5 ± 4	4 ± 4
H7	8 ± 3	7 ± 2	7 ± 2	6 ± 3	5 ± 3	—
Int	8 ± 3	8 ± 3	—	—	—	—

and higher energies in this range. At some energies preferential loss of helium ions causes this ratio to diminish with lower L-shells as the two ion populations diffuse (preferentially inwards) in the radiation belts.

Ion flux comparison at equal ion energy per nucleon is also of interest, and such a comparison of the equatorially mirroring helium ion and proton fluxes is depicted in Fig. 23 covering the energy range ~ 0.125 to 1.25 MeV/nucleon and L-shell range $L = 2.25$ to $L = 5$. This type of flux comparison gives a different perspective, and $j(\text{He})/j(\text{p})$-values in the range from $\sim 10^{-4}$ to $\sim 2 \times 10^{-3}$ are found. The systematic variation with energy is such that a local ratio minimum is found at 0.5 to ~ 0.8 MeV/nucleon at virtually all L-shells. The small energy-dependence of the ion flux ratio at equal energy per nucleon appears to be roughly preserved over this L-shell interval, and the overall magnitude of this ratio also remains fairly constant with L-shell. It is worth noting that the $j(\text{He})/j(\text{p})$-values at equal energy per nucleon are significantly lower than the solar wind average value. In a review of the helium ion abundance in the solar wind, HIRSHBERG (1975) has pointed out that the helium abundance can vary from virtually undetectable ($\ll 1\%$) to perhaps some 20% of the hydrogen ion (proton) flux, with a mean of about 4%. This constitutes evidence against the dominant radiation belt process being one which preserves the source characteristics of the assumed solar wind source for these ions.

Comparison of the fluxes of CNO ions and the proton fluxes is limited to those energies where the unambiguous heavy ion data have been obtained. The data from Explorer 45 also do not overlap very well for comparable energy comparisons. Figure 24 shows spectral comparisons between the fluxes of CNO ions observed at the geomagnetic equator and those of protons on L-shells of $L = 2.5$ to $L = 5$. The proton data for the quasi-integral channel (Int) on Explorer 45, covering proton energies from 3.3 to 22 MeV, have been used with the imposed constraint that above ~ 1 MeV the proton spectrum follows a power law relation $j(E) = j(E_0)\,(E/E_0)^\gamma$ where $J(E_0 = 1.2\ \text{MeV})$ and γ was determined at each L-shell using the H7 and Int channel count rates (Table 1); for details, see FRITZ and SPJELDVIK (1979). The CNO ion data

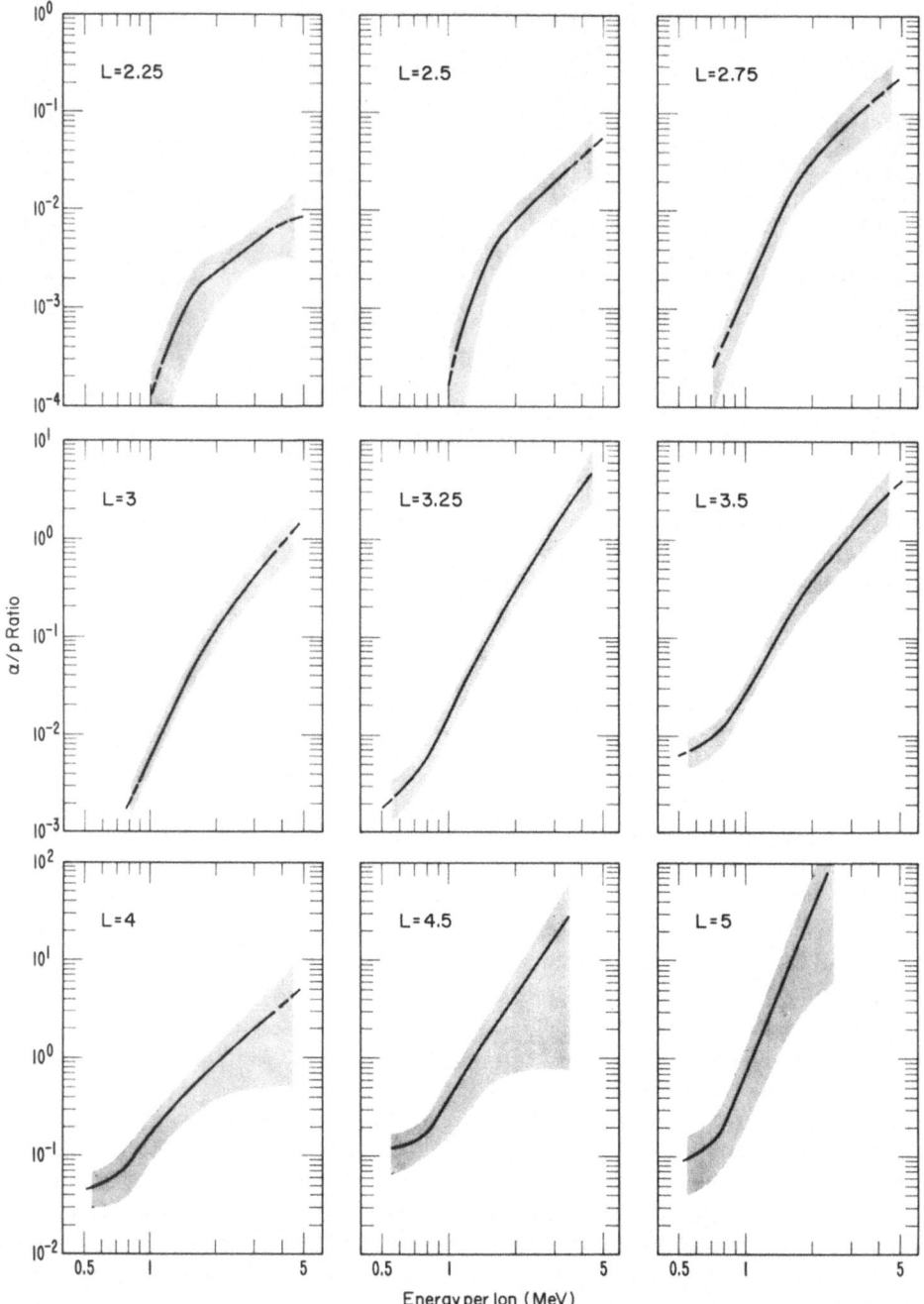

Fig. 22. Experimentally determined quiet time helium ion to proton flux ratio as function of total ion energy from 0.5 to 5 MeV/ion at $L = 2.25$ to $L = 5$. The data are derived from Explover 45 equatorial observations during the first half of June 1972. The shaded areas depict the uncertainties in the derived ratios. Notice that the ratios exceed unity at higher L shells and higher energies.

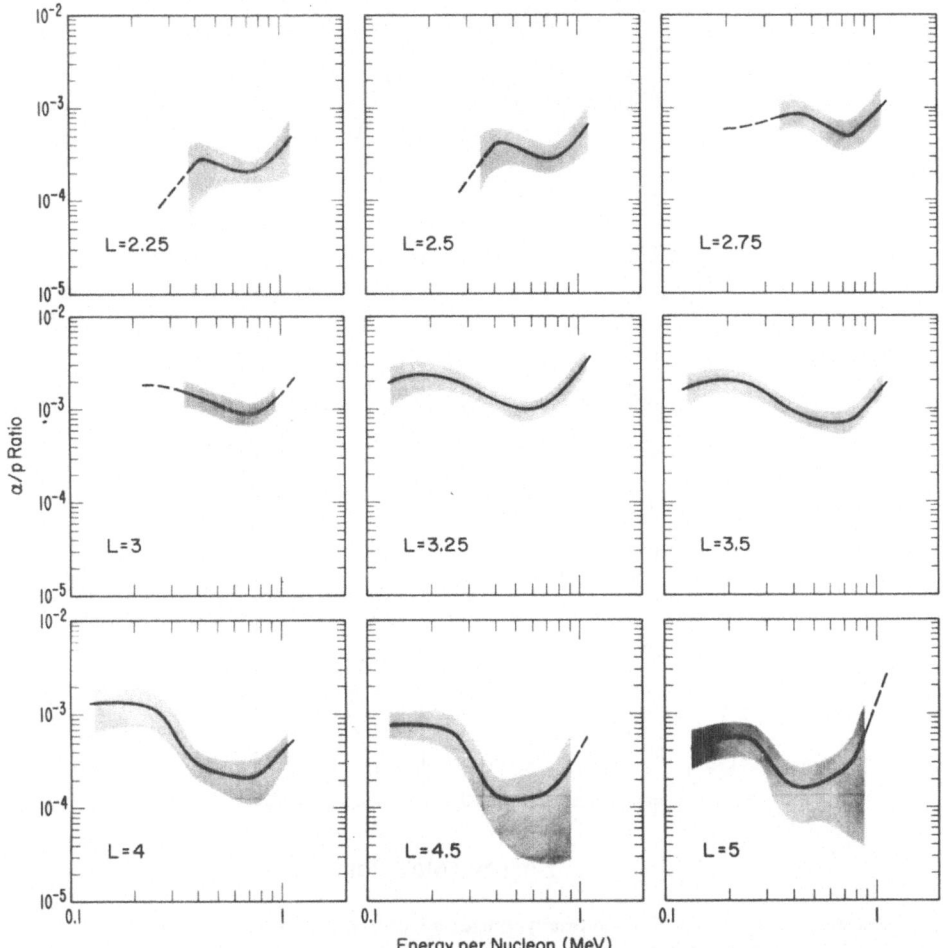

Fig. 23. Experimentally determined quiet time helium ion to proton flux ratio as function of ion energy per nucleon in the range 0.1 to 1 MeV/nucleon. The data were obtained with Explorer 45 and represent the ratios of equatorially mirroring ion fluxes. The shaded areas depict the uncertainties in the derived ratios.

are taken from Fig. 11. These comparisons were made at equal total ion energy, and the CNO ions are assumed to be at least 50% oxygen ions. From this figure it can be seen that the relative fluxes of CNO ions and protons change dramatically from an apparent dominance of the CNO ions (at $E = 1.82–4.8$ MeV/ion) at $L \gtrsim 4$ to a vanishably small ratio ($\lesssim 10^{-4}$) below $L \sim 2.5$. At ~ 3 MeV/ion we deduce 2×10^{-4} at $L = 2.5$, 4×10^{-3} at $L = 3$, 0.2 at $L = 4$, and 20 at $L = 5$. Thus, at equal total ion energy the CNO ions may be lost at an effective rate faster than that of protons.

Comparison of the CNO ion and proton fluxes at equal energy per nucleon generally gives very small values of the $j(\text{CNO})/j(\text{p})$ ion flux ratio. At ~ 100 keV per nucleon we deduce values of 10^{-6} at $L = 5$, 2×10^{-6} at $L = 4$, 9×10^{-6} at $L = 3$,

W. N. SPJELDVIK and T. A. FRITZ

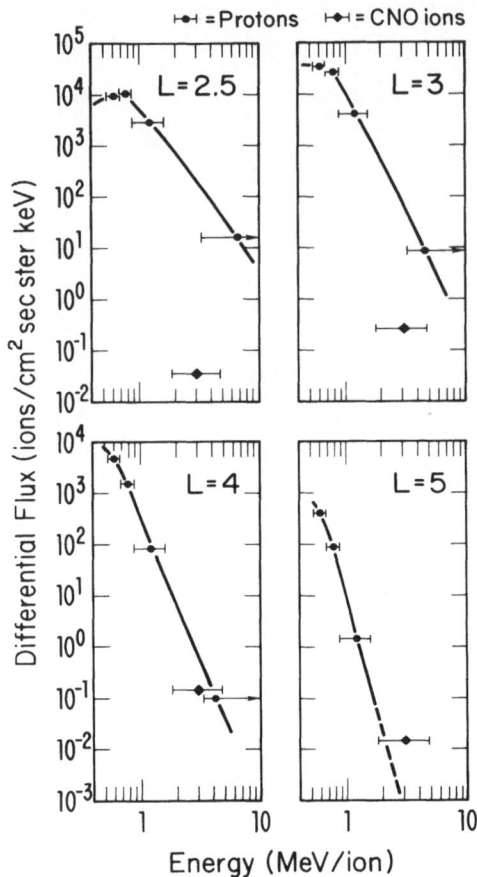

Fig. 24. Spectral comparison of equatorially mirroring proton fluxes (dots) and CNO ion fluxes (diamonds). The horizontal bars depict the energy passbands for the different detectors.

and 4×10^{-6} at $L = 2.5$ under the assumption that oxygen ions make a dominant contribution to the CNO ion flux. The results of HOVESTADT *et al.* (1978) do, however, suggest that at equal energy per nucleon the carbon ion flux may be significantly larger than the oxygen ion flux. Thus if carbon ion fluxes dominate over oxygen ion fluxes, we deduce the ratio $j(CNO)_C/j(p)$ at ~ 276 keV per nucleon to attain the values 2×10^{-6} at $L = 5$, 4×10^{-6} at $L = 4$, 7.7×10^{-6} at $L = 3$, and 3.7×10^{-6} at $L = 2.5$. In either case the $j(CNO)$-values generally fall in the range 10^{-6} to 10^{-5} of the proton flux.

Thus, relative comparison of the different ion species shows that protons are the dominant radiation belt ion species when the comparison is made at equal ion energy per nucleon, and that in such comparisons $j(He)/j(p) \sim 10^{-4}$ to 2×10^{-3} while $j(CNO)/j(p) \sim 10^{-6}$ to 10^{-5}. On the other hand, when the relative comparison is made at equal total ion energy, the $j(He)/j(p)$-values vary strongly from $\lesssim 10^{-4}$ at $E \lesssim 1$ MeV/ion and $L \lesssim 2.5$ to $\gtrsim 10^{+2}$ at $E \gtrsim 3$ MeV/ion and $L \gtrsim 5$; and the

$j(\text{CNO})/j(\text{p})$-values also vary strongly from $\lesssim 2 \times 10^{-4}$ at $E \sim 3$ MeV/ion and $L \lesssim 2.5$ to $\gtrsim 10$ at $E \sim 3$ MeV/ion and $L \gtrsim 5$. Consequently, in these comparisons both helium ions and CNO ions can be the dominant radiation belt ion species at high energies in the outer radiation zone. FRITZ and WILKEN (1976) have also demonstrated that at $L \sim 6.6 \, j(\text{O}) > j(\text{He})$ at multi-MeV energies.

Comparison of quiet time fluxes of the different ion species at $L \gtrsim 1.5$ generally shows increased importance of protons over the heavier ions with lower L-shells. This trend is also generally confirmed by studies of the inner edge of the radiation belt region (e.g., FARLEY and WALT, 1971; HOVESTADT et al., 1972; VALOT, 1972; WHITE, 1973; VALOT and ENGELMANN, 1973; CLAFLIN and WHITE, 1974). In a recent study FISCHER et al. (1977) carried out a comprehensive analysis of data from the DIAL spacecraft at $1.15 \leq L \leq 1.40$. Among the results was the finding that the ion data generally were well ordered as function of the B/B_0-parameter such that the pitch angle anisotropy was found to have a negligible variation with L-shell beyond $L \sim 1.30$. At $L = 1.30$, 1,35 and 1.40 the anisotropy index n was determined to be $n = 12 \pm 2$ for 5.6 to 46 MeV protons. At lower L-shells, the n-index increases strongly. FISCHER et al. (1977) report $n = 16 \pm 2$ at $L = 1.25$, $n = 40 \pm 5$ at $L = 1.20$, and $n = 60 \pm 10$ at $L = 1.15$. From the close correspondence between the theoretically expected proton distribution at these low L-shells (an equilibrium balance between the Cosmic Ray Albedo Neutron Decay (CRAND) source of energetic protons and known losses) it was also concluded that the observed ions at these energies most likely are protons. Further analytic studies have also been carried out in this inner region (e.g., JENTSCH and WIBBERENZ, 1980; JENTSCH, 1981), and WESTPHALEN and SPJELDVIK (1982) have demonstrated the correspondence between the energy-dependence of the radial diffusion coefficient and spectral characteristics of inner zone particles at L-shells beyond the CRAND-source dominated region.

3.   Heavy Ion Observations beyond $L \sim 5$

Much less research effort has been expended in the study of geomagnetically trapped MeV heavy ions beyond L-shells of $L \sim 5$. This segment of the trapping region is not one of characteristic steady-state, but rather one which is significantly influenced by fluctuations in the geomagnetic field and other disturbances such as magnetospheric substorms. Beyond $L \sim 5$ it is not trivial to define "quiet time" conditions, since the absence of major magnetic storms is insufficient to insure time stationary ion fluxes. It is nevertheless customary to distinguish between periods of ion flux enhancements (generally referred to as "injections", although actual transport of ions into this region of space may not necessarily have taken place) and periods of lower flux intensities.

Figure 25 shows the observed flux spectra of helium ions (nuclear charge $Z = 2$), of ions with nuclear charge $Z \geq 3$ and of protons ($Z = 1$) together with a composite panel for spectral comparison of the different ion species. These data were obtained with ion detector telescopes aboard the geostationary spacecraft ATS-6 (at a nominal L-shell of $L = 6.6$) during the magnetospheric substorm that took place at 0–3 UT on June 18, 1974. The intensities of the fluxes at the geostationary altitude were significantly enhanced during this event, and from Fig. 22 it can be seen that helium

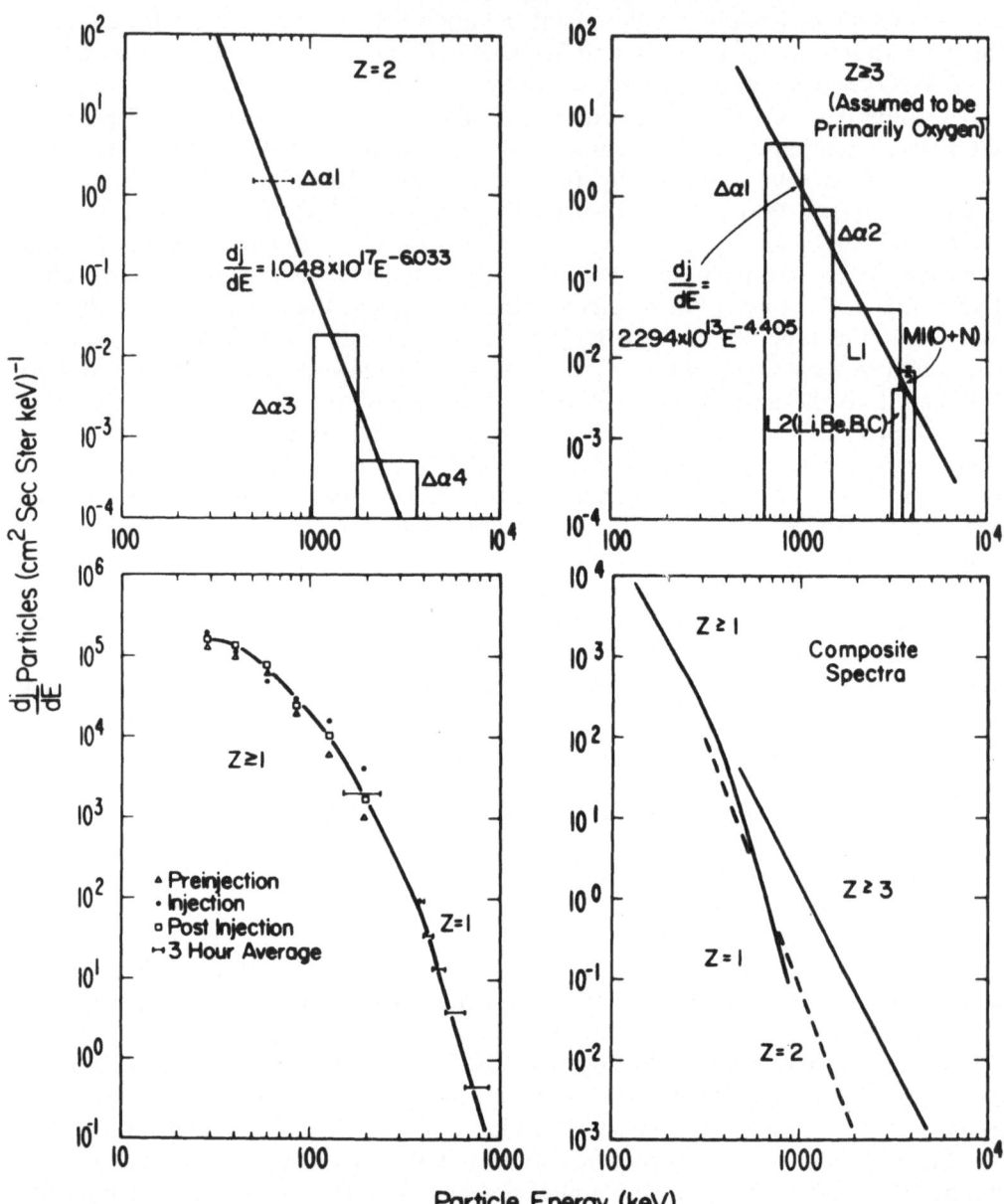

Fig. 25. Equatorial heavy ion spectra monitored with the satellite ATS-6 in the geostationary orbit at $L = 6.6$. The instruments did not permit determination of ion charge states but did discriminate between ion mass. The upper left panel shows heavy ion observations from the $\Delta\alpha 3$ and $\Delta\alpha 4$ detector channels which were sensitive to helium ions only. The upper right panel shows measurements of ions heavier than helium ions (assumed to be primarily atomic oxygen ions) from channels L1, L2, and M1 and measurements of both helium and heavier ions from the channels $\Delta\alpha 1$ and $\Delta\alpha 2$. The lower left panel shows proton observations at $E \gtrsim 100$ keV and total ion measurements at lower energies. The lower right panel summarizes the simultaneous spectral observations (from FRITZ and WILKEN, 1976).

ions are more numerous than protons beyond $E \sim 800$ keV/ion while ions with nuclear charge $Z \geq 3$ (presumably CNO ions) make the predominant contribution at $E \gtrsim 500$ keV/ion (i.e., FRITZ and WILKEN, 1976).

In a later study, SPJELDVIK and FRITZ (1978b) used "quiet time" heavy ion data from ATS–6 to estimate the helium ion and CNO ion (thought of as primarily oxygen ions) spectra at $L = 6.6$. These spectra are depicted in Fig. 26. In this figure extrapolation is made beyond the energy coverage of the ATS–6 data, and in the work of SPJELDVIK and FRITZ (1978b), this was necessitated by the need for outer zone boundary conditions on radiation belt heavy ion model calculations. Apart from the overall lower ion flux level during the "quiet" period of Fig. 26, this figure illustrates that helium ion fluxes can dominate over the CNO ion fluxes at equal total ion energies of $\lesssim 700$ keV/ion, but that even in this non-flux-enhancement situation the CNO ions dominate above ~ 800 keV/ion. Both Fig. 25 and Fig. 26 consistently show the increasing importance of the heavy ions above ~ 1 MeV.

It is of considerable interest to study the outer radiation zone heavy ion pitch angle distribution in great detail. At $L = 3.3 \pm 0.5$ HOVESTADT et al. (1981) have reported that the pitch angle distributions for helium, carbon and oxygen ions with 400–500 keV per nucleon fall off sharply down to an equatorial pitch angle of $\alpha \sim 39°$ ($B/B_0 \sim 2.5$) and that the heavy ion fluxes at $B/B_0 > 2.5$ are subject to large temporal fluctuations (Fig. 16). It is also found that similar conditions exist beyond $L > 5$; the precise details in this outer region are the subject of current research. BLAKE and FENNELL (1981) using data from the SCATHA spacecraft in an outer zone near equatorial orbit (apogee $\sim 8\,R_E$ and perigee $\sim 5\,R_E$) have demonstrated that the helium ion pitch angle distribution at $L = 5.6$ with energies 98 to 240 keV/nucleon (392 to 960 keV/ion) is not fully described by a sinusoidal-type functional relation, but appears to have at least two components. This is exemplified in Fig. 27 which depicts the helium ion pitch angle distribution measured on March 30, 1979 at a local time location of $LT = 15$ hr. This observed pitch angle distribution resembles the typical "bell-shape" which for electrons is known to be caused by pitch angle dependent pitch angle scattering (LYONS et al., 1972). The helium ion data thus suggests that some mode of wave-particle interaction can be effective on the radiation belt heavy ion population at hundreds of keV and perhaps also MeV energies. To further quantify this result, the observed helium ion fluxes are plotted versus the B/B_0-value of observation, and the results are shown in Fig. 28. It can be seen that the B/B_0-dependence is quite strong at $B/B_0 < 2$ (an n-value of $n \sim 8.3$) but that a significant reduction of the anisotropy is found at higher B/B_0-values.

Further work is required to determine storm and substorm effects on the energetic heavy ion fluxes at $L \gtrsim 5$, and to determine the accessibility of interplanetary energetic particles (such as solar cosmic rays and particles from the magnetospheres of other planets) to the outer parts of the radiation belts.

4. Disturbed Time Observations

The effects of major geomagnetic disturbances, such as magnetic storms, on the geomagnetically trapped fluxes of the very energetic (hundreds of keV and MeV) heavy ions have been studied far less than that on electrons and protons. The longevity of

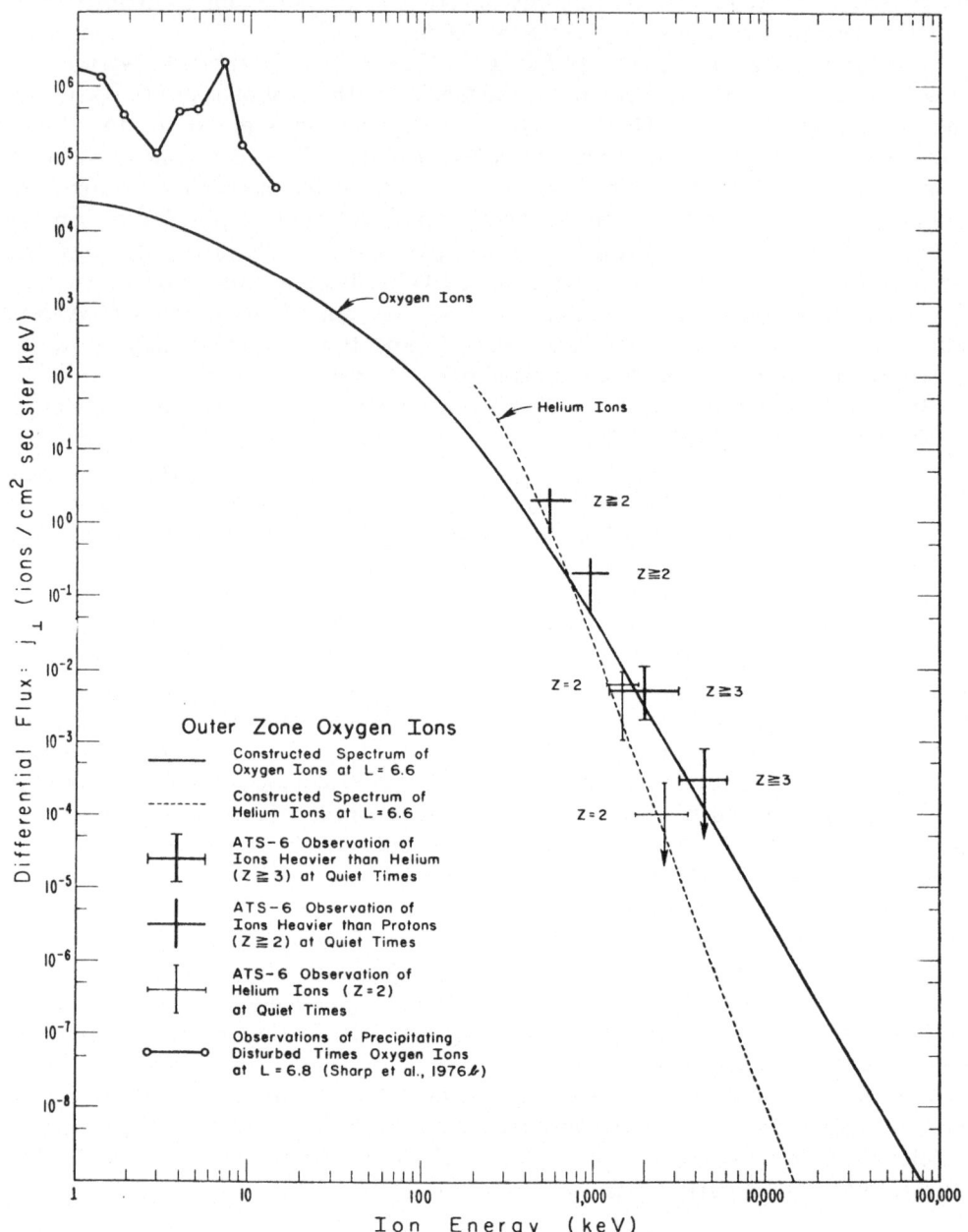

Fig. 26. Estimates of helium ion and oxygen ion spectra at $L = 6.6$ based on ATS-6 observations during a geomagnetically quiet period in 1974 (from Spjeldvik and Fritz, 1978b).

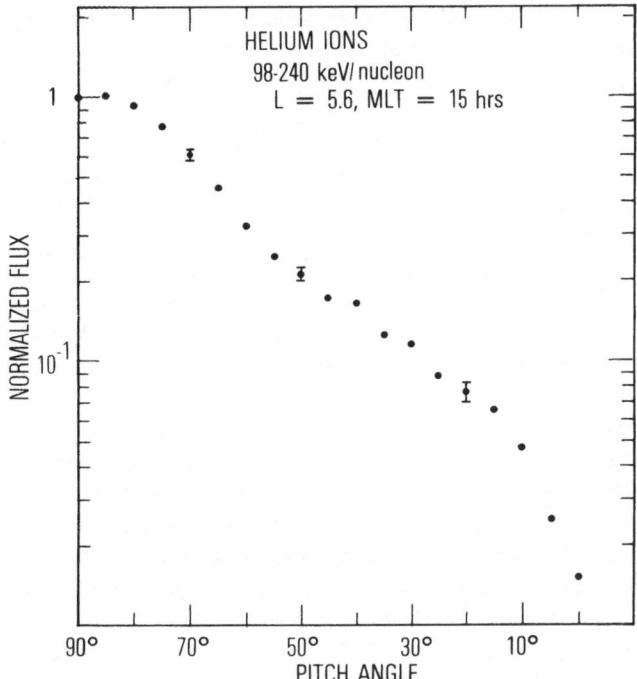

Fig. 27.   Detailed helium ion pitch angle distribution at 98–240 keV/nucleon (392–960 keV/ion) obtained with the SCATHA spacecraft at $L = 5.6$ and MLT = 15 hours (from BLAKE and FENNELL, 1981).

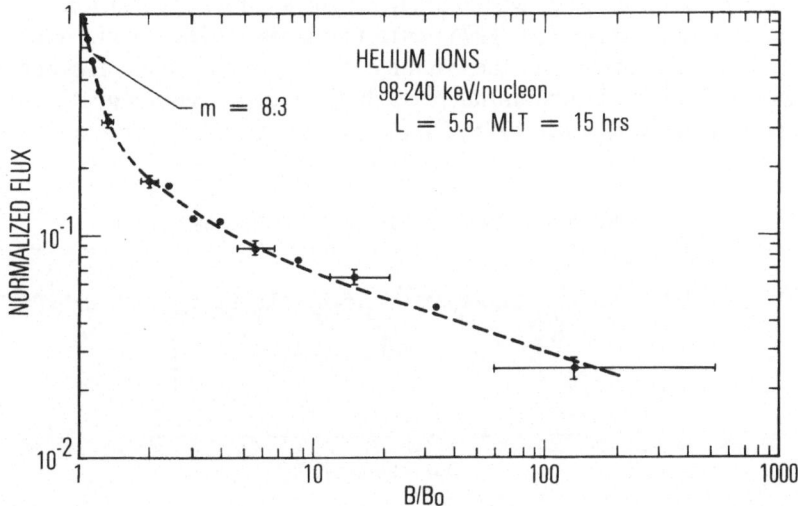

Fig. 28.   Helium ion pitch angle distribution at 98–240 keV/nucleon (392–960 keV/ion) obtained with the SCATHA spacecraft at $L = 5.6$ and MLT = 15 hours (from BLAKE and FENNELL, 1981) plotted versus B/B_0.

these ions, once trapped onto closed magnetic field lines in the inner parts of the magnetosphere, makes them suitable for studies of the long-term behavior of the radiation belts, on time scales of months, much longer than that of magnetospheric substorms ($\lesssim$ hours).

Observations at the geomagnetic equator and at higher latitudes (low altitudes) on the same field line give complementary views of the energetic ion dynamics. While low altitude observations are particularly useful in distinguishing between trapped and precipitated particles, observed variations in the low altitude trapped particle populations $\alpha_{\text{Loss cone}} < \alpha_0 \ll \pi/2$ may result from real overall flux variations (such as an increase also at $\alpha_0 = \pi/2$), or from a particle redistribution process in equatorial pitch angle. Observations made at the geomagnetic equator, on the other hand, permit the entire pitch angle distribution to be studied; and thus this ambiguity may be removed.

In the following we present equatorial heavy ion observations during a sequence of four major magnetic storm periods that ocurred during June-December, 1972. Figure 29 shows the time evolution of the D_{st}-index during this period. The June 17, September 13, and October 31, 1972 magnetic storms were each the classical sudden commencement type of storm with extremum D_{st} reaching $-190, -146,$ and -199 nT respectively. In contrast, the August 4–9, 1972 magnetic storm period was irregular with several separate minima in the D_{st}-values: -188 nT on August 4, -125 nT on August 5, -107 nT on August 6, and -154 nT on August 9, 1972 (SUGIURA and POROS, 1973). This storm period, which may be labeled multiple, overlapping storms, was associated with major solar flare events (e.g., HAKURA, 1976; NAKAGAWA, 1976; BHONSLE et al., 1976; MALITSON et al., 1976). The effects of these flares were observed throughout the solar system: Forbush decreases (e.g., RAO, 1976), solar energetic ion emission (e.g., SIMNETT, 1976), brightening of comet heads (e.g., MILLER, 1976), ionospheric perturbations (e.g., MATSUSHITA, 1976), solar wind variations (e.g., SMITH, 1976; VAISBERG and ZASTENKER, 1976; INTRILLIGATOR, 1976), strong solar energetic particle flux enhancements in the vicinity of the earth (e.g., LANZEROTTI and MACLENNAN, 1974), and substantial perturbations of the geomagnetically trapped radiation (e.g., HOFFMAN et al., 1975; CAHILL, 1976). As will be seen in the following,

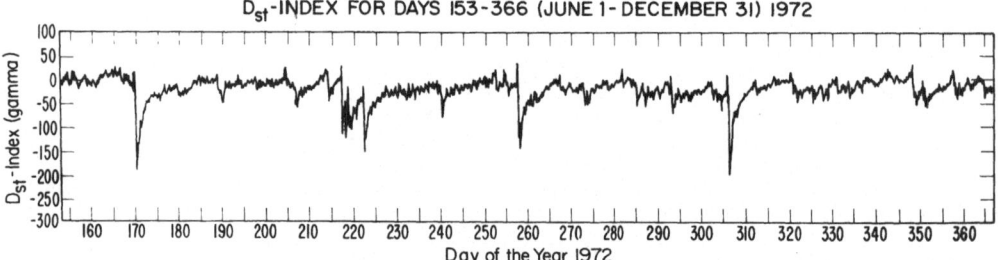

Fig. 29. The D_{st}-index plotted as funtion time for the period June 1–December 31, 1972 (days 153–366). Four magnetic storm periods are indicated with sudden commencements on June 17, August 4, September 13, and October 31, 1972 respectively (from SUGIURA and POROS, 1973).

the August 4–9 magnetic storm period had pronounced effects deep within the trapping region, at least down to $L \sim 2$. The other three magnetic storms had very little effect below $L \sim 3$.

4.1 Helium ions during storms

The equatorial helium ion data from the Explorer 45 $\Delta\alpha3$ channel (1.16–1.74 MeV/ion) is depicted in Fig. 30 as equatorially mirroring ($\alpha_0 = \pi/2$) flux versus time at nominal (undisturbed) L-shells from $L = 2.5$ to $L = 5$ (from SPJELDVIK and FRITZ, 1981a). The time of the sudden commencement for each of the four magnetic storm periods is indicated by the vertical arrows in each panel. This figure illustrates the finding that the effect of the June 17, 1972 magnetic storm was essentially limited to L-shells well beyond $L = 3$. In contrast, the August 4–9 storm period provided an MeV helium ion flux enhancement to L-shells as low as $L \sim 2$; and the enhancement shown at $L = 2.5$ in Fig. 30 is well over an order of magnitude over the pre-storm ion flux intensity. Substantial flux increases are also seen at $L = 3$ and beyond. The third and fourth of these magnetic storms (September 13 and October 31)

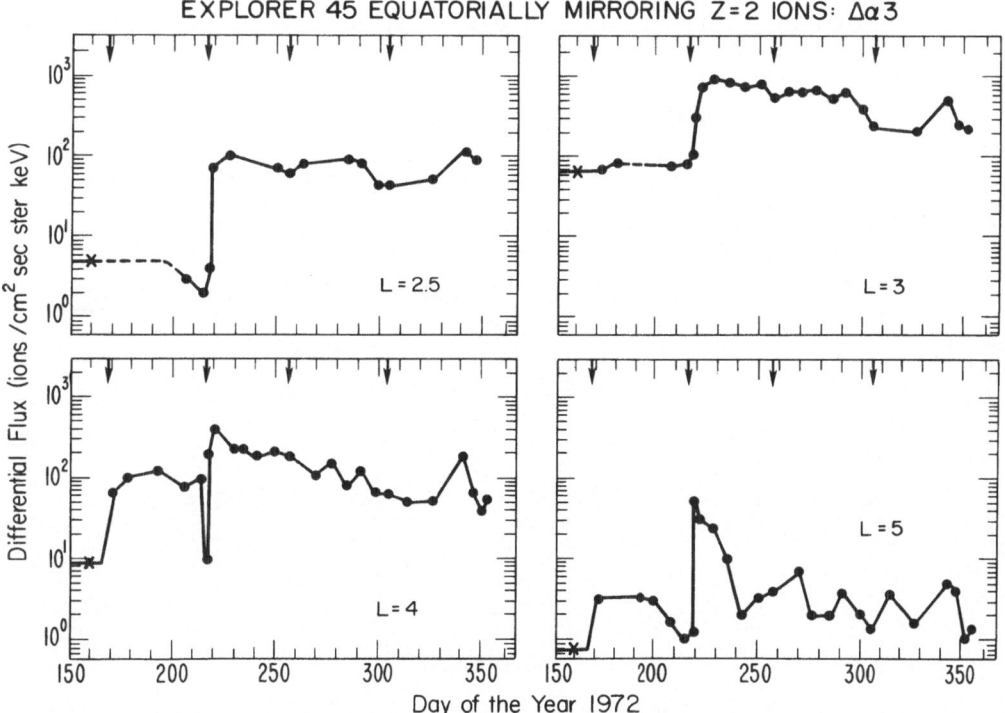

Fig. 30. Time evolution of the Explorer 45 helium ion observations mapped to the geomagnetic equator. The data were obtained with the heavy ion detector telescope channel $\Delta\alpha3$ which covers helium ion energies in the range $E = 1.16$–1.74 MeV/ion (290–435 keV/nucleon) (from SPJELDVIK and FRITZ, 1981a).

provided little discernible helium ion flux enhancements. This provides evidence that different magnetic storms can have significantly differing influences on the trapped population of MeV ions in the radiation belts.

It is also of interest to study spectral changes that can occur during some magnetic storms. During the August 4–9, 1972 magnetic storm period the helium ion spectrum at $L = 2.5$ changed from one characterized by a spectral peak located at $E \sim 1.5$ MeV/ion prior to this storm to a positive spectral slope ($\partial j / \partial E > 0$) with the peak located beyond the Explorer 45 helium ion detector range (i.e., $E > 3.2$ MeV/ion). This result is shown in Fig. 31 which depicts the time evolution of the flux spectrum. A positive spectral slope may lead to an instability involving the growth of plasma waves. From this figure it is seen that the spectral shape eventually (~ 3 months later) is restored to the pre-storm conditions, with the exception that the overall helium ion flux intensity level remained enhanced by more than an order of magnitude even then.

From Fig. 30 we notice that the post-injection decay of the MeV helium ion fluxes depends on the L-shell location. This is explicitly shown in Fig. 32 where the observed decay times for helium ions of energies 0.91–2, 1.16–1.74, and 1.74–3.15 MeV/ion (Explorer 45 detector channels $\Delta\alpha2$, $\Delta\alpha3$, and $\Delta\alpha4$ respectively) are plotted versus L-shell and compared to the characteristic expected time scales for known loss processes (charge exchange loss and Coulomb collision energy degradation). The helium ion

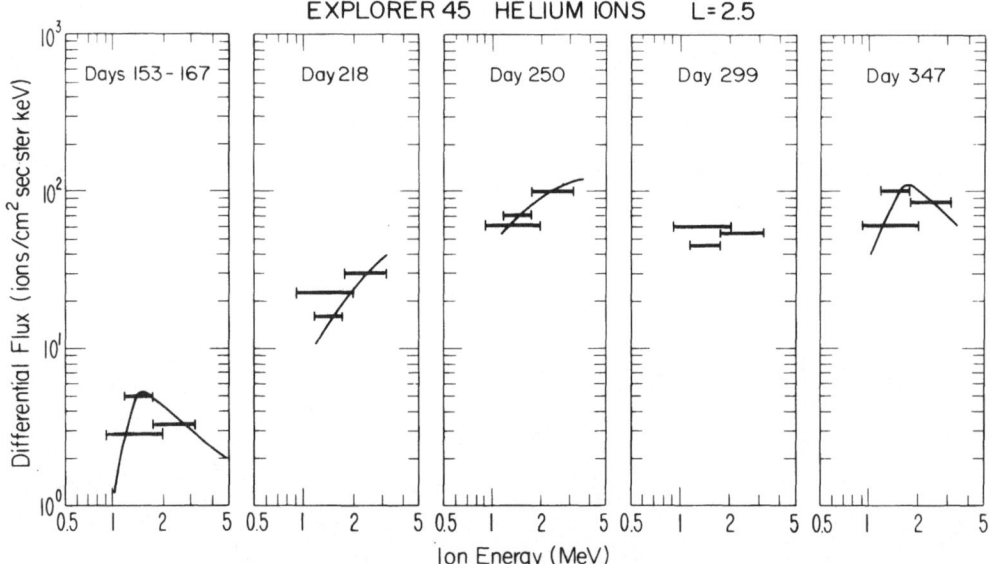

Fig. 31. Equatorially mirroring helium ion spectra measured at $L = 2.5$ using the heavy ion detector telescope on Explorer 45. The data are shown for a geomagnetically quiet period June 1–15 (days 153–167), 1972 (left panel) and for four disturbed and post-disturbance days: the immediate post-injection spectrum on August 5 (day 218) 1972, a late post-storm sample on September 6 (day 250) 1972, the spectrum on October 25 (day 299) 1972, and a sample on December 12 (day 347) 1972.

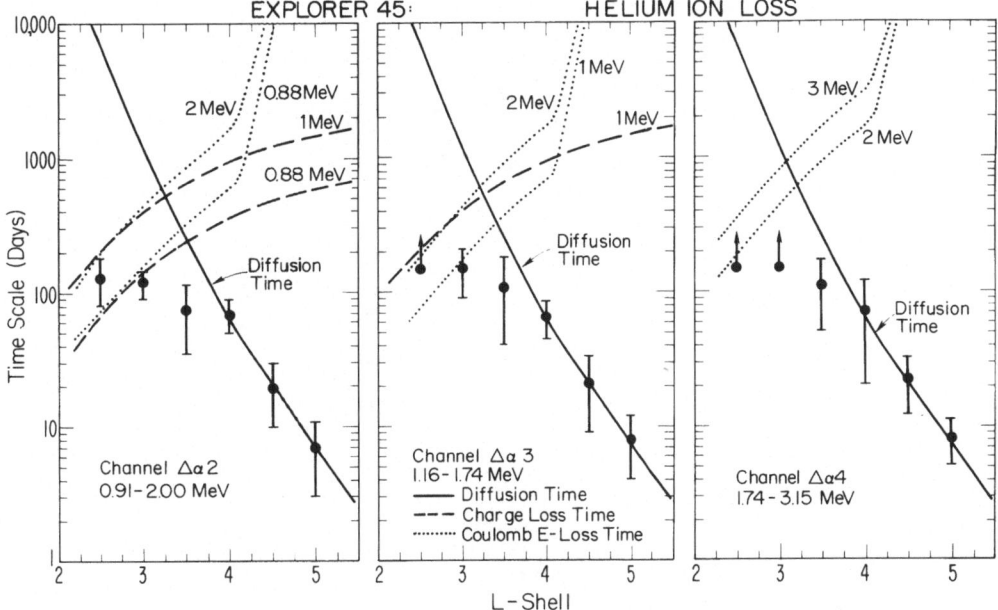

Fig. 32. Time scales of the observed helium ion decay following the August 1972 magnetic storm ion injection. The vertical errorbars show the estimated uncertainty in the time scale determination stemming from statistical variations and the infrequent 'accelerated' mode data sampling. Time scales longer than ~3 months cannot unambiguously be determined because of the statistical spread. Likewise time scales shorter than a few days are difficult to resolve. The solid lines are lines of deduced radial diffusion times following an L^{-10} dependence and normalized to the data; we estimate radial diffusion times of $6.7 \times 10^7 L^{-10}$ days on L shells well beyond $L = 3.5$ with corresponding uncertainty as indicated by the errorbars. Theoretical charge exchange and Coulomb energy degradation time scales are also shown (dashed and dotted lines); for details, see SPJELDVIK and FRITZ (1981a).

charge exchange life times above ~ 1.6 MeV/ion are beyond the range of the graph ($\tau_{\text{CE}} > 10^4$ day). By calculating the helium ion phase space distribution function, f, it was found that $\partial f/\partial L < 0$ on L-shells beyond $L \sim 4$. This is illustrated in Fig. 33 for helium ion magnetic moments ranging from 100 to 700 MeV/Gauss (for further details, see SPJELDVIK and FRITZ, 1981a).

The finding that the post-injection value of $\partial f/\partial L$ can be negative implies that outward radial diffusion following the MeV helium ion injection into the interior of the radiation belts can constitute a major loss mechanism for radiation belt ions. As is evident from Fig. 32, the ion flux decay times decrease essentially monotonically with increasing L-shell. This precludes the possibility that the decay was caused primarily by interactions between the energetic helium ions and the terrestrial exosphere (consisting of neutral thermal hydrogen and plasmaspheric particles), since this mechanism would have caused the longer ion lifetimes to be found at the higher L-shells where the exosphere is less dense. By solving the radial diffusion equation for the diffusion coefficient (when the distribution function and its derivatives are known) SPJELDVIK

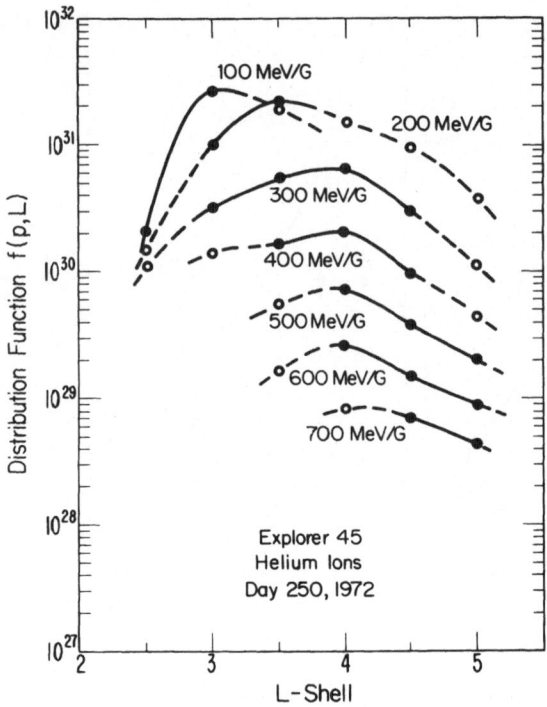

Fig. 33.   Distribution function $f(p, L)$ for day 250 (September 6), 1972, calculated from the MeV helium ion flux observations made by Explorer 45 in the earth's radiation belts. Solid lines and solid circles are from interpolations in the observed data, dashed lines and open circles are from graphic extrapolations in the spectral plots. With the flux $j_\perp(E, L)$ given in ions/cm^{2}s sr keV the distribution function $f(p, L) = p^{-2}j_\perp(E, L)$ is plotted in units of ions s/cm^{4}g^2 sr keV; multiplication with 6.25×10^8 gives cgs units (ions s^3/g^{6}cm^6) (from Spjeldvik and Fritz, 1981a).

and Fritz (1981a) deduced $D_{LL} \sim 1.5 \times 10^{-8}L^{10}$ earth radii^2 per day for this decay period, and the inverse of D_{LL} is also plotted in Fig. 32. The D_{LL} determination was made at $L \sim 4.50 \pm 0.25$. From this result it is reasonable to conclude that the post-injection MeV helium ion flux decay as seen at fixed L-shell is reasonably well understood in terms of known radiation belt physical processes. The initial injection process, however, is *not* understood.

The equatorial radial profiles of energetic helium ion fluxes during this storm period are illustrated in Fig. 34 for five selected spacecraft passes through the radiation belts. The flux profiles for day 217 (August 3) 1972 show typical pre-storm conditions with a narrow L-shell distribution. The data for day 218 (August 4) 1972 demonstrate the effect of this major MeV heavy ion injection event; notice the greatly differing radial structure. Quickly (a few days at most) after the heavy ion injection the radial flux distributions tend to return towards the characteristic pre-storm shapes. This is particularly noticeable on L-shells beyond the quiet time peak flux location (i.e., on $L \gtrsim 3.25$) where the radial diffusion mechanism appears to dominate at these energies.

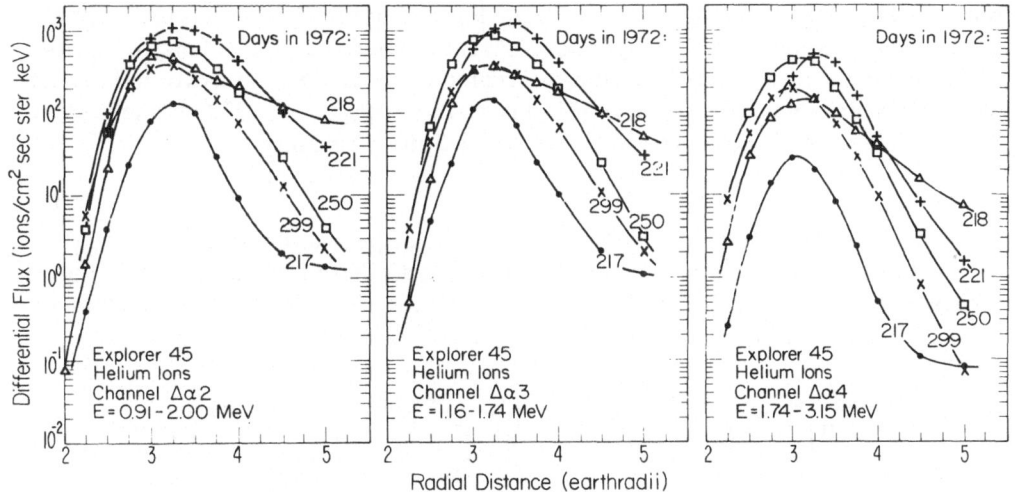

Fig. 34.   Radial profiles of energetic equatorially mirroring helium ions for days 217, 218, 221, 250, and 299, 1972 deduced from Explorer 45 data (from SPJELDVIK and FRITZ, 1981a).

It is remarkable, however, how little the helium ion pitch angle distribution did indeed change from geomagnetically quiet conditions (say, during June 1–15, 1972) prior to the June 17, 1972 magnetic storm to the period following (June 20–30, 1972) this disturbance. The anisotropy index values, n, generally stayed quite similar to the pre-storm values. This is shown in Table 3 (from SPJELDVIK and FRITZ, 1981a) which

Table 3.   Values of the exponent n in the assumed relation $J = J_0 \, (B/B_0)^{-n/2}$ or equivalently $j_\perp = j_{\perp 0} \sin^n \alpha_0$ deduced from Explorer 45 observation during June 20–30, 1972.

Radial distance, R_E	Detector channel	
	$\Delta\alpha 3$	$\Delta\alpha 4$
2.25	9 ± 5	11 ± 6
2.50	8 ± 4	9 ± 4
2.75	8 ± 3	10 ± 4
3.00	8 ± 3	10 ± 3
3.25	8 ± 2	10 ± 2
3.50	9 ± 2	10 ± 2
3.75	8 ± 2	10 ± 3
4.00	8 ± 2	10 ± 3
4.25	8 ± 3	10 ± 4
4.50	8 ± 4	11 ± 5
4.75	7 ± 5	11 ± 5
5.00	9 ± 5	11 ± 6
5.25	10 ± 6	9 ± 7

compares quite well with the quiet time result depicted in Fig. 7. Thus it is reasonable to surmise that the June 17, 1972 magnetic storm cannot have been one of major injection of MeV helium ions into the radiation belt region below $L \sim 5$ from a external particle source. This finding contrasts the conditions for the August 4–9, 1972 magnetic storm period where an external particle source is a definite possibility.

4.2   *CNO ions during storms*

The time evolution of MeV CNO ions during the period of four magnetic storms (June–December, 1972) is even more spectacular than that of the helium ions. Figure 35 depicts the time history of The Explorer 45 equatorial observations from the

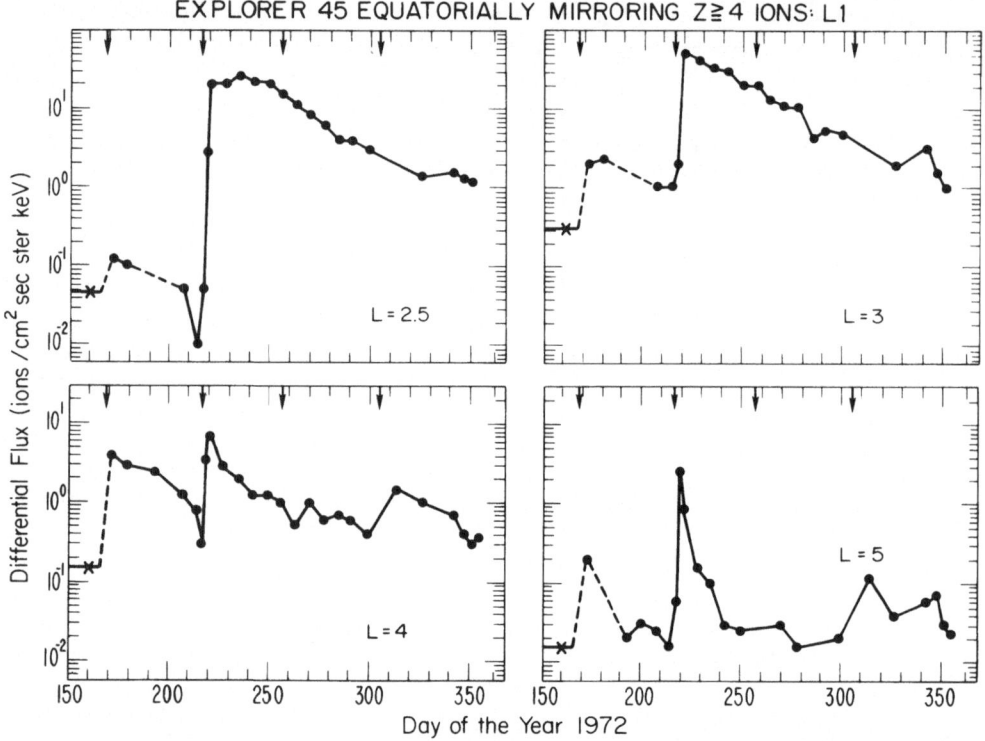

Fig. 35.   Time evolution of the Explorer 45 MeV $Z \geq 4$ ion observations mapped to the geomagnetic equator from the near-equatorial ($1.0 \leq B/B_0 \leq 1.4$) data obtained in the 'accelerated' (2.29-s sampling frequency) data acquisition mode operation of the heavy ion detector. The data were taken from the L1 heavy ion channel, which has the prime sensitivity at 1.82–4.8 MeV/ion for oxygen ions. Use of data from this mode was necessitated by the high counting rates following the August 1972 magnetic storm. The average prestorm flux levels on June 1–15, 1972, are also shown at each of these L shells and are denoted by X. The onsets of the four storm periods are denoted by the vertical arrows in each panel. The figure shows all available accelerated mode data samples which nominally were acquired once a week; more frequent than usual sampling during August 4, 5, and 8 made a crude data resolution possible for the ion injection associated with this storm (from Spjeldvik and Fritz, 1981b).

L1 channel which was sensitive to carbon, nitrogen, oxygen and heavier ions at energies $E = 1.82$–4.8 MeV/ion (114–300 keV/nucleon) for oxygen ions in the principal passband and $E = 40$–160 MeV/ion (2.5–10 MeV/nucleon) in the secondary passband (which presumably does not contribute significantly), and at energies $E = 1.78$–58 MeV/ion (148–$4{,}833$ keV/nucleon) for carbon ions. At the lower L-shells, Fig. 35 shows that there is a small increase in the CNO ion fluxes even during the June 17, 1972 magnetic storm; this amounts to factors of ~ 2 at $L = 2.5$ and ~ 5 at $L = 3$. In contrast to this, the August 4–9, 1972 magnetic storm period provided a spectacular CNO ion flux enhancement at these L-shells; at $L = 2.5$ the observed increase is well over two orders of magnitude, and at $L = 3$ the increase is at least fifty fold. At the higher L-shells, Figure 35 shows that the difference between the two magnetic storms is not particularly large or significant!

The time scales of decay of the newly injected MeV CNO ion fluxes were in all cases as short as—or shorter than—the decay times for the MeV helium ions. For the CNO ions, typical decay times range from a peak of ~ 40 days at $L = 3.5$ to about 6.5 days at $L = 5$. In comparing with the helium ion results, the CNO ions have a positive decay time slope, $\partial \tau_d / \partial L > 0$, below $L \sim 3.5$, which is not found for the helium ions. The CNO ion decay times are depicted in Fig. 36 and compared to the expected time scales of oxygen ion charge exchange decay and the Coulomb collision energy degradation times (from Spjeldvik and Fritz, 1981b). The charge exchange decay times are calculated from

$$\tau_{CE} = \left\{ <\sigma_{10}^0 [\mathrm{H}] V_0> j(0^+) \middle/ \sum_{i=1}^{8} j(0^{i+}) \right\}^{-1}$$

using theoretical values for the normalized oxygen ion charge state distribution (i.e., Spjeldvik and Fritz, 1978), and the Coulomb collision time scales are likewise estimated from a weighted average over the theoretical normalized charge state distribution:

$$\tau_{COUL} = \sum_{i=1}^{8} \tau_{COUL}(0^{i+}) j(0^{i+}) \middle/ \sum_{i=1}^{8} j(0^{i+});$$

σ_{10}^0 is the charge exchange cross section for the $\underline{O}^+ + \mathrm{H} \rightarrow \underline{O} + \mathrm{H}^+$ process, [H] is the number density of exospheric hydrogen, V_0 is the oxygen ion velocity and τ_{COUL} is the Coulomb collision time scale; for details, see Spjeldvik and Fritz (1981b). From Fig. 36 it can be seen that the charge exchange loss time scales at $E \lesssim 2$ MeV are shorter than the observed CNO ion flux decay times on L-shells below $L \sim 4$. From theoretical considerations it follows that the MeV oxygen ion spectra have maximum flux at the lower energies (1–2 MeV/ion) on the higher L-shells ($L \sim 4$–5), but that the theoretical spectra turn over to exhibit peak fluxes at $E \lesssim 4$ MeV/ion at $L \lesssim 3$. Applied to the observed CNO ion flux data, this shift in spectral shape with lower L-shells presumably makes the count contributions to the Explorer 45 L1 data channel come from the higher energies (within the primary passband: 1.82–4.8 MeV/ion) with lower L-shells, and this would account for the observed time scale features in Fig. 36 (Spjeldvik and Fritz, 1981b).

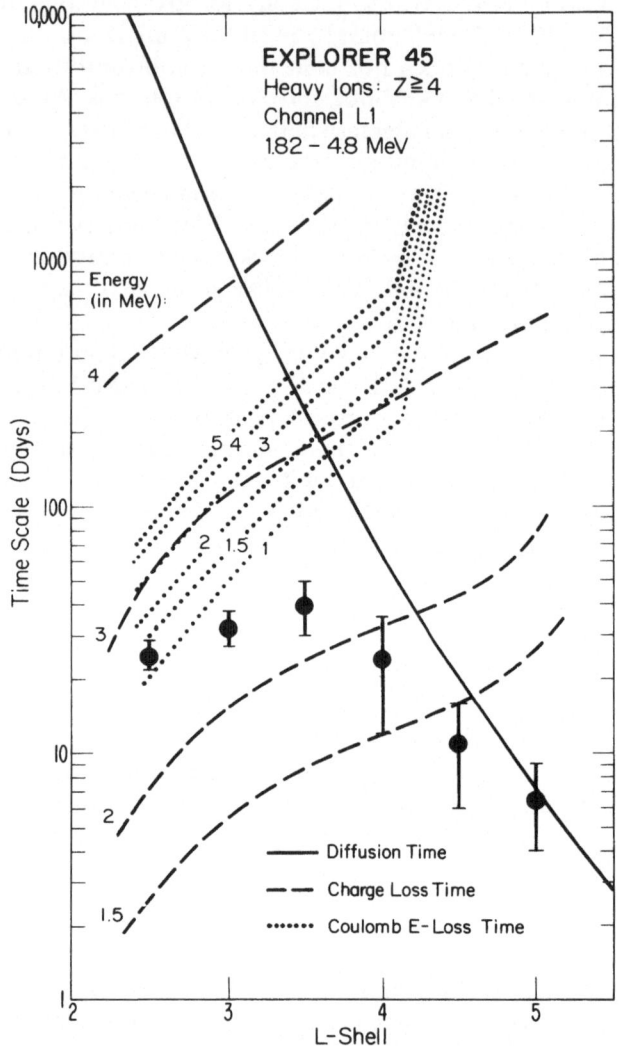

Fig. 36.   Time scales of the observed MeV $Z \geq 4$ ion decay following the August 1972 magnetic storm ion injection. The data are taken from the L1 channel, which has the prime sensitivity at 1.82–4.8 MeV/ion for oxygen ions. The vertical errorbars show the estimated uncertainty in the time scale determination stemming from statistical variations and the infrequent accelerated mode data sampling. Comparison is made with the characteristic time scales of transport and collisional processes. The solid lines depict radial diffusion times deduced from helium ion observations (SPJELDVIK and FRITZ, 1981a) assuming an L^{-10} dependence: $\tau_{\text{transport}} = 6.7 \times 10^7 L^{-10}$ days. Theoretical charge exchange loss times and Coulomb collision energy degradation times are also shown; for details, see SPJELDVIK and FRITZ (1981b).

The MeV CNO ion flux pitch angle distributions remain steep with an n-index value of typically $n \sim 10$ after the June 17, 1972 magnetic storm. This is only slightly different from the pre-storm period where values around $n = 12$ were observed (Fig. 13), and there is a substantial overlap of the errorbars for the two data sets (Table 2 of SPJELDVIK and FRITZ (1981b) and Fig. 13). This lends further support to the idea of little heavy ion injection during the June 19, 1972 magnetic storm.

The evolution of the radial profiles of CNO ions is given in Fig. 37 which demonstrates a qualitatively similar evolution of the MeV CNO ion flux L-shell distribution to that of the MeV helium ions. This result suggests that the radial trapped ion flux redistribution mechanism found to operate on the helium ions also is effective on the CNO ions at these energies following the particle injection during the August 4–9, 1972 magnetic storm period. At the lower L-shells ($L \lesssim 3.5$) interactions between the CNO ions and the earth's exosphere becomes increasingly important, and because of the larger interaction cross sections for MeV oxygen ions than that of helium ions at the same total ion energy, the MeV CNO ion lifetimes become significantly shorter than the collisional lifetimes of the MeV helium ions. This is the likely cause of the decrease in CNO ion lifetimes with decreasing L-shell below $L \sim 3.5$.

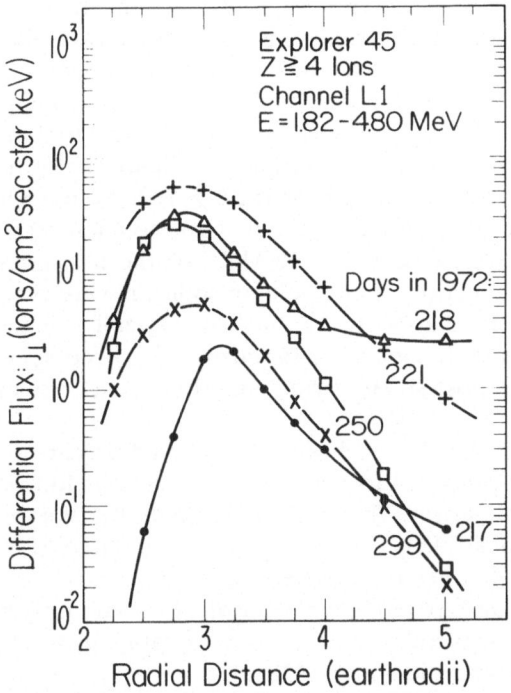

Fig. 37. Radial profiles of energetic equatorially mirroring MeV $Z \geq 4$ ions obtained with Explorer 45 for days 217, 218, 221, 250, and 299 of 1972. The data are from the L1 channel, which has the prime sensitivity at 1.82–4.8 MeV/ion for oxygen ions. Notice the large increase at low L, the occurrence of the peak on day 221 (August 8) 1972, and the rapid restoration of the shape of the radial profile at $L \gtrsim 4$ following the storm (from SPJELDVIK and FRITZ, 1981b).

4.3   Ions heavier than fluorine during storms

Equatorial measurements made with the heavy ion detector telescope aboard Explorer 45 have shown that ions with nuclear charge $Z \gtrsim 9$ generally are not present in large quantities during quiet conditions. During November 1971–July 1972 an upper limit of 0.1 ions/cm^{2}sec ster at energies beyond $E \sim 10$ Mev/ion in the interior of the radiation belts ($L \lesssim 5.2$) was established (Spjeldvik and Fritz, 1981c). A more sensitive ion detector on ISEE-1 gave more detailed results on these rare ions, as outlined above. The August 4–9, 1972 magnetic storm period, which resulted in a large MeV helium ion and CNO ion flux injection to low L-shells, injected also significant, measurable amounts of $Z \gtrsim 9$ ions into the L-shell region, $L \sim 2$–4. The observed time evolution of the equatorial MeV $Z \gtrsim 9$ ion fluxes is depicted in Fig. 38 covering $L = 2.5$, $L = 2.75$, $L = 3$, and $L = 3.25$ where the count rates were sufficiently high to permit meaningful count statistics. Notice the sudden appearance and fairly systematic ion flux decay during the post-injection period. The effects of the September 13 and October 31, 1972 magnetic storms are rather minor compared to the substantial injection during the August 4–9, 1972 event period. Also, the preceding magnetic storm, on June 17, 1972, had no discernible effect on the MeV $Z \gtrsim 9$ ion flux observations (which then were virtually zero).

The post-injection decay of the MeV $Z \gtrsim 9$ ion fluxes was found to vary systematically with L-shell in this range, such that the shorter lifetimes are seen at the higher L-shells. This result is depicted in Fig. 39. Unfortunately, not much theoretical work has been done for radiation belt ions heavier than oxygen ions. The passband sensitivity for the Explorer 45 detector channel A5 sensitive to the $Z \gtrsim 9$ ions is given in Table 4. Notice that ions of magnesium, silicon, sulphur, chlorine, potassium, argon, iron, and heavier ions would all be recorded at $E \gtrsim 13$ MeV/ion while sodium and neon ions could be detected only above 17.2 and ~ 20 MeV/ion respectively.

For radiation belt $Z \geq 9$ ions at $E \gtrsim 10$ MeV/ion the collisional lifetimes (resulting from charge exchange reactions and Coulomb collisions) are very long, and probably well exceeding the observed decay times which are typically a few tens of days. Also, at $L \sim 3$ the time scale characteristic of the radial diffusion process ($\sim D_{LL}^{-1}$) is of the order of ~ 3 years during this post-storm period, and that is also very much longer than the oserved decay times.

The radial profiles of the observed MeV $Z \gtrsim 9$ ion fluxes are depicted in Fig. 40 for nine selected spacecraft passes through the radiation belts following the onset of the August 4, 1972 magnetic storm. The irregularities in the fluxes shown in this figure (and those in Fig. 38) are most likely due to the low count statistics of these ions where in general only a few counts per minute were observed.

The MeV $Z \gtrsim 9$ ion flux pitch angle anisotropy could only be crudely estimated from these data, and it was found that the n-index values generally exceeded $n \sim 10$ at $L = 2.7$ to 3.3 (Spjeldvik and Fritz, 1981c).

The question of the proper identity of the MeV $Z \gtrsim 9$ ions observed following the August 4–9, 1972 heavy ion injection event period cannot be answered with certainty. There are at least two major possibilities to consider: (1) Molecular ions such as NO^+, O_2^+ etc., from the earth's lower ionosphere ($h \lesssim 100$ km) and (2) heavy (atomic) ions such as Neon, Sodium, Magnesium, Aluminum, Silicon, or Iron in various charge

Explorer 45 Observations of Z ≧ 9 Ions

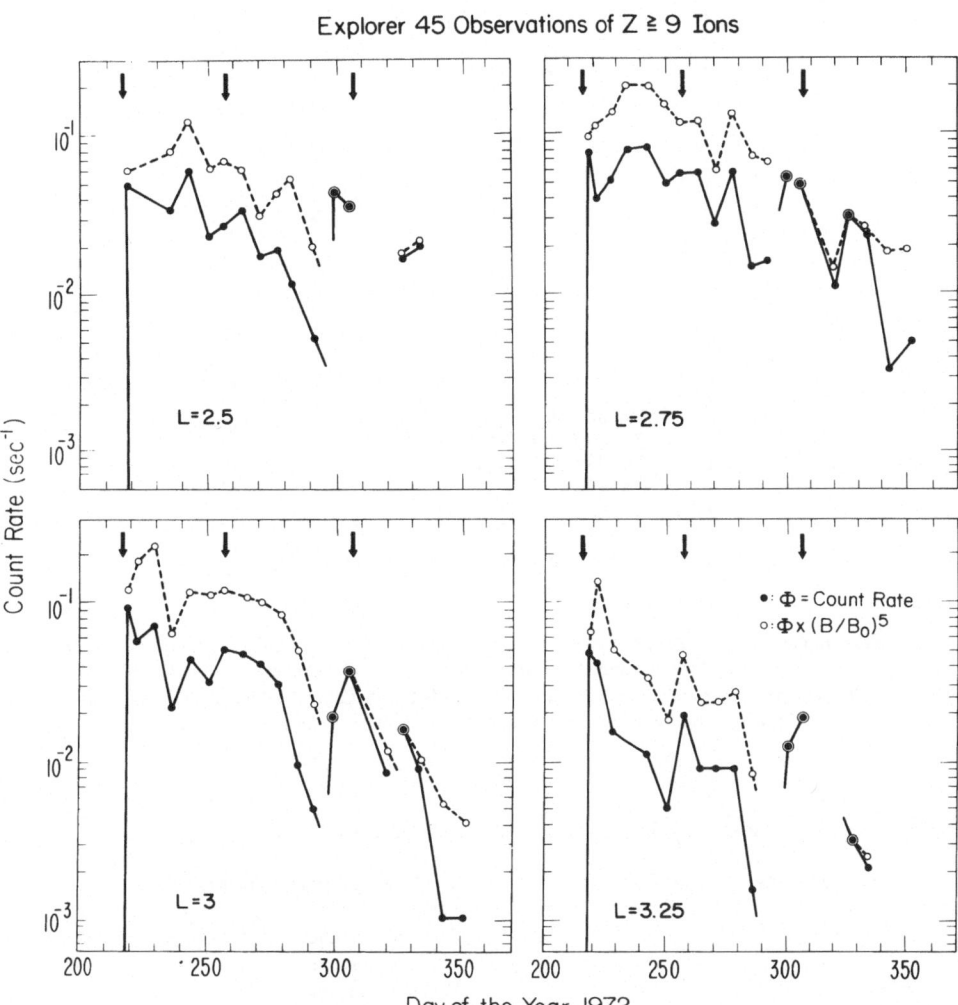

Fig. 38.   Explorer 45 observations of MeV ions with nuclear charge $Z \gtrsim 9$ (i.e., ions heavier than fluorine). The energy range sensitivities for the various heavy ions are given in Table 4. The data are given as detector count rate versus time where the solid circles indicate the actual observed rates (at the spacecraft location which varied in the range $1.0 \leqq B/B_0 \lesssim 1.4$) and the open circles represent data mapped to the geomagnetic equator using pitch angle distribution information (see the text). The onsets of the August, September, and October/November magnetic storms are indicated by the vertical arrows (from SPJELDVIK and FRITZ, 1981c).

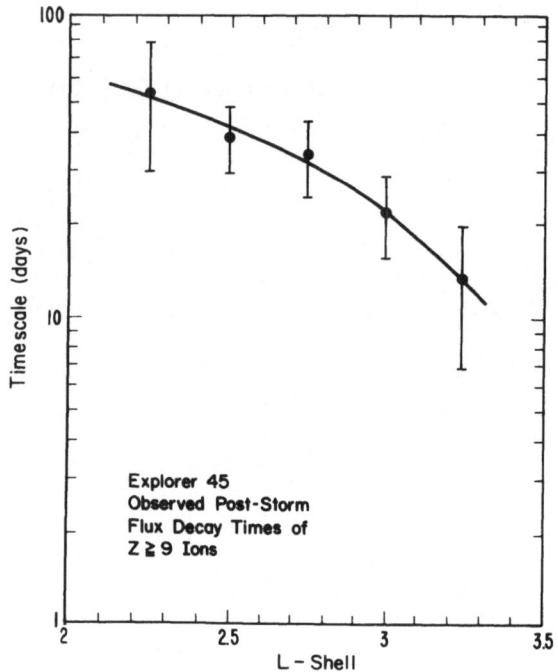

Fig. 39.   Ion flux decay times for the post-storm period following the August 1972 magnetic storm observed with the Explorer 45 heavy ion data channel A5 that responded to ions of nuclear charge $Z \gtrsim 9$ and energies typically $E \gtrsim 10$ MeV per ion (passbands of specific ions are listed in Table 4). The vertical errorbars depict the estimated uncertainties in the deduced time scales (from SPJELDVIK and FRITZ, 1981c).

Table 4.   Explorer 45 detector channel 45 sensitivity to ions.

Ion (in any charge state)	Energy passband, MeV*
[1]Proton	insensitive
[4]Helium	insensitive
[12]Carbon	insensitive
[14]Nitrogen	insensitive
[16]Oxygen	insensitive
[19]Fluorine	marginally insensitive
[20]Neon	~20–22
[23]Natrium (sodium)	17.2–28.4
[24]Magnesium	12.8–51.2
[27]Aluminum	11.9–80.0
[28]Silicon	11.0–108
[32]Sulphur	10.5–186
[35]Chlorine	10.4–240
[39]Kalium (potassium)	10.1–355
[40]Argon	10.5–320
[56]Ferrum (iron)	9.4–> 700

*Theoretically calculated.

Explorer 45 Z ≳ 9 Ion Observations in 1972

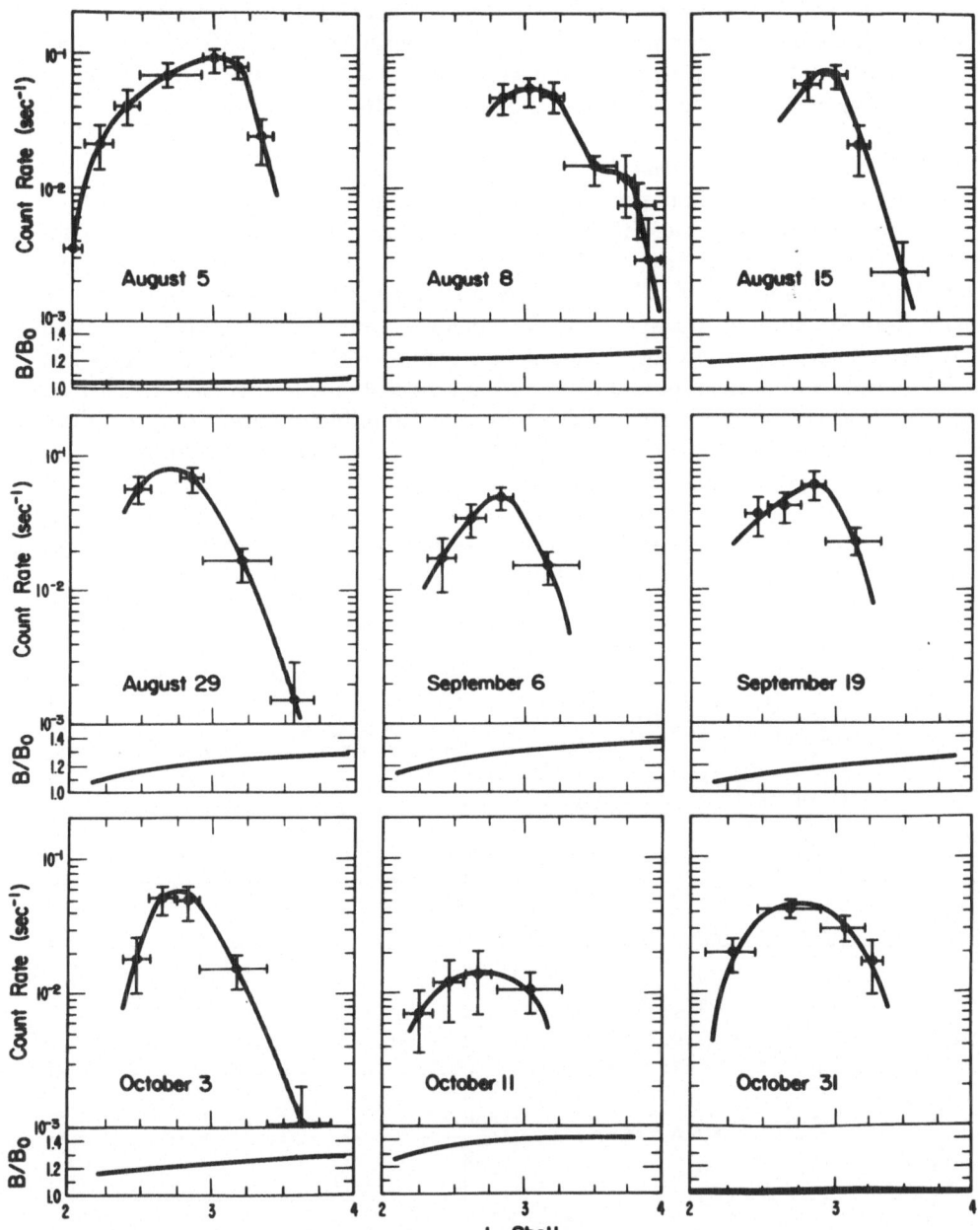

Fig. 40. Observed radial profiles of the Explorer 45 heavy ion data channel A5 count rates (which are proportional to the ion flux) versus L shell as determined from a quiet time mean geomagnetic field model. The data are shown for nine orbital passes through the radiation belts during (August 5) and following (August 8–October 31) the August 1972 magnetic storm period. Horizontal errorbars indicate L shell range of the data sampling (because of the satellite motion), and the vertical errorbars indicate probable count rate uncertainty. The orbital variation of the B/B_0 values is also given in each panel; there is a tendency of observing higher count rates when B/B_0 is small (from SPJELDVIK and FRITZ, 1981c).

states, from an extraterrestrial source.

The data set alone does not permit a distinction between these possible candidates, but from the elemental abundance of multi-MeV ions in the August 1972 solar cosmic ray emission events at energies above 7.5 MeV/nucleon ($\gtrsim$ 210 MeV/ion for Silicon) all the heavy ions noted above may have viable contributions (e.g., Webber et al., 1975).

4.4 Low altitude observations during storms

Observations carried out on L-shells in the radiation belts at altitudes near the top of the atmosphere are valuable because the atmospheric loss cone then spans a significant section of pitch angle space. Randall (1973) has reported ion observations from the Injun-5 spacecraft at low altitudes during a geomagnetic storm period when the ratios of the different ion species varied strongly; the j (He)/j (p) ratio was found to increase significantly following the magnetic storm period. More recently, Scholer et al. (1979) report that in a segment of the radiation belt L-shell range, $L \sim 2.6$ to 4.7, the low altitude ion observations from the S3-2 spacecraft indicate significant precipitation of protons ($E > 0.4$ MeV to $E > 2$ MeV in four integral channels), helium ions ($E = 0.95$ to 2.89 MeV/ion in three differential pass-bands), and CNO ions ($E > 0.25$ MeV/nucleon). This is illustrated in Fig. 41 which depicts the count rates in these channels versus time, B-field and L-shell of observation during March 26, 1976. Notice that both upward and downward loss cones (seen because of the satellite spin

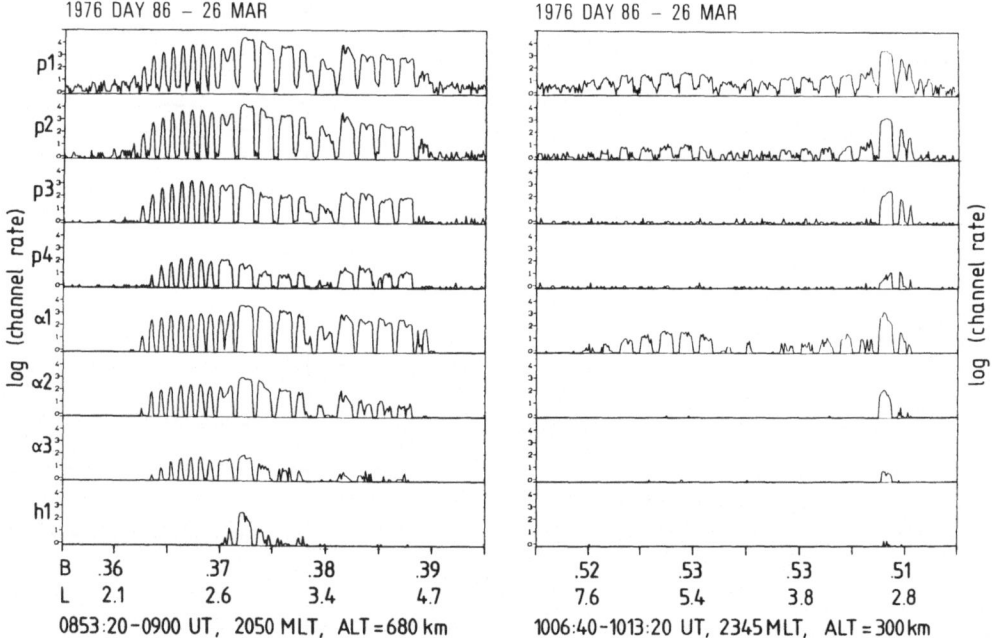

Fig. 41.   Proton, helium ion and heavier ion count rates versus time for two nightside passes of the S3-2 spacecraft through the radiation belts on March 26, 1976. The energy coverage is in the lower MeV range; for details, see Scholer et al. (1979).

modulation) are essentially empty at $L \lesssim 2.6$ while at the higher L-shells only the upward loss cone is empty. This might imply that there exists a band of plasma waves which can interact and scatter ions efficiently in different mass ranges at energies ranging from a fraction of an MeV to several MeV total ion energy. The data in Fig. 41 were obtained during the late main phase of a geomagnetic storm with extremum $D_{st} = -229$ nT.

5. Summary and Conclusions

In the foregoing sections we have presented selected data on radiation belt heavy ions during geomagnetically quiet periods, and during and following magnetic storms. We are able to positively identify ions of helium, carbon, oxygen, and even heavier ions. The charge states of these ions (with a few exceptions) have not been experimentally established, but the theory predicts that at energies exceeding ~ 1 MeV the higher charge states available to a given ionic species will dominate the charge state distribution. Comparison between theory and measurement generally give an overall agreement within the energy range where the experimental data are available. There is, however, a significant gap in our present ability to detect and positively identify the radiation belt ion species. This gap extends from tens of keV total ion energies, below which ion mass spectrometers are effective, to hundreds of keV or even MeV energies, depending on ion species. This observational mass-gap encompasses the energetically dominant part of the extraterrestrial ring current and the dynamically very interesting energy range where the heavy ions are predicted to have sharp spectral gradients (e.g., SPJELDVIK, 1979).

There are a few areas where the agreement between theory and data is less than perfect. It is recognized (e.g., Scholer, private communication, 1981) that the helium ion distribution function, $f(\mu, L, Q)$ (Q denoting the ion charge state), can exhibit a small negative spectral gradient, $\partial f / \partial L < 0$, around $L \sim 3$ at a few MeV total ion energies even during geomagnetically undisturbed conditions. A similar feature is also deducible from the Explorer 45 helium ion data (unpublished). Since conditions of pure radial diffusion under steady state conditions require that $\phi = -D_{LL} L^{-2} (\partial f / \partial L)$ is constant with L-shell, the fact that $f > 0$ at $L > 1$ and $f = 0$ at $L = 1$ implies that the theoretical value of $\partial f / \partial L > 0$ at all L-shells in this case. Thus, there must either be a source of MeV helium ions at $L \sim 3$ to 4, a significant rate of energy degradation from higher to lower energies at these L-shells, or the ideal situation of perfect time independence of the MeV helium ion fluxes is not satisfied.

A similar testing of the CNO and heavier ions has not yet been carried out. Comparison with the CNO ions at $E = 1.82$–4.8 MeV/ion at the equator has, however, shown that the heavy ion theory is capable of reproducing the overall flux intensities observed at the different L-shells in one data channel. Future work using better spectral, angular and spatial resolution will be undertaken with data from several experiments on the ISEE-1 spacecraft.

This paper is the written form of an invited review given at the International Association of Geomagnetism and Aeronomy Symposium on Ion Composition in Edinburgh, Scotland, August 3–14, 1981. One of us (W.N.S.) was supported by a NASA grant S-50897G.

418   W. N. Spjeldvik and T. A. Fritz

REFERENCES

Bhonsle, R. V., S. S. Degaonkar, and S. K. Alurkar, Ground-based solar radio observations of the August 1972 events, *Space Sci. Rev.*, **19**, 475, 1976.

Blake, J. B., Experimental test to determine the origin of geomagnetically trapped radiation, *J. Geophys, Res.*, **78**, 5822, 1973.

Blake, J. B. and G. A. Paulikas, Measurements of trapped α-particles: $2 \leq L \leq 4.5$, in *Particles and Fields in the Magnetosphere*, edited by B. M. McCormac, D. Reidel, Dordrecht-Holland, 1970.

Blake, J. B. and G. A. Paulikas, Geomagnetically trapped alpha particles, 1. Off-equator particles in the outer zone, *J. Geophys. Res.*, **77**, 3431, 1972.

Blake, J. B. and J. F. Fennell, Heavy ion measurements in the synchronous altitude region, *Planet. Space Sci*, **29**, 1205, 1981.

Blake, J. B., J. F. Fennell, and D. Hovestadt, Measurements of heavy ions in the low-altitude regions of the outer zone, *J. Geophys. Res.*, **85**, 5992, 1980.

Blake, J. B., J. F. Fennell, M. Schulz, and G. A. Paulikas, Geomagnetically trapped alpha particles: 2, The inner zone, *J. Geophys. Res.*, **78**, 5498, 1973.

Cahill, L. J., Jr., Response of the magnetosphere during early August 1972, *Space Sci. Res.*, **19**, 703, 1976.

Claflin, E. S. and R. S. White, A study of equatorial inner belt protons from 2 to 200 MeV, *J. Geophys. Res.*, **79**, 959, 1974.

Cornwall, J. M., Radial diffusion of ionized helium and protons: A probe for magnetospheric dynamics, *J. Geophys. Res.*, **77**, 1756, 1972.

Farley, T. A. and M. Walt, Source and loss processes of protons of the inner radiation belt, *J. Geophys. Res.*, **76**, 8223, 1971.

Fennell, J. F. and J. B. Blake, Geomagnetically trapped alpha particles, in *Magnetospheric Particles and Fields*, edited by B. M. McCormac, pp. 149–156, D. Reidel, Dordrecht-Holland, 1976.

Fennell, J. F., J. B. Blake, and G. A. Paulikas, Geomagnetically trapped alpha particles: 3. Low-altitude alpha-proton comparisons, *J. Geophys. Res.*, **79**, 521, 1974.

Fischer, H. M., V. W. Auschrat, and G. Wibberenz, Augular distribution and energy spectra of protons of energy $5 \leq E \leq 50$ MeV at the lower edge of the radiation belts in equatorial latitudes, *J. Geophys. Res.*, **82**, 537, 1977.

Fritz, T. A., Ion composition, in *Physics of the Solar Planetary Environments*, edited by D. J. Williams, p. 716, American Geophysical Union, Washington, D.C., 1976.

Fritz, T. A. and S. M. Krimigis, Initial observations of geomagnetically trapped protons and alpha particles with OGO-4, *J. Geophys. Res.*, **74**, 5132, 1969.

Fritz, T. A. and D. J. Williams, Initial observations of geomagnetically trapped alpha particles at the equator, *J. Geophys, Res.*, **78**, 4719, 1973.

Fritz, T. A. and B. Wilken, Substorm generated fluxes of heavy ions at the geostationary orbit, in *Magnetospheric Particles and Fields*, edited by B. M. McCormac, p. 171, D. Reidel Reidel Publishing, Dordrecht-Holland, 1976.

Fritz, T. A. and W. N. Spjeldvik, Observations of energetic radiation belt helium ions at the geomagnetic equator during quiet conditions, *J. Geophys. Res.*, **83**, 2579, 1978.

Fritz, T. A. and W. N. Spjeldvik, Simultaneous quiet time observations of energetic radiation belt protons and helium ions: The equatorial α/p-ratio near 1 MeV, *J. Geophys, Res.* **84**, 2608, 1979.

Fritz, T. A. and W. N. Spjeldvik, Pitch angle distributions of geomagnetically trapped helium ions during quiet times, *J. Geophys. Res.*, **87**, 5095, 1982.

Fritz, T. A., C. W. Arthur, J. B. Blake, P. J. Coleman, Jr., J. P. Corrigan, W. D. Cummings, S. E. DeForest, K. N. Erickson, A. Konradi, W. Lennartsson, A. J. Masley, B. H. Mauk, C. E. McIlwain, R. L. McPherron, G. A. Paulikas, K. A. Pfitzer, D. L. Reasoner, P. R. Satterblom, S. Y. Su, R. J. Walker, E. C. Whipple, Jr., B. Wilken, and J. R. Winckler, Significant initial results from ATS-6, NASA Technical Paper 1101, Goddard Space Flight Center, Maryland, 1977.

Hakura, Y. Interdisciplinary summary of solar/interplanetary events during August 1972, *Space Sci. Rev.*, **19**, 411, 1976.

Hauge, O. and O. Engvold, The chemical composition of the solar atmosphere, Institute for Theoretical Astrophysics, Reprint 31, p. 18, Oslo-Blindern, Norway, 1970.

HIRSHBERG, J., Composition of the solar wind: Present and past, *Rev. Geophys. Space Phys.*, **13**, 1059, 1975.

HOFFMAN, R. A., L. J. CAHILL, Jr., R. R. ANDERSON, N. C. MAYNARD, P. H. SMITH, T. A. FRITZ, D. J. WILLIAMS, A. KONRADI, and D. A. GURNETT, Explorer 45 (S^3–A) observations of the magnetosphere and magnetopause during the August 4–6, 1972 magnetic storm period, *J. Geophys. Res.*, **80**, 4287, 1975.

HOVESTADT, D., E. ACHTERMANN, B. EBEL, B. HÄUSLER, and G. PASCHMANN, New observations of the proton population of the radiation belt between 1.5 and 104 MeV, in *Earth's Magnetospheric Processes*, edited by B. M. McCormac, p. 115, D. Reidel, Dordrecht, Holland, 1972.

HOVESTADT, D., B. KLECKER, E. MITCHELL, J. F. FENNELL, G. GLOECKLER, and C. Y. FAN, Spatial distribution of $Z \geq 2$ ions in the outer radiation belt during quiet conditions, *Adv. Space Res.*, **1**, 305, 1981.

HOVESTADT, D., G. GLOECKLER, C. Y. FAN, L. A. FISK, F. M. IPAVICH, B. KLECKER, J. J. O'GALLAGHER, and M. SCHOLER, Evidence for solar wind origin of energetic heavy ions in the earth's radiation belts, *Geophys. Res. Lett.*, **5**, 1055, 1978.

INTRILIGATOR, D. S., The August 1972 solar-terrestrial events: Solar wind plasma observations, *Space Sci. Rev.*, **19**, 629, 1976.

JENTSCH, V., On the role of external and internal source in generating energy and pitch angle distributions of inner zone protons, *J. Geophys. Res.*, **86**, 701, 1981.

JENTSCH, V. and G. WIBBERENZ, An analytic study of the energy and pitch angle distribution of inner zone protons, *J. Geophys. Res.*, **85**, 1, 1980.

JOHNSON, R. G., R. D. SHARP, and E. G. SHELLEY, The discovery of energetic He$^+$ ions in the magnetosphere, *J. Geophys. Res.*, **79**, 3135, 1974.

JOHNSON, R. G., R. D. SHARP, and E. G. SHELLEY, Composition of the hot plasmas in the magnetosphere, in *Physics of the Hot Plasma in the Magnetosphere*, edited by B. Hultqvist and L. Stenflo, p. 45, Plenum, New York, 1975.

JOHNSON, R. G., R. D. SHARP, and E. G. SHELLEY, Observations of the ring current composition during the 29 July 1977 magnetic storm, *EOS, Trans. Am. Geophys. Union*, **58**, 1217, 1977.

JOHNSON, R. G., R. D. SHARP, and E. G. SHELLEY, Observations of ions of ionospheric origin in the storm-time ring current, *Geophys. Res. Lett.*, **5**, 59, 1978.

KRIMIGIS, S. M., Alpha particles trapped in the earth's magnetic field, in *Particles and Fields in the Magnetosphere*, edited by B. M. McCormac, pp. 364–379, D. Reidel Pub., Dordrecht-Holland, 1970.

KRIMIGIS, S. M. and J. A. VAN ALLEN, Geomagnetically trapped alpha particles, *J. Geophys. Res.*, **72**, 5779, 1967.

KRIMIGIS, S. M. and P. VERZARIU, Measurements of geomagnetically trapped alpha particles, 1968–1970: 1, Quiet-time distributions, *J. Geophys. Res.*, **78**, 7275, 1973.

KRIMIGIS, S. M., P. VERZARIU, J. A. VAN ALLEN, T. P. ARMSTRONG, T. A. FRITZ, and B. A. RANDALL, Trapped energetic nucleii $Z \geq 3$ in the earth's outer radiation zone, *J. Geophys. Res.*, **75**, 4210, 1970.

LANZEROTTI, L. J. and C. G. MACLENNAN, Solar particle observations during the August 1972 event, in *Correlated Interplanetary and Magnetospheric Observations*, edited by D. E. Page, p. 587, D. Reidel, Dordrecht-Holland, 1974.

LUNDIN, R., L. R. LYONS, and N. PISSARENKO, Observations of the ring current composition at L-values less than 4, *Geophys. Res. Lett.*, **7**, 425, 1980

LYONS, L. R., R. M. THORNE, and C. F. KENNEL, Pitch angle diffusion of radiation belt electrons within the plasmasphere, *J. Geophys. Res.*, **77**, 3455, 972.

MALITSON, H. H., J. FAINBERG, and R. G. STONE, Hectometric and kilometric solar radio emission observed from satellites in August 1972, *Space Sci. Rev.*, **19**, 511, 1976.

MASUNG, L. S., G. GLOECKLER, C. Y. FAN, and D. HOVESTADT, Observations of the mean ionization states of energetic particles in the vicinity of the earth's magnetosphere, *J. Geophys. Res.*, **85**, 5983, 1980.

MATSUSHITA, S., Ionospheric and thermospheric responses during August 1972 storm—A review, *Space Sci. Rev.*, **19**, 713, 1976.

MILLER, F. D., Solar-Cometary relations and the events of June–August 1972, *Space Sci. Rev.*, **19**, 739, 1976.

MOGRO-CAMPERO, A., Geomagnetically trapped carbon, nitrogen and oxygen nucleii, *J. Geophys. Res.*, **77**, 2799, 1972.

NAKAGAWA, Y., Flares of August 1972: Analysis of dynamics, *Space Sci. Rev.*, **19**, 459, 1976.

PANASYUK, M. I., Charge states of energetic ions, *Cosmic Res.*, **18**, 64, 1980.

PANASYUK, M. I. and N. A. VLASOVA, Anisotropy of fluxes of protons and alpha particles with energies greater than 4 MeV in the radiation belts, *Cosmic Res.*, **19**, 52, 1981.

PANASYUK, M. I., S. Ya. REIZMAN, E. N. SOSUOVETS, and V. N. FILATOV, Experimental results of proton and alpha-particle measurements at energies more than 1 MeV/nucleon in the radiation belts, *Geomagn. Aeron.*, **15**, 887, 1977.

RANDALL, B. A., Time variations of magnetospheric intensities of outer zone protons, alpha particles and ions ($Z > 2$), University of Iowa, Report 73–3, Iowa City, 1973.

RAO, U. R., High energy cosmic ray observations during August 1972, *Space Sci. Rev.*, **19**, 533, 1976.

ROEDERER, J. G., *Dynamics of Geomagnetically Trapped Radiation*, Springer Verlag, New York, 1970.

RUBIN, A. G., R. C. FILZ, P. L. ROTHWELL, and B. SELLERS, Geomagnetically trapped alpha particles from 18 to 70 MeV, *J. Geophys. Res.*, **82**, 1938, 1977.

SCHOLER, M., D. HOVESTADT, G. HARTMANN, J. B. BLAKE, J. F. FENNELL, and G. GLOECKLER, Low-altitude measurements of precipitating protons, alpha particles, and heavy ions during the geomagnetic storm on March 26–27, 1976, *J. Geophys. Res.*, **84**, 79, 1979.

SCHULZ, M., Geomagnetically trapped radiation, *Space Sci. Rev.*, **17**, 481, 1975.

SCHULZ, M. and L. J. LANZEROTTI, *Particle Diffusion in the Radiation Belts*, Springer Verlag, New York, 1974.

SHARP, R. D., R. G. JOHNSON, E. G. SHELLEY, and K. K. HARRIS, Energetic O^+ ions in the magnetosphere, *J. Geophys. Res.*, **79**, 1844, 1974a.

SHARP, R. D., R. G. JOHNSON, and E. G. SHELLEY, Satellite measurements of auroral alpha particles, *J. Geophys. Res.*, **79**, 5167, 1974b.

SHARP, R. D., R. G. JOHNSON, and E. G. SHELLEY, The morphology of energetic O^+ ions during two magnetic storms: Temporal variations, *J. Geophys. Res.*, **81**, 3283, 1976a.

SHARP, R. D., R. G. JOHNSON, and E. G. SHELLEY, The morphology of energetic O^+ ions during two magnetic storms: Latitudinal variations, *J. Geophys. Res.*, **81**, 3292, 1976b.

SHARP, R. D., E. G. SHELLEY, and R. G. JOHNSON, A search for helium ions in the recovery phase of a magnetic storm, *J. Geophys. Res.*, **82**, 2361, 1977a.

SHARP, R. D., R. G. JOHNSON, and E. G. SHELLEY, Observation of an ionospheric acceleration mechanism producing energetic (keV) ions primarily normal to the geomagnetic field direction, *J. Geophys. Res.*, **82**, 3324, 1977b.

SHELLEY, E. G., R. G. JOHNSON, and R. D. SHARP, Satellite observations of energetic heavy ions during a geomagnetic storm, *J. Geophys. Res.*, **77**, 6104, 1972.

SHELLEY, E. G., R. G. JOHNSON, and R. D. SHARP, Morphology of energetic O^+ in the magnetosphere, in *Magnetospheric Physics*, edited by B. M. McCormac, pp. 135–139, D. Reidel, Dordrecht-Holland, 1974.

SHELLEY, E. G., R. D. SHARP, and R. G. JOHNSON, Satellite observations of an ionospheric acceleration mechanism, *Geophys. Res. Lett.* **3**, 654, 1976a.

SHELLEY, E. G., R. D. SHARP, and R. G. JOHNSON, He^{++} and H^+ flux measurements in the dayside cusp: Estimates of convection electric field, *J. Geophys. Res.*, **81**, 2363, 1976b.

SHELLEY, E. G., R. D. SHARP, and R. G. JOHNSON, Ion composition in the quiet time magnetosphere, *EOS, Trans. Am. Geophys. Union*, **58**, 1217, 1977.

SIMNETT, G. M. Solar cosmic radiation during August 1972, *Space Sci. Rev.*, **19**, 579, 1976.

SMITH, E. J., The August 1972 solar-terrestrial events: Interplanetary magnetic field observations, *Space Sci. Rev.*, **19**, 661, 1976.

SPJELDVIK, W. N., Equilibrium structure of equatorially mirroring radiation belt protons, *J. Geophys. Res.*, **82**, 2801, 1977.

SPJELDVIK, W. N., Expected charge states of energetic ions in the magnetosphere, *Space Sci. Rev.*, **23**, 499, 1979.

SPJELDVIK, W. N. and T. A. FRITZ, Energetic ionized helium in the quiet time radiation belts: Theory and comparison with observation, *J. Geophys. Res.*, **83**, 654, 1978a.

SPJELDVIK, W. N. and T. A. FRITZ, Theory for charge states of energetic oxygen ions in the earth's radiation belts, *J. Geophys. Res.*, **83**, 1583, 1978b.

SPJELDVIK, W. N. and T. A. FRITZ, Quiet time observations of equatorially trapped megaelectronvolt radiation belt ions with nuclear charge $Z \geq 4$, *J. Geophys. Res.*, **83**, 4401, 1978c.

SPJELDVIK, W. N. and T. A. FRITZ, Observations of energetic helium ions in the earth's radiation belts during a sequence of geomagnetic storms, *J. Geophys. Res.*, **86**, 2317, 1981a.

SPJELDVIK, W. N. and T. A. FRITZ, Energetic heavy ions with nuclear charge $Z \geq 4$ in the equatorial radiation belts of the earth: Magnetic storms, *J. Geophys, Res.*, **86**, 2349, 1981b.

SPJELDVIK, W. N. and T. A. FRITZ, Observations of ions with nuclear charge $Z \geq 9$ in the inner magnetosphere, *J. Geophys. Res.*, **86**, 7749, 1981c.

SUGIURA, M. and D. J. POROS, Hourly values of equatorial D_{st} for 1971 and 1972, World Data Center A for Rockets and Satellites/National Space Science Data Center, NASA/GSFC, Greenbelt, Maryland, 1973.

TEEGARDEN, B. J., T. T. VON ROSENWINGE, and F. B. MACDONALD, Satellite measurements of the charge compositon of solar cosmic rays in the $6 \leq Z \leq 26$ interval, *Astrophys. J.*, **180**, 571, 1973.

VAISBERG, O. L. and G. N. ZASTENKER, Solar wind and magnetosheath observations at earth during August 1972, *Space Sci. Rev.*, **19**, 687, 1976.

VALOT, P., Differential energy spectrum of geomagnetically trapped protons with the ESRO–2 satellite, *J. Geophys. Res.*, **77**, 2309, 1972.

VALOT, P. and J. ENGELMANN, Pitch angle distribution of geomagnetically trapped protons for $1.2 \leq L \leq 2.1$, *Space Res.*, XIII, 675, 1973.

VAN ALLEN, J. A., B. A. RANDALL, and S. M. KRIMIGIS, Energetic C, N, O nuclei in the earth's outer radiation zone, *J. Geophys. Res.*, **75**, 6085, 1970.

WEBBER, W. R., E. C. ROELOF, F. B. MCDONALD, B. J. TEEGARDEN, and J. TRAINOR, Pioneer 10 measurements of the charge and energy spectrum of solar cosmic rays during 1972 August, *Astrophys. J.*, **199**, 482, 1975.

WESTPHALEN, H. and W. N. SPJELDVIK, On the energy dependence of the radial diffusion coefficient and spectra of inner radiation belt particles: Analytic solutions and comparison with numerical results, *J. Geophys. Res.*, **87**, 8321, 1982.

WHITE, R. S., The high energy proton radiation belt, *Rev. Geophys. Space Phys.*, **11**, 595, 1973.

WITHBROE, G. L., The chemical composition of the photosphere and the corona, in *The Menzel Symposium on Solar Physics, Atomic Spectra and Gaseous Nebulae*, Spec. Publ. 353, edited by K. B. Gebbie, p. 353, National Bureau of Standards, Washington, D.C., 1971.

Energetic Ion Composition in the Earth's Magnetosphere, edited by R. G. Johnson, 423–438.

Atmospheric and Ionospheric Effects of Precipitated Energetic O^+ Ions

Marsha R. TORR

Center for Atmospheric and Space Sciences, Department of Physics,
Utah State University, Logan, Utah 84322, U.S.A.

(Received February 26, 1982)

The impact on the atmosphere of heavy ion precipitation has been slowly recognized over the past few years. At present the quantification of this impact is very dependent on assumptions made concerning the relevant collision cross-sections and morphological details of the precipitation events. Nonetheless, in view of the considerable magnitude that measured O^+ fluxes reach on occasion (0.4 ergs $cm^{-2} sec^{-1} sr^{-1}$), we can draw the following conclusions. A major effect is the transfer of the bulk of the incoming energy flux to the neutral atmosphere at F region altitudes in the form of heat. This heating source in turn has a significant effect on global thermospheric winds and temperatures. The nocturnal ionization rate could reach a few tens of ion pairs $cm^{-3} sec^{-1}$ in the region of maximum precipitation, while direct excitation in the same region could be as much as ~ 300 R of oxygen emissions. The precipitation events produce a large backsplash of energetic neutral oxygen atoms. The flux of oxygen capable of escape from the Earth is estimated to be comparable to that for hydrogen. The portion of the splash atoms that could be ionized would be of a magnitude ten times that of the incoming O^+ ions. Given appropriate acceleration mechanisms, the events may be self-sustaining.

1. Introduction

Since the first reports of the precipitation of energetic O^+ fluxes into the ionosphere (SHELLEY *et al.*, 1972), there has been interest in the potential atmospheric effects in view of the magnitude of the measured energy fluxes associated with these events. An initial study (TORR *et al.*, 1974) showed the ionospheric/atmospheric impact indeed do produce some significant effects. Following the initial report of the energetic O^+ fluxes, further morphological details were published (SHELLEY *et al.*, 1974; SHARP *et al.*, 1974, 1976a, b). Subsequent studies (TORR and TORR, 1979; TORR *et al.*, 1982) have improved the early model and have extended the analysis. While such studies of the impact of the energetic O^+ fluxes are hampered by lack of accurate knowledge of the relevant collision cross-sections, the resulting effects are large enough that they remain significant over broad likely ranges for such parameters.

In the sections below, we summarize the main conclusions drawn in the earlier studies, and the reader is referred to the original papers for further details.

A study of the ionospheric /atmospheric impact of the energetic O^+ events starts at 800 km, the altitude at which the observations were made. The measurements were made over the energy range 0.7 to 12 keV and the energetic ions were found to occur in every storm studied over a one-year period, reaching energy fluxes of 0.4 ergs cm^{-2} sec^{-1} sr^{-1} with 0.1 ergs cm^{-2} sec^{-1} sr^{-1} being more typical. Little information is available concerning the pitch angle distribution. The measurements were made at pitch angles near the edge of the loss cone, 55°–58°. There is, however, some indication that the distribution is isotropic, and that is what has been assumed in the calculations discussed in this paper.

At an altitude of 800 km, the primary atmospheric constituent is atomic oxygen, so the studies referenced above have investigated the fate of energetic O^+ ions encountering an atomic oxygen atmosphere. From the cross-section and collision information that is available, the two main energetic ion processes over the energy range of the observations are charge exchange and momentum transfer:

$$O_f^+ + O \rightarrow O_f + O^+ \tag{1}$$

$$O_f^+ + O \rightarrow O_f^+ + O_f \tag{2}$$

where the subscript f denotes a fast particle. Because the cross-sections for these two processes are large ($\sim 10^{15}$ cm^2; Lo and FITE, 1970; SOLOV'EV et al., 1972) we neglect other processes such as ionization and excitation. Via these two processes, the energetic O^+ ions are rapidly converted into energetic O atoms. This process is illustrated in Fig. 1. Figure 1a shows the O^+ flux energy spectra measured by SHELLEY et al., (1972) during the December 16/17, 1971, magnetic storm. Figure 1b shows the production rates of energetic oxygen atoms as a function of altitude that would result from these energetic O^+ spectra at 800 km. In this computation the MSIS model atmosphere appropriate for the likely conditions prevailing at the time of the storm has been used. We have used the model (HEDIN, 1979) for 45°N, solar maximum ($F_{10.7} = 200$) for magnetically disturbed conditions ($Ap = 100$) for December. Because the energetic O^+ events are predominantly a nightside phenomenon, the model is for 0300 local time.

The energetic O atoms formed via these collision processes then undergo further collisions with the ambient atomic oxygen. In each collision, they transfer part of their energy to the ambient atoms, until they thermalize. This process is interesting in that each energetic O^+ ion subsequently results in a large number of thermalizing O atoms. The studies by TORR et al., (1974, 1981) and TORR and TORR (1979) model these energetic O atoms as they scatter in altitude, energy and angle. In the following section the results of this modeling are summarized. Additional aspects are discussed in the final section.

2. Results

2.1 Heating of the thermosphere

One of the most significant effects of the precipitating energetic O^+ fluxes is that approximately 60–70% of the incoming energy is transferred to the neutral atmosphere as heat. Figure 2 shows the heating rate profile that would result from the O^+ energy spectrum measured by SHELLEY *et al.*, (1972) during the December 1971 magnetic storm. The O^+ precipitation was observed to occur throughout the two days of the storm, as is shown in Fig. 3, with the calculations reported here corresponding to the time of peak flux of 0.4 ergs cm^{-2} sec^{-1} sr^{-1}. This energy flux is large and in the initial attempts to include it as a heating source in the NCAR global circulation model

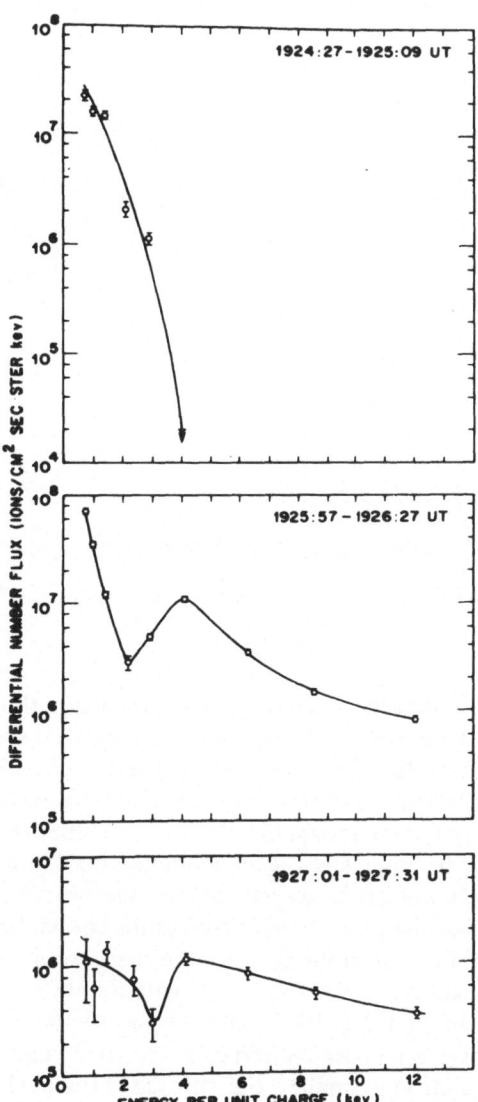

Fig. 1. (a) Flux energy spectra observed during the December 16/17, 1971, magnetic storm. From top to bottom the spectra correspond to L values of ~ 2.7, 3.4, and 4.2 respectively (from SHELLEY *et al.*, 1972).

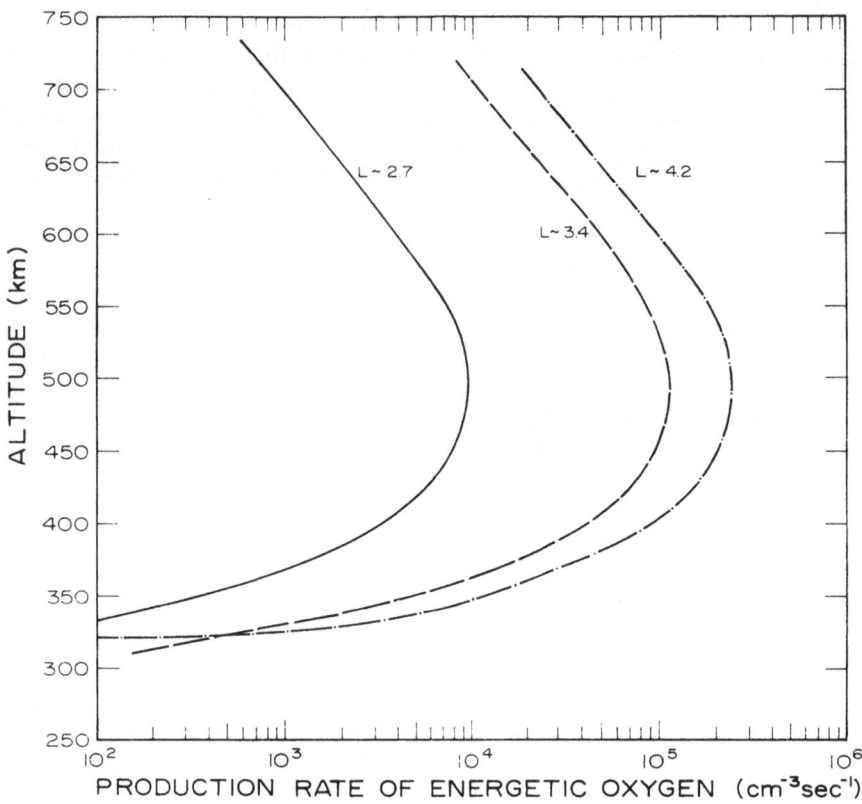

Fig. 1. (b) Production rates of energetic oxygen from the three observed O^+ spectra shown in (a) (from TORR *et al.*, 1981).

described below, some difficulty was experienced in getting the program to accept the large perturbation. The measured energy flux was therefore scaled down by 2π in the heating rate computations described here and by TORR *et al.*(1982). The same was done by KOZYRA *et al.* (1982). As a result, the case modeled here and by TORR *et al.*, (1982) is representative of a fairly typical storm O^+ flux event, rather than the maximum fluxes observed. Also shown in Fig. 2 is the dependence on the altitude of peak energy deposition to both the collision cross-section and the model atmosphere used in the calculation. The model atmosphere used in the results shown below has been discussed in the previous section. The collisions between the energetic and ambient oxygen atoms are assumed to be hard sphere and elastic. A measurement for oxygen on N_2 by AMDUR *et al.* (1957) varied from $\sim 1.2 \times 10^{-15}$ cm^2 at 10 keV to $\sim 2 \times 10^{-15}$ cm^2. In the thermalization calculations a cross-section of 1.7×10^{-15} cm^2 was used, independent of energy. If the cross-section is smaller, the altitude of the peak of the heating rate will be lowered. An additional factor might cause the peak heating rate to be lower in the atmosphere, namely a forward peak to the angular dependence of the O–O collisions. In the calculations shown in Figure 2, hard sphere collisions were assumed, in which

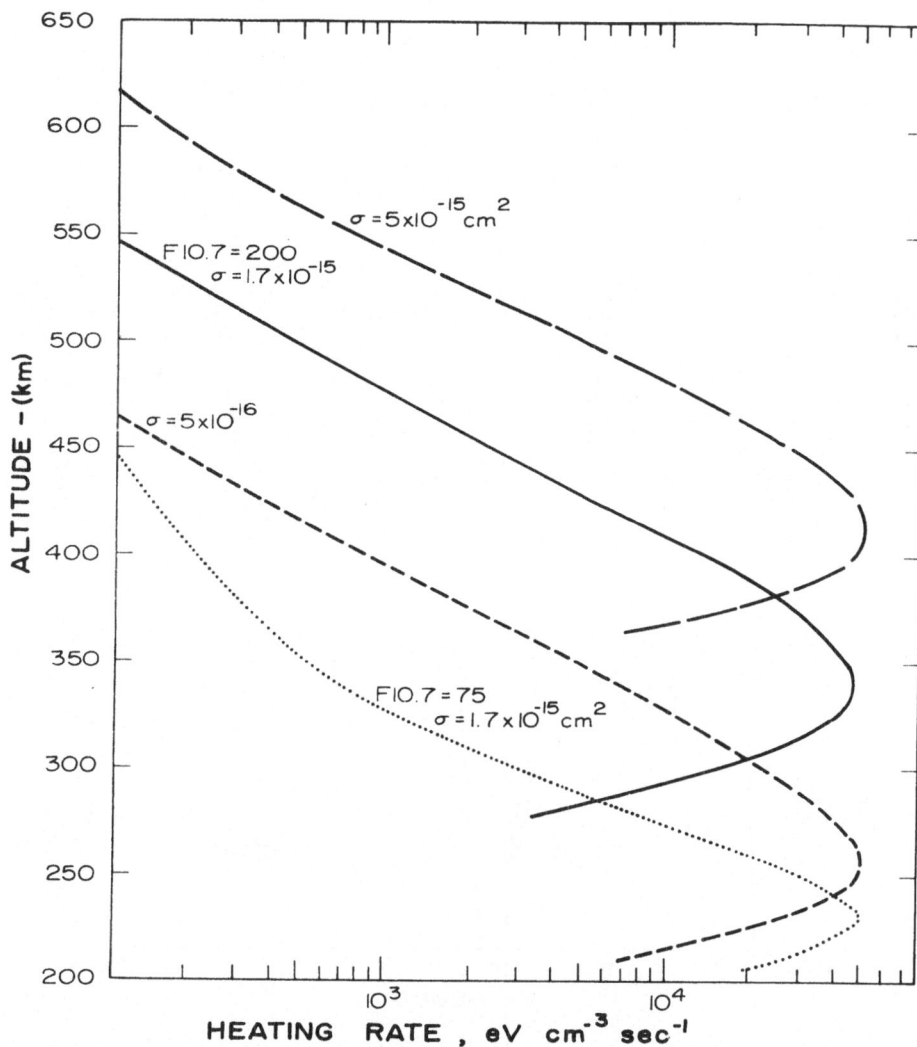

Fig. 2. Heating rate profile due to the energtic oxygen. The results represented by the full line curve are for the cross-section used in the rest of the modelling, namely $\sigma = 1.7 \times 10^{-15}$ cm^2 for a solar maximum, storm time model atmosphere. The dotted curve shows the effect of using a model atmosphere appropriate to solar minimum. The two dashed curves show the effect of varying σ from 5×10^{-15} cm^2 to 5×10^{-16} cm^2 (from TORR *et al.*, 1982). The input energy flux used in the calculation is the peak flux measured during the December 16/17 storm, divided by 2π.

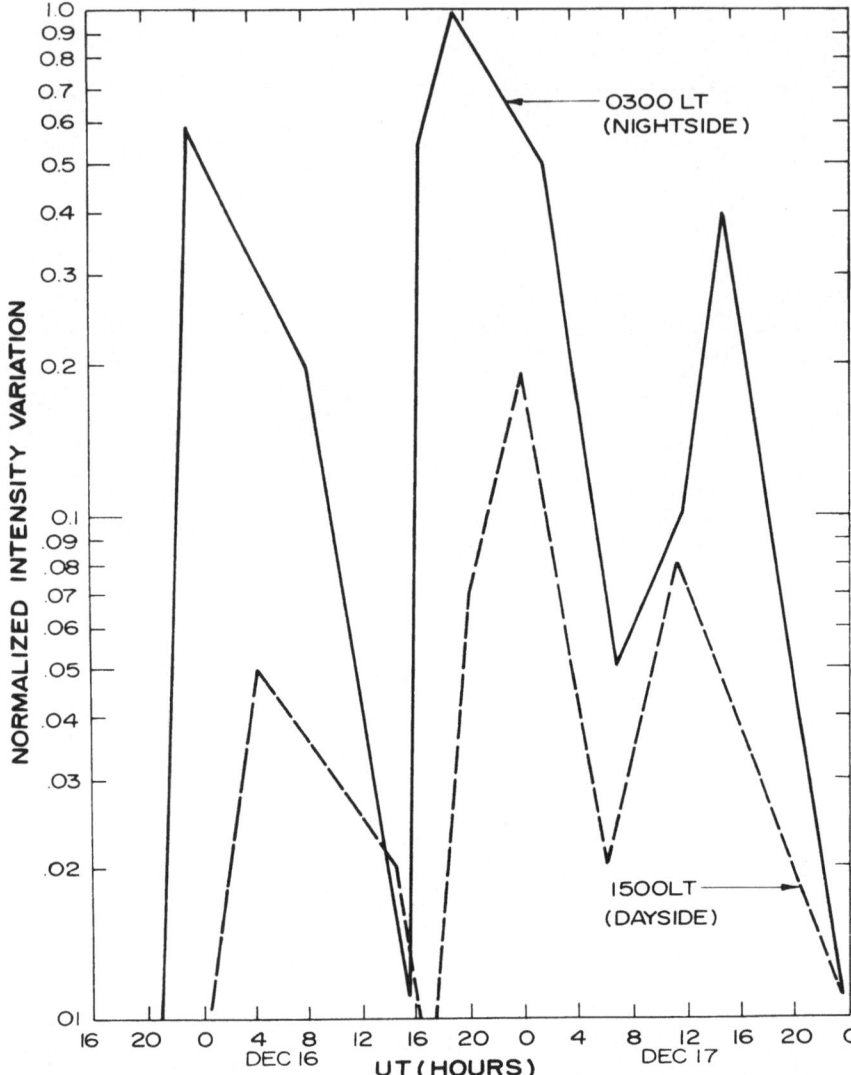

Fig. 3. Smoothed representation of the observed variation in O^+ flux over the duration of the December 16/17, 1971 magnetic storm. The normalization factor is used to scale the heating input shown in Fig. 5 (from TORR *et al.*, 1982).

the energy of the collision is on the average equally shared between the target and the projectile atoms. KOZYRA *et al.* (1982) have investigated the effects of using the more forward peaked elastic scattering. The results of this comparison are shown in Fig. 4 in which it can be seen that the effect of elastic scattering over hard sphere is to broaden the bottom side of the profile.

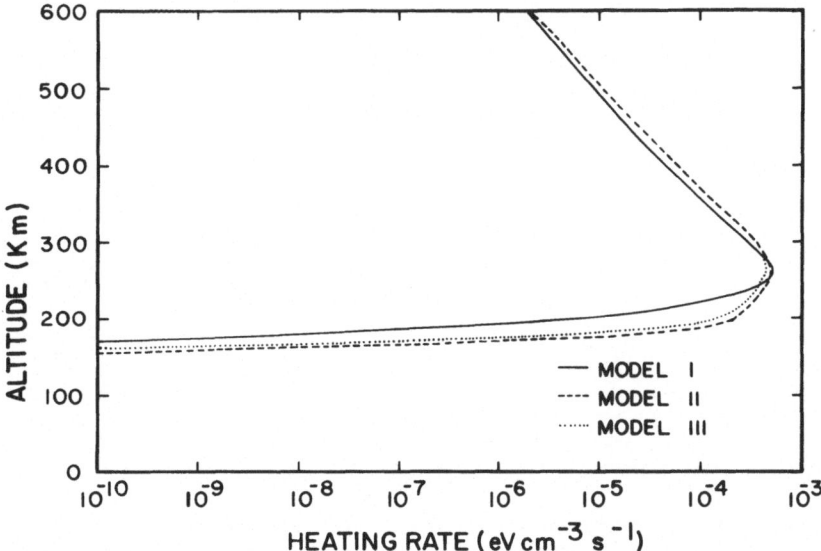

Fig. 4.   Equivalent heating rate due to energetic oxygen as computed by KOZYRA *et al.* (1982) using a two-stream model. Model I uses hard sphere collisions while Model II uses true elastic collisions. (Model III includes inelastic collisions as well).

The heating rate profile shown in Fig. 2 constitutes a significant source of energy to the nightside thermosphere. This is illustrated in Fig. 5 in which a comparison is shown between this nocturnal heating rate and that due to solar EUV heating on the dayside at solar maximum. Even if the peak altitude of the energetic O+ heating rate is lowered by 100 km, it will still constitute a nocturnal stormtime source within a factor of two of the daytime solar heating. It should also be borne in mind that the O+ energy flux used in Fig. 5 is 6 times less than the maximum observed flux.

In view of the magnitude of the energetic O+ heating rate on occasion, and the fact that the precipitation can occur for hours, the events might be expected to produce some effect on the thermospheric dynamics. The fluxes have been found to occur over a wide latitudinal range extending from $L \sim 2$ to $L \sim 9$, and were typically 5–10 times lower on the dayside than on the nightside. In order to examine the impact on the thermospheric dynamics, the results of the energetic oxygen heating rate calculation have been parameterized in terms of altitude, latitude and local time, and this source has been used as input to the NCAR global thermospheric circulation model (DICKINSON *et al.*, 1981). The results of this study are reported by TORR *et al.*, (1982) and show that the energetic O+ events can significantly perturb the thermosphere above about 300 km.

The model is time dependent and runs from 2000 UT on December 15 till 2000 UT on December 17. During this time the energetic O+ heating rate is varied according to the temporal modulation shown in Fig. 3. The assumption is made that the heating occurs symmetrically in both hemispheres.

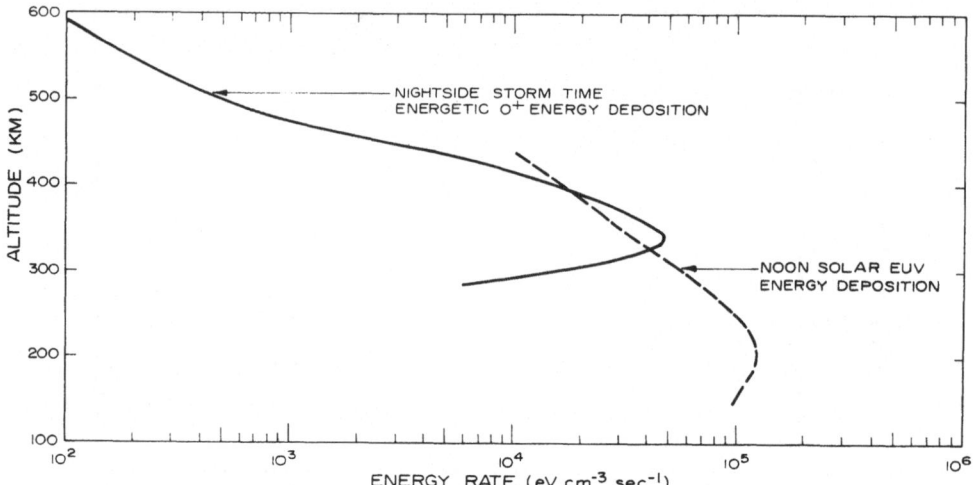

Fig. 5.   Comparison of nocturnal energetic oxygen heating rate profile, with dayside solar EUV energy deposition rate (dashed curve) (from TORR *et al.*, 1982). As in Fig. 2, the O$^+$ energy flux has been reduced by 2π from the peak observed flux.

In Figs. 6–9 are shown the modelled temperature and neutral wind perturbation caused by the energetic O$^+$ heating source. In each case the results of the model for solar heating alone are followed by the results obtained when the energetic O$^+$ heating is included in the model. The temperature effects are shown as the perturbation to the departure from the mean global temperature at any location.

Figure 6a shows the global temperature perturbation and velocity vector pattern at an altitude of 450 km for solar heating only. Figure 6b shows the results of including the particle heating source. There is an overall temperature increase with the maximum perturbations occuring in the regions of O$^+$ precipitation. The maximum midnight winds at high latitudes are decreased from 205 m s^{-1} to 190 m s^{-1}. The decrease in winds affects the energy transport, thereby increasing the overall temperature in the polar cap.

Figures 7 and 8 depict an altitude-longitude cross-section at a latitude of 52.5°N. Figure 7a and b show the temperature perturbation, without and with the energetic O$^+$ source respectively. With no particle source there is a temperature minimum near midnight which becomes a maximum after the inclusion of the O$^+$ source. The maximum perturbation is 400°K. In the case of the meridional wind component, the particle heating causes a decrease in the magnitude of the poleward winds from 150 m s^{-1} to 100 m s^{-1}, while increasing the nighttime equatorward winds from 130 m s^{-1} to 240 m s^{-1}. The major effect on the zonal winds is a reduction in magnitude by up to 100 ms^{-1}. The effect on the vertical wind component is illustrated in Fig. 8, from which it can be seen that the energetic O$^+$ source results in a reversal at night.

A cross-section in latitude is shown in Fig. 9 for the temperature perturbation at midnight.

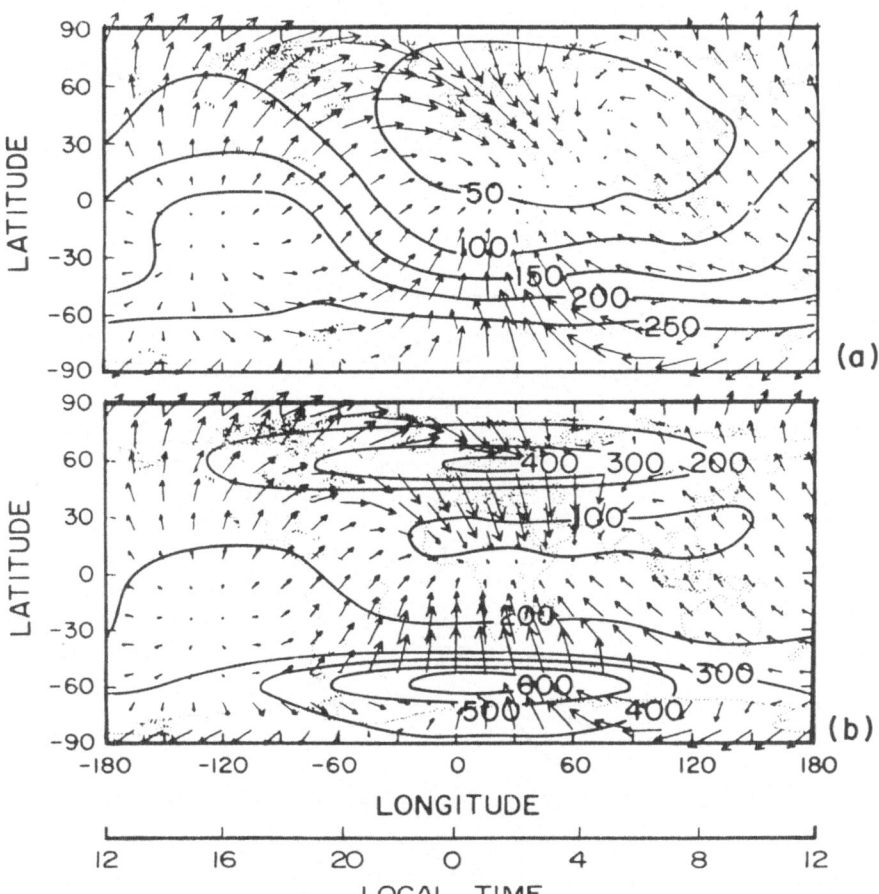

Fig. 6. Calculated global distribution of perturbation temperature (°K) and winds at 0000UT along a constant pressure surface in the TGCM of $Z = +3$ (approximately 450 km) for December solstice conditions. (a) basic case that includes solar heating and heat and momentum sources associated with magnetospheric convection with a cross-tail potential of 20 kV. (b) same as (a) except heating by fast O⁺ ions at 0000UT on 17 December is included. The maximum wind vector is 205 ms⁻¹ in (a) and 188 ms⁻¹ in (b). Local midnight is 0° longitude and local noon is at 180° longitude (from TORR *et al.*, 1982).

The effects shown above follow the corresponding temporal variations in the O⁺ energy influx with a very short time constant. At the conclusion of the storm there is a net overall thermospheric temperature increase of about 20°K which slowly decreases as the storm energy is thermally conducted to the lower thermosphere.

It would be very interesting to compare the global thermospheric circulation effects described above with those of an auroral substorm. Unfortunately, such a model has not yet been attempted. However, a study by ROBLE and DICKINSON (1970) on SAR-arcs showed that the response of the atmosphere in this region is linear. Thus these effects at the time of energetic O⁺ precipitation are expected to be superimposed on those due to the substorm effects.

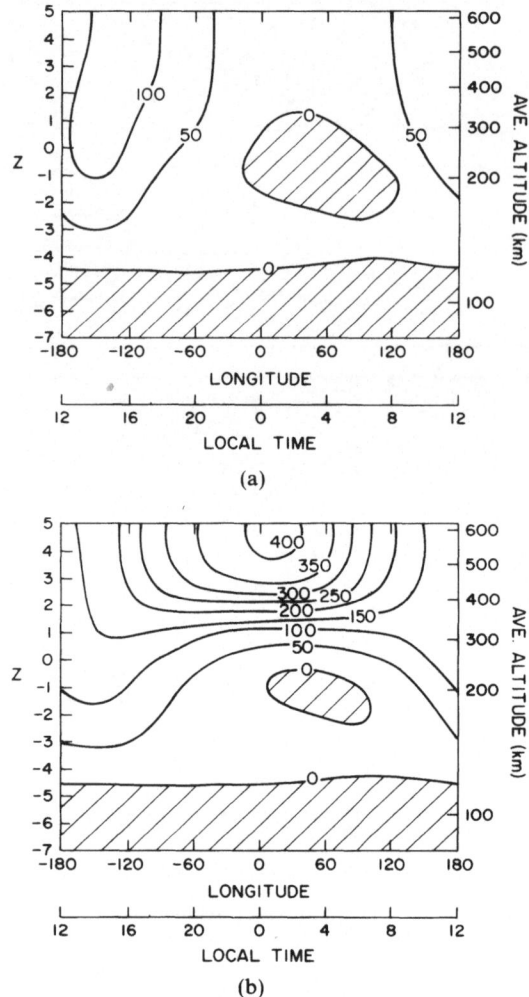

Fig. 7.   Calculated contours of the diurnal variation of the perturbation temperature at 0000UT around the 52.5°N latitude circle for (a) basic case and (b) heating by energetic O^+ included.

2.2 Escape of oxygen from the earth

The precipitation of the energetic O^+ ions into the atmosphere generates a large backsplash of atomic oxygen out of the atmosphere (TORR et al., 1974, 1981). The oxygen atoms with energies in excess of 10 eV are potentially capable of escaping from the Earth. Approximately 30% of the total incoming energy is converted into energetic atoms in this category. At the very peak of the storm a flux of 4×10^{10} cm^{-2} sec^{-1} is potentially capable of escape.

By computing the escape number flux from the model, and by making some

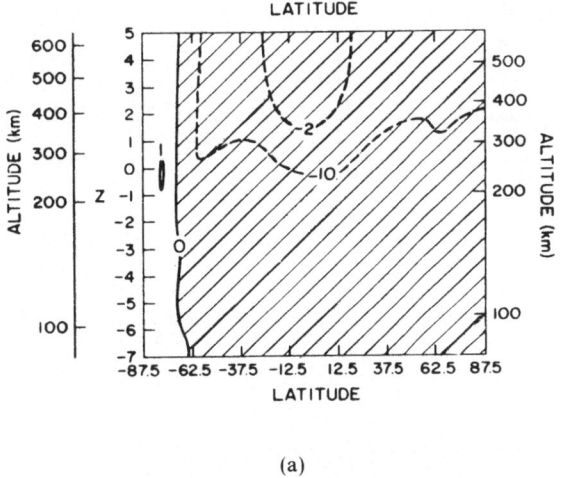

(a)

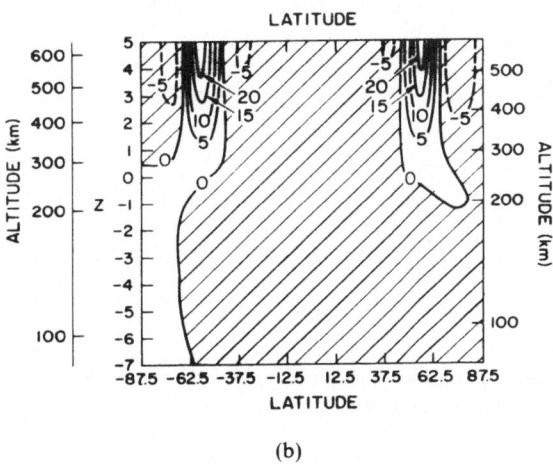

(b)

Fig. 8.   Same as (7) except that the contours represent the vertical wind component.

assumptions concerning the magnitude and frequency of the events, the lifetime for the escape of oxygen from the Earth can be determined. If it is assumed that the precipitation occurs between $50°$ to $70°$ latitude whenever $K_p > 4$, the events would take place over 17% of the Earth for 22% of the time. If in addition, it is assumed that the typical nocturnal energy flux is 0.1 ergs cm^{-2} sec^{-1} sr^{-1}, with that on the dayside to be an order of magnitude less, the average energy flux would be 0.05 ergs cm^{-2}sec^{-1}sr^{-1}. The average global escape flux would thus be $\sim 2 \times 10^8$ cm^{-2} sec^{-1}, which yields an escape time for oxygen comparable with the lifetime of the Earth.

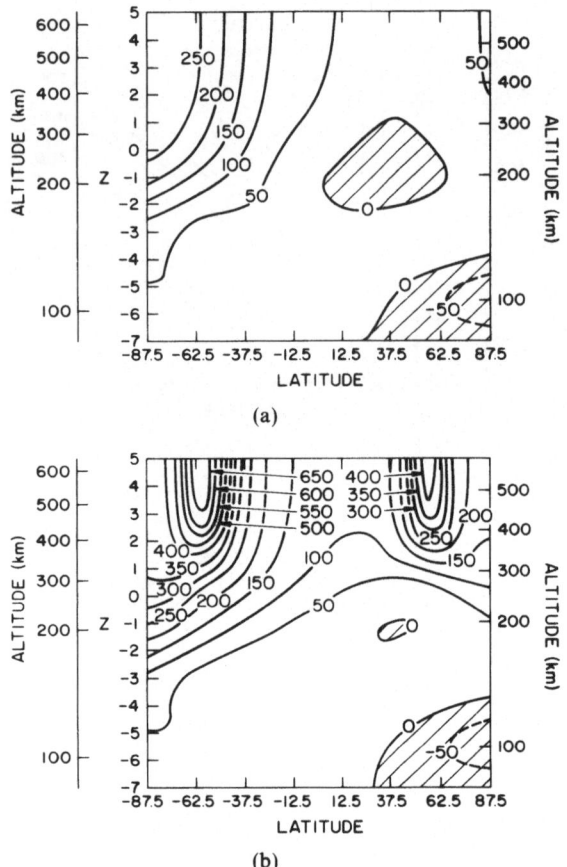

Fig. 9. Contours of calculated latitudinal distribution of temperature perturbation along the 0° longitude at 0000UT for (a) basic case and (b) heating by energetic O^+ included.

2.3   Ionization and excitation

For the O^+ energies considered in this study the velocity of the orbital electron in the atom is about 10^9 cm sec^{-1}, compared with oxygen ion velocities of less than 4×10^7 cm sec^{-1}. Thus direct ionization and excitation in the O–O collisions would not be expected to be an important process (MASSEY, 1949). DRAWIN (1968, 1969) has published formulas which yield an ionization cross-section at 1 keV of 1×10^{-16} cm^2. This is small by comparison with the values assumed for the momentum exchange process. If, however, there is a sufficient number of collisions, the ionization produced might be significant on certain occasions. By computing the total number of collisions involving an energetic oxygen atom capable of causing ionization (say, > 50 eV) a rough estimate of the ionization rate can be made. For this purpose the ionization cross-section has been assumed to be constant with energy at 2×10^{-17} cm^2 yielding a peak ionization rate of $60 \, \text{cm}^{-3} \, \text{s}^{-1}$ at ~ 400 km for the peak O^+ energy flux. KOZYRA *et al.*, (1982) repeated the estimate using the forward scattering $O-O$ collisions, the energy dependence of the Drawin formula and an initial flux six times smaller. They obtained a peak ionization of $\sim 2 \, \text{cm}^{-3} \, \text{sec}^{-1}$ near 300 km.

It can thus be concluded that for an incoming O^+ energy flux of ~ 0.4 $\text{ergs} \, \text{cm}^{-2} \, \text{s}^{-1} \, \text{sr}^{-1}$, the peak ionization rate would be $10-60$ ion pairs $\text{cm}^{-3} \, \text{sec}^{-1}$. At night in the topside F-region and in the absence of strong electron precipitation, there are no comparable sources of ionization. The effect would be an apparent slowing down of the recombination process at these altitudes in the precipitation region.

2.4   Excitation

If the excitation cross-section is also at least $2 \times 10^{-17} \, \text{cm}^2$, a similar rough estimate of the resulting emission can be made. For the $L \sim 3.3$ spectrum shown in Fig. 1, there are a total of $\sim 3 \times 10^8$ ions $\text{cm}^{-2} \, \text{sec}^{-1}$ with an average energy of $\sim 4,000 \, \text{eV}$. This will result in a total of $\sim 3 \times 10^{11}$ collisions to cascade in energy down to $\sim 8 \, \text{eV}$. Using the above cross-section together with ambient oxygen scale height and concentration, the event would yield ~ 300 R at topside altitudes.

3.   Low Latitude Effects

TINSLEY (1979a, b) has presented arguments for the charge exchange of ring current ions resulting in the precipitation of heavy neutral atoms into the thermosphere with associated emissions and heating. The concept of the precipitation of the heavy neutrals is illustrated in Fig. 10.

It was originally anticipated that the heavy neutrals would be protons and H Balmer emissions at low latitudes have been reported (LEVASSEUR and BLAMONT, 1973; TINSLEY and BURNSIDE, 1981). More recently additional information has been obtained on the composition of the ring current. The first measurements of trapped ring current ions (made at energies $< 50 \, \text{keV}$ and $L < 4$) were reported by JOHNSON *et al.*, (1977). It was found that O^+ and H^+ ions at energies less than 16 keV were approximately equal during the main phase of three magnetic storms, with He^+ an order of magnitude less. During a recovery phase they found O^+ to dominate, He^+ to follow in magnitude, and H^+ to be the smallest. Recently evidence has been presented (LUNDIN *et al.*, 1980) from the PROGNOZ-7 satellite, which shows that a large fraction of the ring current ($\sim 82\%$ on occasion) at ion energies less than 17 keV is O^+ ions for $L < 4$. Earlier, LYONS and EVANS (1976) presented theoretical arguments as to why the ring current ions at energies $< 50 \, \text{keV}$ and $L < 4$ would not be dominated by protons during the recovery phase of large geomagnetic storms. Both O^+ and He^+ would meet the requirements for a longer lifetime for charge exchange with hydrogen than that of protons.

Under these circumstances the heavy neutrals precipitating at low latitudes generally would be expected to be energetic oxygen atoms. From calculations such as those reported in the previous section, it would therefore also be expected that these particles would deposit their energy at higher altitudes than would protons where the major constituent with which they would collide again being atomic oxygen. TINSLEY

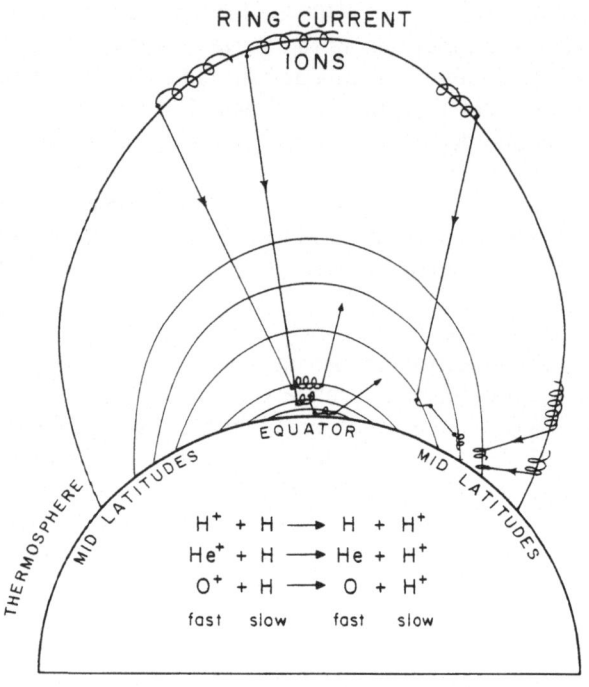

RING CURRENT
IONS

EQUATOR

THERMOSPHERE

MID LATITUDES

MID LATITUDES

$H^+ + H \longrightarrow H + H^+$
$He^+ + H \longrightarrow He + H^+$
$O^+ + H \longrightarrow O + H^+$

fast   slow   fast   slow

POLAR SECTION
THROUGH MAGNETOSPHERE

Fig. 10. Geometry of energetic neutral atom production from ring current ions. The neutrals travel on straight line trajectories, mostly outwards, but a fraction intersect the thermosphere, and depending on species and energy, some will then reionize, and near the equator become temporarily trapped.

(1981) has presented optical evidence to support the constituent being energetic atomic oxygen. It must be noted however that the ring current composition has still not been directly measured in the energy range 30–150 keV which dominates the energy content of the ring current.

4.   Discussion

The precipitation of energetic O^+ ions into the atmosphere during magnetic storms gives rise to large fluxes of energetic oxygen atoms which in turn cause significant effects. The largest of these effects is the heating of the neutral atmosphere with the associated perturbations to the global thermospheric temperatures and circulation patterns. While the ionization effects resulting from the heavy ion precipitation may not be significant, the excitation may result in several hundred Rayleighs of oxygen emission. The O^+ precipitation results in a large splash flux of energetic oxygen atoms. The flux capable of escaping from the Earth is comparable with that for hydrogen. The flux of atomic oxygen atoms entering long trajectory ballistic orbits can produce a reionized flux which is 10 times the magnitude of the original O^+ flux (TORR and TORR, 1979). This raises the intriguing possibility that the events may be self-sustaining.

At lower latitudes, the charge exchange of hydrogen atoms with ring current ions may result in a light rain of heavy neutrals which could constitute a nocturnal ionization source larger in magnitude than those previously considered, such as EUV and electron sources. In addition the heating effects need to be evaluated.

Studies of these effects have been hampered by the lack of sufficient observations of the energetic ions, and to an even larger extent by the lack of observations of the energetic neutrals. Instrumentation currently in operation and under development should provide valuable data in this regard.

This work was supported by a NASA grant to the Utah State University NAGW-105, and a grant of computer resources from the National Center for Atmospheric Research which is sponsored by the National Science Foundation.

REFERENCES

AMDUR, I., E. A. MASON, and J. E. JORDAN, Scattering of high velocity neutral particles, 10, $He-N_2$; $A-N_2$: The N_2-N_2 interaction, *J. Chem. Phys.*, **27**, 527, 1957.

DICKINSON, R. E., E. C. RIDLEY, and R. G. ROBLE, A three-dimensional general circulation model of the thermosphere, *J. Geophys. Res.*, **86**, 1499, 1981.

DRAWIN, H. W., Zur formelmässigen Darstellung des Ionisierungsquerschnitts für den Atom-atomstoss und uber die Ionen-Elektronen-Rekombination im dichten Naturalgas, *Z. Physik*, **211**, 404, 1968.

DRAWIN, H. W. Influence of atom-atom collisions on the collisional-radioactive ionization and recombination coefficients of hydrogen plasmas, *Z. Physik* **225**, 483, 1969.

HEDIN, A. E., Tables of thermospheric temperature, density and composition derived from satellite and ground based measurements, Goddard Space Flight Center Special Report, Vol. 3, 1979.

JOHNSON, R. G., R. D. SHARP, and E. G. SHELLEY, Observations of ions of ionospheric origin in the storm time ring current, *Geophys. Res. Lett.*, **4**, 403, 1977.

KOZYRA, J. U., T. E. CRAVENS, and A. F. NAGY, Energetic O^+ precipitation *J. Geophys. Res.*, **87**, 2481, 1982.

LEVASSEUR, A.-C. and J.E. BLAMONT, Satellite observations of strong Balmer Atmospheric emissions around the magnetic equator, *J. Geophys. Res.*, **78**, 3881, 1973.

LO, H. H. and W. L. FITE, Electron-capture and loss cross sections for fast heavy particles passing through gases, *Atomic Data*, **1**, 305, 1970.

LUNDIN, R., L. R. LYONS, and N. PISSARENKO, Observations of the ring current composition at L-values less than 4, *Geophys. Res. Lett.*, **7**, 425, 1980.

LYONS, L. R. and D. S. EVANS, The inconsistency between proton charge exchange and the observed ring current decay, *J. Geophys. Res.*, **81**, 6197, 1976.

ROBLE, R. G. and R. E. DICKINSON, Atmospheric response to heating within a stable mid-latitude red arc, *Planet. Space Sci.*, **18**, 1489, 1970.

SHARP, R. D., R. G. JOHNSON, E. G. SHELLEY, and K. K. HARRIS, Energetic O^+ ions in the magnetosphere, *J. Geophys. Res.*, **79**, 144, 1974.

SHARP, R. D., R. G. JOHNSON, and E. G. SHELLEY, The morphology of energetic O^+ ions during two magnetic storms: temporal variations, *J. Geophys. Res.*, **81**, 3283, 1976a.

SHARP, R. D., R. G. JOHNSON, and E. G. SHELLEY, The morphology of energetic O^+ ions during two magnetic storms: latitudinal variations, *J. Geophys. Res.*, **81**, 3292, 1976b.

SHELLEY, E. G., R. G. JOHNSON, and R. D. SHARP, Satellite observations of energetic heavy ions during a geomagnetic storm, *J. Geophys. Res.*, **77**, 6104, 1972.

Shelley, E. G., R. G. Johnson, and R. D. Sharp, Morphology of energetic O^+ in the magnetosphere, in *Magnetospheric Physics*, pp. 135–139, Reidel Publishing Company, Dordrecht-Holland, 1974.

Solov'ev, E. S., R. N. Il'in, V. A. Oparin, I. T. Serenkov, and N. V. Fedorenko, Capture and loss of electrons by fast oxygen atoms and ions in air, nitrogen and oxygen atoms in air, nitrogen and oxygen, *Sov. Phys. Tech. Phys.*, **17**, 267, 1972.

Tinsley, B. A., Energetic neutral atom precipitation during magnetic storms; optical emission, ionization and energy deposition at low and middle latitudes, *J. Geophys. Res.*, **84**, 1855, 1979a.

Tinsley, B. A., Energetic neutral atom precipitation as a possible source of midlatitude F region winds, *Geophys. Res. Lett.*, **6**, 291, 1979b.

Tinsley, B. A. and R. G. Burnside, Precipitation of energetic neutral hydrogen atoms at Arecibo during a magnetic equator, *J. Geophys, Res.*, **8**, 87, 1981.

Tinsley, B. A., R. P. Rohrbaugh, Y. Sahai, and N. Teixeira, Observations of optical emissions from particle precipitation at low latitudes, *EOS*, (abstract) **62**, 978, 1981.

Torr, Marsha R. and D. G. Torr, Energetic oxygen: a direct coupling mechanism between the magnetosphere and thermosphere, *Geophys. Res. Lett.*, 6, 700, 1979.

Torr, Marsha R., J. C. G. Walker, and D. G. Torr, Escape of fast oxygen from the atmosphere during geomagnetic storms, *J. Geophys. Res.*, **79**, 5267, 1974.

Torr, Marsha R., D. G. Torr, and R. E. Roble, Energetic O^+ precipitation; a significant energy source for the thermosphere, in *Physical Basis of the Ionosphere in the Solar-Terrestrial System*, AGARD, Hartford House, 1981.

Torr, Marsha R., D. G. Torr, R. E. Roble, and E. C. Ridley, The dynamic response of the thermosphere to the energy influx resulting from energetic O^+ ions, *J. Geophys. Res.*, **87**, 5290, 1982.